MECHANICS OF MATERIALS

A… [illegible] …lastic and

Plastic [illegible] …Structural Components

Also of interest

ASHBY
Materials Selection in Mechanical Design

ASHBY & JONES
Engineering Materials 1
Engineering Materials 2

GIBSON & ASHBY
Cellular Solids

HULL & BACON
Introduction to Dislocations, 3rd Edition

JONO & INOUE
Mechanical Behaviour of Materials—VI (ICM 6)

KETTUNEN et al
Strength of Metals and Alloys (ICSMA 8)

SALAMA et al
Advances in Fracture Research (ICF 7)

TAYA & ARSENAULT
Metal Matrix Composites: Thermomechanical Behavior

YAN et al
Mechanical Behaviour of Materials (ICM 5)

MECHANICS OF MATERIALS

An Introduction to the Mechanics of Elastic and Plastic Deformation of Solids and Structural Components

SECOND EDITION

In Two Volumes

VOLUME I

E. J. HEARN

Ph.D., B.Sc. (Eng.) Hons., C.Eng., F.I.Mech.E., F.I.Prod.E., F.I.Diag.E.

Associate Director (Industrial Liaison & External Relations)
Hong Kong Polytechnic
Hong Kong

BUTTERWORTH HEINEMANN

Butterworth-Heinemann Ltd
Linacre House, Jordan Hill, Oxford OX2 8DP

A member of the Reed Elsevier plc group

OXFORD LONDON BOSTON
MUNICH NEW DELHI SINGAPORE SYDNEY
TOKYO TORONTO WELLINGTON

First published 1977
Reprinted with corrections 1980, 1981, 1982
Second edition 1985
Reprinted with corrections 1988
Reprinted 1989, 1991, 1993, 1995

British Library Cataloguing in Publication Data
Hearn, E. J.
Mechanics of materials: an introduction to the mechanics of elastic and plastic deformation of solids and structural components
1. Materials 2. Mechanics, applied
I. Title II. Series
620.1'12 TA404.8

ISBN 0 7506 2540 6

Library of Congress Cataloguing in Publication Data
Hearn, E. J.
Mechanics of materials (international series on materials science and technology; v.19)
Includes bibliographies and indexes
1. Strength of materials I. Title II. Series
III. Series: International series on materials science and technology; v.19)
TA405.H3 1985 620.'123 84-25470

Printed and bound in Great Britain

CONTENTS OF VOLUME 1

CONTENTS OF VOLUME 2

INTRODUCTION

This text is a substantially revised and extended version of the highly successful first edition initially published in 1977 and reprinted in 1980, 1981 and 1982. Volumes 1 and 2 are intended to cover the material normally contained in degree and honours degree courses in mechanics of materials and in courses leading to exemption from the academic requirements of the Engineering Council. They should also serve as a valuable reference medium for industry and for post-graduate courses. Whilst both volumes should also prove valuable for students following BTEC* Higher Certificate and Higher Diploma courses and studying mechanical science, stress analysis, solid mechanics or similar modules, it is expected that Volume 1 will be more appropriate for the Higher Certificate level.

The study of mechanics of materials is the study of the behaviour of solid bodies under load. The way in which they react to applied forces, the deflections resulting and the stresses and strains set up within the bodies, are all considered in an attempt to provide sufficient knowledge to enable any component to be designed such that it will not fail within its service life. Typical components considered in detail in the first edition of this text included beams, shafts, cylinders, struts, diaphragms and springs and, in most simple loading cases, theoretical expressions are derived to cover the mechanical behaviour of these components. Because of the reliance of such expressions on certain basic assumptions the text also included a chapter devoted to the important experimental stress and strain measurement techniques in use today with recommendations for further reading.

This second edition retains all the contents of the first edition suitably modified and updated as appropriate, and incorporating all corrections and amendments made since the work was first introduced in 1977. In addition, major changes have been made in order to modernise the text to take account of various movements within the educational system, changes in syllabi and courses, and changes in the technology of the subject itself. In particular, new chapters have been introduced covering the topics of (a) fatigue, creep and fracture, and (b) contact stresses, residual stresses and stress concentrations, together with detailed coverage of Airy stress functions and an introduction to finite element stress analysis.

Full details of the new material contained in each volume is given below:–

Volume 1

Expansion of introductory chapter on simple stress and strain to include a wider coverage of materials, the application of Poisson's ratio effects to two-dimensional stress systems and the expansion of partially constrained bars.

Limitations of the simple bending theory as related to the assumptions used in its derivation.

* Business and Technician Education Council

Deflections of beams due to temperature effects; finite difference method for beam deflections.

Derivation of equation of shear stress distribution in flanges of beams subjected to bending.

Power transmitted by shafts; combined stress systems – bending and torsion; bending, torsion and internal pressure. Thin rotating rings and cylinders.

Application of "unit-load" method for slopes or angular rotations of beams, etc.

Limitations of spring theories; Wahl and other factors; extension springs – initial tension; allowable stresses.

Alternative representations of stress at a point; cartesian and polar plots for uniaxial and biaxial stress conditions.

Alternative representations of strain distribution at a point – cartesian and polar plots.

Revised table for evaluation of strain gauge rosette data.

Limitations of failure theories; effect of stress concentrations; safety factors; modes of failure.

Additional worked examples and problems for solution.

Additional appendices for basic mechanical properties of metals and non-metallics.

Volume 2

Three new chapters:

(a) *Miscellaneous Topics* including bending of beams with initial curvature; bending of wide beams; visco-elasticity and finite element stress analysis.

(b) *Contact stresses, residual stresses and stress concentrations.*

(c) *Fatigue, Creep and Fracture.*

all three written from a practical design viewpoint related closely to the needs of industry.

A major extension of the Elasticity chapter including further work on "stress at a point"; tensor notations; stress and strain invariants; eigen values; reduced stresses; and an extensive section on Airy stress functions.

Extended coverage of inelastic bending and torsion and a full treatment of the plastic yielding (autofrettage) of thick cylinders.

Use of an equivalent "J" for torsion of non-circular sections and treatment of torsion of thin-walled stiffened sections.

Additional worked examples and problems for solution.

Additional appendices for the basic mechanical properties of metals and non-metallics.

The style of the text remains unchanged with, normally, each chapter containing a summary of essential formulae which are developed within the chapter and a large number of worked examples. The examples have been selected to provide progression in terms of complexity of problem and to illustrate the logical way in which the solution to a difficult problem can be developed. Graphical solutions have been introduced where appropriate. In order to provide clarity of working in the worked examples there is inevitably more detailed explanation of individual steps than would be expected in the model answer to an examination problem.

All chapters (with the exception of Chapter 21) conclude with an extensive list of problems for solution by students together with answers. These have been collected from various sources and include questions from past examination papers in imperial units which have been converted to the equivalent SI values. Each problem is graded according to its degree of difficulty as follows:

A	Relatively easy problem of an introductory nature.
A/B	Generally suitable for first-year studies.
B	Generally suitable for second or third year studies.
C	More difficult problems generally suitable for third year studies.

Gratitude is expressed to the following examination boards, universities and colleges who have kindly given permission for questions to be reproduced:

City University	C.U.
East Midland Educational Union	E.M.E.U.
Engineering Institutions Examination	E.I.E. and C.E.I.
Institution of Mechanical Engineers	I.Mech.E.
Institution of Structural Engineers	I.Struct.E.
Union of Educational Institutions	U.E.I.
Union of Lancashire and Cheshire Institutes	U.L.C.I.
University of Birmingham	U.Birm.
University of London	U.L.

In all, the text contains 150 worked examples and more than 500 problems for solution, and whilst it is hoped that no errors are present it is perhaps inevitable that some errors will be detected. In this event any comment, criticism or correction will be gratefully acknowledged.

The symbols and abbreviations throughout the text are in accordance with the latest recommendations of BS 1991 and PD 5686†.

As mentioned above, graphical methods of solution have been introduced where appropriate since it is the author's experience that these are more readily accepted and understood by students than some of the more involved analytical procedures; substantial time saving can also result. Extensive use has also been made of diagrams throughout the text since in the words of the old adage "a single diagram is worth 1000 words".

Finally, the author is indebted to all those who have assisted in the production of this volume; to Professor H. G. Hopkins, Mr R. Brettell, Mr R. J. Phelps for their work associated with the first edition and to Dr A. S. Tooth[1], Dr N. Walker[2], Mr R. Winters[2] and Mr M. Daniels[3] for their contributions to the new material of volume 2. Thanks also go to the publishers for their advice and assistance, especially in the preparation of the diagrams and editing, to Prof. C. C. Perry (USA) for his most valuable critique of the first edition, and to Mrs J. Beard and Miss S. Benzing for typing the manuscript.

July 1985 E. J. HEARN

† Relevant Standards for use in Great Britain: BS 1991; PD 5686: Other useful SI Guides: *The International System of Units*, N.P.L. Ministry of Technology, H.M.S.O. (Britain). Mechty, *The International System of Units (Physical Constants and Conversion Factors)*, NASA, No. SP-7012, 3rd edn. 1973 (U.S.A.) *Metric Practice Guide*, A.S.T.M. Standard E380-72 (U.S.A.).

[1] §23.27. Dr. A. S. Tooth. University of Strathclyde, Glasgow.

[2] §26. Dr. N. Walker and Mr. R. Winters, City of Birmingham Polytechnic.

[3] §24.4. Mr. M. Daniels, City of Birmingham Polytechnic.

NOTATION

Quantity	*Symbol*	*SI Unit*
Angle	$\alpha, \beta, \theta, \gamma, \phi$	rad (radian)
Length	L, s	m (metre) mm (millimetre)
Area	A	m^2
Volume	V	m^3
Time	t	s (second)
Angular velocity	ω	rad/s
Velocity	v	m/s
Weight	W	N (newton)
Mass	m	kg (kilogram)
Density	ρ	kg/m^3
Force	F or P or W	N
Moment	M	N m
Pressure	P	Pa (Pascal) N/m^2 bar ($= 10^5\ N/m^2$)
Stress	σ	N/m^2
Strain	ε	—
Shear stress	τ	N/m^2
Shear strain	γ	—
Young's modulus	E	N/m^2
Shear modulus	G	N/m^2
Bulk modulus	K	N/m^2
Poisson's ratio	ν	—
Modular ratio	m	—
Power	—	W (watt)
Coefficient of linear expansion	α	m/m °C
Coefficient of friction	μ	—
Second moment of area	I	m^4
Polar moment of area	J	m^4
Product moment of area	I_{xy}	m^4
Temperature	T	°C
Direction cosines	l, m, n	—
Principal stresses	$\sigma_1, \sigma_2, \sigma_3$	N/m^2
Principal strains	$\varepsilon_1, \varepsilon_2, \varepsilon_3$	—
Maximum shear stress	τ_{max}	N/m^2
Octahedral stress	σ_{oct}	N/m^2

Quantity	*Symbol*	*SI Unit*
Deviatoric stress	σ'	N/m^2
Deviatoric strain	ε'	—
Hydrostatic or mean stress	$\bar{\sigma}$	N/m^2
Volumetric strain	Δ	—
Stress concentration factor	K	—
Strain energy	U	J
Displacement	δ	m
Deflection	δ or y	m
Radius of curvature	ρ	m
Photoelastic material fringe value	f	N/m^2/fringe/m
Number of fringes	n	—
Body force stress	X, Y, Z F_R, F_θ, F_Z	N/m^3
Radius of gyration	k	m
Slenderness ratio	L/k	—
Gravitational acceleration	g	m/s^2
Cartesian coordinates	x, y, z	—
Cylindrical coordinates	r, θ, z	—
Eccentricity	e	m
Number of coils or leaves of spring	n	—
Equivalent J or effective polar moment of area	J_{eq} or J_E	m^4
Autofrettage pressure	P_A	N/m^2 or bar
Radius of elastic–plastic interface	R_p	m
Thick cylinder radius ratio R_2/R_1	K	—
Ratio elastic–plastic interface radius to internal radius of thick cylinder R_p/R_1	m	—
Resultant stress on oblique plane	p_n	N/m^2
Normal stress on oblique plane	σ_n	N/m^2
Shear stress on oblique plane	τ_n	N/m^2
Direction cosines of plane	l, m, n	—
Direction cosines of line of action of resultant stress	l', m', n'	—
Direction cosines of line of action of shear stress	l_s, m_s, n_s	—
Components of resultant stress on oblique plane	p_{xn}, p_{yn}, p_{zn}	N/m^2
Shear stress in any direction ϕ on oblique plane	τ_ϕ	N/m^2
Invariants of stress	I_1 I_2 I_3	N/m^2 $(N/m^2)^2$ $(N/m^2)^3$
Invariants of reduced stresses	J_1, J_2, J_3	
Airy stress function	ϕ	—

Quantity	*Symbol*	*SI Unit*
'Operator' for Airy stress function biharmonic equation	∇	—
Strain rate	$\dot{\varepsilon}$	s^{-1}
Coefficient of viscosity	η	
Retardation time (creep strain recovery)	t'	s
Relaxation time (creep stress relaxation)	t''	s
Creep contraction or lateral strain ratio	$J(t)$	—
Maximum contact pressure (Hertz)	p_0	N/m^2
Contact formulae constant	Δ	$(N/m^2)^{-1}$
Contact area semi-axes	a, b	m
Maximum contact stress	$\sigma_c = -p_0$	N/m^2
Spur gear contact formula constant	K	N/m^2
Helical gear profile contact ratio	m_p	—
Elastic stress concentration factor	K_t	—
Fatigue stress concentration factor	K_f	—
Plastic flow stress concentration factor	K_p	—
Shear stress concentration factor	Kt_s	—
Endurance limit for n cycles of load	S_n	N/m^2
Notch sensitivity factor	q	—
Fatigue notch factor	K_f	—
Strain concentration factor	K_ε	—
Griffith's critical strain energy release	G_c	
Surface energy of crack face	γ	N m
Plate thickness	B	m
Strain energy	U	N m
Compliance	C	$m N^{-1}$
Fracture stress	σ_f	N/m^2
Stress Intensity Factor	K or K_I	$N/m^{3/2}$
Compliance function	Y	—
Plastic zone dimension	r_p	m
Critical stress intensity factor	K_{IC}	$N/m^{3/2}$
"J" Integral	J	
Fatigue crack dimension	a	m
Coefficients of Paris Erdogan law	C, m	—
Fatigue stress range	σ_r	N/m^2
Fatigue mean stress	σ_m	N/m^2
Fatigue stress amplitude	σ_a	N/m^2
Fatigue stress ratio	R_s	—
Cycles to failure	N_f	—
Fatigue strength for N cycles	σ_N	N/m^2
Tensile strength	σ_{TS}	N/m^2
Factor of safety	F	—

Quantity	*Symbol*	*SI Unit*
Elastic strain range	$\Delta\varepsilon_e$	—
Plastic strain range	$\Delta\varepsilon_p$	—
Total strain range	$\Delta\varepsilon_t$	—
Ductility	D	
Secondary creep rate	ε_s^0	s^{-1}
Activation energy	H	N m
Universal Gas Constant	R	J/kg K
Absolute temperature	T	°K
Arrhenius equation constant	A	—
Larson–Miller creep parameter	P_1	
Sherby–Dorn creep parameter	P_2	
Manson–Haford creep parameter	P_3	
Initial stress	σ_i	N/m^2
Time to rupture	t_r	s
Constants of power law equation	β, n	—

CHAPTER 1

SIMPLE STRESS AND STRAIN

1.1. Load

In any engineering structure or mechanism the individual components will be subjected to external forces arising from the service conditions or environment in which the component works. If the component or member is in equilibrium, the resultant of the external forces will be zero but, nevertheless, they together place a load on the member which tends to deform that member and which must be reacted by internal forces which are set up within the material.

If a cylindrical bar is subjected to a direct pull or push along its axis as shown in Fig. 1.1, then it is said to be subjected to *tension* or *compression.* Typical examples of tension are the forces present in towing ropes or lifting hoists, whilst compression occurs in the legs of your chair as you sit on it or in the support pillars of buildings.

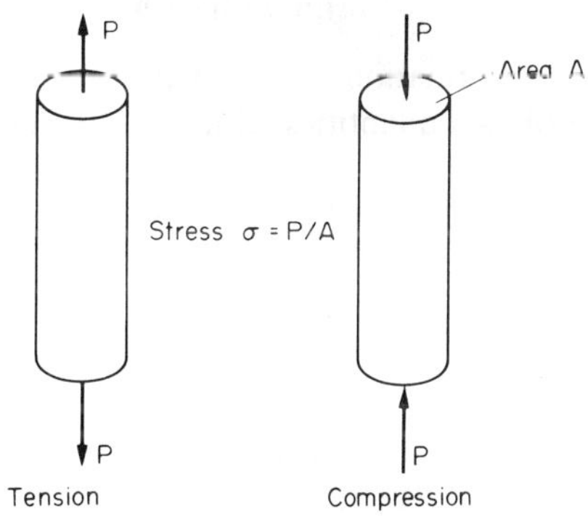

Fig. 1.1. Types of direct stress.

In the SI system of units load is measured in *newtons*, although a single newton, in engineering terms, is a very small load. In most engineering applications, therefore, loads appear in SI multiples, i.e. kilonewtons (kN) or meganewtons (MN).

There are a number of different ways in which load can be applied to a member. Typical loading types are:

(a) *Static* or *dead* loads, i.e. non-fluctuating loads, generally caused by gravity effects.
(b) *Live* loads, as produced by, for example, lorries crossing a bridge.
(c) *Impact* or *shock* loads caused by sudden blows.
(d) *Fatigue, fluctuating* or *alternating* loads, the magnitude and sign of the load changing with time.

1.2. Direct or normal stress (σ)

It has been noted above that external force applied to a body in equilibrium is reacted by internal forces set up within the material. If, therefore, a bar is subjected to a uniform tension or compression, i.e. a direct force, which is uniformly or equally applied across the cross-section, then the internal forces set up are also distributed uniformly and the bar is said to be subjected to a uniform *direct or normal stress*, the stress being defined as

$$\textbf{stress } (\sigma) = \frac{\textbf{load}}{\textbf{area}} = \frac{P}{A}$$

Stress σ may thus be compressive or tensile depending on the nature of the load and will be measured in units of newtons per square metre (N/m^2) or multiples of this.

In some cases the loading situation is such that the stress will vary across any given section, and in such cases the stress at any point is given by the limiting value of $\delta P/\delta A$ as δA tends to zero.

1.3. Direct strain (ε)

If a bar is subjected to a direct load, and hence a stress, the bar will change in length. If the bar has an original length L and changes in length by an amount δL, the *strain* produced is defined as follows:

$$\textbf{strain } (\varepsilon) = \frac{\textbf{change in length}}{\textbf{original length}} = \frac{\delta L}{L}$$

Strain is thus a measure of the deformation of the material and is non-dimensional, i.e. it has no units; it is simply a ratio of two quantities with the same unit (Fig. 1.2).

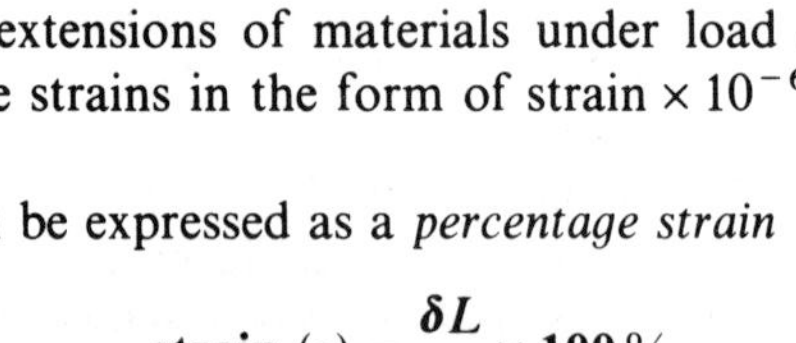

Fig. 1.2.

Since, in practice, the extensions of materials under load are very small, it is often convenient to measure the strains in the form of strain $\times 10^{-6}$, i.e. *microstrain*, when the symbol used becomes $\mu\varepsilon$.

Alternatively, strain can be expressed as a *percentage strain*

i.e. $$\textbf{strain } (\varepsilon) = \frac{\delta L}{L} \times 100\%$$

1.4. Sign convention for direct stress and strain

Tensile stresses and strains are considered POSITIVE in sense producing an *increase* in length. Compressive stresses and strains are considered NEGATIVE in sense producing a *decrease* in length.

1.5. Elastic materials – Hooke's law

A material is said to be *elastic* if it returns to its original, unloaded dimensions when load is removed. A particular form of elasticity which applies to a large range of engineering materials, at least over part of their load range, produces deformations which are proportional to the loads producing them. Since loads are proportional to the stresses they produce and deformations are proportional to the strains, this also implies that, whilst materials are elastic, stress is proportional to strain. *Hooke's law*, in its simplest form*, therefore states that

$$\text{stress } (\sigma) \propto \text{strain } (\varepsilon)$$

i.e.

$$\frac{\textbf{stress}}{\textbf{strain}} = \textbf{constant*}$$

It will be seen in later sections that this law is obeyed within certain limits by most ferrous alloys and it can even be assumed to apply to other engineering materials such as concrete, timber and non-ferrous alloys with reasonable accuracy. Whilst a material is elastic the deformation produced by any load will be *completely* recovered when the load is removed; there is no permanent deformation.

Other classifications of materials with which the reader should be acquainted are as follows:

A material which has a uniform structure throughout without any flaws or discontinuities is termed a *homogeneous* material. *Non-homogeneous* or *inhomogeneous* materials such as concrete and poor-quality cast iron will thus have a structure which varies from point to point depending on its constituents and the presence of casting flaws or impurities.

If a material exhibits uniform properties throughout in all directions it is said to be *isotropic*; conversely one which does not exhibit this uniform behaviour is said to be *non-isotropic* or *anisotropic*.

An *orthotropic* material is one which has different properties in different planes. A typical example of such a material is wood, although some composites which contain systematically orientated "inhomogeneities" may also be considered to fall into this category.

1.6. Modulus of elasticity – Young's modulus

Within the elastic limits of materials, i.e. within the limits in which Hooke's law applies, it has been shown that

$$\frac{\text{stress}}{\text{strain}} = \text{constant}$$

This constant is given the symbol E and termed the *modulus of elasticity* or *Young's modulus*.

Thus

$$E = \frac{\textbf{stress}}{\textbf{strain}} = \frac{\sigma}{\varepsilon} \tag{1.1}$$

$$= \frac{P}{A} \div \frac{\delta L}{L} = \frac{PL}{A\,\delta L} \tag{1.2}$$

* Readers should be warned that in more complex stress cases this simple form of Hooke's law will not apply and mis-application could prove dangerous; see §14.1, page 361.

Young's modulus E is generally assumed to be the same in tension or compression and for most engineering materials has a high numerical value. Typically, $E = 200 \times 10^9\ \text{N/m}^2$ for steel, so that it will be observed from (1.1) that strains are normally very small since

$$\varepsilon = \frac{\sigma}{E} \tag{1.3}$$

In most common engineering applications strains do not often exceed 0.003 or 0.3% so that the assumption used later in the text that deformations are small in relation to original dimensions is generally well founded.

The actual value of Young's modulus for any material is normally determined by carrying out a standard tensile test on a specimen of the material as described below.

1.7. Tensile test

In order to compare the strengths of various materials it is necessary to carry out some standard form of test to establish their relative properties. One such test is the standard tensile test in which a circular bar of uniform cross-section is subjected to a gradually increasing tensile load until failure occurs. Measurements of the change in length of a selected *gauge length* of the bar are recorded throughout the loading operation by means of extensometers (see page 610) and a graph of load against extension or stress against strain is produced as shown in Fig. 1.3; this shows a typical result for a test on a mild (low carbon) steel bar; other materials will exhibit different graphs but of a similar general form see Figs 1.5 to 1.7.

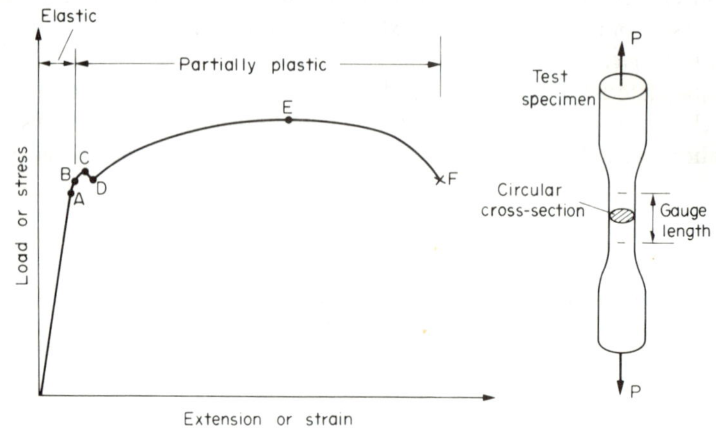

Fig. 1.3. Typical tensile test curve for mild steel.

For the first part of the test it will be observed that Hooke's law is obeyed, i.e. the material behaves elastically and stress is proportional to strain, giving the straight-line graph indicated. Some point A is eventually reached, however, when the linear nature of the graph ceases and this point is termed the *limit of proportionality*.

For a short period beyond this point the material may still be elastic in the sense that deformations are completely recovered when load is removed (i.e. strain returns to zero) but

Hooke's law does not apply. The limiting point B for this condition is termed the *elastic limit*. For most practical purposes it can often be assumed that points A and B are coincident.

Beyond the elastic limit *plastic deformation* occurs and strains are not totally recoverable. There will thus be some permanent deformation or *permanent set* when load is removed. After the points C, termed the *upper yield point*, and D, the *lower yield point*, relatively rapid increases in strain occur without correspondingly high increases in load or stress. The graph thus becomes much more shallow and covers a much greater portion of the strain axis than does the elastic range of the material. The capacity of a material to allow these large plastic deformations is a measure of the so-called *ductility* of the material, and this will be discussed in greater detail below.

For certain materials, for example, high carbon steels and non-ferrous metals, it is not possible to detect any difference between the upper and lower yield points and in some cases no yield point exists at all. In such cases a *proof stress* is used to indicate the onset of plastic strain or as a comparison of the relative properties with another similar material. This involves a measure of the permanent deformation produced by a loading cycle; the 0.1 % proof stress, for example, is that stress which, when removed, produces a permanent strain or "set" of 0.1 % of the original gauge length – see Fig. 1.4(a).

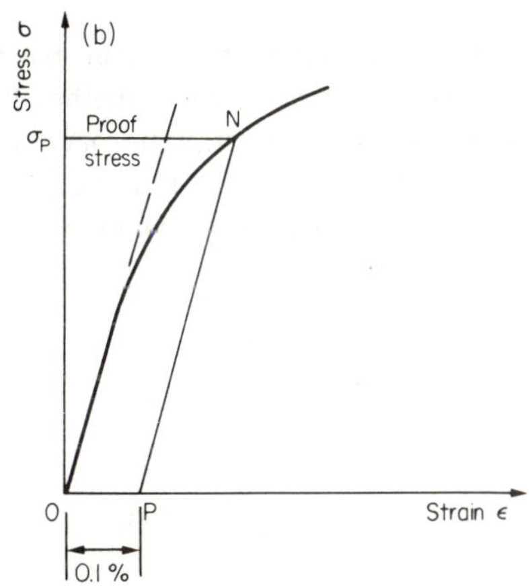

Fig. 1.4. (a) Determination of 0.1 % proof stress.

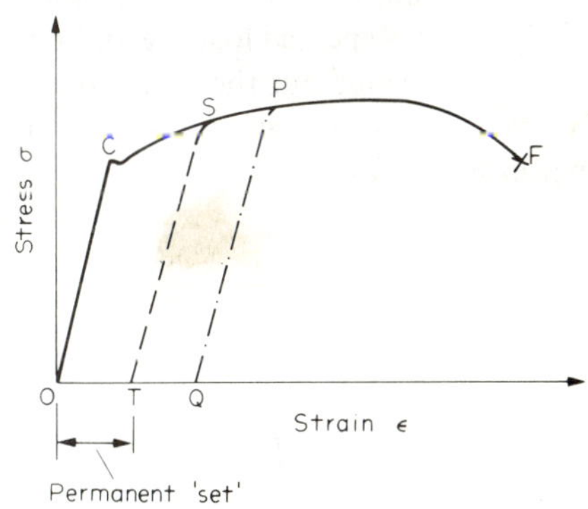

Fig. 1.4. (b) Permanent deformation or "set" after straining beyond the yield point.

The 0.1 % proof stress value may be determined from the tensile test curve for the material in question as follows:

Mark the point P on the strain axis which is equivalent to 0.1 % strain. From P draw a line parallel with the initial straight line portion of the tensile test curve to cut the curve in N. The stress corresponding to N is then the *0.1 % proof stress*. A material is considered to satisfy its specification if the permanent set is no more than 0.1 % after the proof stress has been applied for 15 seconds and removed.

Beyond the yield point some increase in load is required to take the strain to point E on the graph. Between D and E the material is said to be in the *elastic–plastic* state, some of the section remaining elastic and hence contributing to recovery of the original dimensions if load is removed, the remainder being plastic. Beyond E the cross-sectional area of the bar begins to reduce rapidly over a relatively small length of the bar and the bar is said to *neck*. This necking takes place whilst the load reduces, and fracture of the bar finally occurs at point F.

The nominal stress at failure, termed the *maximum* or *ultimate tensile stress*, is given by the load at E divided by the original cross-sectional area of the bar. (This is also known as the *tensile strength* of the material of the bar.) Owing to the large reduction in area produced by the necking process the actual stress at fracture is often greater than the above value. Since, however, designers are interested in maximum loads which can be carried by the complete cross-section, the stress at fracture is seldom of any practical value.

If load is removed from the test specimen after the yield point C has been passed, e.g. to some position S, Fig. 1.4(b), the unloading line ST can, for most practical purposes, be taken to be linear. Thus, despite the fact that loading to S comprises both elastic (OC) and partially plastic (CS) portions, the unloading procedure is totally elastic. A second load cycle, commencing with the permanent elongation associated with the strain OT, would then follow the line TS and continue along the previous curve to failure at F. It will be observed, however, that the repeated load cycle has the effect of increasing the elastic range of the material, i.e. raising the effective yield point from C to S, while the tensile strength is unaltered. The procedure could be repeated along the line PQ, etc., and the material is said to have been *work hardened*.

In fact, careful observation shows that the material will no longer exhibit true elasticity since the unloading and reloading lines will form a small *hysteresis loop*, neither being precisely linear. Repeated loading and unloading will produce a yield point approaching the ultimate stress value but the elongation or strain to failure will be much reduced.

Typical stress–strain curves resulting from tensile tests on other engineering materials are shown in Figs 1.5 to 1.7.

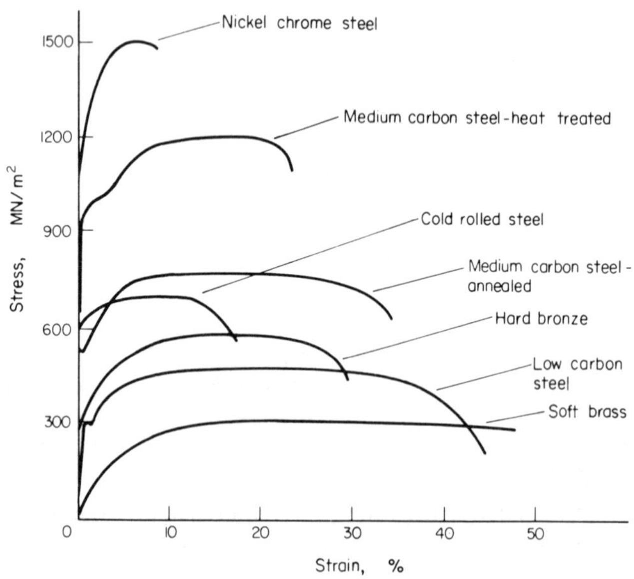

Fig. 1.5. Tensile test curves for various metals.

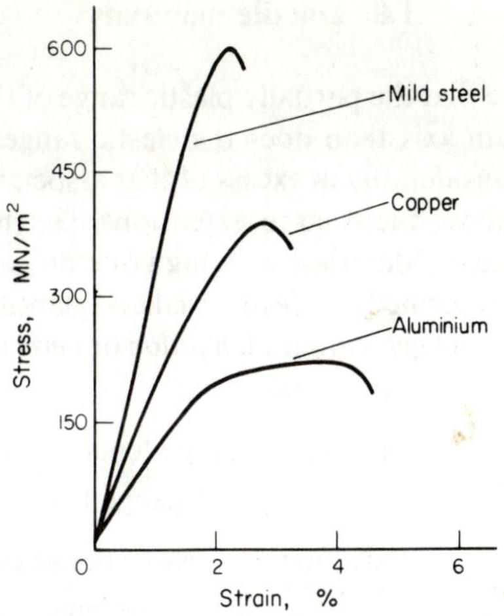

Fig. 1.6. Typical stress–strain curves for hard drawn wire materials – note large reduction in strain values from those of Fig. 1.5.

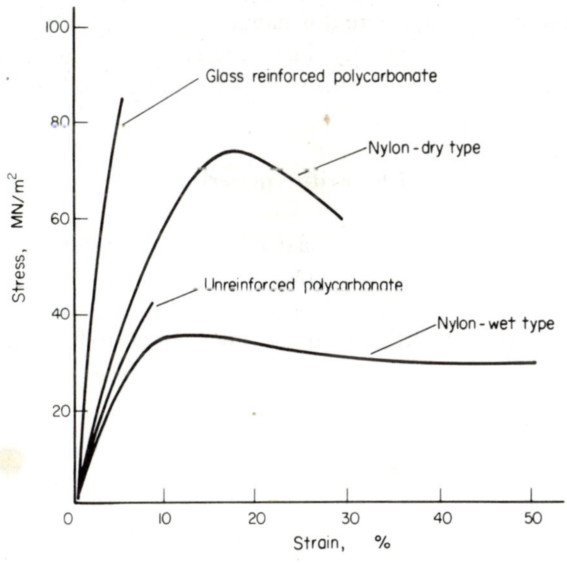

Fig. 1.7. Typical tension test results for various types of nylon and polycarbonate.

After completing the standard tensile test it is usually necessary to refer to some "British Standard Specification" or "Code of Practice" to ensure that the material tested satisfies the requirements, for example:

BS 4360	British Standard Specification for Weldable Structural Steels.
BS 970	British Standard Specification for Wrought Steels.
BS 153	British Standard Specification for Steel Girder Bridges.
BS 449	British Standard Specification for the use of Structural Steel in Building, etc.

1.8. Ductile materials

It has been observed above that the partially plastic range of the graph of Fig. 1.3 covers a much wider part of the strain axis than does the elastic range. Thus the extension of the material over this range is considerably in excess of that associated with elastic loading. The capacity of a material to allow these large extensions, i.e. the ability to be drawn out plastically, is termed its *ductility*. Materials with high ductility are termed *ductile* materials, members with low ductility are termed *brittle* materials. A quantitative value of the ductility is obtained by measurements of the *percentage elongation* or *percentage reduction in area*, both being defined below.

$$\text{Percentage elongation} = \frac{\text{increase in gauge length to fracture}}{\text{original gauge length}} \times 100$$

$$\text{Percentage reduction in area} = \frac{\text{reduction in cross-sectional area of necked portion}}{\text{original area}} \times 100$$

The latter value, being independent of any selected gauge length, is generally taken to be the more useful measure of ductility for reference purposes.

A property closely related to ductility is *malleability*, which defines a material's ability to be hammered out into thin sheets. A typical example of a malleable material is lead. This is used extensively in the plumbing trade where it is hammered or beaten into corners or joints to provide a weatherproof seal. Malleability thus represents the ability of a material to allow permanent extensions in all lateral directions under compressive loadings.

1.9. Brittle materials

A brittle material is one which exhibits relatively small extensions to fracture so that the partially plastic region of the tensile test graph is much reduced (Fig. 1.8). Whilst Fig. 1.3 referred to a low carbon steel, Fig. 1.8 could well refer to a much higher strength steel with a higher carbon content. There is little or no necking at fracture for brittle materials.

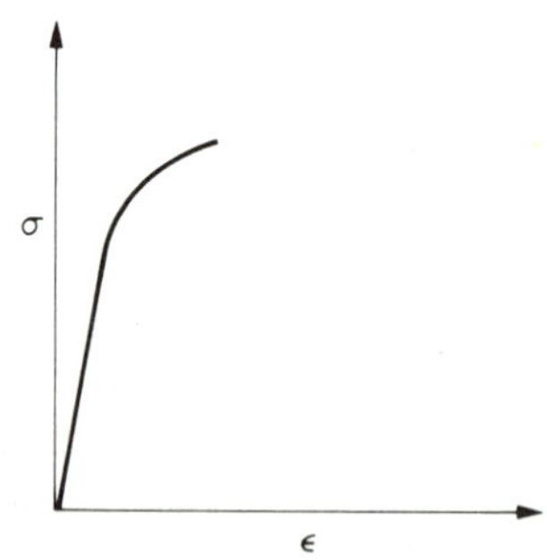

Fig. 1.8. Typical tensile test curve for a brittle material.

Typical variations of mechanical properties of steel with carbon content are shown in Fig. 1.9.

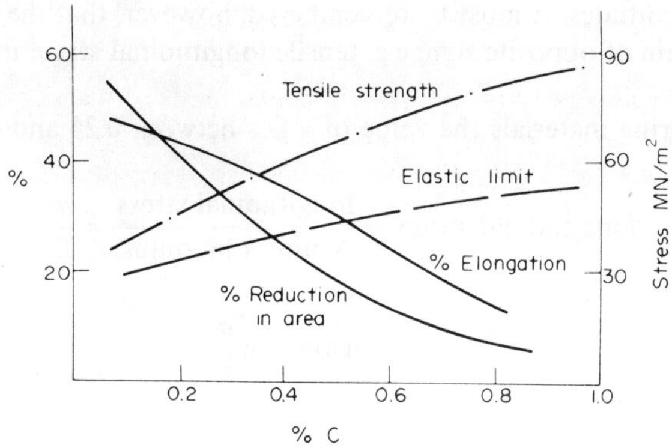

Fig. 1.9. Variation of mechanical properties of steel with carbon content.

1.10. Poisson's ratio

Consider the rectangular bar of Fig. 1.10 subjected to a tensile load. Under the action of this load the bar will increase in length by an amount δL giving a longitudinal strain in the bar of

$$\varepsilon_L = \frac{\delta L}{L}$$

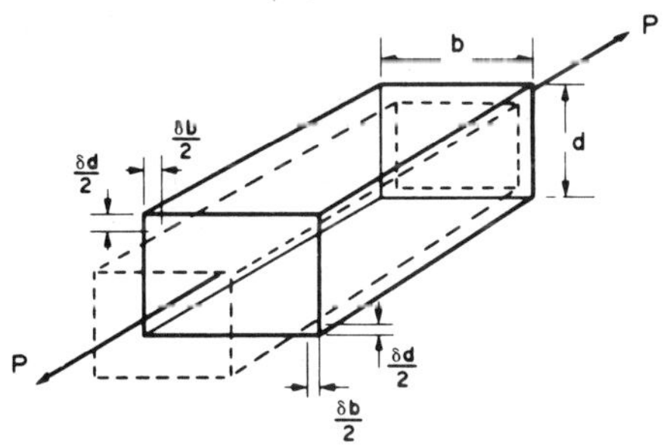

Fig. 1.10.

The bar will also exhibit, however, a *reduction* in dimensions laterally, i.e. its breadth and depth will both reduce. The associated lateral strains will both be equal, will be of opposite sense to the longitudinal strain, and will be given by

$$\varepsilon_{lat} = -\frac{\delta b}{b} = -\frac{\delta d}{d}$$

Provided the load on the material is retained within the elastic range the ratio of the lateral and longitudinal strains will always be constant. This ratio is termed *Poisson's ratio.*

i.e. $$\textbf{Poisson's ratio } (\nu) = \frac{\textbf{lateral strain}}{\textbf{longitudinal strain}} = \frac{(-\delta d/d)}{\delta L/L} \qquad (1.4)$$

The negative sign of the lateral strain is normally ignored to leave Poisson's ratio simply as a ratio of strain magnitudes. It must be remembered, however, that the longitudinal strain induces a lateral strain of opposite sign, e.g. tensile longitudinal strain induces compressive lateral strain.

For most engineering materials the value of ν lies between 0.25 and 0.33.

Since

$$\text{longitudinal strain} = \frac{\text{longitudinal stress}}{\text{Young's modulus}} = \frac{\sigma}{E} \tag{1.4a}$$

Hence

$$\textbf{lateral strain} = \nu\frac{\sigma}{E} \tag{1.4b}$$

1.11. Application of Poisson's ratio to a two-dimensional stress system

A two-dimensional stress system is one in which all the stresses lie within one plane such as the X–Y plane. From the work of §1.10 it will be seen that if a material is subjected to a tensile stress σ on one axis producing a strain σ/E and hence an extension on that axis, it will be subjected simultaneously to a lateral strain of ν times σ/E on any axis at right angles. This lateral strain will be compressive and will result in a compression or reduction of length on this axis.

Consider, therefore, an element of material subjected to two stresses at right angles to each other and let both stresses, σ_x and σ_y, be considered tensile, see Fig. 1.11.

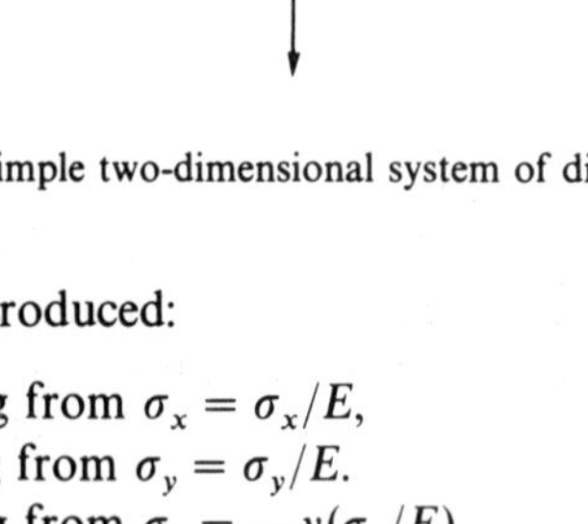

Fig. 1.11. Simple two-dimensional system of direct stresses.

The following strains will be produced:

(a) in the X direction resulting from $\sigma_x = \sigma_x/E$,
(b) in the Y direction resulting from $\sigma_y = \sigma_y/E$.
(c) in the X direction resulting from $\sigma_y = -\nu(\sigma_y/E)$,
(d) in the Y direction resulting from $\sigma_x = -\nu(\sigma_x/E)$.

strains (c) and (d) being the so-called *Poisson's ratio strain*, opposite in sign to the applied strains, i.e. compressive.

The total strain in the X direction will therefore be given by:

$$\varepsilon_x = \frac{\sigma_x}{E} - \nu\frac{\sigma_y}{E} = \frac{1}{E}(\sigma_x - \nu\sigma_y) \tag{1.5}$$

and the total strain in the Y direction will be:

$$\varepsilon_y = \frac{\sigma_y}{E} - \nu\frac{\sigma_x}{E} = \frac{1}{E}(\sigma_y - \nu\sigma_x) \qquad (1.6)$$

If any stress is, in fact, compressive its value must be substituted in the above equations together with a negative sign following the normal sign convention.

1.12. Shear stress

Consider a block or portion of material as shown in Fig. 1.12a subjected to a set of equal and opposite forces Q. (Such a system could be realised in a bicycle brake block when contacted with the wheel.) There is then a tendency for one layer of the material to slide over another to produce the form of failure shown in Fig. 1.12b. If this failure is restricted, then a *shear stress* τ is set up, defined as follows:

$$\textbf{shear stress } (\tau) = \frac{\textbf{shear load}}{\textbf{area resisting shear}} = \frac{Q}{A}$$

This shear stress will always be *tangential* to the area on which it acts; direct stresses, however, are always *normal* to the area on which they act.

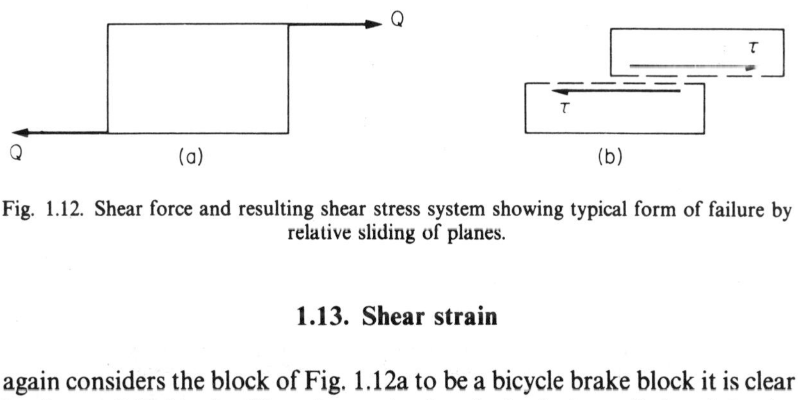

Fig. 1.12. Shear force and resulting shear stress system showing typical form of failure by relative sliding of planes.

1.13. Shear strain

If one again considers the block of Fig. 1.12a to be a bicycle brake block it is clear that the rectangular shape of the block will not be retained as the brake is applied and the shear forces introduced. The block will in fact change shape or "strain" into the form shown in Fig. 1.13. The angle of deformation γ is then termed the *shear strain.*

Shear strain is measured in radians and hence is non-dimensional, i.e. it has no units.

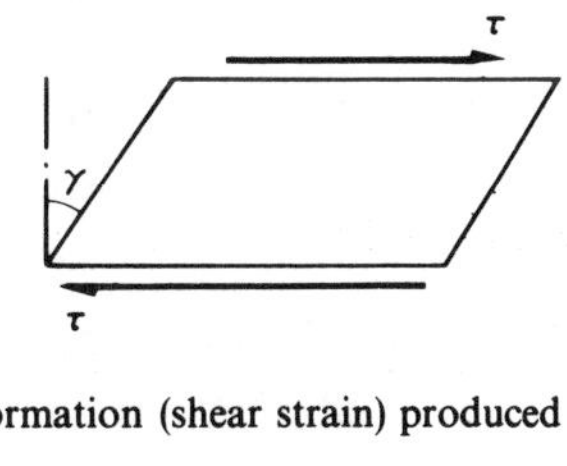

Fig. 1.13. Deformation (shear strain) produced by shear stresses.

1.14. Modulus of rigidity

For materials within the elastic range the shear strain is proportional to the shear stress producing it,

i.e.

$$\frac{\textbf{shear stress}}{\textbf{shear strain}} = \frac{\tau}{\gamma} = \textbf{constant} = G \tag{1.7}$$

The constant G is termed the *modulus of rigidity* or *shear modulus* and is directly comparable to the modulus of elasticity used in the direct stress application. The term *modulus* thus implies a ratio of stress to strain in each case.

1.15. Double shear

Consider the simple riveted lap joint shown in Fig. 1.14a. When load is applied to the plates the rivet is subjected to shear forces tending to shear it on one plane as indicated. In the butt joint with two cover plates of Fig. 1.14b, however, each rivet is subjected to possible shearing on two faces, i.e. *double shear*. In such cases twice the area of metal is resisting the applied forces so that the shear stress set up is given by

$$\textbf{shear stress } \tau \textbf{ (in double shear)} = \frac{P}{2A} \tag{1.8}$$

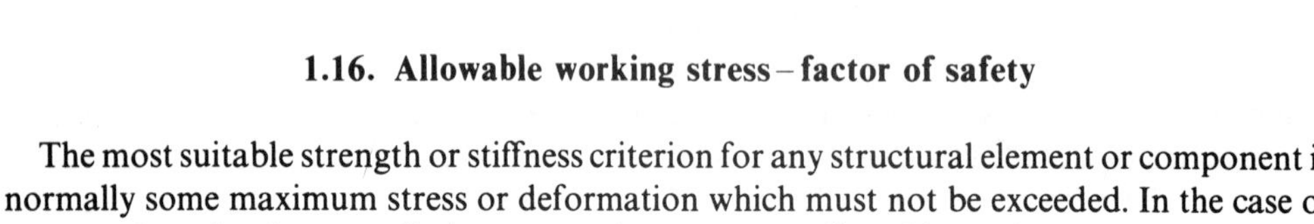

Fig. 1.14. (a) Single shear. (b) Double shear.

1.16. Allowable working stress – factor of safety

The most suitable strength or stiffness criterion for any structural element or component is normally some maximum stress or deformation which must not be exceeded. In the case of stresses the value is generally known as the *maximum allowable working stress*.

Because of uncertainties of loading conditions, design procedures, production methods, etc., designers generally introduce a *factor of safety* into their designs, defined as follows:

$$\textbf{factor of safety} = \frac{\textbf{maximum stress}}{\textbf{allowable working stress}} \tag{1.9}$$

However, in view of the fact that plastic deformations are seldom accepted this definition is sometimes modified to

$$\text{factor of safety} = \frac{\text{yield stress (or proof stress)}}{\text{allowable working stress}}$$

In the absence of any information as to which definition has been used for any quoted value of safety factor the former definition must be assumed. In this case a factor of safety of 3 implies that the design is capable of carrying three times the maximum stress to which it is expected the structure will be subjected in any normal loading condition. There is seldom any realistic basis for the selection of a particular safety factor and values vary significantly from one branch of engineering to another. Values are normally selected on the basis of a consideration of the social, human safety and economic consequences of failure. Typical values range from 2.5 (for relatively low consequence, static load cases) to 10 (for shock load and high safety risk applications)—see §15.12.

1.17. Load factor

In some loading cases, e.g. buckling of struts, neither the yield stress nor the ultimate strength is a realistic criterion for failure of components. In such cases it is convenient to replace the safety factor, based on stresses, with a different factor based on loads. The *load factor* is therefore defined as:

$$\textbf{load factor} = \frac{\textbf{load at failure}}{\textbf{allowable working load}} \tag{1.10}$$

This is particularly useful in applications of the so-called plastic limit design procedures, and further discussion of load factors is referred to §18.8.

1.18. Temperature stresses

When the temperature of a component is increased or decreased the material respectively expands or contracts. If this expansion or contraction is not resisted in any way then the processes take place free of stress. If, however, the changes in dimensions are restricted then stresses termed *temperature stresses* will be set up within the material.

Consider a bar of material with a linear coefficient of expansion α. Let the original length of the bar be L and let the temperature *increase* be t. If the bar is free to expand the change in length would be given by

$$\Delta L = L\alpha t \tag{1.11}$$

and the new length

$$L' = L + L\alpha t = L(1 + \alpha t)$$

If this extension were totally prevented, then a compressive stress would be set up equal to that produced when a bar of length $L(1 + \alpha t)$ is compressed through a distance of $L\alpha t$. In this case the bar experiences a compressive strain

$$\varepsilon = \frac{\Delta L}{L} = \frac{L\alpha t}{L(1 + \alpha t)}$$

In most cases αt is very small compared with unity so that

$$\varepsilon = \frac{L\alpha t}{L} = \alpha t$$

But $$\frac{\sigma}{\varepsilon} = E$$

$$\therefore \qquad \textbf{stress } \sigma = E\varepsilon = E\alpha t \qquad (1.12)$$

This is the stress set up owing to total restraint on expansions or contractions caused by a temperature rise, or fall, t. In the former case the stress is compressive, in the latter case the stress is tensile.

If the expansion or contraction of the bar is *partially* prevented then the stress set up will be less than that given by eqn. (1.10). Its value will be found in a similar way to that described above except that instead of being compressed through the total free expansion distance of $L\alpha t$ it will be compressed through some proportion of this distance depending on the amount of restraint.

Assuming some fraction n of $L\alpha t$ is *allowed*, then the extension which is prevented is $(1-n)L\alpha t$. This will produce a compressive strain, as described previously, of magnitude

$$\varepsilon = \frac{(1-n)\,L\alpha t}{L(1+\alpha t)}$$

or, approximately,

$$\varepsilon = (1-n)L\alpha t/L = (1-n)\alpha t.$$

The stress set up will then be E times ε.

i.e. $$\sigma = (1-n)\,E\alpha t \qquad (1.13)$$

Thus, for example, if one-third of the free expansion is prevented the stress set up will be two-thirds of that given by eqn. (1.12).

1.19. Stress concentrations – stress concentration factor

If a bar of uniform cross-section is subjected to an axial tensile or compressive load the stress is assumed to be uniform across the section. However, in the presence of any sudden change of section, hole, sharp corner, notch, keyway, material flaw, etc., the local stress will rise significantly. The ratio of this stress to the nominal stress at the section in the absence of any of these so-called *stress concentrations* is termed the *stress concentration* factor.

1.20. Toughness

Toughness is defined as the ability of a material to withstand cracks, i.e. to prevent the transfer or propagation of cracks across its section hence causing failure. Two distinct types of toughness mechanism exist and in each case it is appropriate to consider the crack as a very high local stress concentration.

The first type of mechanism relates particularly to ductile materials which are generally regarded as tough. This arises because the very high stresses at the end of the crack produce local yielding of the material and local plastic flow at the crack tip. This has the action of blunting the sharp tip of the crack and hence reduces its stress-concentration effect considerably (Fig. 1.15).

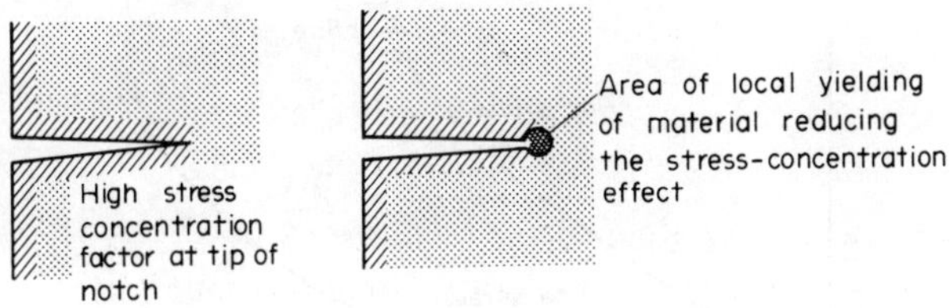

Fig. 1.15. Toughness mechanism – type 1.

The second mechanism refers to fibrous, reinforced or resin-based materials which have weak interfaces. Typical examples are glass-fibre reinforced materials and wood. It can be shown that a region of local tensile stress always exists at the front of a propagating crack and provided that the adhesive strength of the fibre/resin interface is relatively low (one-fifth the cohesive strength of the complete material) this tensile stress opens up the interface and produces a crack sink, i.e. it blunts the crack by effectively increasing the radius at the crack tip, thereby reducing the stress-concentration effect (Fig. 1.16).

This principle is used on occasions to stop, or at least delay, crack propagation in engineering components when a temporary "repair" is carried out by drilling a hole at the end of a crack, again reducing its stress-concentration effect.

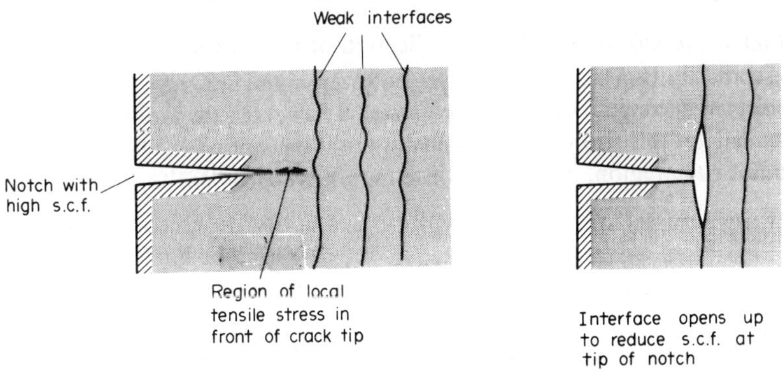

Fig. 1.16. Toughness mechanism – type 2.

1.21. Creep and fatigue

In the preceding paragraphs it has been suggested that failure of materials occurs when the ultimate strengths have been exceeded. Reference has also been made in §1.15 to cases where excessive deformation, as caused by plastic deformation beyond the yield point, can be considered as a criterion for effective failure of components. This chapter would not be complete, therefore, without reference to certain loading conditions under which materials can fail at stresses much less than the yield stress, namely *creep* and *fatigue*.

Creep is the gradual increase of plastic strain in a material with time at constant load. Particularly at elevated temperatures some materials are susceptible to this phenomenon and even under the constant load mentioned strains can increase continually until fracture. This form of fracture is particularly relevant to turbine blades, nuclear reactors, furnaces, rocket motors, etc.

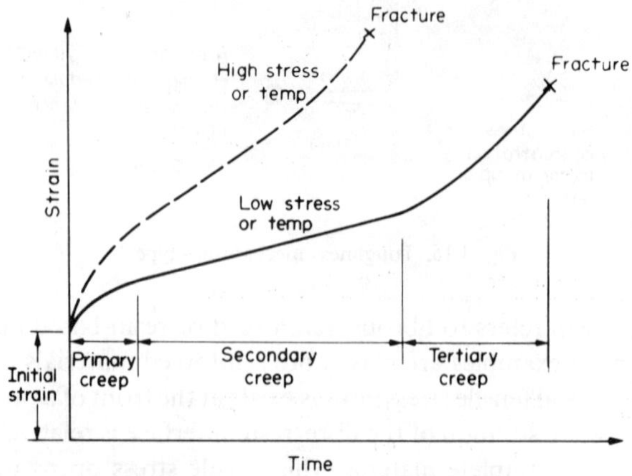

Fig. 1.17. Typical creep curve.

The general form of the strain versus time graph or *creep curve* is shown in Fig. 1.17 for two typical operating conditions. In each case the curve can be considered to exhibit four principal features.

(a) An *initial strain*, due to the initial application of load. In most cases this would be an elastic strain.
(b) A *primary creep* region, during which the creep rate (slope of the graph) diminishes.
(c) A *secondary creep* region, when the creep rate is sensibly constant.
(d) A *tertiary creep* region, during which the creep rate accelerates to final fracture.

It is clearly imperative that a material which is susceptible to creep effects should only be subjected to stresses which keep it in the secondary (straight line) region throughout its service life. This enables the amount of creep extension to be estimated and allowed for in design.

Fatigue is the failure of a material under fluctuating stresses each of which is believed to produce minute amounts of plastic strain. Fatigue is particularly important in components subjected to repeated and often rapid load fluctuations, e.g. aircraft components, turbine blades, vehicle suspensions, etc. Fatigue behaviour of materials is usually described by a *fatigue life* or *S–N curve* in which the number of stress cycles N to produce failure with a stress peak of S is plotted against S. A typical S–N curve for mild steel is shown in Fig. 1.18. The particularly relevant feature of this curve is the limiting stress S_n since it is assumed that stresses below this value will not produce fatigue failure however many cycles are applied, i.e. there is *infinite life*. In the simplest design cases, therefore, there is an aim to keep all stresses below this limiting level. However, this often implies an over-design in terms of physical size and material usage, particularly in cases where the stress may only occasionally exceed the limiting value noted above. This is, of course, particularly important in applications such as aerospace structures where component weight is a premium. Additionally the situation is complicated by the many materials which do not show a defined limit, and modern design procedures therefore rationalise the situation by aiming at a prescribed, long, but *finite life*, and accept that service stresses will occasionally exceed the value S_n. It is clear that the number of occasions on which the stress exceeds S_n, and by how

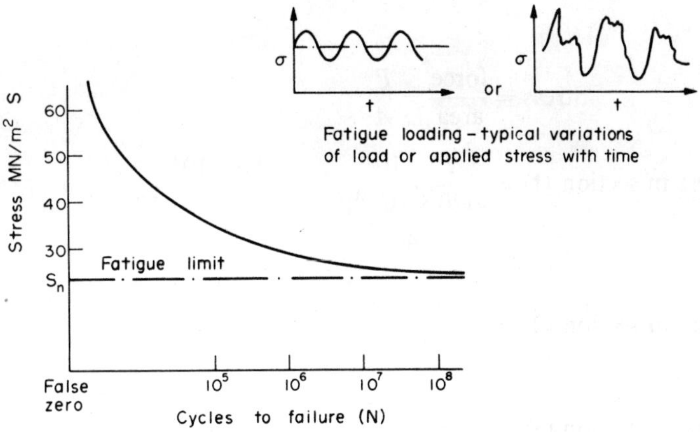

Fig. 1.18. Typical *S–N* fatigue curve for mild steel.

much, will have an important bearing on the prescribed life and considerable specimen, and often full-scale, testing is required before sufficient statistics are available to allow realistic life assessment.

The importance of the creep and fatigue phenomena cannot be over-emphasised and the comments above are only an introduction to the concepts and design philosophies involved. For detailed consideration of these topics and of the other materials testing topics introduced earlier the reader is referred to Chapter 26 and to the texts listed at the end of this chapter.

Examples

Example 1.1

Determine the stress in each section of the bar shown in Fig. 1.19 when subjected to an axial tensile load of 20 kN. The central section is 30 mm square cross-section; the other portions are of circular section, their diameters being indicated. What will be the total extension of the bar? For the bar material $E = 210\,\text{GN/m}^2$.

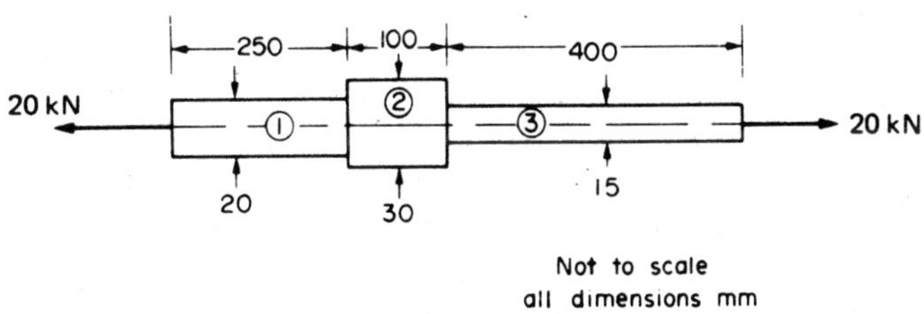

Fig. 1.19.

Solution

$$\text{Stress} = \frac{\text{force}}{\text{area}} = \frac{P}{A}$$

$$\text{Stress in section (1)} = \frac{20 \times 10^3}{\dfrac{\pi(20 \times 10^{-3})^2}{4}} = \frac{80 \times 10^3}{\pi \times 400 \times 10^{-6}} = 63.66\ \text{MN/m}^2$$

$$\text{Stress in section (2)} = \frac{20 \times 10^3}{30 \times 30 \times 10^{-6}} = 22.2\ \text{MN/m}^2$$

$$\text{Stress in section (3)} = \frac{20 \times 10^3}{\dfrac{\pi(15 \times 10^{-3})^2}{4}} = \frac{80 \times 10^3}{\pi \times 225 \times 10^{-6}} = 113.2\ \text{MN/m}^2$$

Now the extension of a bar can always be written in terms of the stress in the bar since

$$E = \frac{\text{stress}}{\text{strain}} = \frac{\sigma}{\delta/L}$$

i.e.
$$\delta = \frac{\sigma L}{E}$$

$\therefore$
$$\text{extension of section (1)} = 63.66 \times 10^6 \times \frac{250 \times 10^{-3}}{210 \times 10^9} = 75.8 \times 10^{-6}\ \text{m}$$

$$\text{extension of section (2)} = 22.2 \times 10^6 \times \frac{100 \times 10^{-3}}{210 \times 10^9} = 10.6 \times 10^{-6}\ \text{m}$$

$$\text{extension of section (3)} = 113.2 \times 10^6 \times \frac{400 \times 10^{-3}}{210 \times 10^9} = 215.6 \times 10^{-6}\ \text{m}$$

$\therefore$
$$\text{total extension} = (75.8 + 10.6 + 215.6)10^{-6}$$
$$= 302 \times 10^{-6}\ \text{m}$$
$$= \mathbf{0.302\ mm}$$

Example 1.2

(a) A 25 mm diameter bar is subjected to an axial tensile load of 100 kN. Under the action of this load a 200 mm gauge length is found to extend 0.19×10^{-3} mm. Determine the modulus of elasticity for the bar material.

(b) If, in order to reduce weight whilst keeping the external diameter constant, the bar is bored axially to produce a cylinder of uniform thickness, what is the maximum diameter of bore possible given that the maximum allowable stress is 240 MN/m^2? The load can be assumed to remain constant at 100 kN.

(c) What will be the change in the outside diameter of the bar under the limiting stress quoted in (b)? ($E = 210\ \text{GN/m}^2$ and $v = 0.3$).

Solution

(a) From eqn. (1.2),

$$\text{Young's modulus } E = \frac{PL}{A\,\delta L}$$

$$= \frac{100\times10^3\times200\times10^{-3}}{\dfrac{\pi(25\times10^{-3})^2}{4}\times0.19\times10^{-3}}$$

$$= \mathbf{214\,GN/m^2}$$

(b) Let the required bore diameter be d mm; the cross-sectional area of the bar will then be reduced to

$$A = \left[\frac{\pi\times25^2}{4} - \frac{\pi d^2}{4}\right]10^{-6} = \frac{\pi}{4}(25^2 - d^2)10^{-6}\,\text{m}^2$$

$$\therefore \qquad \text{stress in bar} = \frac{P}{A} = \frac{4\times100\times10^3}{\pi(25^2-d^2)10^{-6}}$$

But this stress is restricted to a maximum allowable value of $240\,\text{MN/m}^2$.

$$\therefore \qquad 240\times10^6 = \frac{4\times100\times10^3}{\pi(25^2-d^2)10^{-6}}$$

$$\therefore \qquad 25^2 - d^2 = \frac{4\times100\times10^3}{240\times10^6\times\pi\times10^{-6}} = 530.5$$

$$\therefore \qquad d^2 = 94.48 \quad \text{and} \quad d = 9.72\,\text{mm}$$

The maximum bore possible is thus **9.72 mm**.

(c) The change in the outside diameter of the bar will be obtained from the lateral strain,

$$\text{i.e.} \qquad \text{lateral strain} = \frac{\delta d}{d}$$

$$\text{But} \qquad \text{Poisson's ratio } \nu = \frac{\text{lateral strain}}{\text{longitudinal strain}}$$

$$\text{and} \qquad \text{longitudinal strain} = \frac{\sigma}{E} = \frac{240\times10^6}{210\times10^9}$$

$$\therefore \qquad \frac{\delta d}{d} = -\nu\frac{\sigma}{E} = -\frac{0.3\times240\times10^6}{210\times10^9}$$

$$\therefore \qquad \text{change in outside diameter} = -\frac{0.3\times240\times10^6}{210\times10^9}\times25\times10^{-3}$$

$$= \mathbf{-8.57\times10^{-6}\,m} \text{ (a reduction)}$$

Example 1.3

The coupling shown in Fig. 1.20 is constructed from steel of rectangular cross-section and is designed to transmit a tensile force of 50 kN. If the bolt is of 15 mm diameter calculate:

(a) the shear stress in the bolt;
(b) the direct stress in the plate;
(c) the direct stress in the forked end of the coupling.

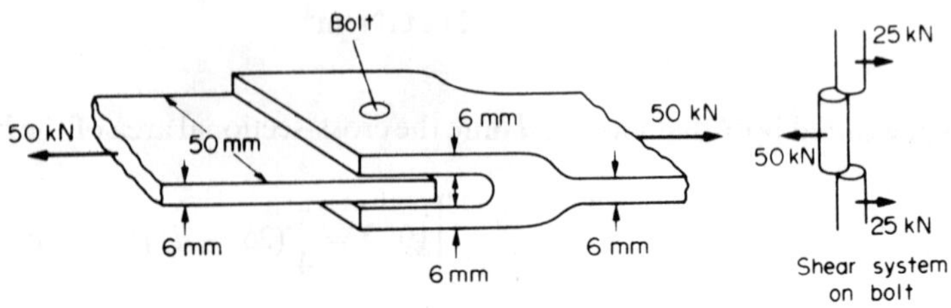

Fig. 1.20.

Solution

(a) The bolt is subjected to double shear, tending to shear it as shown in Fig. 1.14b. There is thus twice the area of the bolt resisting the shear and from eqn. (1.8)

$$\text{shear stress in bolt} = \frac{P}{2A} = \frac{50 \times 10^3 \times 4}{2 \times \pi(15 \times 10^{-3})^2}$$

$$= \frac{100 \times 10^3}{\pi(15 \times 10^{-3})^2} = \mathbf{141.5\,MN/m^2}$$

(b) The plate will be subjected to a direct tensile stress given by

$$\sigma = \frac{P}{A} = \frac{50 \times 10^3}{50 \times 6 \times 10^{-6}} = \mathbf{166.7\,MN/m^2}$$

(c) The force in the coupling is shared by the forked end pieces, each being subjected to a direct stress

$$\sigma = \frac{P}{A} = \frac{25 \times 10^3}{50 \times 6 \times 10^{-6}} = \mathbf{83.3\,MN/m^2}$$

Example 1.4

Derive an expression for the total extension of the tapered bar of circular cross-section shown in Fig. 1.21 when it is subjected to an axial tensile load W.

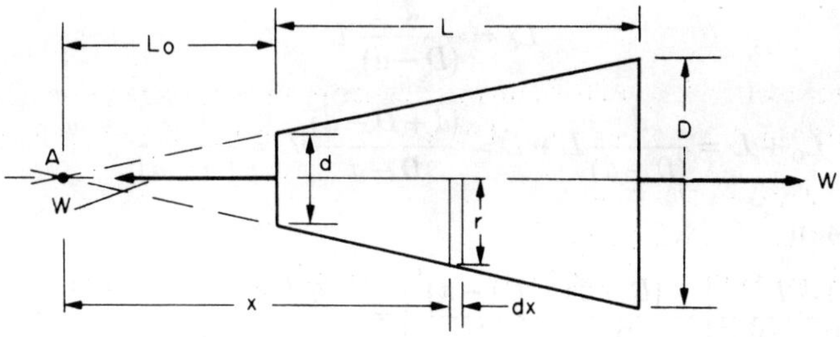

Fig. 1.21.

Solution

From the proportions of Fig. 1.21,

$$\frac{d/2}{L_0} = \frac{(D-d)/2}{L}$$

$$\therefore \qquad L_0 = \frac{d}{(D-d)} L$$

Consider an element of thickness dx and radius r, distance x from the point of taper A.

$$\text{Stress on the element} = \frac{W}{\pi r^2}$$

But

$$\frac{r}{x} = \frac{d}{2L_0}$$

$$\therefore \qquad r = d\left(\frac{D-d}{2dL}\right)x = \frac{x(D-d)}{2L}$$

$$\therefore \qquad \text{stress on the element} = \frac{4WL^2}{\pi(D-d)^2x^2}$$

$$\therefore \qquad \text{strain on the element} = \frac{\sigma}{E}$$

$$\text{and extension of the element} = \frac{\sigma dx}{E}$$

$$= \frac{4WL^2}{\pi(D-d)^2x^2E}dx$$

$$\therefore \qquad \text{total extension of bar} = \int_{L_0}^{L_0+L} \frac{4WL^2}{\pi(D-d)^2E}\frac{dx}{x^2}$$

$$= \frac{4WL^2}{\pi(D-d)^2E}\left[-\frac{1}{x}\right]_{L_0}^{L_0+L}$$

$$= \frac{4WL^2}{\pi(D-d)^2E}\left[-\frac{1}{(L_0+L)} - \left(-\frac{1}{L_0}\right)\right]$$

But $$L_0 = \frac{d}{(D-d)}L$$

$$\therefore \quad L_0 + L = \frac{d}{(D-d)}L + L = \frac{(d+D-d)}{D-d}L = \frac{DL}{(D-d)}$$

$\therefore$ total extension

$$= \frac{4WL^2}{\pi(D-d)^2E}\left[-\frac{(D-d)}{DL} + \frac{(D-d)}{dL}\right] = \frac{4WL}{\pi(D-d)E}\left[\frac{(-d+D)}{Dd}\right]$$

$$= \frac{4WL}{\pi DdE}$$

Example 1.5

The following figures were obtained in a standard tensile test on a specimen of low carbon steel:

diameter of specimen, 11.28 mm;
gauge length, 56 mm;
minimum diameter after fracture, 6.45 mm.

Using the above information and the table of results below, produce:

(1) a load/extension graph over the complete test range;
(2) a load/extension graph to an enlarged scale over the elastic range of the specimen.

Load (kN)	2.47	4.97	7.4	9.86	12.33	14.8	17.27	19.74	22.2	24.7
Extension ($m \times 10^{-6}$)	5.6	11.9	18.2	24.5	31.5	38.5	45.5	52.5	59.5	66.5
Load (kN)	27.13	29.6	32.1	33.3	31.2	32	31.5	32	32.2	34.5
Extension ($m \times 10^{-6}$)	73.5	81.2	89.6	112	224	448	672	840	1120	1680
Load (kN)	35.8	37	38.7	39.5	40	39.6	35.7	28		
Extension ($m \times 10^{-6}$)	1960	2520	3640	5600	7840	11200	13440	14560		

Using the two graphs and other information supplied, determine the values of:

(a) Young's modulus of elasticity;
(b) the ultimate tensile stress;
(c) the stress at the upper and lower yield points;
(d) the percentage reduction of area;
(e) the percentage elongation;
(f) the nominal and actual stress at fracture.

Solution

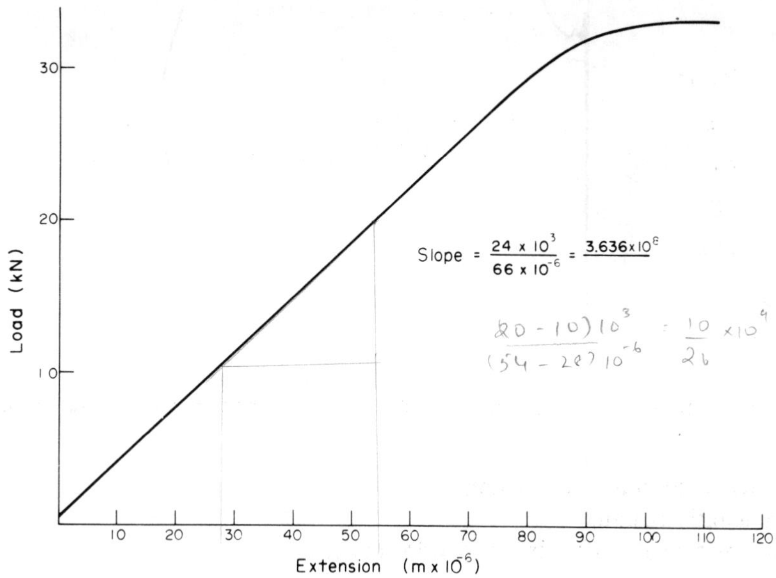

Fig. 1.22. Load–extension graph for elastic range.

(a) $$\text{Young's modulus } E = \frac{\sigma}{\varepsilon} = \frac{\text{load}}{\text{area}} \times \frac{\text{gauge length}}{\text{extension}}$$

$$= \frac{\text{load}}{\text{extension}} \times \frac{\text{gauge length}}{\text{area}}$$

i.e $$E = \text{slope of graph} \times \frac{L}{A} = 3.636 \times 10^8 \times \frac{56 \times 10^{-3}}{100 \times 10^{-6}}$$

$$= 203.6 \times 10^9 \text{ N/m}^2$$

$\therefore$ $$\mathbf{E = 203.6\ GN/m^2}$$

(b) $$\text{Ultimate tensile stress} = \frac{\text{maximum load}}{\text{cross-section area}} = \frac{40.2 \times 10^3}{100 \times 10^{-6}} = \mathbf{402\ MN/m^2}$$

(see Fig. 1.23).

(c) $$\text{Upper yield stress} = \frac{33.3 \times 10^3}{100 \times 10^{-6}} = \mathbf{333\ MN/m^2}$$

$$\text{Lower yield stress} = \frac{31.2 \times 10^3}{100 \times 10^{-6}} = \mathbf{312\ MN/m^2}$$

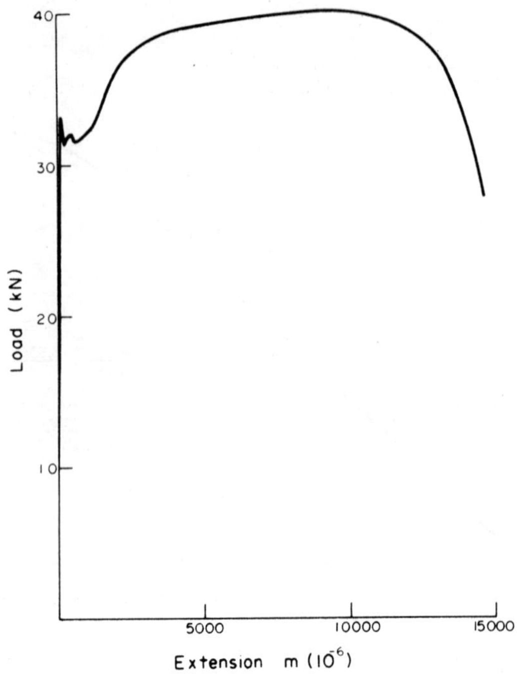

Fig. 1.23. Load–extension graph for complete load range.

(d) $$\text{Percentage reduction of area} = \frac{\left(\frac{\pi}{4}D^2 - \frac{\pi}{4}d^2\right)}{\frac{\pi}{4}D^2} \times 100$$

$$= \frac{(D^2 - d^2)}{D^2} \times 100$$

$$= \frac{(11.28^2 - 6.45^2)}{11.28^2} = \mathbf{67.3\%}$$

(e) $$\text{Percentage elongation} = \frac{(70.56 - 56)}{56} \times 100$$

$$= \mathbf{26\%}$$

(f) $$\text{Nominal stress at fracture} = \frac{28 \times 10^3}{100 \times 10^{-6}} = \mathbf{280\ MN/m^2}$$

$$\text{Actual stress at fracture} = \frac{28 \times 10^3}{\frac{\pi}{4}(6.45)^2 \times 10^{-6}} = \mathbf{856.9\ MN/m^2}$$

Problems

1.1 (A). A 25 mm square-cross-section bar of length 300 mm carries an axial compressive load of 50 kN. Determine the stress set up in the bar and its change of length when the load is applied. For the bar material $E = 200\,\text{GN/m}^2$. [80 MN/m^2; 0.12 mm.]

1.2 (A). A steel tube, 25 mm outside diameter and 12 mm inside diameter, carries an axial tensile load of 40 kN. What will be the stress in the bar? What further increase in load is possible if the stress in the bar is limited to 225 MN/m^2? [106 MN/m^2; 45 kN.]

1.3 (A). Define the terms *shear stress* and *shear strain*, illustrating your answer by means of a simple sketch.
Two circular bars, one of brass and the other of steel, are to be loaded by a shear load of 30 kN. Determine the necessary diameter of the bars (a) in single shear, (b) in double shear, if the shear stress in the two materials must not exceed 50 MN/m^2 and 100 MN/m^2 respectively. [27.6, 19.5, 19.5, 13.8 mm.]

1.4 (A). Two fork-end pieces are to be joined together by a single steel pin of 25 mm diameter and they are required to transmit 50 kN. Determine the minimum cross-sectional area of material required in one branch of either fork if the stress in the fork material is not to exceed 180 MN/m^2. What will be the maximum shear stress in the pin? [1.39×10^{-4} m^2; 50.9 MN/m^2.]

1.5 (A). A simple turnbuckle arrangement is constructed from a 40 mm outside diameter tube threaded internally at each end to take two rods of 25 mm outside diameter with threaded ends. What will be the nominal stresses set up in the tube and the rods, ignoring thread depth, when the turnbuckle carries an axial load of 30 kN? Assuming a sufficient strength of thread, what maximum load can be transmitted by the turnbuckle if the maximum stress is limited to 180 MN/m^2? [39.2, 61.1 MN/m^2, 88.4 kN.]

1.6 (A). An I-seetion girder is constructed from two 80 mm × 12 mm flanges joined by an 80 mm × 12 mm web. Four such girders are mounted vertically one at each corner of a horizontal platform which the girders support. The platform is 4 m above ground level and weighs 10 kN. Assuming that each girder supports an equal share of the load, determine the maximum compressive stress set up in the material of each girder when the platform supports an additional load of 15 kN. The weight of the girders may *not* be neglected. The density of the cast iron from which the girders are constructed is 7470 kg/m^3. [2.46 MN/m^2.]

1.7 (A). A bar *ABCD* consists of three sections: *AB* is 25 mm square and 50 mm long, *BC* is of 20 mm diameter and 40 mm long and *CD* is of 12 mm diameter and 50 mm long. Determine the stress set up in each section of the bar when it is subjected to an axial tensile load of 20 kN. What will be the total extension of the bar under this load? For the bar material, $E = 210\,\text{GN/m}^2$. [32, 63.7, 176.8 MN/m^2, 0.062 mm.]

1.8 (A). A steel bar *ABCD* consists of three sections: *AB* is of 20 mm diameter and 200 mm long, *BC* is 25 mm square and 400 mm long, and *CD* is of 12 mm diameter and 200 mm long. The bar is subjected to an axial compressive load which induces a stress of 30 MN/m^2 on the largest cross-section. Determine the total decrease in the length of the bar when the load is applied. For steel $E = 210\,\text{GN/m}^2$. [0.272 mm.]

1.9 (A). During a tensile test on a specimen the following results were obtained:

Load (kN)	15	30	40	50	55	60	65
Extension (mm)	0.05	0.094	0.127	0.157	1.778	2.79	3.81
Load (kN)	70	75	80	82	80		70
Extension (mm)	5.08	7.62	12.7	16.0	19.05		22.9

Diameter of gauge length = 19 mm Gauge length = 100 mm
Diameter at fracture = 16.49 mm Gauge length at fracture = 121 mm

Plot the complete load extension graph and the straight line portion to an enlarged scale. Hence determine:

(a) the modulus of elasticity;
(b) the percentage elongation;
(c) the percentage reduction in area;
(d) the nominal stress at fracture;
(e) the actual stress at fracture;
(f) the tensile strength.

[116 GN/m^2; 21 %; 24.7 %; 247 MN/m^2; 328 MN/m^2; 289 MN/m^2.]

1.10 Figure 1.24 shows a special spanner used to tighten screwed components. A torque is applied at the tommy-bar and is transmitted to the pins which engage into holes located into the end of a screwed component.

(a) Using the data given in Fig. 1.24 calculate:
  (i) the diameter D of the shank if the shear stress is not to exceed 50 N/mm^2,
  (ii) the stress due to bending in the tommy-bar,
  (iii) the shear stress in the pins.
(b) Why is the tommy-bar a preferred method of applying the torque?

[C.G.] [9.14 mm; 254.6 MN/m^2; 39.8 MN/m^2.]

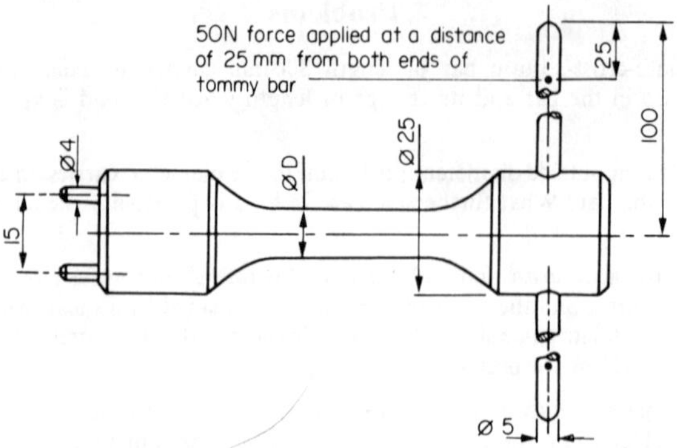

Fig. 1.24.

1.11 (**a**) A test piece is cut from a brass bar and subjected to a tensile test. With a load of 6.4 kN the test piece, of diameter 11.28 mm, extends by 0.04 mm over a gauge length of 50 mm. Determine:
(i) the stress, (ii) the strain, (iii) the modulus of elasticity.

(**b**) A spacer is turned from the same bar. The spacer has a diameter of 28 mm and a length of 250 mm, both measurements being made at 20°C. The temperature of the spacer is then increased to 100°C, the natural expansion being entirely prevented. Taking the coefficient of linear expansion to be 18×10^{-6}/°C determine:
(i) the stress in the spacer, (ii) the compressive load on the spacer.

[C.G.] [64 MN/m^2, 0.0008, 80 GN/m^2, 115.2 MN/m^2, 71 kN.]

Bibliography

1. J. G. Tweedale, *Mechanical Properties of Metal*, George Allen & Unwin Ltd., 1964.
2. E. N. Simons, *The Testing of Metals*, David & Charles, Newton Abbot, 1972.
3. J. Y. Mann, *Fatigue of Materials – An Introductory Text*, Melbourne University Press, 1967.
4. P. G. Forrest, *Fatigue of Metals*, Pergamon, 1970.
5. R. B. Heywood, *Designing against Fatigue*, Chapman & Hall, 1962.
6. *Fatigue – An Interdisciplinary Approach*, 10th Sagamore Army Materials Research Conference Proceedings, Syracuse University Press, 1964.
7. A. J. Kennedy, *Processes of Creep and Fatigue in Metals*, Oliver & Boyd, Edinburgh and London, 1962.
8. R. K. Penny and D. L. Marriott, *Design for Creep*, McGraw-Hill (U.K.), 1971.
9. A. I. Smith and A. M. Nicolson, *Advances in Creep Design*, Applied Science Publishers, London, 1971.
10. J. F. Knott, *Fundamentals of Fracture Mechanics*, Butterworths, London, 1973.
11. H. Liebowitz, *Fracture – An Advanced Treatise*, vols. 1 to 7, Academic Press, New York and London, 1972.
12. W. D. Biggs, *The Brittle Fracture of Steel*, MacDonald & Evans Ltd., 1960.
13. D. Broek, *Elementary Engineering Fracture Mechanics*, Noordhoff International Publishing, Holland, 1974.
14. J. E. Gordon, *The New Science of Strong Materials*, Pelican 213, Penguin, 1970.
15. R. J. Roark and W. C. Young, *Formulas for Stress and Strain*, 5th Edition, McGraw-Hill, 1975.

CHAPTER 2

COMPOUND BARS

Summary

When a compound bar is constructed from members of different materials, lengths and areas and is subjected to an external tensile or compressive load W the load carried by any single member is given by

$$F_1 = \frac{\dfrac{E_1 A_1}{L_1}}{\Sigma \dfrac{EA}{L}} W$$

where suffix 1 refers to the single member and $\Sigma \dfrac{EA}{L}$ is the sum of all such quantities for all the members.

Where the bars have a common length the compound bar can be reduced to a single equivalent bar with an equivalent Young's modulus, termed a *combined E.*

$$\text{Combined } E = \frac{\Sigma EA}{\Sigma A}$$

The free expansion of a bar under a temperature change from T_1 to T_2 is

$$\alpha(T_2 - T_1)L$$

where α is the coefficient of linear expansion and L is the length of the bar.

If this expansion is prevented a stress will be induced in the bar given by

$$\alpha(T_2 - T_1)E$$

To determine the stresses in a compound bar composed of two members of different free lengths two principles are used:

(1) The tensile force applied to the short member by the long member is equal in magnitude to the compressive force applied to the long member by the short member.
(2) The extension of the short member plus the contraction of the long member equals the difference in free lengths.

This difference in free lengths may result from the tightening of a nut or from a temperature change in two members of different material (i.e. different coefficients of expansion) but of equal length initially.

If such a bar is then subjected to an additional external load the resultant stresses may be obtained by using the *principle of superposition.* With this method the stresses in the members arising from the separate effects are obtained and the results added, taking account of sign, to give the resultant stresses.

N.B.: Discussion in this chapter is concerned with compound bars which are symmetrically proportioned such that no bending results.

2.1. Compound bars subjected to external load

In certain applications it is necessary to use a combination of elements or bars made from different materials, each material performing a different function. In overhead electric cables, for example, it is often convenient to carry the current in a set of copper wires surrounding steel wires, the latter being designed to support the weight of the cable over large spans. Such combinations of materials are generally termed *compound bars*. Discussion in this chapter is concerned with compound bars which are symmetrically proportioned such that no bending results.

When an external load is applied to such a compound bar it is shared between the individual component materials in proportions depending on their respective lengths, areas and Young's moduli.

Consider, therefore, a compound bar consisting of n members, each having a different length and cross-sectional area and each being of a different material; this is shown diagrammatically in Fig. 2.1. Let all members have a common extension x, i.e. the load is positioned to produce the same extension in each member.

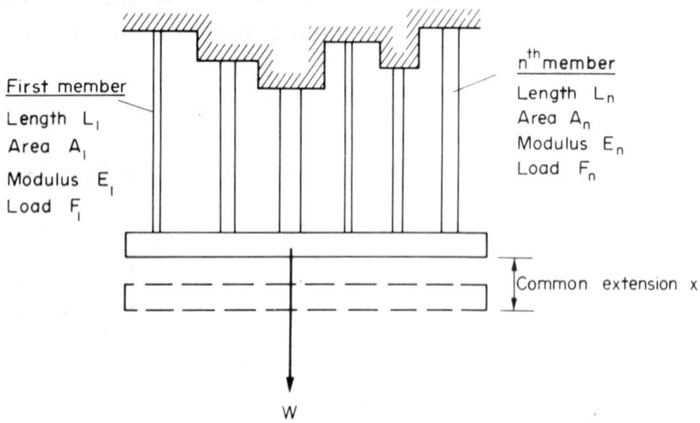

Fig. 2.1. Diagrammatic representation of a compound bar formed of different materials with different lengths, cross-sectional areas and Young's moduli.

For the nth member,

$$\frac{\text{stress}}{\text{strain}} = E_n = \frac{F_n L_n}{A_n x_n}$$

$$\therefore \qquad F_n = \frac{E_n A_n x}{L_n} \qquad (2.1)$$

where F_n is the force in the nth member and A_n and L_n are its cross-sectional area and length.

The total load carried will be the sum of all such loads for all the members

i.e.
$$W = \sum \frac{E_n A_n x}{L_n} = x \sum \frac{E_n A_n}{L_n} \tag{2.2}$$

Now from eqn. (2.1) the force in member 1 is given by

$$F_1 = \frac{E_1 A_1 x}{L_1}$$

But, from eqn. (2.2),

$$x = \frac{W}{\sum \frac{E_n A_n}{L_n}}$$

∴
$$F_1 = \frac{\frac{E_1 A_1}{L_1}}{\sum \frac{EA}{L}} W \tag{2.3}$$

i.e. *each member carries a portion of the total load W proportional to its EA/L value.*

If the wires are all of equal length the above equation reduces to

$$F_1 = \frac{E_1 A_1}{\Sigma EA} W \tag{2.4}$$

The stress in member 1 is then given by

$$\sigma_1 = \frac{F_1}{A_1} \tag{2.5}$$

2.2. Compound bars – "equivalent" or "combined" modulus

In order to determine the common extension of a compound bar it is convenient to consider it as a single bar of an imaginary material with an *equivalent* or *combined* modulus E_c. Here it is necessary to assume that both the extension and the original lengths of the individual members of the compound bar are the same; the strains in all members will then be equal.

Now total load on compound bar $= F_1 + F_2 + F_3 + \ldots + F_n$ where F_1, F_2, etc., are the loads in members 1, 2, etc.

But force = stress × area

∴
$$\sigma(A_1 + A_2 + \ldots + A_n) = \sigma_1 A_1 + \sigma_2 A_2 + \ldots + \sigma_n A_n$$

where σ is the stress in the equivalent single bar.

Dividing through by the common strain ε,

$$\frac{\sigma}{\varepsilon}(A_1 + A_2 + \ldots + A_n) = \frac{\sigma_1}{\varepsilon} A_1 + \frac{\sigma_2}{\varepsilon} A_2 + \ldots + \frac{\sigma_n}{\varepsilon} A_n$$

i.e.
$$E_c(A_1 + A_2 + \ldots + A_n) = E_1 A_1 + E_2 A_2 + \ldots + E_n A_n$$

where E_c is the *equivalent* or *combined* E of the single bar.

$$\therefore \qquad \text{combined } E = \frac{E_1A_1 + E_2A_2 + \ldots + E_nA_n}{A_1 + A_2 + \ldots + A_n}$$

i.e.

$$E_c = \frac{\Sigma EA}{\Sigma A} \tag{2.6}$$

With an external load W applied,

$$\text{stress in the equivalent bar} = \frac{W}{\Sigma A}$$

and

$$\text{strain in the equivalent bar} = \frac{W}{E_c \Sigma A} = \frac{x}{L}$$

$\therefore$ since

$$\frac{\text{stress}}{\text{strain}} = E$$

$$\textbf{common extension } x = \frac{WL}{E_c \Sigma A} \tag{2.7}$$

$$= \text{extension of single bar}$$

2.3. Compound bars subjected to temperature change

When a material is subjected to a change in temperature its length will change by an amount

$$\alpha Lt$$

where α is the coefficient of linear expansion for the material, L is the original length and t the temperature change. (An increase in temperature produces an increase in length and a decrease in temperature a decrease in length except in very special cases of materials with zero or negative coefficients of expansion which need not be considered here.)

If, however, the free expansion of the material is prevented by some external force, then a stress is set up in the material. This stress is equal in magnitude to that which would be produced in the bar by initially allowing the free change of length and then applying sufficient force to return the bar to its original length.

Now

$$\text{change in length} = \alpha Lt$$

$$\therefore \qquad \text{strain} = \frac{\alpha Lt}{L} = \alpha t$$

Therefore, the stress created in the material by the application of sufficient force to remove this strain

$$= \text{strain} \times E$$

$$= \boldsymbol{E\alpha t}$$

Consider now a compound bar constructed from two different materials rigidly joined together as shown in Fig. 2.2 and Fig. 2.3(a). For simplicity of description consider that the materials in this case are steel and brass.

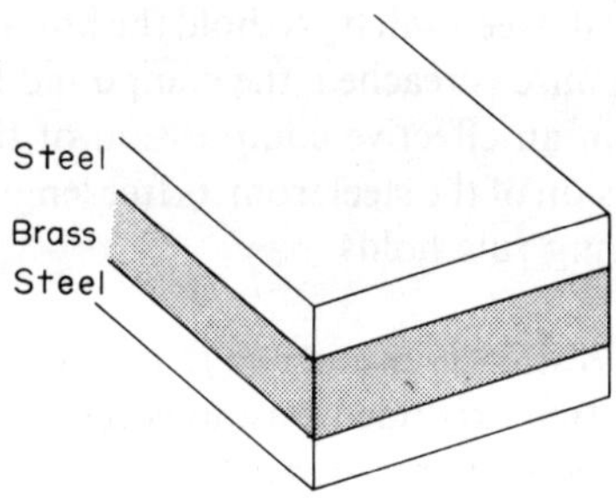

Fig. 2.2.

In general, the coefficients of expansion of the two materials forming the compound bar will be different so that as the temperature rises each material will attempt to expand by different amounts. Figure 2.3b shows the positions to which the individual materials will extend if they are completely free to expand (i.e. not joined rigidly together as a compound bar). The extension of any length L is given by

$$\alpha L t$$

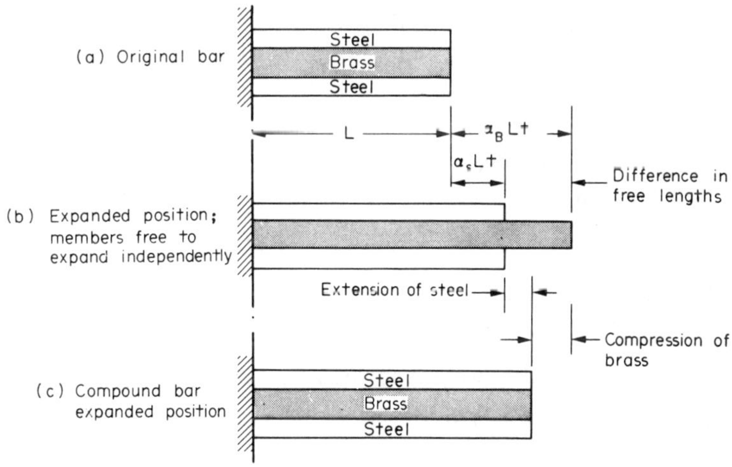

Fig. 2.3. Thermal expansion of compound bar.

Thus the difference of "free" expansion lengths or so-called *free lengths*

$$= \alpha_B Lt - \alpha_S Lt = (\alpha_B - \alpha_S) Lt$$

since in this case the coefficient of expansion of the brass α_B is greater than that for the steel α_S. The initial lengths L of the two materials are assumed equal.

If the two materials are now rigidly joined as a compound bar and subjected to the same temperature rise, each material will attempt to expand to its free length position but each will be affected by the movement of the other. The higher coefficient of expansion material (brass) will therefore seek to pull the steel up to its free length position and conversely the lower coefficient of expansion material (steel) will try to hold the brass back to the steel "free length" position. In practice a compromise is reached, the compound bar extending to the position shown in Fig. 2.3c, resulting in an effective compression of the brass from its free length position and an effective extension of the steel from its free length position. From the diagram it will be seen that the following rule holds.

Rule 1.

Extension of steel + compression of brass = difference in "free" lengths.

Referring to the bars in their free expanded positions the rule may be written as

Extension of "short" member + compression of "long" member = difference in free lengths.

Applying Newton's law of equal action and reaction the following second rule also applies.

Rule 2.

The tensile force applied to the short member by the long member is equal in magnitude to the compressive force applied to the long member by the short member.

Thus, in this case,

tensile force in steel = compressive force in brass

Now, from the definition of Young's modulus

$$E = \frac{\text{stress}}{\text{strain}} = \frac{\sigma}{\delta/L}$$

where δ is the change in length.

$$\therefore \qquad \delta = \frac{\sigma L}{E}$$

Also $$\text{force} = \text{stress} \times \text{area} = \sigma A$$

where A is the cross-sectional area.

Therefore Rule 1 becomes

$$\frac{\sigma_S L}{E_S} + \frac{\sigma_B L}{E_B} = (\alpha_B - \alpha_S)Lt \tag{2.8}$$

and Rule 2 becomes

$$\sigma_S A_S = \sigma_B A_B \tag{2.9}$$

We thus have two equations with two unknowns σ_S and σ_B and it is possible to evaluate the magnitudes of these stresses (see Example 2.2).

2.4. Compound bar (tube and rod)

Consider now the case of a hollow tube with washers or endplates at each end and a central threaded rod as shown in Fig. 2.4. At first sight there would seem to be no connection with the work of the previous section, yet, in fact, the method of solution to determine the stresses set up in the tube and rod when one nut is tightened is identical to that described in §2.3.

The compound bar which is formed after assembly of the tube and rod, i.e. with the nuts tightened, is shown in Fig. 2.4c, the rod being in a state of tension and the tube in compression. Once again Rule 2 applies, i.e.

compressive force in tube = tensile force in rod

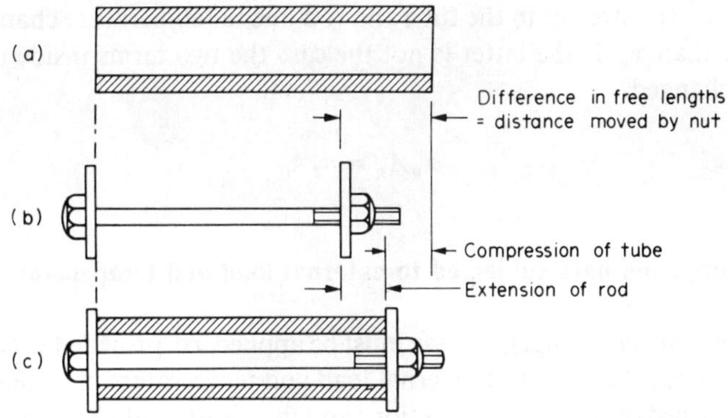

Fig. 2.4. Equivalent "mechanical" system to that of Fig. 2.3.

Figure 2.4a and b show, *diagrammatically*, the effective positions of the tube and rod before the nut is tightened and the two components are combined. As the nut is turned there is a simultaneous compression of the tube and tension of the rod leading to the final state shown in Fig. 2.4c. As before, however, the diagram shows that Rule 1 applies:

compression of tube + extension of rod = difference in free lengths = axial advance of nut

i.e. the axial movement of the nut (= number of turns n × threads per metre) is taken up by combined compression of the tube and extension of the rod.

Thus, with suffix t for tube and R for rod,

$$\frac{\sigma_t L}{E_t} + \frac{\sigma_R L}{E_R} = n \times \textbf{threads/metre} \tag{2.10}$$

also

$$\sigma_R A_R = \sigma_t A_t \tag{2.11}$$

If the tube and rod are now subjected to a change of temperature they may be treated as a normal compound bar of § 2.3 and Rules 1 and 2 again apply (Fig. 2.5),

i.e.

$$\frac{\sigma'_t L}{E_t} + \frac{\sigma'_R L}{E_R} = (\alpha_t - \alpha_R) L t \tag{2.12}$$

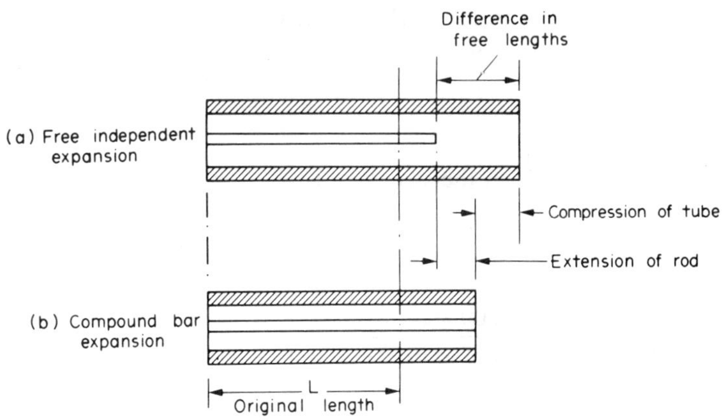

Fig. 2.5.

where σ'_t and σ'_R are the stresses in the tube and rod due to temperature change only and α_t is assumed greater than α_R. If the latter is not the case the two terms inside the final bracket should be interchanged.

Also

$$\sigma'_R A_R = \sigma'_t A_t$$

2.5. Compound bars subjected to external load and temperature effects

In this case the *principle of superposition* must be applied, i.e. provided that stresses remain within the elastic limit the effects of external load and temperature change may be assessed separately as described in the previous sections and the results added, taking account of sign, to determine the resultant total effect;

i.e. *total strain = sum of strain due to external loads and temperature strain*

2.6. Compound thick cylinders subjected to temperature changes

The procedure described in §2.3 has been applied to compound cylinders constructed from tubes of different materials on page 230.

Examples

Example 2.1

(a) A compound bar consists of four brass wires of 2.5 mm diameter and one steel wire of 1.5 mm diameter. Determine the stresses in each of the wires when the bar supports a load of 500 N. Assume all of the wires are of equal lengths.

(b) Calculate the "equivalent" or "combined" modulus for the compound bar and determine its total extension if it is initially 0.75 m long. Hence check the values of the stresses obtained in part (a).

For brass $E = 100\ \text{GN/m}^2$ and for steel $E = 200\ \text{GN/m}^2$.

Solution

(a) From eqn. (2.3) the force in the steel wire is given by

$$F_s = \frac{E_s A_s}{\Sigma EA} W$$

$$= \left[\frac{200 \times 10^9 \times \frac{\pi}{4} \times 1.5^2 \times 10^{-6}}{200 \times 10^9 \times \frac{\pi}{4} \times 1.5^2 \times 10^{-6} + 4(100 \times 10^9 \times \frac{\pi}{4} \times 2.5^2 \times 10^{-6})}\right] 500$$

$$= \left[\frac{2 \times 1.5^2}{(2 \times 1.5^2) + (4 \times 2.5^2)}\right] 500 = 76.27\ \text{N}$$

$\therefore$ total force in brass wires = 500 − 76.27 = 423.73 N

$$\therefore \quad \text{stress in steel} = \frac{\text{load}}{\text{area}} = \frac{76.27}{\frac{\pi}{4} \times 1.5^2 \times 10^{-6}} = \mathbf{43.2\ MN/m^2}$$

and
$$\text{stress in brass} = \frac{\text{load}}{\text{area}} = \frac{423.73}{4 \times \frac{\pi}{4} \times 2.5^2 \times 10^{-6}} = \mathbf{21.6\ MN/m^2}$$

(b) From eqn. (2.6)

$$\text{combined } E = \frac{\Sigma EA}{\Sigma A} = \frac{200 \times 10^9 \times \frac{\pi}{4} \times 1.5^2 \times 10^{-6} + 4(100 \times 10^9 \times \frac{\pi}{4} \times 2.5^2 \times 10^{-6})}{\frac{\pi}{4}(1.5^2 + 4 \times 2.5^2)^{10-6}}$$

$$= \frac{(200 \times 1.5^2 + 400 \times 2.5^2)}{(1.5^2 + 4 \times 2.5^2)} 10^9 = \mathbf{108.26\ GN/m^2}$$

Now
$$E = \frac{\text{stress}}{\text{strain}}$$

and the stress in the equivalent bar

$$= \frac{500}{\Sigma A} = \frac{500}{\frac{\pi}{4}(1.5^2 + 4 \times 2.5^2)10^{-6}} = 23.36\ \text{MN/m}^2$$

$$\therefore \quad \text{strain in the equivalent bar} = \frac{\text{stress}}{E} = \frac{23.36 \times 10^6}{108.26 \times 10^9} = 0.216 \times 10^{-3}$$

$$\therefore \quad \text{common extension} = \text{strain} \times \text{original length}$$
$$= 0.216 \times 10^{-3} \times 0.75 = 0.162 \times 10^{-3}$$
$$= \mathbf{0.162\ mm}$$

This is also the extension of any single bar, giving a strain in any bar

$$= \frac{0.162 \times 10^{-3}}{0.75} = 0.216 \times 10^{-3} \text{ as above}$$

$$\therefore \quad \text{stress in steel} = \text{strain} \times E_s = 0.216 \times 10^{-3} \times 200 \times 10^9$$
$$= \mathbf{43.2\ MN/m^2}$$

and
$$\text{stress in brass} = \text{strain} \times E_B = 0.216 \times 10^{-3} \times 100 \times 10^9$$
$$= \mathbf{21.6\ MN/m^2}$$

These are the same values as obtained in part (a).

Example 2.2

(a) A compound bar is constructed from three bars 50 mm wide by 12 mm thick fastened together to form a bar 50 mm wide by 36 mm thick. The middle bar is of aluminium alloy for which $E = 70\ \text{GN/m}^2$ and the outside bars are of brass with $E = 100\ \text{GN/m}^2$. If the bars are initially fastened at 18°C and the temperature of the whole assembly is then raised to 50°C, determine the stresses set up in the brass and the aluminium.

$$\alpha_B = 18 \times 10^{-6} \text{ per °C} \quad \text{and} \quad \alpha_A = 22 \times 10^{-6} \text{ per °C}$$

(b) What will be the changes in these stresses if an external compressive load of 15 kN is applied to the compound bar at the higher temperature?

Solution

With any problem of this type it is convenient to let the stress in one of the component members or materials, e.g. the brass, be x.

Then, since

$$\text{force in brass} = \text{force in aluminium}$$

and

$$\text{force} = \text{stress} \times \text{area}$$

$$x \times 2 \times 50 \times 12 \times 10^{-6} = \sigma_A \times 50 \times 12 \times 10^{-6}$$

i.e.

$$\text{stress in aluminium } \sigma_A = 2x$$

Now, from eqn. (2.8),

$$\text{extension of brass} + \text{compression of aluminium} = \text{difference in free lengths}$$

$$= (\alpha_A - \alpha_B)(T_2 - T_1)L$$

$$\frac{xL}{100 \times 10^9} + \frac{2xL}{70 \times 10^9} = (22-18)10^{-6}(50-18)L$$

$$\frac{(7x + 20x)}{700 \times 10^9} = 4 \times 10^{-6} \times 32$$

$$27x = 4 \times 10^{-6} \times 32 \times 700 \times 10^9$$

$$x = \mathbf{3.32\ MN/m^2}$$

The stress in the brass is thus **3.32 MN/m²** (**tensile**) and the stress in the aluminium is $2 \times 3.32 =$ **6.64 MN/m² (compressive)**.

(b) With an external load of 15 kN applied each member will take a proportion of the total load given by eqn. (2.3).

$$\text{Force in aluminium} = \frac{E_A A_A}{\Sigma EA} W$$

$$= \left[\frac{70 \times 10^9 \times 50 \times 12 \times 10^{-6}}{(70 \times 50 \times 12 + 2 \times 100 \times 50 \times 12)10^9 \times 10^{-6}}\right] 15 \times 10^3$$

$$= \left[\frac{70}{(70+200)}\right] 15 \times 10^3$$

$$= 3.89 \text{ kN}$$

$$\therefore \quad \text{force in brass} = 15 - 3.89 = 11.11 \text{ kN}$$

$$\therefore \quad \text{stress in brass} = \frac{\text{load}}{\text{area}} = \frac{11.11 \times 10^3}{2 \times 50 \times 12 \times 10^{-6}}$$

$$= \mathbf{9.26\ MN/m^2\ (compressive)}$$

$$\text{Stress in aluminium} = \frac{\text{load}}{\text{area}} = \frac{3.89 \times 10^3}{50 \times 12 \times 10^{-6}}$$

$$= \mathbf{6.5\ MN/m^2\ (compressive)}$$

These stresses represent the *changes* in the stresses owing to the applied load. The total or resultant stresses owing to combined applied loading plus temperature effects are, therefore,

$$\text{stress in aluminium} = -6.64 - 6.5 = -13.14\ \text{MN/m}^2$$

$$= \mathbf{13.14\ MN/m^2\ (compressive)}$$

$$\text{stress in brass} = +3.32 - 9.26 = -5.94\ \text{MN/m}^2$$

$$= \mathbf{5.94\ MN/m^2\ (compressive)}$$

Example 2.3

A 25 mm diameter steel rod passes concentrically through a bronze tube 400 mm long, 50 mm external diameter and 40 mm internal diameter. The ends of the steel rod are threaded and provided with nuts and washers which are adjusted initially so that there is no end play at 20°C.

(a) Assuming that there is no change in the thickness of the washers, find the stress produced in the steel and bronze when one of the nuts is tightened by giving it one-tenth of a turn, the pitch of the thread being 2.5 mm.
(b) If the temperature of the steel and bronze is then raised to 50°C find the changes that will occur in the stresses in both materials.

The coefficient of linear expansion per °C is 11×10^{-6} for steel and 18×10^{-6} for bronze. E for steel $= 200\ \text{GN/m}^2$. E for bronze $= 100\ \text{GN/m}^2$.

Solution

(a) Let x be the stress in the tube resulting from the tightening of the nut and σ_R the stress in the rod.

Then, from eqn. (2.11),

$$\text{force (stress} \times \text{area) in tube} = \text{force (stress} \times \text{area) in rod}$$

$$x \times \tfrac{\pi}{4}(50^2 - 40^2)10^{-6} = \sigma_R \times \tfrac{\pi}{4} \times 25^2 \times 10^{-6}$$

$$\sigma_R = \frac{(50^2 - 40^2)}{25^2} x = 1.44x$$

And since compression of tube + extension of rod = axial advance of nut, from eqn. (2.10),

$$\frac{x \times 400 \times 10^{-3}}{100 \times 10^9} + \frac{1.44x \times 400 \times 10^{-3}}{200 \times 10^9} = \frac{1}{10} \times 2.5 \times 10^{-3}$$

$$400 \frac{(2x + 1.44x)}{200 \times 10^9} 10^{-3} = 2.5 \times 10^{-4}$$

$$\therefore \quad 6.88x = 2.5 \times 10^8$$

$$x = 36.3\ \text{MN/m}^2$$

The stress in the tube is thus **36.3 MN/m² (compressive)** and the stress in the rod is $1.44 \times 36.3 =$ **52.3 MN/m² (tensile)**.

(b) Let p be the stress in the tube resulting from temperature change. The relationship between the stresses in the tube and the rod will remain as in part (a) so that the stress in the rod is then $1.44p$. In this case, if free expansion were allowed in the independent members, the bronze tube would expand more than the steel rod and from eqn. (2.8)

$$\text{compression of tube} + \text{extension of rod} = \text{difference in free length}$$

$$\therefore \quad \frac{pL}{100 \times 10^9} + \frac{1.44pL}{200 \times 10^9} = (\alpha_B - \alpha_S)(T_2 - T_1)L$$

$$\frac{(2p + 1.44p)}{200 \times 10^9} = (18 - 11)10^{-6}(50 - 20)$$

$$3.44p = 7 \times 10^{-6} \times 30 \times 200 \times 10^9$$

$$p = 12.21 \text{ MN/m}^2$$

and

$$1.44p = 17.6 \text{ MN/m}^2$$

The changes in the stresses resulting from the temperature effects are thus 12.2 MN/m² (compressive) in the tube and 17.6 MN/m² (tensile) in the rod.

The final, resultant, stresses are thus:

stress in tube $= -36.3 - 12.2 =$ **48.5 MN/m² (compressive)**

stress in rod $= 52.3 + 17.6 =$ **69.9 MN/m² (tensile)**

Example 2.4

A composite bar is constructed from a steel rod of 25 mm diameter surrounded by a copper tube of 50 mm outside diameter and 25 mm inside diameter. The rod and tube are joined by two 20 mm diameter pins as shown in Fig. 2.6. Find the shear stress set up in the pins if, after pinning, the temperature is raised by 50°C.

For steel $E = 210$ GN/m² and $\alpha = 11 \times 10^{-6}$ per °C.

For copper $E = 105$ GN/m² and $\alpha = 17 \times 10^{-6}$ per °C.

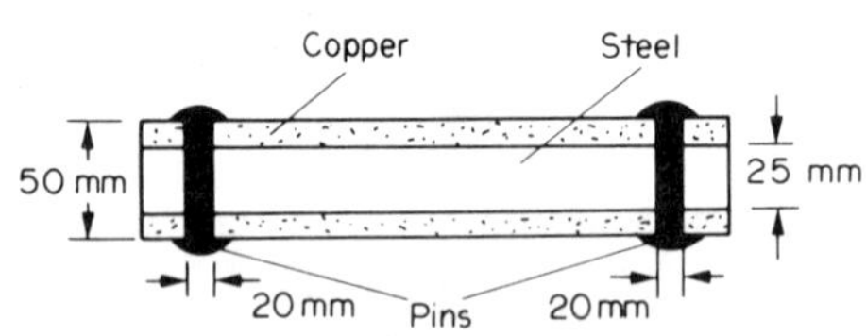

Fig. 2.6.

Solution

In this case the copper attempts to expand more than the steel, thus tending to shear the pins joining the two.

Let the stress set up in the steel be x, then, since

$$\text{force in steel} = \text{force in copper}$$

$$x \times \tfrac{\pi}{4} \times 25^2 \times 10^{-6} = \sigma_c \times \tfrac{\pi}{4}(50^2 - 25^2)10^{-6}$$

i.e.

$$\text{stress in copper } \sigma_c = \frac{x \times 25^2}{(50^2 - 25^2)} = 0.333x = \frac{x}{3}$$

Now the extension of the steel from its freely expanded length to its forced length in the compound bar is given by

$$\frac{\sigma L}{E} = \frac{xL}{210 \times 10^9}$$

where L is the original length.

Similarly, the compression of the copper from its freely expanded position to its position in the compound bar is given by

$$\frac{\sigma L}{E} = \frac{x}{3} \times \frac{L}{105 \times 10^9}$$

Now the extension of steel + compression of copper

$$= \text{difference in "free" lengths}$$

$$= (\alpha_2 - \alpha_1)(T_2 - T_1)L$$

$$\therefore \quad \frac{xL}{210 \times 10^9} + \frac{xL}{3 \times 105 \times 10^9} = (17 - 11)10^{-6} \times 50 \times L$$

$$\frac{3x + 2x}{6 \times 105 \times 10^9} = 6 \times 10^{-6} \times 50$$

$$5x = 6 \times 10^{-6} \times 50 \times 6 \times 105 \times 10^9$$

$$x = 37.8 \times 10^6 = 37.8 \text{ MN/m}^2$$

$\therefore$ load carried by the steel = stress × area

$$= 37.8 \times 10^6 \times \tfrac{\pi}{4} \times 25^2 \times 10^{-6}$$

$$= 18.56 \text{ kN}$$

The pins will be in a state of double shear (see §1.15), the shear stress set up being given by

$$\tau = \frac{\text{load}}{2 \times \text{area}} = \frac{18.56 \times 10^3}{2 \times \frac{\pi}{4} \times 20^2 \times 10^{-6}}$$

$$= \mathbf{29.5\ MN/m^2}$$

Problems

2.1 (A). A power transmission cable consists of ten copper wires each of 1.6 mm diameter surrounding three steel wires each of 3 mm diameter. Determine the combined E for the compound cable and hence determine the extension of a 30 m length of the cable when it is being laid with a tension of 2 kN.

For steel, $E = 200$ GN/m^2; for copper, $E = 100$ GN/m^2. [151.3 GN/m^2; 9.6 mm.]

2.2 (A). If the maximum stress allowed in the copper of the cable of problem 2.1 is 60 MN/m^2, determine the maximum tension which the cable can support. [3.75 kN.]

2.3 (A). What will be the stress induced in a steel bar when it is heated from 15°C to 60°C, all expansion being prevented?

For mild steel, $E = 210\ GN/m^2$ and $\alpha = 11 \times 10^{-6}$ per °C. [104 MN/m^{2}°.]

2.4 (A). A 75 mm diameter compound bar is constructed by shrinking a circular brass bush onto the outside of a 50 mm diameter solid steel rod. If the compound bar is then subjected to an axial compressive load of 160 kN determine the load carried by the steel rod and the brass bush and the compressive stress set up in each material.

For steel, $E = 210\ GN/m^2$; for brass, $E = 100\ GN/m^2$. [I. Struct. E.] [100.3, 59.7 kN; 51.1, 24.3 MN/m^2.]

2.5 (B). A steel rod of cross-sectional area 600 mm^2 and a coaxial copper tube of cross-sectional area 1000 mm^2 are firmly attached at their ends to form a compound bar. Determine the stress in the steel and in the copper when the temperature of the bar is raised by 80°C and an axial tensile force of 60 kN is applied.

For steel, $E = 200\ GN/m^2$ with $\alpha = 11 \times 10^{-6}$ per °C.

For copper, $E = 100\ GN/m^2$ with $\alpha = 16.5 \times 10^{-6}$ per °C. [E.I.E.] [94.6, 3.3 MN/m^2.]

2.6 (B). A stanchion is formed by buttwelding together four plates of steel to form a square tube of outside cross-section 200 mm × 200 mm. The constant metal thickness is 10 mm. The inside is then filled with concrete.

(**a**) Determine the cross-sectional area of the steel and concrete

(**b**) If E for steel is 200 GN/m^2 and this value is twenty times that for the concrete find, when the stanchion carries a load of 368.8 kN,

(i) The stress in the concrete

(ii) The stress in the steel

(iii) The amount the stanchion shortens over a length of 2*m*. [C.G.] [2, 40 MN/m^2; 40 mm]

CHAPTER 3

SHEARING FORCE AND BENDING MOMENT DIAGRAMS

Summary

At any section in a beam carrying transverse loads the shearing force is defined as the algebraic sum of the forces taken on either side of the section.

Similarly, the bending moment at any section is the algebraic sum of the moments of the forces about the section, again taken on either side.

In order that the shearing-force and bending-moment values calculated on either side of the section shall have the same magnitude and sign, a convenient sign convention has to be adopted. This is shown in Figs. 3.1 and 3.2 (see page 42).

Shearing-force (S.F.) and bending-moment (B.M.) diagrams show the variation of these quantities along the length of a beam for any fixed loading condition.

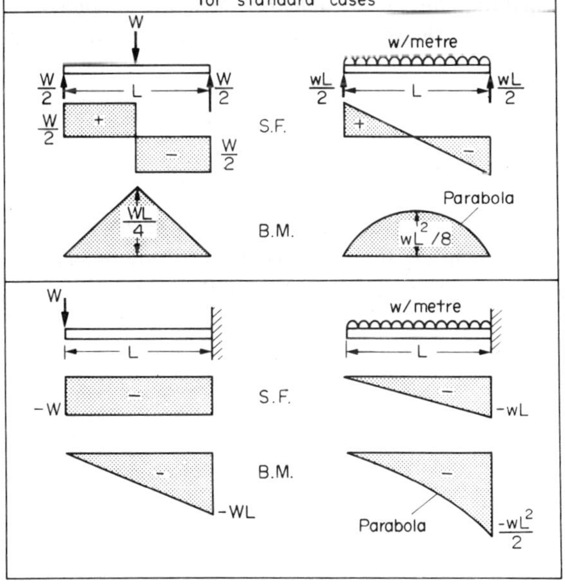

Shearing force and bending moment diagrams for standard cases

3.1. Shearing force and bending moment

At every section in a beam carrying transverse loads there will be resultant forces on either side of the section which, for equilibrium, must be equal and opposite, and whose combined action tends to shear the section in one of the two ways shown in Fig. 3.1a and b. *The shearing force (S.F.) at the section is defined therefore as the algebraic sum of the forces taken on one side of the section.* Which side is chosen is purely a matter of convenience but in order that the value obtained on both sides shall have the same magnitude and sign a convenient sign convention has to be adopted.

3.1.1. Shearing force (S.F.) sign convention

Forces upwards to the left of a section or downwards to the right of the section are positive. Thus Fig. 3.1a shows a positive S.F. system at X–X and Fig. 3.1b shows a negative S.F. system.

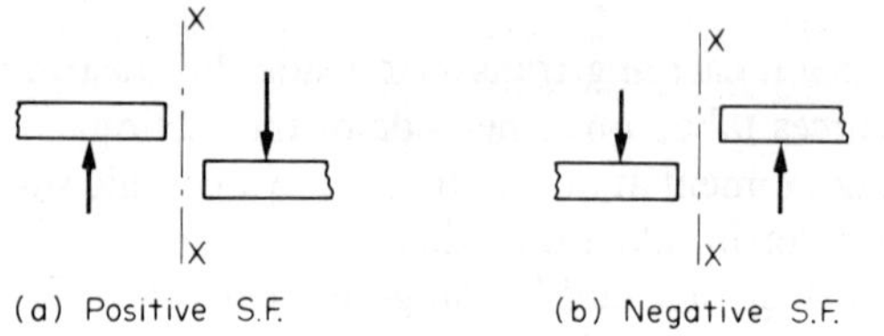

Fig. 3.1. S.F. sign convention.

In addition to the shear, every section of the beam will be subjected to bending, i.e. to a resultant B.M. which is the net effect of the moments of each of the individual loads. Again, for equilibrium, the values on either side of the section must have equal values. *The bending moment (B.M.) is defined therefore as the algebraic sum of the moments of the forces about the section, taken on either side of the section.* As for S.F., a convenient sign convention must be adopted.

3.1.2. Bending moment (B.M.) sign convention

Clockwise moments to the left and counterclockwise to the right are positive. Thus Fig. 3.2a shows a positive bending moment system resulting in *sagging* of the beam at X–X and Fig. 3.2b illustrates a negative B.M. system with its associated *hogging* beam.

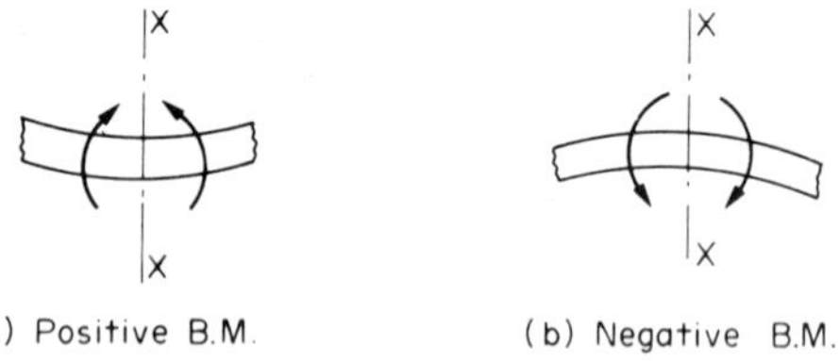

Fig. 3.2. B.M. sign convention.

It should be noted that whilst the above sign conventions for S.F. and B.M. are somewhat arbitrary and could be completely reversed, the systems chosen here are the only ones which yield the mathematically correct signs for slopes and deflections of beams in subsequent work and therefore are highly recommended.

Diagrams which illustrate the variation in the B.M. and S.F. values along the length of a beam or structure for any fixed loading condition are termed *B.M. and S.F. diagrams.* They are therefore graphs of B.M. or S.F. values drawn on the beam as a base and they clearly illustrate in the early design stages the positions on the beam which are subjected to the greatest shear or bending stresses and hence which may require further consideration or strengthening.

At this point it is imperative to note that there are two general forms of loading to which structures may be subjected, namely, concentrated and distributed loads. The former are assumed to act at a point and immediately introduce an oversimplification since all practical loading systems must be applied over a finite area. Nevertheless, for calculation purposes this area is assumed to be so small that the load can be justly assumed to act at a point. Distributed loads are assumed to act over part, or all, of the beam and in most cases are assumed to be equally or uniformly distributed; they are then termed uniformly distributed loads (u.d.l.). Occasionally, however, the distribution is not uniform but may vary linearly across the loaded portion or have some more complex distribution form.

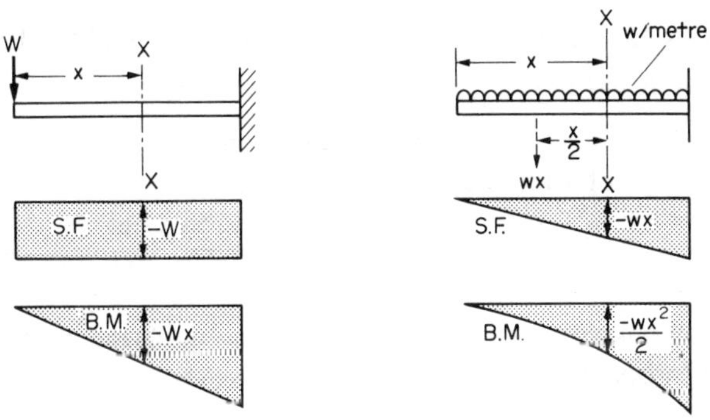

Fig. 3.3. S.F.–B.M. diagrams for standard cases.

Thus in the case of a cantilever carrying a concentrated load W at the end (Fig. 3.3), the S.F. at any section X–X, distance x from the free end, is S.F. $= -W$. This will be true whatever the value of x, and so the S.F. diagram becomes a rectangle. The B.M. at the same section X–X is $-Wx$ and this will increase linearly with x. The B.M. diagram is therefore a triangle.

If the cantilever now carries a uniformly distributed load, the S.F. at X–X is the net load to one side of X–X, i.e. $-wx$. In this case, therefore, the S.F. diagram becomes triangular, increasing to a maximum value of $-wL$ at the support. The B.M. at X–X is obtained by treating the load to the left of X–X as a concentrated load of the same value acting at the centre of gravity,

i.e. $$\text{B.M. at } X\text{–}X = -wx\frac{x}{2} = -\frac{-wx^2}{2}$$

Plotted against x this produces the parabolic B.M. diagram shown.

3.2. S.F. and B.M. diagrams for beams carrying concentrated loads only

In order to illustrate the procedure to be adopted for the determination of S.F. and B.M. values for more complicated load conditions, consider the simply supported beam shown in

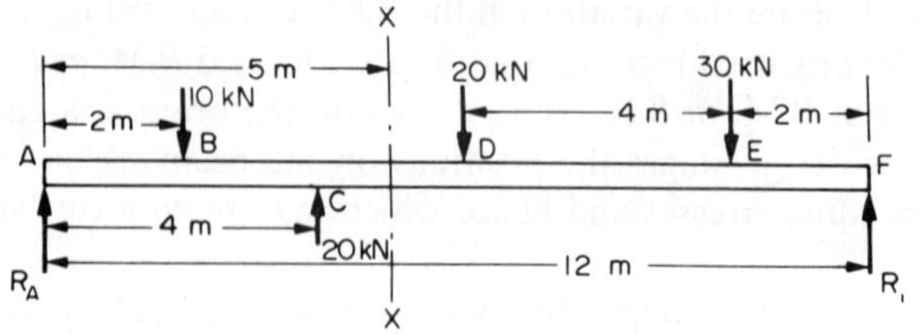

Fig. 3.4.

Fig. 3.4 carrying concentrated loads only. (The term *simply supported* means that the beam can be assumed to rest on knife-edges or roller supports and is free to bend at the supports without any restraint.)

The values of the reactions at the ends of the beam may be calculated by applying normal equilibrium conditions, i.e. by taking moments about F.

Thus $$R_A \times 12 = (10 \times 10) + (20 \times 6) + (30 \times 2) - (20 \times 8) = 120$$
$$R_A = 10\,\text{kN}$$

For vertical equilibrium

$$\text{total force up} = \text{total load down}$$
$$R_A + R_F = 10 + 20 + 30 - 20 = 40$$
$$R_F = 30\,\text{kN}$$

At this stage it is advisable to check the value of R_F by taking moments about A.

Summing up the forces on either side of X–X we have the result shown in Fig. 3.5. Using the sign convention listed above, the shear force at X–X is therefore $+20\,\text{kN}$, i.e. the resultant force at X–X tending to shear the beam is 20 kN.

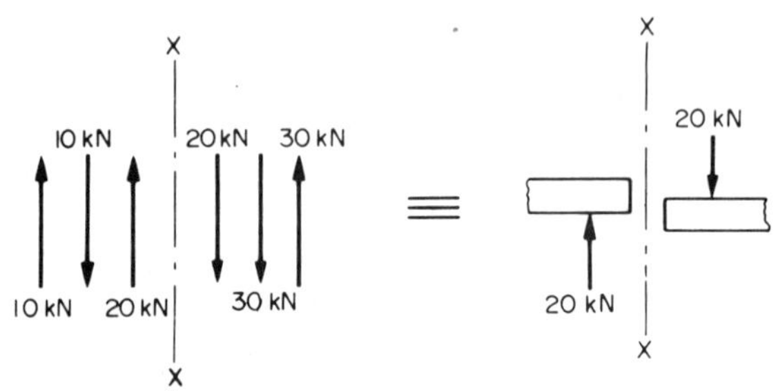

Fig. 3.5. Total S.F. at X–X.

Similarly, Fig. 3.6 shows the summation of the moments of the forces at X–X, the resultant B.M. being 40 kN m.

In practice only one side of the section is normally considered and the summations involved can often be completed by mental arithmetic. The complete S.F. and B.M. diagrams for the beam are shown in Fig. 3.7, and the B.M. values used to construct the diagram are derived on page 45.

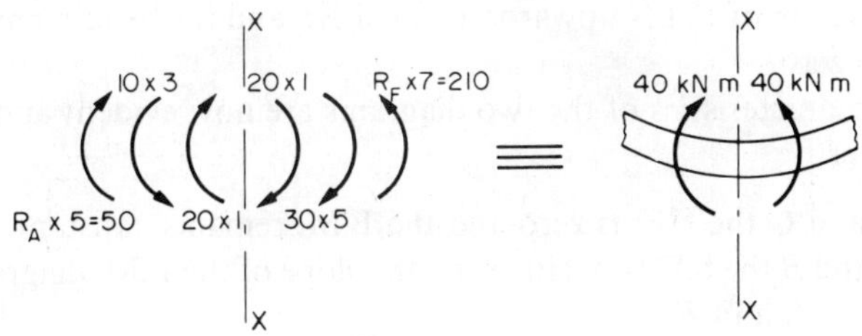

Fig. 3.6. Total B.M. at X–X.

$$\begin{aligned}
&\text{B.M. at } A && = 0\\
&\text{B.M. at } B = +(10\times 2) && = +20\,\text{kN m}\\
&\text{B.M. at } C = +(10\times 4)-(10\times 2) && = +20\,\text{kN m}\\
&\text{B.M. at } D = +(10\times 6)+(20\times 2)-(10\times 4) && = +60\,\text{kN m}\\
&\text{B.M. at } E = +(30\times 2) && = +60\,\text{kN m}\\
&\text{B.M. at } F && = 0
\end{aligned}$$

All the above values have been calculated from the moments of the forces to the *left* of each section considered except for E where forces to the right of the section are taken.

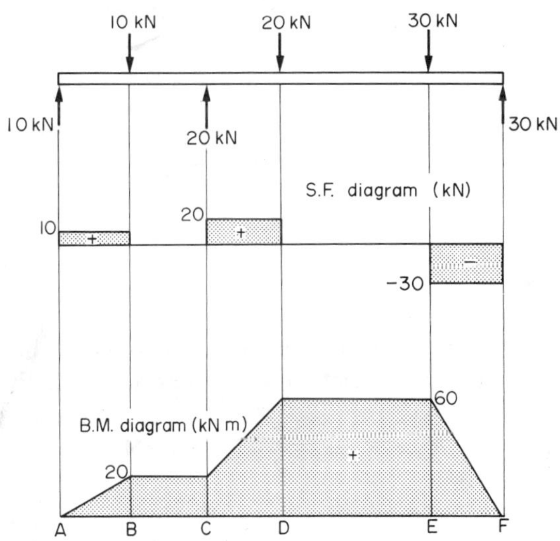

Fig. 3.7.

It may be observed at this stage that the S.F. diagram can be obtained very quickly when working from the left-hand side, since after plotting the S.F. value at the support all subsequent steps are in the direction of and equal in magnitude to the applied loads, e.g. 10 kN up at A, down 10 kN at B, up 20 kN at C, etc., with horizontal lines joining the steps to show that the S.F. remains constant between points of application of concentrated loads.

The S.F. and B.M. values at the left-hand support are determined by considering a section an infinitely small distance to the right of the support. The only load to the left (and hence the

S.F.) is then the reaction of 10 kN upwards, i.e. positive, and the bending moment = reaction × zero distance = zero.

The following characteristics of the two diagrams are now evident and will be explained later in this chapter:

(a) between *B* and *C* the S.F. is zero and the B.M. remains constant;
(b) between *A* and *B* the S.F. is positive and the slope of the B.M. diagram is positive; vice versa between *E* and *F*;
(c) the difference in B.M. between *A* and *B* = 20 kN m = area of S.F. diagram between *A* and *B*.

3.3. S.F. and B.M. diagrams for uniformly distributed loads

Consider now the simply supported beam shown in Fig. 3.8 carrying a u.d.l. $w = 25$ kN/m across the complete span.

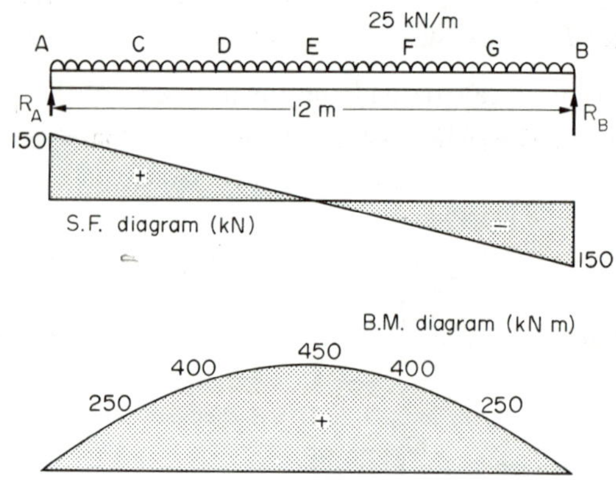

Fig. 3.8.

Here again it is necessary to evaluate the reactions, but in this case the problem is simplified by the symmetry of the beam. Each reaction will therefore take half the applied load,

i.e.
$$R_A = R_B = \frac{25 \times 12}{2} = 150 \text{ kN}$$

The S.F. at *A*, using the usual sign convention, is therefore +150 kN.

Consider now the beam divided into six equal parts 2 m long. The S.F. at any other point *C* is, therefore,

$$150 - \text{load downwards between } A \text{ and } C$$
$$= 150 - (25 \times 2) = +100 \text{ kN}$$

The whole diagram may be constructed in this way, or much more quickly by noticing that the S.F. at *A* is +150 kN and that between *A* and *B* the S.F. decreases uniformly, producing the required sloping straight line, shown in Fig. 3.7. Alternatively, the S.F. at *A* is +150 kN and between *A* and *B* this decreases gradually by the amount of the applied load (i.e. by 25 × 12 = 300 kN) to −150 kN at *B*.

When evaluating B.M.'s it is assumed that a u.d.l. can be replaced by a concentrated load of equal value acting at the middle of its spread. When taking moments about C, therefore, the portion of the u.d.l. between A and C has an effect equivalent to that of a concentrated load of $25 \times 2 = 50\,\text{kN}$ acting the centre of AC, i.e. 1 m from C.

$$\text{B.M. at } C = (R_A \times 2) - (50 \times 1) = 300 - 50 = 250\,\text{kN m}$$

Similarly, for moments at D the u.d.l. on AD can be replaced by a concentrated load of

$$25 \times 4 = 100\,\text{kN at the centre of } AD, \text{ i.e. at } C.$$

$$\text{B.M. at } D = (R_A \times 4) - (100 \times 2) = 600 - 200 = 400\,\text{kN m}$$

Similarly,

$$\text{B.M. at } E = (R_A \times 6) - (25 \times 6)3 = 900 - 450 = 450\,\text{kN m}$$

The B.M. diagram will be symmetrical about the beam centre line; therefore the values of B.M. at F and G will be the same as those at D and C respectively. The final diagram is therefore as shown in Fig. 3.8 and is parabolic.

Point (a) of the summary is clearly illustrated here, since the B.M. is a maximum when the S.F. is zero. Again, the reason for this will be shown later.

3.4. S.F. and B.M. diagrams for combined concentrated and uniformly distributed loads

Consider the beam shown in Fig. 3.9 loaded with a combination of concentrated loads and u.d.l.s.

Taking moments about E

$$(R_A \times 8) + (40 \times 2) = (10 \times 2 \times 7) + (20 \times 6) + (20 \times 3) + (10 \times 1) + (20 \times 3 \times 1.5)$$

$$8R_A + 80 = 420$$

$$R_A = 42.5\,\text{kN} \,(= \text{S.F. at } A)$$

Now

$$R_A + R_E = (10 \times 2) + 20 + 20 + 10 + (20 \times 3) + 40 = 170$$

$$R_E = 127.5\,\text{kN}$$

Working from the left-hand support it is now possible to construct the S.F. diagram, as indicated previously, by following the direction arrows of the loads. In the case of the u.d.l.'s the S.F. diagram will decrease gradually by the amount of the total load until the end of the u.d.l. or the next concentrated load is reached. Where there is no u.d.l. the S.F. diagram remains horizontal between load points.

In order to plot the B.M. diagram the following values must be determined:

B.M. at A $= 0$

B.M. at $B = (42.5 \times 2) - (10 \times 2 \times 1) = 85 - 20 = 65\,\text{kN m}$

B.M. at $C = (42.5 \times 5) - (10 \times 2 \times 4) - (20 \times 3) = 212.5 - 80 - 60 = 72.5\,\text{kN m}$

B.M. at $D = (42.5 \times 7) - (10 \times 2 \times 6) - (20 \times 5) - (20 \times 2)$
$- (20 \times 2 \times 1) = 297.5 - 120 - 100 - 40 - 40 = 297.5 - 300 = -2.5\,\text{kN m}$

B.M. at $E = (-40 \times 2)$ working from r.h.s. $= -80\,\text{kN m}$

B.M. at F $= 0$

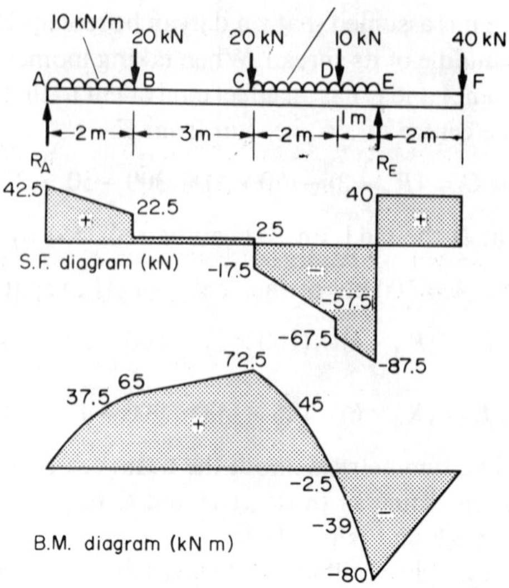

Fig. 3.9.

For complete accuracy one or two intermediate values should be obtained along each u.d.l. portion of the beam,

e.g. $\qquad$ B.M. midway between A and $B = (42.5 \times 1) - (10 \times 1 \times \frac{1}{2})$

$$= 42.5 - 5 = 37.5 \text{ kN m}$$

Similarly, B.M. midway between C and $D = 45$ kN m

B.M. midway between D and $E = -39$ kN m

The B.M. and S.F. diagrams are then as shown in Fig. 3.9.

3.5. Points of contraflexure

A point of contraflexure is a point where the curvature of the beam changes sign. It is sometimes referred to as a *point of inflexion* and will be shown later to occur at the point, or points, on the beam where the B.M. is zero.

For the beam of Fig. 3.9, therefore, it is evident from the B.M. diagram that this point lies somewhere between C and D (B.M. at C is positive, B.M. at D is negative). If the required point is a distance x from C then at that point

$$\text{B.M.} = (42.5)(5+x) - (10 \times 2)(4+x) - 20(3+x) - 20x - \frac{20x^2}{2}$$

$$= 212.5 + 42.5x - 80 - 20x - 60 - 20x - 20x - 10x^2$$

$$= 72.5 - 17.5x - 10x^2$$

Thus the B.M. is zero where

$$0 = 72.5 - 17.5x - 10x^2$$

i.e. where $\qquad x = 1.96$ or -3.7

Since the last answer can be ignored (being outside the beam), the point of contraflexure must be situated at 1.96 m to the right of C.

3.6. Relationship between shear force Q, bending moment M and intensity of loading w

Consider the beam AB shown in Fig. 3.10 carrying a uniform loading intensity (uniformly distributed load) of w kN/m. By symmetry, each reaction takes half the total load, i.e., $wL/2$.

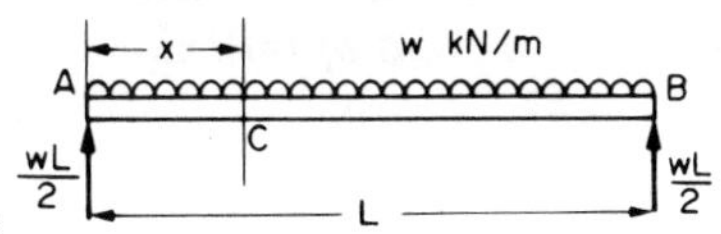

Fig. 3.10.

The B.M. at any point C, distance x from A, is given by

$$M = \frac{wL}{2}x - (wx)\frac{x}{2}$$

i.e.

$$M = \tfrac{1}{2}wLx - \tfrac{1}{2}wx^2$$

Differentiating,

$$\frac{dM}{dx} = \tfrac{1}{2}wL - wx$$

Now

$$\text{S.F. at } C = \tfrac{1}{2}wL - wx = Q \qquad (3.1)$$

$\therefore$

$$\frac{dM}{dx} = Q \qquad (3.2)$$

Differentiating eqn. (3.1),

$$\frac{dQ}{dx} = -w \qquad (3.3)$$

These relationships are the basis of the rules stated in the summary, the proofs of which are as follows:

(a) The maximum or minimum B.M. occurs where $dM/dx = 0$

But

$$\frac{dM}{dx} = Q$$

Thus where S.F. is zero B.M. is a maximum or minimum.

(b) The slope of the B.M. diagram $= dM/dx = Q$.

Thus where $Q = 0$ the slope of the B.M. diagram is zero, and the B.M. is therefore constant.

(c) Also, since Q represents the slope of the B.M. diagram, it follows that ***where the S.F. is positive the slope of the B.M. diagram is positive, and where the S.F. is negative the slope of the B.M. diagram is also negative.***

(d) The area of the S.F. diagram between any two points, from basic calculus, is

$$\int Q\,dx$$

But $$\frac{dM}{dx} = Q \quad \text{or} \quad M = \int Q dx$$

i.e. ***the B.M. change between any two points is the area of the S.F. diagram between these points.***

This often provides a very quick method of obtaining the B.M. diagram once the S.F. diagram has been drawn.

(e) With the chosen sign convention, when the B.M. is positive the beam is *sagging* and when it is negative the beam is *hogging*. Thus when the curvature of the beam changes from *sagging* to *hogging*, as at X–X in Fig. 3.11, or vice versa, the B.M. changes sign, i.e. becomes instantaneously zero. This is termed a ***point of inflexion*** or ***contraflexure. Thus a point of contraflexure occurs where the B.M. is zero.***

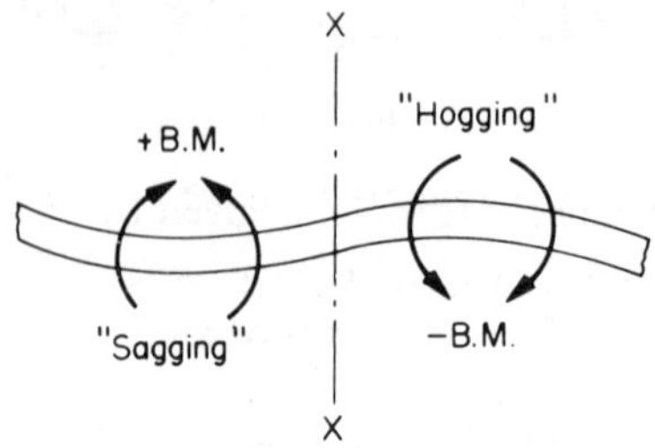

Fig. 3.11. Beam with point of contraflexure at X–X.

3.7. S.F. and B.M. diagrams for an applied couple or moment

In general there are two ways in which the couple or moment can be applied: (a) with horizontal loads and (b) with vertical loads, and the method of solution is different for each.

Type (a): couple or moment applied with horizontal loads

Consider the beam *AB* shown in Fig. 3.12 to which a moment *F.d* is applied by means of horizontal loads at a point *C*, distance *a* from *A*.

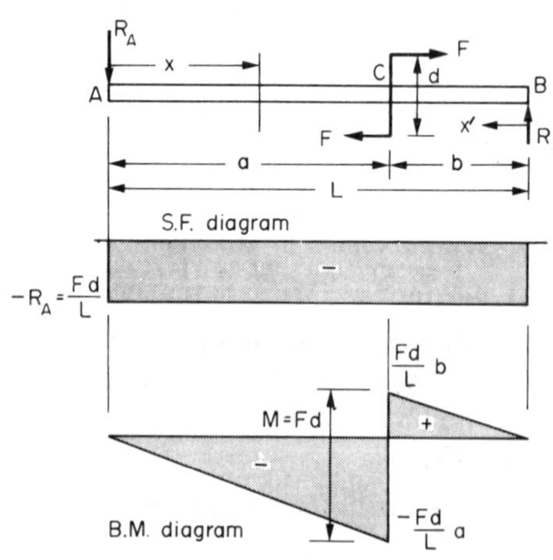

Fig. 3.12.

Since this will tend to lift the beam at A, R_A acts downwards.

Moments about B: $$R_A \times L = Fd$$

$$\therefore \quad R_A = \frac{Fd}{L}$$

and for vertical equilibrium $$R_B = R_A = \frac{Fd}{L}$$

The S.F. diagram can now be drawn as the horizontal loads have no effect on the vertical shear.

The B.M. at any section between A and C is

$$M = -R_A x = \frac{-Fd}{L}x$$

Thus the value of the B.M. increases linearly from zero at A to $\frac{-Fd}{L}a$ at C.

Similarly, the B.M. at any section between C and B is

$$M = -R_A x + Fd = R_B x' = \frac{Fd}{L}x'$$

i.e. the value of the B.M. again increases linearly from zero at B to $\frac{Fd}{L}b$ at C. The B.M. diagram is therefore as shown in Fig. 3.12.

Type (b): moment applied with vertical loads

Consider the beam AB shown in Fig. 3.13; taking moments about B:

$$R_A L = F(d+b)$$

$$\therefore \quad R_A = \frac{F(d+b)}{L}$$

Similarly, $$R_B = \frac{F(a-d)}{L}$$

The S.F. diagram can therefore be drawn as in Fig. 3.13 and it will be observed that in this case F does affect the diagram.

For the B.M. diagram an equivalent system is used. The offset load F is replaced by a moment and a force acting at C, as shown in Fig. 3.13. Thus

$$\text{B.M. between } A \text{ and } C = R_A x = \frac{F(d+b)}{L}x$$

i.e. increasing linearly from zero to $\frac{F(d+b)}{L}a$ at C.

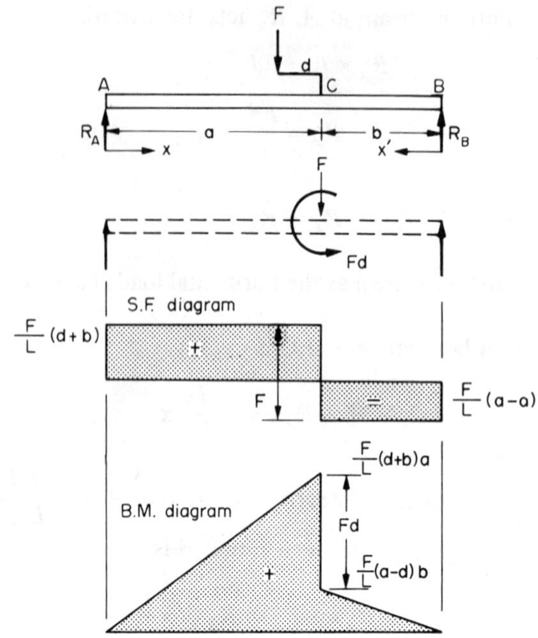

Fig. 3.13.

Similarly,

$$\text{B.M. between } C \text{ and } B = R_B x'$$

$$= \frac{F(a-d)}{L} x'$$

i.e. increasing linearly from zero to $\dfrac{F(a-d)}{L} b$ at C.

The difference in values at C is equal to the applied moment Fd, as with type (a).

Consider now the beam shown in Fig. 3.14 carrying concentrated loads in addition to the applied moment of 30 kN m (which can be assumed to be of type (a) unless otherwise stated). The principle of superposition states that the total effect of the combined loads will be the same as the algebraic sum of the effects of the separate loadings, i.e. the final diagram will be the combination of the separate diagrams representing applied moment and those representing concentrated loads. The final diagrams are therefore as shown shaded, all values quoted being measured from the normal base line of each diagram. In each case, however, the applied-moment diagrams have been inverted so that the negative areas can easily be subtracted. Final values are now measured from the dotted lines: e.g. the S.F. and B.M. at any point G are as indicated in Fig. 3.14.

3.8. S.F. and B.M. diagrams for inclined loads

If a beam is subjected to inclined loads as shown in Fig. 3.15 each of the loads must be resolved into its vertical and horizontal components as indicated. The vertical components

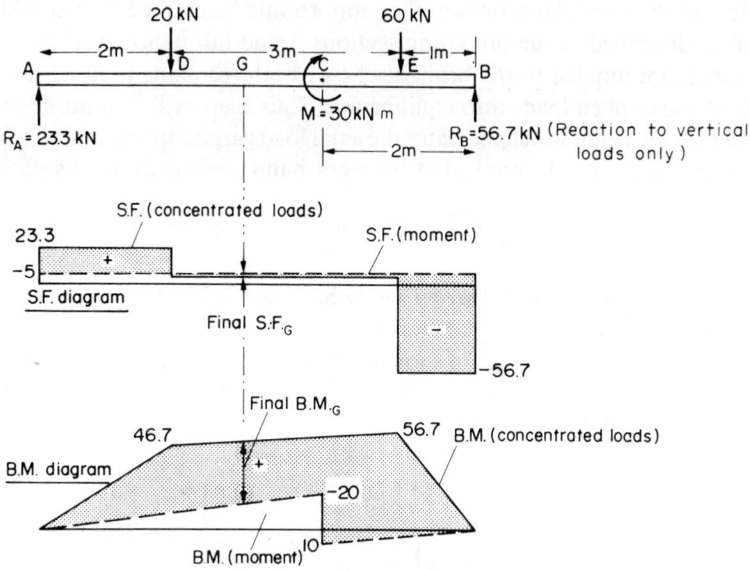

Fig. 3.14.

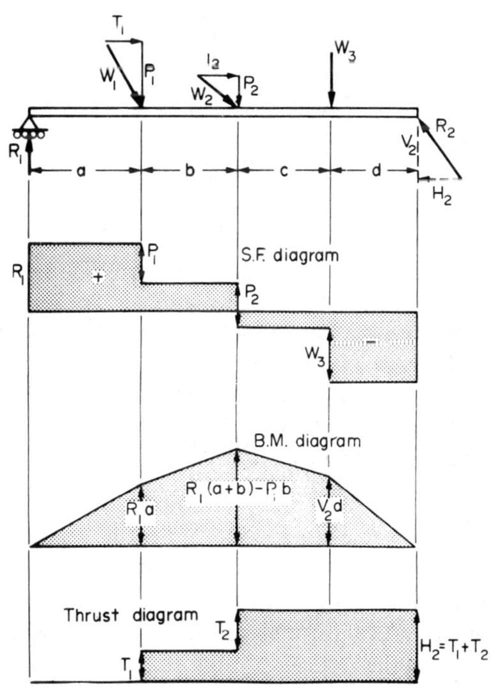

Fig. 3.15. S.F., B.M. and thrust diagrams for system of inclined loads.

yield the values of the vertical reactions at the supports and hence the S.F. and B.M. diagrams are obtained as described in the preceding sections. In addition, however, there must be a horizontal constraint applied to the beam at one or both reactions to bring the horizontal components of the applied loads into equilibrium. Thus there will be a horizontal force or *thrust diagram* for the beam which indicates the axial load carried by the beam at any point. If the constraint is assumed to be applied at the right-hand end the thrust diagram will be as indicated.

3.9. Graphical construction of S.F. and B.M. diagrams

Consider the simply supported beam shown in Fig. 3.16 carrying three concentrated loads of different values. The procedure to be followed for graphical construction of the S.F. and B.M. diagrams is as follows.

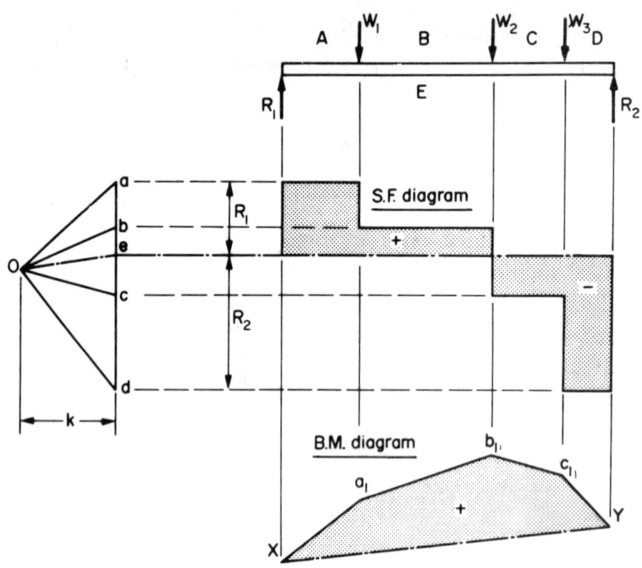

Fig. 3.16. Graphical construction of S.F. and B.M. diagrams.

(a) Letter the spaces between the loads and reactions *A, B, C, D* and *E*. Each force can then be denoted by the letters of the spaces on either side of it.
(b) To one side of the beam diagram construct a force vector diagram for the applied loads, i.e. set off a vertical distance *ab* to represent, in magnitude and direction, the force W_1 dividing spaces *A* and *B* to some scale, *bc* to represent W_2 and *cd* to represent W_3.
(c) Select any point *O*, known as a *pole point*, and join *Oa*, *Ob*, *Oc* and *Od*.
(d) Drop verticals from all loads and reactions.
(e) Select any point *X* on the vertical through reaction R_1 and from this point draw a line in space *A* parallel to *Oa* to cut the vertical through W_1 in a_1. In space *B* draw a line from a_1 parallel to *ob*, continue in space *C* parallel to *Oc*, and finally in space *D* parallel to *Od* to cut the vertical through R_2 in *Y*.

(f) Join XY and through the pole point O draw a line parallel to XY to cut the force vector diagram in e. The distance ea then represents the value of the reaction R_1 in magnitude and direction and de represents R_2.
(g) Draw a horizontal line through e to cut the vertical projections from the loading points and to act as the base line for the S.F. diagram. Horizontal lines from a in gap A, b in gap B, c in gap C, etc., produce the required S.F. diagram to the same scale as the original force vector diagram.
(h) The diagram $Xa_1b_1c_1Y$ is the B.M. diagram for the beam, vertical distances from the inclined base line XY giving the bending moment at any required point to a certain scale.

If the original beam diagram is drawn to a scale 1 cm = L metres (say), the force vector diagram scale is 1 cm = W newton, and, if the horizontal distance from the pole point O to the vector diagram is k cm, then the scale of the B.M. diagram is

$$1\text{ cm} = kLW\text{ newton metre}$$

The above procedure applies for beams carrying concentrated loads only, but an approximate solution is obtained in a similar way for u.d.l.s. by considering the load divided into a convenient number of concentrated loads acting at the centres of gravity of the divisions chosen.

3.10. S.F. and B.M. diagrams for beams carrying distributed loads of increasing value

For beams which carry distributed loads of varying intensity as in Fig. 3.18 a solution can be obtained from eqn. (3.3) provided that the loading variation can be expressed in terms of the distance x along the beam span, i.e. as a function of x.

$$\frac{dQ}{dx} = -w = -f(x)$$

Integrating once yields the shear force Q in terms of a constant of integration A since

$$\frac{dM}{dx} = Q$$

Integration again yields an expression for the B.M. M in terms of A and a second constant of integration B. Known conditions of B.M. or S.F., usually at the supports or ends of the beam, yield the values of the constants and hence the required distributions of S.F. and B.M. A typical example of this type has been evaluated on page 57.

3.11. S.F. at points of application of concentrated loads

In the preceding sections it has been assumed that concentrated loads can be applied precisely at a point so that S.F. diagrams are shown to change value suddenly from one value to another, and sometimes one sign to another, at the loading points. It would appear from the S.F. diagrams drawn previously, therefore, that two possible values of S.F. exist at any one loading point and this is obviously not the case. In practice, loads can only be applied over finite areas and the S.F. must change gradually from one value to another across these areas. The vertical line portions of the S.F. diagrams are thus highly idealised versions of what actually occurs in practice and should be replaced more accurately by lines slightly inclined to the vertical. All sharp corners of the diagrams should also be rounded. Despite these minor inaccuracies, B.M. and S.F. diagrams remain a highly convenient, powerful and useful representation of beam loading conditions for design purposes.

Examples

Example 3.1

Draw the S.F. and B.M. diagrams for the beam loaded as shown in Fig. 3.17, and determine (a) the position and magnitude of the maximum B.M., and (b) the position of any point of contraflexure.

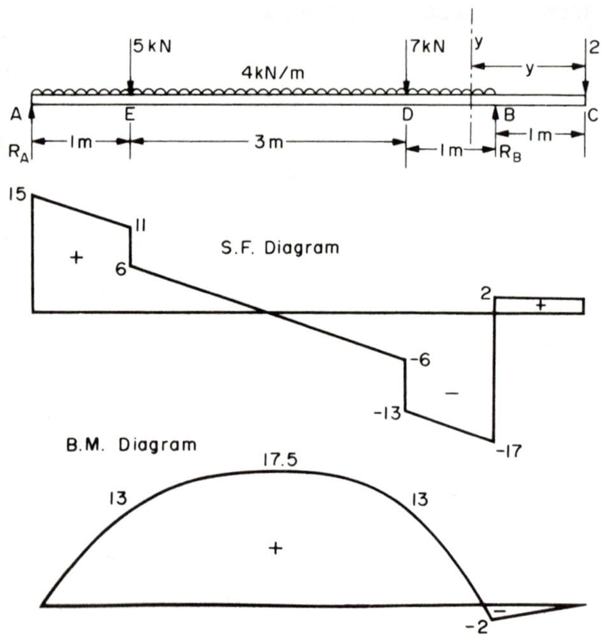

Fig. 3.17.

Solution

Taking the moments about A,

$$5R_B = (5 \times 1) + (7 \times 4) + (2 \times 6) + (4 \times 5) \times 2.5$$

$$\therefore \qquad R_B = \frac{5 + 28 + 12 + 50}{5} = \mathbf{19\,kN}$$

and since $$R_A + R_B = 5 + 7 + 2 + (4 \times 5) = 34$$
$$R_A = 34 - 19 = \mathbf{15\,kN}$$

The S.F. diagram may now be constructed as described in §3.4 and is shown in Fig. 3.17.

Calculation of bending moments

B.M. at A and $C = 0$

B.M. at $B \quad = -2 \times 1 = -2\,\text{kN m}$

B.M. at $D \quad = -(2 \times 2) + (19 \times 1) - (4 \times 1 \times \tfrac{1}{2}) = +13\,\text{kN m}$

B.M. at $E \quad = +(15 \times 1) - (4 \times 1 \times \tfrac{1}{2}) = +13\,\text{kN m}$

The maximum B.M. will be given by the point (or points) at which dM/dx (i.e. the shear force) is zero. By inspection of the S.F. diagram this occurs midway between D and E, i.e. at 1.5 m from E.

$$\text{B.M. at this point} = (2.5 \times 15) - (5 \times 1.5) - \left(4 \times 2.5 \times \frac{2.5}{2}\right)$$
$$= \mathbf{+17.5\,kN\,m}$$

There will also be local maxima at the other points where the S.F. diagram crosses its zero axis, i.e. at point B.

Owing to the presence of the concentrated loads (reactions) at these positions, however, these will appear as discontinuities in the diagram; there will not be a smooth contour change. The value of the B.M.s at these points should be checked since the position of maximum stress in the beam depends upon the numerical maximum value of the B.M.; this does not necessarily occur at the mathematical maximum obtained above.

The B.M. diagram is therefore as shown in Fig. 3.17. Alternatively, the B.M. at any point between D and E at a distance of x from A will be given by

$$M_{xx} = 15x - 5(x-1) - \frac{4x^2}{2} = 10x + 5 - 2x^2$$

The maximum B.M. position is then given where $\dfrac{dM}{dx} = 0$.

$$\frac{dM}{dx} = 10 - 4x = 0 \qquad \therefore \qquad x = 2.5\,\text{m}$$

i.e. **1.5 m from E,** as found previously.

(b) Since the B.M. diagram only crosses the zero axis once there is only one point of contraflexure, i.e. between B and D. Then, B.M. at distance y from C will be given by

$$M_{yy} = -2y + 19(y-1) - 4(y-1)\tfrac{1}{2}(y-1)$$
$$= -2y + 19y - 19 - 2y^2 + 4y - 2 = 0$$

The point of contraflexure occurs where B.M. = 0, i.e. where $M_{yy} = 0$,

$$\therefore \qquad 0 = -2y^2 + 21y - 21$$

i.e. $$2y^2 - 21y + 21 = 0$$

Then $$y = \frac{21 \pm \sqrt{(21^2 - 4 \times 2 \times 21)}}{4} = 1.12\,\text{m}$$

i.e. point of contraflexure occurs **0.12 m to the left of *B*.**

Example 3.2

A beam ABC is 9 m long and supported at B and C, 6 m apart as shown in Fig. 3.18. The beam carries a triangular distribution of load over the portion BC together with an applied counterclockwise couple of moment 80 kN m at B and a u.d.l. of 10 kN/m over AB, as shown. Draw the S.F. and B.M. diagrams for the beam.

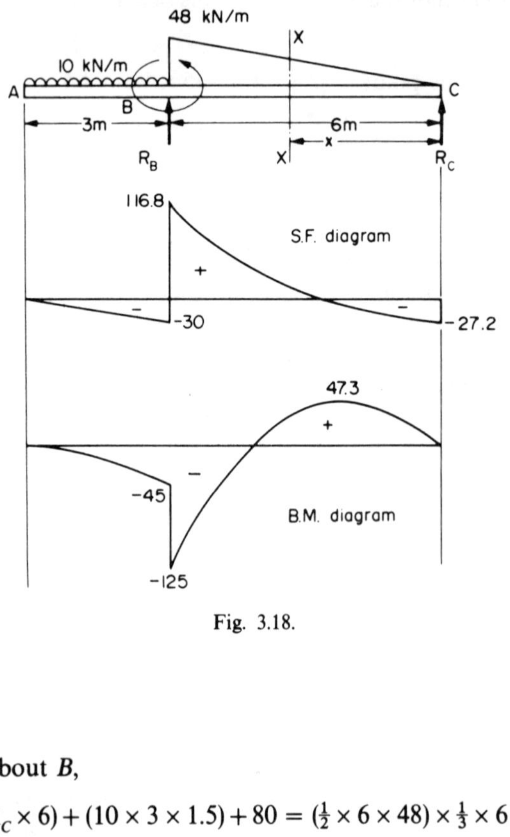

Fig. 3.18.

Solution

Taking moments about B,

$$(R_C \times 6) + (10 \times 3 \times 1.5) + 80 = (\tfrac{1}{2} \times 6 \times 48) \times \tfrac{1}{3} \times 6$$

$$6R_C + 45 + 80 = 288$$

$$R_C = 27.2\,\text{kN}$$

and $$R_C + R_B = (10 \times 3) + (\tfrac{1}{2} \times 6 \times 48)$$

$$= 30 + 144 = 174$$

$\therefore$ $$R_B = 146.8\,\text{kN}$$

At any distance x from C between C and B the shear force is given by

$$\text{S.F.}_{XX} = -\tfrac{1}{2}wx + R_C$$

and by proportions $$\frac{w}{x} = \frac{48}{6} = 8$$

i.e. $$w = 8x \text{ kN/m}$$

$\therefore$ $$\text{S.F.}_{XX} = -(R_C - \tfrac{1}{2} \times 8x \times x)$$
$$= -R_C + 4x^2$$
$$= -27.2 + 4x^2$$

The S.F. diagram is then as shown in Fig. 3.18.

Also $$\text{B.M.}_{XX} = -(\tfrac{1}{2}wx)\frac{x}{3} + R_C x$$

$$= 27.2x - \frac{4x^3}{3}$$

For a maximum value, $$\frac{d\,(\text{B.M.})}{dx} = \text{S.F.} = 0$$

i.e.,where $$4x^2 = 27.2$$

or $$x = 2.61 \text{ m from } C$$

$$\text{B.M.}_{max} = 27.2(2.61) - \frac{4}{3}(2.61)^3$$

$$= 47.3 \text{ kN m}$$

$$\text{B.M. at } A \text{ and } C = 0$$

$$\text{B.M. immediately to left of } B = -(10 \times 3 \times 1.5) = -45 \text{ kN m}$$

At the point of application of the applied moment there will be a sudden change in B.M. of 80 kN m. (There will be no such discontinuity in the S.F. diagram; the effect of the moment will merely be reflected in the values calculated for the reactions.)

The B.M. diagram is therefore as shown in Fig. 3.18.

Problems

3.1 (A). A beam *AB*, 1.2 m long, is simply-supported at its ends *A* and *B* and carries two concentrated loads, one of 10 kN at *C*, the other 15 kN at *D*. Point *C* is 0.4 m from *A*, point *D* is 1 m from *A*. Draw the S.F. and B.M. diagrams for the beam inserting principal values. [9.17, −0.83, −15.83 kN; 3.67, 3.17 kNm.]

3.2 (A). The beam of question 3.1 carries an additional load of 5 kN *upwards* at point *E*, 0.6 m from *A*. Draw the S.F. and B.M. diagrams for the modified loading. What is the maximum B.M.? [6.67, −3.33, 1.67, −13.33 kN; 2.67, 2, 2.67 kN m.]

3.3 (A). A cantilever beam *AB*, 2.5 m long is rigidly built in at *A* and carries vertical concentrated loads of 8 kN at *B* and 12 kN at *C*, 1 m from *A*. Draw S.F. and B.M. diagrams for the beam inserting principal values. [−8, −20 kN; −11.2, −31.2 kNm.]

3.4 (A). A beam AB, 5 m long, is simply-supported at the end B and at a point C, 1 m from A. It carries vertical loads of 5 kN at A and 20 kN at D, the centre of the span BC. Draw S.F. and B.M. diagrams for the beam inserting principal values. [−5, 11.25, −8.75 kN; −5, 17.5 kNm.]

3.5 (A). A beam AB, 3 m long, is simply-supported at A and B. It carries a 16 kN concentrated load at C, 1.2 m from A, and a u.d.l. of 5 kN/m over the remainder of the beam. Draw the S.F. and B.M. diagrams and determine the value of the maximum B.M. [12.3, −3.7, −12.7 kN; 14.8 kNm.]

3.6 (A). A simply supported beam has a span of 4 m and carries a uniformly distributed load of 60 kN/m together with a central concentrated load of 40 kN. Draw the S.F. and B.M. diagrams for the beam and hence determine the maximum B.M. acting on the beam. [S.F. 140, ±20, −140 kN; B.M. 0, 160, 0 kN m.]

3.7 (A). A 2 m long cantilever is built-in at the right-hand end and carries a load of 40 kN at the free end. In order to restrict the deflection of the cantilever within reasonable limits an upward load of 10 kN is applied at mid-span. Construct the S.F. and B.M. diagrams for the cantilever and hence determine the values of the reaction force and moment at the support. [30 kN, 70 kN m.]

3.8 (A). A beam 4.2 m long overhangs each of two simple supports by 0.6 m. The beam carries a uniformly distributed load of 30 kN/m between supports together with concentrated loads of 20 kN and 30 kN at the two ends. Sketch the S.F. and B.M. diagrams for the beam and hence determine the position of any points of contraflexure. [S.F. −20, +43, −47, +30 kN; B.M. −12, 18.75, −18 kN m; 0.313 and 2.553 m from l.h. support.]

3.9 (A/B). A beam $ABCDE$, with A on the left, is 7 m long and is simply supported at B and E. The lengths of the various portions are $AB = 1{\cdot}5$ m, $BC = 1{\cdot}5$ m, $CD = 1$ m and $DE = 3$ m. There is a uniformly distributed load of 15 kN/m between B and a point 2 m to the right of B and concentrated loads of 20 kN act at A and D with one of 50 kN at C.

(a) Draw the S.F. diagrams and hence determine the position from A at which the S.F. is zero.
(b) Determine the value of the B.M. at this point.
(c) Sketch the B.M. diagram approximately to scale, quoting the principal values.

[3.32 m; 69.8 kN m; 0, −30, 69.1, 68.1, 0 kN m.]

3.10 (A/B). A beam $ABCDE$ is simply supported at A and D. It carries the following loading: a distributed load of 30 kN/m between A and B; a concentrated load of 20 kN at B; a concentrated load of 20 kN at C; a concentrated load of 10 kN at E; a distributed load of 60 kN/m between D and E. Span $AB = 1.5$ m, $BC = CD = DE = 1$ m. Calculate the value of the reactions at A and D and hence draw the S.F. and B.M. diagrams. What are the magnitude and position of the maximum B.M. on the beam? [41.1, 113.9 kN; 28.15 kN m; 1.37 m from A.]

3.11 (B). A beam, 12 m long, is to be simply supported at 2 m from each end and to carry a u.d.l. of 30 kN/m together with a 30 kN point load at the right-hand end. For ease of transportation the beam is to be jointed in two places, one joint being situated 5 m from the left-hand end. What load (to the nearest kN) must be applied to the left-hand end to ensure that there is no B.M. at the joint (i.e. the joint is to be a point of contraflexure)? What will then be the best position on the beam for the other joint? Determine the position and magnitude of the maximum B.M. present on the beam. [114 kN, 1.6 m from r.h. reaction; 4.7 m from l.h. reaction; 43.35 kN m.]

3.12 (B). A horizontal beam AB is 4 m long and of constant flexural rigidity. It is rigidly built-in at the left-hand end A and simply supported on a non-yielding support at the right-hand end B. The beam carries uniformly distributed vertical loading of 18 kN/m over its whole length, together with a vertical downward load of 10 kN at 2.5 m from the end A. Sketch the S.F. and B.M. diagrams for the beam, indicating all main values. [I. Struct. E.] [S.F. 45, −10, −37.6 kN; B.M. −18.6, +36.15 kN m.]

3.13 (B). A beam ABC, 6 m long, is simply-supported at the left-hand end A and at B 1 m from the right-hand end C. The beam is of weight 100 N/metre run.

(a) Determine the reactions at A and B.
(b) Construct to scales of 20 mm = 1 m and 20 mm = 100 N, the shearing-force diagram for the beam, indicating thereon the principal values.
(c) Determine the magnitude and position of the maximum bending moment. (You may, if you so wish, deduce the answers from the shearing force diagram without constructing a full or partial bending-moment diagram.)

[C.G.] [240 N, 360 N, 288 Nm, 2.4 m from A.]

3.14 (B). A beam $ABCD$, 6 m long, is simply-supported at the right-hand end D and at a point B 1m from the left-hand end A. It carries a vertical load of 10 kN at A, a second concentrated load of 20 kN at C, 3 m from D, and a uniformly distributed load of 10 kN/m between C and D. Determine:

(a) the values of the reactions at B and D,
(b) the position and magnitude of the maximum bending moment.

[33 kN, 27 kN, 2.7 m from D, 36.45 k Nm.]

3.15 (B). A beam $ABCD$ is simply supported at B and C with $AB = CD = 2$ m; $BC = 4$ m. It carries a point load of 60 kN at the free end A, a uniformly distributed load of 60 kN/m between B and C and an anticlockwise moment of

80 kN m in the plane of the beam applied at the free end *D*. Sketch and dimension the S.F. and B.M. diagrams, and determine the position and magnitude of the maximum bending moment.

[E.I.E.] [S.F. −60, +170, −70 kN; B.M. −120, +120.1, +80 kN m; 120.1 kN m at 2.83 m to right of *B*.]

3.16 (B). A beam *ABCDE* is 4.6 m in length and loaded as shown in Fig. 3.19. Draw the S.F. and B.M. diagrams for the beam, indicating all major values.

[I.E.I.] [S.F. 28.27, 7.06, −12.94, −30.94, +18, 0; B.M. 28.27, 7.06, 15.53, −10.8.]

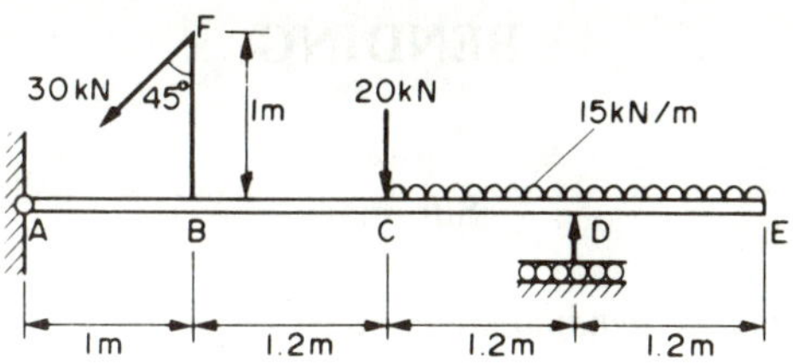

Fig. 3.19.

3.17 (B). A simply supported beam has a span of 6 m and carries a distributed load which varies in a linear manner from 30 kN/m at one support to 90 kN/m at the other support. Locate the point of maximum bending moment and calculate the value of this maximum. Sketch the S.F. and B.M. diagrams.

[U.L.] [3.25 m from l.h. end; 272 kN m.]

3.18 (B). Obtain the relationship between the bending moment, shearing force, and intensity of loading of a laterally loaded beam. A simply supported beam of span *L* carries a distributed load of intensity kx^2/L^2, where x is measured from one support towards the other. Determine: (a) the location and magnitude of the greatest bending moment, (b) the support reactions. [U. Birm.] [$0.0394\ kL^2$ at 0.63 of span; $kL/12$ and $kL/4$.]

3.19 (B). A beam *ABC* is continuous over two spans. It is built-in at *A*, supported on rollers at *B* and *C* and contains a hinge at the centre of the span *AB*. The loading consists of a uniformly distributed load of total weight 20 kN on the 7 m span *AB* and a concentrated load of 30 kN at the centre of the 3 m span *BC*. Sketch the S.F. and B.M. diagrams, indicating the magnitudes of all important values.

[I.E.I.] [S.F. 5, −15, 26.67, −3.33 kN; B.M. 4.38, −35, +5 kN m.]

3.20 (B). A log of wood 225 mm square cross-section and 5 m in length is rendered impervious to water and floats in a horizontal position in fresh water. It is loaded at the centre with a load just sufficient to sink it completely. Draw S.F. and B.M. diagrams for the condition when this load is applied, stating their maximum values. Take the density of wood as 770 kg/m^3 and of water as 1000 kg/m^3. [S.F. 0, ±0.285, 0 kN; B.M. 0, 0.356, 0 kN m.]

3.21 (B). A simply supported beam is 3 m long and carries a vertical load of 5 kN at a point 1 m from the left-hand end. At a section 2 m from the left-hand end a clockwise couple of 3 kN m is exerted, the axis of the couple being horizontal and perpendicular to the longitudinal axis of the beam. Draw to scale the B.M. and S.F. diagrams and mark on them the principal dimensions. [I.Mech.E.] [S.F. 2.33, −2.67 kN; B.M. 2.33, −0.34, +2.67 kN m.]

CHAPTER 4

BENDING

Summary

The *simple theory of elastic bending* states that

$$\frac{M}{I} = \frac{\sigma}{y} = \frac{E}{R}$$

where M is the applied bending moment (B.M.) at a transverse section, I is the second moment of area of the beam cross-section about the neutral axis (N.A.), σ is the stress at distance y from the N.A. of the beam cross-section, E is the Young's modulus of elasticity for the beam material, and R is the radius of curvature of the N.A. at the section.

Certain assumptions and conditions must obtain before this theory can strictly be applied: see page 64.

In some applications the following relationship is useful:

$$M = Z\sigma_{max}$$

where $Z = I/y_{max}$ and is termed the *section modulus*; σ_{max} is then the stress at the maximum distance from the N.A.

The most useful standard values of the second moment of area I for certain sections are as follows (Fig. 4.1):

$$\text{rectangle about axis through centroid} = \frac{bd^3}{12} = I_{N.A.}$$

$$\text{rectangle about axis through side} = \frac{bd^3}{3} = I_{XX}$$

$$\text{circle about axis through centroid} = \frac{\pi D^4}{64} = I_{N.A.}$$

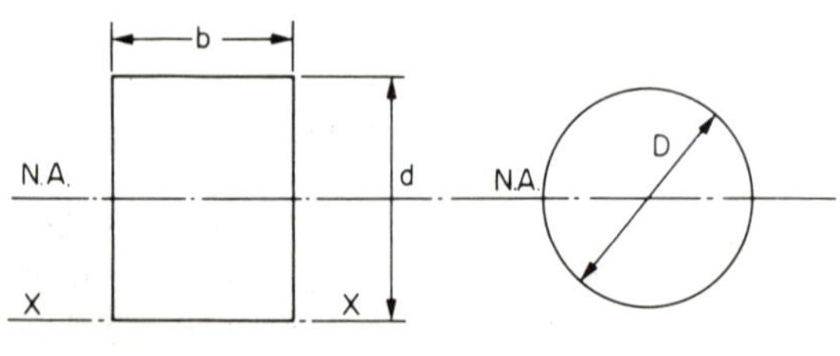

Fig. 4.1.

The *centroid* is the centre of area of the section through which the N.A., or axis of zero stress, is always found to pass.

In some cases it is convenient to determine the second moment of area about an axis other than the N.A. and then to use the *parallel axis theorem.*

$$I_{\text{N.A.}} = I_G + Ah^2$$

For *composite beams* one material is replaced by an equivalent width of the other material given by

$$t' = \frac{E}{E'}t$$

where E/E' is termed the *modular ratio.* The relationship between the stress in the material and its equivalent area is then given by

$$\sigma = \frac{E}{E'}\sigma'$$

For *skew loading of symmetrical sections* the stress at any point (x, y) is given by

$$\sigma = \frac{M_{xx}}{I_{xx}}y \pm \frac{M_{yy}}{I_{yy}}x$$

the angle of the N.A. being given by

$$\tan\theta = \pm\frac{M_{yy}}{M_{xx}}\frac{I_{xx}}{I_{yy}}$$

For *eccentric loading on one axis,*

$$\sigma = \frac{P}{A} \pm \frac{Pey}{I}$$

the N.A. being positioned at a distance

$$y_N = \pm\frac{I}{Ae}$$

from the axis about which the eccentricity is measured.

For *eccentric loading on two axes,*

$$\sigma = \frac{P}{A} \pm \frac{Ph}{I_{yy}}x \pm \frac{Pk}{I_{xx}}y$$

For *concrete or masonry rectangular or circular section columns*, the load must be retained within the middle third or middle quarter areas respectively.

Introduction

If a piece of rubber, most conveniently of rectangular cross-section, is bent between one's fingers it is readily apparent that one surface of the rubber is stretched, i.e. put into tension, and the opposite surface is compressed. The effect is clarified if, before bending, a regular set of lines is drawn or scribed on each surface at a uniform spacing and perpendicular to the axis of the rubber which is held between the fingers. After bending, the spacing between the set of lines on one surface is clearly seen to increase and on the other surface to reduce. The thinner the rubber, i.e. the closer the two marked faces, the smaller is the effect for the same applied moment. The change in spacing of the lines on each surface is a measure of the strain and hence the stress to which the surface is subjected and it is convenient to obtain a formula relating the stress in the surface to the value of the B.M. applied and the amount of curvature produced. In order for this to be achieved it is necessary to make certain simplifying assumptions, and for this reason the theory introduced below is often termed the simple theory of bending. The assumptions are as follows:

(1) The beam is initially straight and unstressed.
(2) The material of the beam is perfectly homogeneous and isotropic, i.e. of the same density and elastic properties throughout.
(3) The elastic limit is nowhere exceeded.
(4) Young's modulus for the material is the same in tension and compression.
(5) Plane cross-sections remain plane before and after bending.
(6) Every cross-section of the beam is symmetrical about the plane of bending, i.e. about an axis perpendicular to the N.A.
(7) There is no resultant force perpendicular to any cross-section.

4.1. Simple bending theory

If we now consider a beam initially unstressed and subjected to a constant B.M. along its length, i.e. pure bending, as would be obtained by applying equal couples at each end, it will bend to a radius R as shown in Fig. 4.2b. As a result of this bending the top fibres of the beam will be subjected to tension and the bottom to compression. It is reasonable to suppose, therefore, that somewhere between the two there are points at which the stress is zero. The locus of all such points is termed the *neutral axis*. The radius of curvature R is then measured to this axis. For symmetrical sections the N.A. is the axis of symmetry, but whatever the section the N.A. will always pass through the centre of area or centroid.

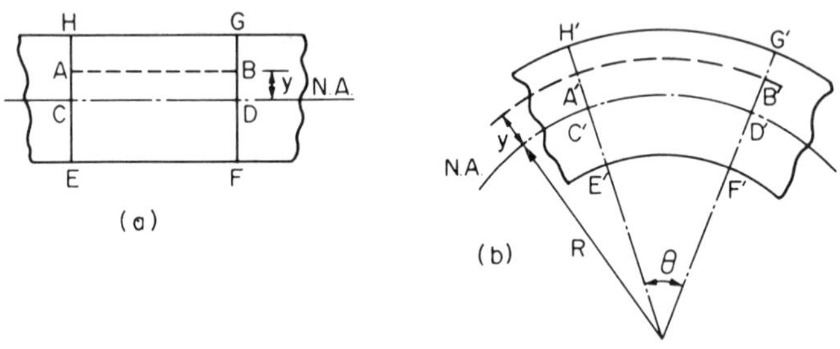

Fig. 4.2. Beam subjected to pure bending (a) before, and (b) after, the moment M has been applied.

Consider now two cross-sections of a beam, HE and GF, originally parallel (Fig. 4.2a). When the beam is bent (Fig. 4.2b) it is assumed that these sections remain plane; i.e. $H'E'$ and $G'F'$, the final positions of the sections, are still straight lines. They will then subtend some angle θ.

Consider now some fibre AB in the material, distance y from the N.A. When the beam is bent this will stretch to $A'B'$.

$$\text{Strain in fibre } AB = \frac{\text{extension}}{\text{original length}} = \frac{A'B' - AB}{AB}$$

But $AB = CD$, and, since the N.A. is unstressed, $CD = C'D'$.

$$\therefore \qquad \text{strain} = \frac{A'B' - C'D'}{C'D'} = \frac{(R+y)\theta - R\theta}{R\theta} = \frac{y}{R}$$

But $$\frac{\text{stress}}{\text{strain}} = \text{Young's modulus } E$$

$$\therefore \qquad \text{strain} = \frac{\sigma}{E}$$

Equating the two equations for strain,

$$\frac{\sigma}{E} = \frac{y}{R}$$

or $$\frac{\boldsymbol{\sigma}}{\boldsymbol{y}} = \frac{\boldsymbol{E}}{\boldsymbol{R}} \tag{4.1}$$

Consider now a cross-section of the beam (Fig. 4.3). From eqn. (4.1) the stress on a fibre at distance y from the N.A. is

$$\sigma = \frac{E}{R} y$$

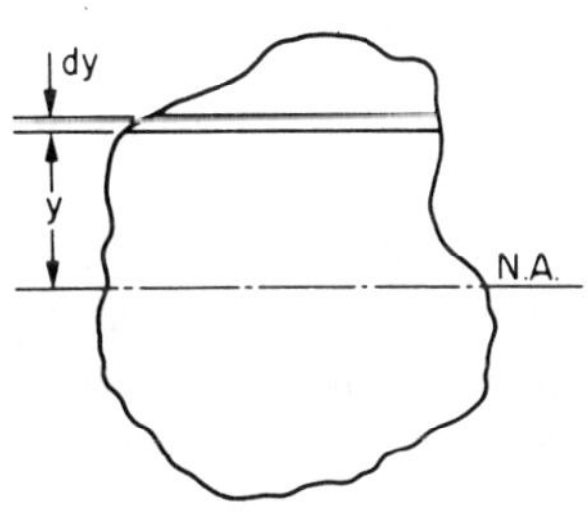

Fig. 4.3. Beam cross-section.

If the strip is of area δA the force on the strip is

$$F = \sigma \delta A = \frac{E}{R} y \delta A$$

This has a moment about the N.A. of

$$Fy = \frac{E}{R} y^2 \delta A$$

The total moment for the whole cross-section is therefore

$$M = \sum \frac{E}{R} y^2 \delta A$$

$$= \frac{E}{R} \Sigma y^2 \delta A$$

since E and R are assumed constant.

The term $\Sigma y^2 \delta A$ is called the *second moment of area* of the cross-section and given the symbol I.

$$\therefore \qquad M = \frac{E}{R} I \quad \text{and} \quad \frac{M}{I} = \frac{E}{R} \tag{4.2}$$

Combining eqns. (4.1) and (4.2) we have the bending theory equation

$$\frac{M}{I} = \frac{\sigma}{y} = \frac{E}{R} \tag{4.3}$$

From eqn. (4.2) it will be seen that if the beam is of uniform section, the material of the beam is homogeneous and the applied moment is constant, the values of I, E and M remain constant and hence the radius of curvature of the bent beam will also be constant. Thus for pure bending of uniform sections, beams will deflect into circular arcs and for this reason the term *circular bending* is often used. From eqn. (4.2) the radius of curvature to which any beam is bent by an applied moment M is given by:

$$R = \frac{EI}{M}$$

and is thus directly related to the value of the quantity EI. Since the radius of curvature is a direct indication of the degree of flexibility of the beam (the larger the value of R, the smaller the deflection and the greater the rigidity) the quantity EI is often termed the *flexural rigidity* or *flexural stiffness* of the beam. The relative stiffnesses of beam sections can then easily be compared by their EI values.

It should be observed here that the above proof has involved the assumption of pure bending without any shear being present. From the work of the previous chapter it is clear that in most practical beam loading cases shear and bending occur together at most points. Inspection of the S.F. and B.M. diagrams, however, shows that when the B.M. is a maximum the S.F. is, in fact, always zero. It will be shown later that bending produces by far the greatest magnitude of stress in all but a small minority of special loading cases so that beams designed on the basis of the maximum B.M. using the simple bending theory are generally more than adequate in strength at other points.

4.2. Neutral axis

As stated above, it is clear that if, in bending, one surface of the beam is subjected to tension and the opposite surface to compression there must be a region within the beam cross-section at which the stress changes sign, i.e. where the stress is zero, and this is termed the *neutral axis*.

Further, eqn. (4.3) may be re-written in the form

$$\sigma = \frac{M}{I}y \tag{4.4}$$

which shows that at any section the stress is directly proportional to y, the distance from the N.A., i.e. σ varies linearly with y, the maximum stress values occurring in the outside surface of the beam where y is a maximum.

Consider again, therefore, the general beam cross-section of Fig. 4.3 in which the N.A. is located at some arbitrary position. The force on the small element of area is σdA acting perpendicular to the cross-section, i.e. parallel to the beam axis. The total force parallel to the beam axis is therefore $\int \sigma dA$.

Now one of the basic assumptions listed earlier states that when the beam is in equilibrium there can be no resultant force across the section, i.e. the tensile force on one side of the N.A. must exactly balance the compressive force on the other side.

$$\therefore \qquad \int \sigma dA = 0$$

Substituting from eqn. (4.1)

$$\int \frac{E}{R} y dA = 0 \quad \text{and hence} \quad \frac{E}{R}\int y dA = 0$$

This integral is the *first moment of area* of the beam cross-section about the N.A. since y is always measured from the N.A. Now the only first moment of area for the cross-section which is zero is that about an axis through the centroid of the section since this is the basic condition required of the centroid. It follows therefore that *the neutral axis must always pass through the centroid.*

It should be noted that this condition only applies with stresses maintained within the elastic range and different conditions must be applied when stresses enter the plastic range of the materials concerned.

Typical stress distributions in bending are shown in Fig. 4.4. It is evident that the material near the N.A. is always subjected to relatively low stresses compared with the areas most removed from the axis. In order to obtain the maximum resistance to bending it is advisable therefore to use sections which have large areas as far away from the N.A. as possible. For this reason beams with I- or T-sections find considerable favour in present engineering applications, such as girders, where bending plays a large part. Such beams have large moments of area about one axis and, provided that it is ensured that bending takes place about this axis, they will have a high resistance to bending stresses.

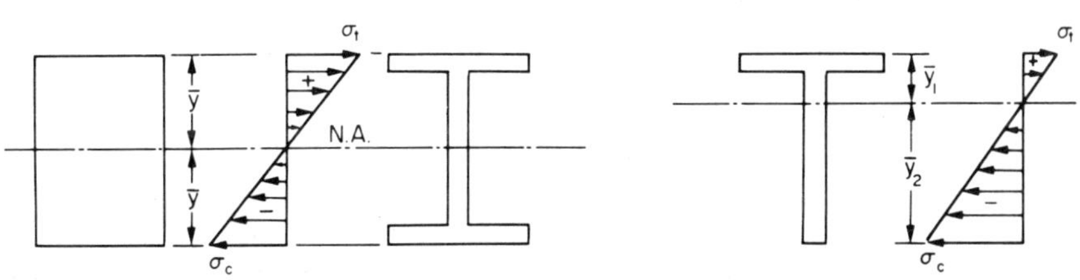

Fig. 4.4. Typical bending stress distributions.

4.3. Section modulus

From eqn. (4.4) the maximum stress obtained in any cross-section is given by

$$\sigma_{max} = \frac{M}{I} y_{max} \qquad (4.5)$$

For any given allowable stress the maximum moment which can be accepted by a particular shape of cross-section is therefore

$$M = \frac{I}{y_{max}} \sigma_{max}$$

For ready comparison of the strength of various beam cross-sections this is sometimes written in the form

$$M = Z\sigma_{max} \qquad (4.6)$$

where $Z(= I/y_{max})$ is termed the *section modulus*. The higher the value of Z for a particular cross-section the higher the B.M. which it can withstand for a given maximum stress.

In applications such as cast-iron or reinforced concrete where the properties of the material are vastly different in tension and compression two values of maximum allowable stress apply. This is particularly important in the case of unsymmetrical sections such as T-sections where the values of y_{max} will also be different on each side of the N.A. (Fig. 4.4) and here two values of section modulus are often quoted,

$$Z_1 = I/y_1 \quad \text{and} \quad Z_2 = I/y_2 \qquad (4.7)$$

each being then used with the appropriate value of allowable stress.

Standard handbooks† are available which list section modulus values for a range of girders, etc; to enable appropriate beams to be selected for known section modulus requirements.

4.4. Second moment of area

Consider the rectangular beam cross-section shown in Fig. 4.5 and an element of area dA, thickness dy, breadth B and distance y from the N.A. which by symmetry passes through the

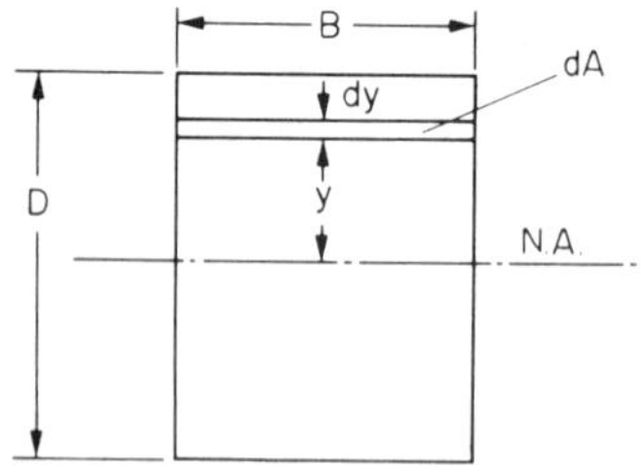

Fig. 4.5.

† *Handbook on Structural Steelwork*. BCSA/CONSTRADO. London, 1971, Supplement 1971, 2nd Supplement 1976 (in accordance with BS449, 'The use of structural steel in building'). *Structural Steelwork Handbook for Standard Metric Sections*. CONSTRADO. London, 1976 (in accordance with BS4848, 'Structural hollow sections').

centre of the section. The *second moment of area I* has been defined earlier as

$$I = \int y^2 dA$$

Thus for the rectangular section the second moment of area about the N.A., i.e. an axis through the centre, is given by

$$I_{\text{N.A.}} = \int_{-D/2}^{D/2} y^2 B dy = B \int_{-D/2}^{D/2} y^2 dy$$

$$= B\left[\frac{y^3}{3}\right]_{-D/2}^{D/2} = \frac{BD^3}{12} \qquad (4.8)$$

Similarly, the second moment of area of the rectangular section about an axis through the lower edge of the section would be found using the same procedure but with integral limits of 0 to D,

$$I = B\left[\frac{y^3}{3}\right]_0^D = \frac{BD^3}{3} \qquad (4.9)$$

These standard forms prove very convenient in the determination of $I_{\text{N.A.}}$ values for built-up sections which can be conveniently divided into rectangles. For *symmetrical sections* as, for instance, the I-section shown in Fig. 4.6,

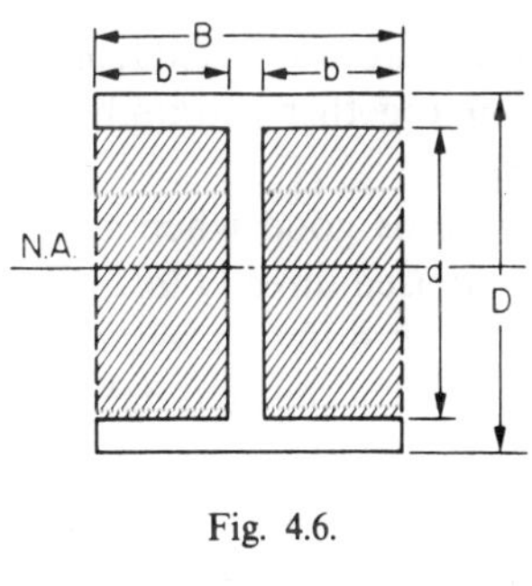

Fig. 4.6.

$$I_{\text{N.A.}} = I \text{ of dotted rectangle} - I \text{ of shaded portions}$$

$$= \frac{BD^3}{12} - 2\left(\frac{bd^3}{12}\right) \qquad (4.10)$$

It will be found that any symmetrical section can be divided into convenient rectangles with the N.A. running through each of their centroids and the above procedure can then be employed to effect a rapid solution.

For *unsymmetrical sections* such as the T-section of Fig. 4.7 it is more convenient to divide the section into rectangles with their *edges* in the N.A., when the second type of standard form may be applied.

$$I_{\text{N.A.}} = \underset{(\text{about } DC)}{I_{ABCD}} - \underset{(\text{about } DC)}{I_{\text{shaded areas}}} + \underset{(\text{about } HG)}{I_{EFGH}}$$

(each of these quantities may be written in the form $BD^3/3$).

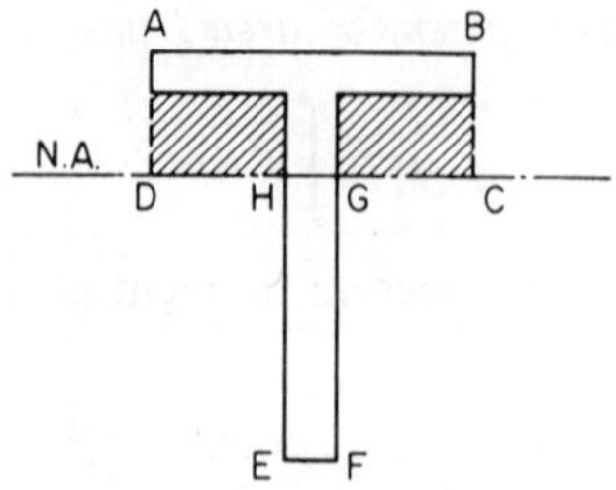

Fig. 4.7.

As an alternative procedure it is possible to determine the second moment of area of each rectangle about an axis through its own centroid ($I_G = BD^3/12$) and to "shift" this value to the equivalent value about the N.A. by means of the *parallel axis theorem*.

$$I_{\text{N.A.}} = I_G + Ah^2 \tag{4.11}$$

where A is the area of the rectangle and h the distance of its centroid G from the N.A. Whilst this is perhaps not so quick or convenient for sections built-up from rectangles, it is often the only procedure available for sections of other shapes, e.g. rectangles containing circular holes.

4.5. Bending of composite or flitched beams

(a) A composite beam is one which is constructed from a combination of materials. If such a beam is formed by rigidly bolting together two timber joists and a reinforcing steel plate, then it is termed a *flitched beam*.

Since the bending theory only holds good when a constant value of Young's modulus applies across a section it cannot be used directly to solve composite-beam problems where two different materials, and therefore different values of E, are present. The method of solution in such a case is to replace one of the materials by an *equivalent section* of the other.

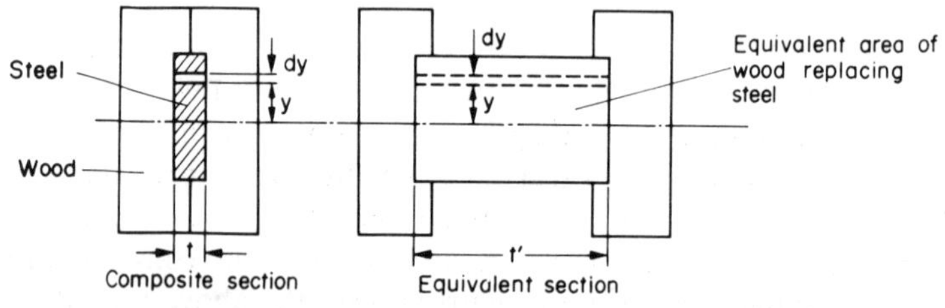

Fig. 4.8. Bending of composite or flitched beams: original beam cross-section and equivalent of uniform material (wood) properties.

Consider, therefore, the beam shown in Fig. 4.8 in which a steel plate is held centrally in an appropriate recess between two blocks of wood. Here it is convenient to replace the steel by an equivalent area of wood, retaining the same bending strength, i.e. the moment at any section must be the same in the equivalent section as in the original so that the force at any given dy in the equivalent beam must be equal to that at the strip it replaces.

$$\therefore \qquad \sigma t dy = \sigma' t' dy$$

$$\boldsymbol{\sigma t = \sigma' t'} \tag{4.12}$$

$$\varepsilon E t = \varepsilon' E' t'$$

since $$\frac{\sigma}{\varepsilon} = E$$

Again, for true similarity the strains must be equal,

$$\therefore \qquad \varepsilon = \varepsilon'$$

$$\therefore \qquad Et = E't' \quad \text{or} \quad \frac{t'}{t} = \frac{E}{E'} \tag{4.13}$$

i.e. $$t' = \frac{E}{E'} t \tag{4.14}$$

Thus to replace the steel strip by an equivalent wooden strip the thickness must be multiplied by the modular ratio E/E′.

The equivalent section is then one of the same material throughout and the simple bending theory applies. The stress in the wooden part of the original beam is found directly and that in the steel found from the value at the same point in the equivalent material as follows:

from eqn. (4.12) $$\frac{\sigma}{\sigma'} = \frac{t'}{t}$$

and from eqn. (4.13) $$\frac{\sigma}{\sigma'} = \frac{E}{E'} \quad \text{or} \quad \boldsymbol{\sigma = \frac{E}{E'}\sigma'} \tag{4.15}$$

i.e. ***stress in steel = modular ratio × stress in equivalent wood***

The above procedure, of course, is not limited to the two materials treated above but applies equally well for any material combination. The wood and steel flitched beam was merely chosen as a convenient example.

4.6. Reinforced concrete beams – simple tension reinforcement

Concrete has a high compressive strength but is very weak in tension. Therefore in applications where tension is likely to result, e.g. bending, it is necessary to reinforce the concrete by the insertion of steel rods. The section of Fig. 4.9a is thus a compound beam and can be treated by reducing it to the equivalent concrete section, shown in Fig. 4.9b.

In calculations, the concrete is assumed to carry no tensile load; hence the gap below the N.A. in Fig. 4.9b. The N.A. is then fixed since it must pass through the centroid of the area assumed in this figure: i.e. moments of area about the N.A. must be zero.

Let t = tensile stress in the steel,
c = compressive stress in the concrete,
A = total area of steel reinforcement,
m = modular ratio, $E_{\text{steel}}/E_{\text{concrete}}$,

other symbols representing the dimensions shown in Fig. 4.9.

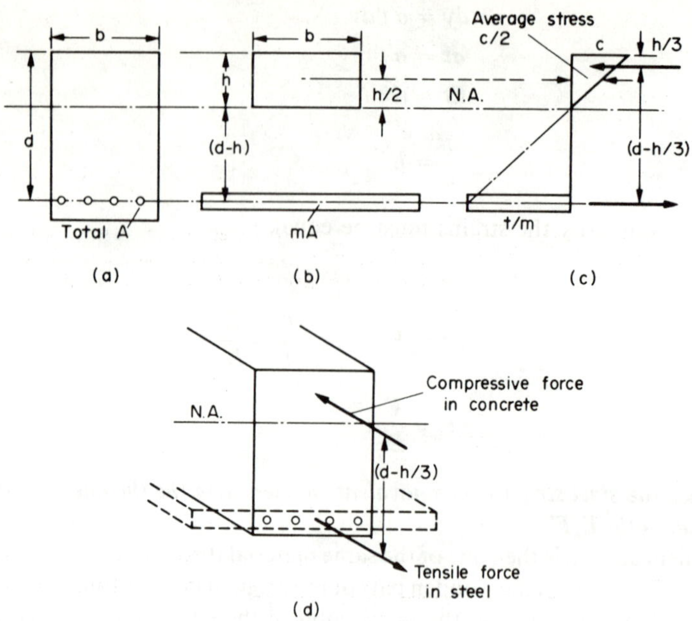

Fig. 4.9. Bending of reinforced concrete beams with simple tension reinforcement.

Then
$$bh\frac{h}{2} = mA(d-h) \tag{4.16}$$

which can be solved for h.

The moment of resistance is then the moment of the couple in Fig. 4.9c and d. Therefore

moment of resistance (based on compressive forces)

$$M = \underset{\text{area}}{(bh)} \times \underset{\substack{\text{average}\\ \text{stress}}}{\frac{c}{2}} \times \underset{\substack{\text{lever}\\ \text{arm}}}{\left(d-\frac{h}{3}\right)} = \frac{\boldsymbol{bhc}}{\mathbf{2}}\left(\boldsymbol{d}-\frac{\boldsymbol{h}}{\mathbf{3}}\right) \tag{4.17}$$

Similarly,

moment of resistance (based on tensile forces)

$$M = \underset{\text{area}}{mA} \times \underset{\text{stress}}{\frac{t}{m}} \times \underset{\substack{\text{lever}\\ \text{arm}}}{\left(d-\frac{h}{3}\right)} = \boldsymbol{At}\left(\boldsymbol{d}-\frac{\boldsymbol{h}}{\mathbf{3}}\right) \tag{4.18}$$

Both t and c are usually given as the maximum allowable values, which may or may not be reached at the same time. Equations (4.17) and (4.18) must both be worked out, therefore, and the lowest value taken, since the larger moment would give a stress greater than the allowed maximum stress in the other material.

In design applications where the dimensions of reinforced concrete beams are required which will carry a known B.M. the above equations generally contain too many unknowns, and certain simplifications are necessary. It is usual in these circumstances to assume a *balanced* section, i.e. one in which the maximum allowable stresses in the steel and concrete occur simultaneously. There is then no wastage of materials, and for this reason the section is also known as an *economic* or *critical* section.

For this type of section the N.A. is positioned by proportion of the stress distribution (Fig. 4.9c).

Thus by similar triangles

$$\frac{c}{h} = \frac{t/m}{(d-h)}$$

$$\boldsymbol{mc(d-h) = th} \tag{4.19}$$

Thus d can be found in terms of h, and since the moment of resistance is known this relationship can be substituted in eqn. (4.17) to solve for the unknown depth d.

Also, with a balanced section,

moment of resistance (compressive) = moment of resistance (tensile)

$$bh\frac{c}{2}\left(d-\frac{h}{3}\right) = mA\frac{t}{m}\left(d-\frac{h}{3}\right)$$

$$\boldsymbol{\frac{bhc}{2} = At} \tag{4.20}$$

By means of eqn. (4.20) the required total area of reinforcing steel A can thus be determined.

4.7. Skew loading (bending of symmetrical sections about axes other than the axes of symmetry)

Consider the simple rectangular-section beam shown in Fig. 4.10 which is subjected to a load inclined to the axes of symmetry. In such cases bending will take place about an inclined axis, i.e. the N.A. will be inclined at some angle θ to the XX axis and deflections will take place perpendicular to the N.A.

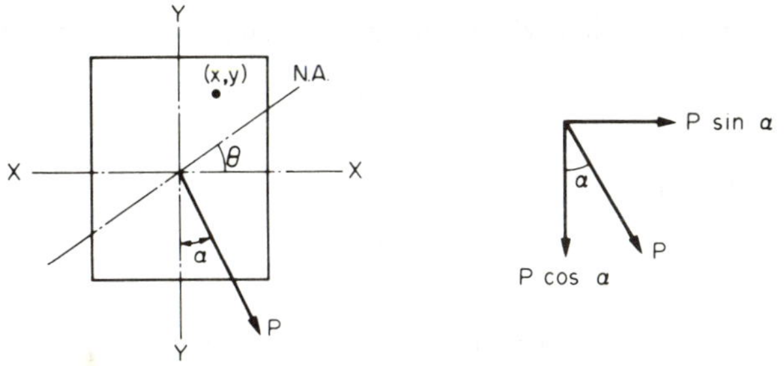

Fig. 4.10. Skew loading of symmetrical section.

In such cases it is convenient to resolve the load P, and hence the applied moment, into its components parallel with the axes of symmetry and to apply the simple bending theory to the resulting bending about both axes. It is thus assumed that simple bending takes place simultaneously about both axes of symmetry, the total stress at any point (x, y) being given by combining the results of the separate bending actions algebraically using the normal conventions for the signs of the stress, i.e. tension-positive, compression-negative.

Thus
$$\sigma = \frac{M_{xx}}{I_x} y \pm \frac{M_{yy}}{I_{yy}} x \qquad (4.21)$$

The equation of the N.A. is obtained by setting eqn. (4.21) to zero,

i.e.
$$\tan\theta = \frac{y}{x} = \pm \frac{M_{yy}}{M_{xx}} \frac{I_{xx}}{I_{yy}} \qquad (4.22)$$

4.8. Combined bending and direct stress – eccentric loading

(a) *Eccentric loading on one axis*

There are numerous examples in engineering practice where tensile or compressive loads on sections are not applied through the centroid of the section and which thus will introduce not only tension or compression as the case may be but also considerable bending effects. In concrete applications, for example, where the material is considerably weaker in tension than in compression, any bending and hence tensile stresses which are introduced can often cause severe problems. Consider, therefore, the beam shown in Fig. 4.11 where the load has been applied at an eccentricity e from one axis of symmetry. The stress at any point is determined by calculating the bending stress at the point on the basis of the simple bending theory and combining this with the direct stress (load/area), taking due account of sign,

i.e.
$$\sigma = \frac{P}{A} \pm \frac{My}{I} \qquad (4.23)$$

where
$$M = \text{applied moment} = Pe$$

$\therefore$
$$\sigma = \frac{P}{A} \pm \frac{Pey}{I} \qquad (4.24)$$

The positive sign between the two terms of the expression is used when both parts have the same effect and the negative sign when one produces tension and the other compression.

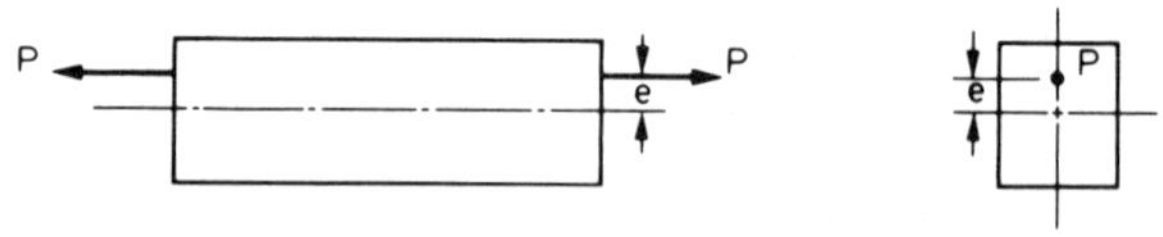

Fig. 4.11. Combined bending and direct stress – eccentric loading on one axis.

It should now be clear that any eccentric load can be treated as precisely equivalent to a direct load acting through the centroid plus an applied moment about an axis through the centroid equal to load × eccentricity. The distribution of stress across the section is then given by Fig. 4.12.

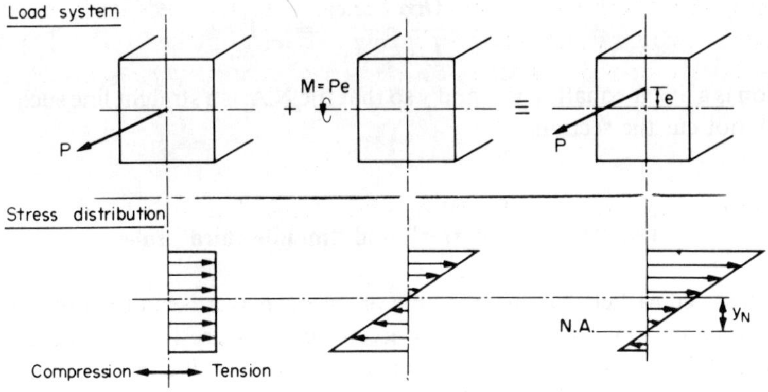

Fig. 4.12. Stress distributions under eccentric loading.

The *equation of the N.A.* can be obtained by setting σ equal to zero in eqn. (4.24),

i.e.
$$y = \pm \frac{I}{Ae} = y_N \tag{4.25}$$

Thus with the load eccentric to one axis the N.A. will be parallel to that axis and a distance y_N from it. The larger the eccentricity of the load the nearer the N.A. will be to the axis of symmetry through the centroid for given values of A and I.

(*b*) *Eccentric loading on two axes*

It some cases the applied load will not be applied on either of the axes of symmetry so that there will now be a direct stress effect plus simultaneous bending about both axes. Thus, for the section shown in Fig. 4.13, with the load applied at P with eccentricities of h and k, the total stress at any point (x, y) is given by

$$\sigma = \frac{P}{A} \pm \frac{Phx}{I_{yy}} \pm \frac{Pky}{I_{xx}} \tag{4.26}$$

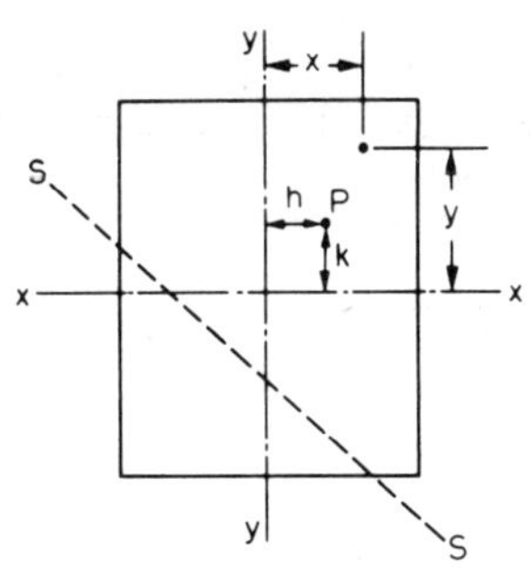

Fig. 4.13. Eccentric loading on two axes showing possible position of neutral axis SS.

Again the *equation of the N.A.* is obtained by equating eqn. (4.26) to zero, when

$$\frac{P}{A} \pm \frac{Phx}{I_{yy}} \pm \frac{Pky}{I_{xx}} = 0$$

or
$$\frac{Ahx}{I_{yy}} \pm \frac{Aky}{I_{xx}} = \pm 1 \qquad (4.27)$$

This equation is a linear equation in x and y so that the N.A. is a straight line such as SS which may or may not cut the section.

4.9. "Middle-quarter" and "middle-third" rules

It has been stated earlier that considerable problems may arise in the use of cast-iron or concrete sections in applications in which eccentric loads are likely to occur since both materials are notably weaker in tension than in compression. It is convenient, therefore, that for rectangular and circular cross-sections, provided that the load is applied within certain defined areas, no tension will be produced whatever the magnitude of the applied compressive load. (Here we are solely interested in applications such as column and girder design which are principally subjected to compression.)

Consider, therefore, the rectangular cross-section of Fig. 4.13. The stress at any point (x, y) is given by eqn. (4.26) as

$$\sigma = \frac{P}{A} \pm \frac{Phx}{I_{yy}} \pm \frac{Pky}{I_{xx}}$$

Thus, with a compressive load applied, the most severe tension stresses are introduced when the last two terms have their maximum value and are tensile in effect,

i.e.
$$\sigma = \frac{P}{A} - \frac{Ph}{I_{yy}} \times \frac{b}{2} - \frac{Pk}{I_{xx}} \times \frac{d}{2}$$

$$= \frac{P}{bd} - \frac{Phb}{2} \times \frac{12}{db^3} - \frac{Pkd}{2} \times \frac{12}{bd^3}$$

For no tension to result in the section, σ must be equated to zero,

$$0 = \frac{1}{bd} - \frac{6h}{db^2} - \frac{6k}{bd^2}$$

or
$$\frac{bd}{6} = dh + bk$$

This is a linear expression in h and k producing the line SS in Fig. 4.13. If the load is now applied in each of the other three quadrants the total limiting area within which P must be applied to produce zero tension in the section is obtained. This is the diamond area shown shaded in Fig. 4.14 with diagonals of $b/3$ and $d/3$ and hence termed the *middle third.*

For circular sections of diameter d, whatever the position of application of P, an axis of symmetry will pass through this position so that the problem reduces to one of eccentricity about a single axis of symmetry.

Now from eqn. (4.23)

$$\sigma = \frac{P}{A} \pm \frac{Pe}{I} y$$

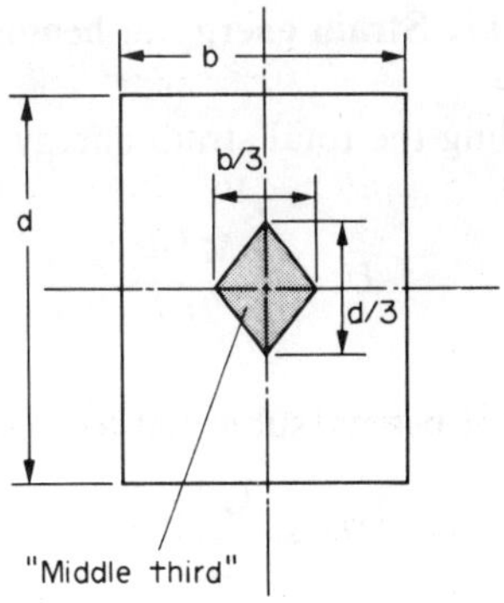

Fig. 4.14. Eccentric loading of rectangular sections – "middle third".

Therefore for zero tensile stress in the presence of an eccentric compressive load

$$\frac{P}{A} = \frac{Pey}{I}$$

$$\frac{4}{\pi d^2} = e \times \frac{d}{2} \times \frac{64}{\pi d^4}$$

$$e = \frac{d}{8}$$

Thus the limiting region for application of the load is the shaded circular area of diameter $d/4$ (shown in Fig. 4.15) which is termed the *middle quarter.*

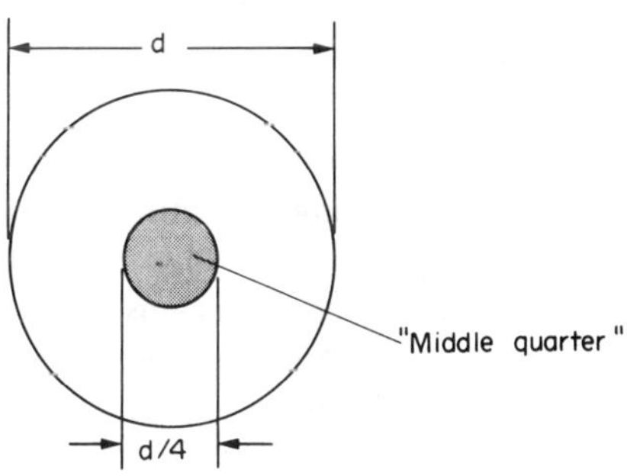

Fig. 4.15. Eccentric loading of circular sections – "middle quarter".

4.10. Shear stresses owing to bending

It can be shown that any cross-section of a beam subjected to bending by transverse loads experiences not only direct stresses as given by the bending theory but also shear stresses. The magnitudes of these shear stresses at a particular section is always such that they sum up to the total shear force Q at that section. A full treatment of the procedures used to determine the distribution of the shear stresses is given in Chapter 7.

4.11. Strain energy in bending

For beams subjected to bending the total strain energy of the system is given by

$$U = \int_0^L \frac{M^2 ds}{2EI}$$

For uniform beams, or parts of beams, subjected to a constant B.M. M, this reduces to

$$U = \frac{M^2 L}{2EI}$$

In most beam-loading cases the strain energy due to bending far exceeds that due to other forms of loading, such as shear or direct stress, and energy methods of solution using *Castigliano* or *unit load* procedures based on the above equations are extremely powerful methods of solution. These are covered fully in Chapter 11.

4.12. Limitations of the simple bending theory

It has been observed earlier that the theory introduced in preceding sections is often termed the "simple theory of bending" and that it relies on a number of assumptions which either have been listed on page 64 or arise in the subsequent proofs. It should thus be evident that in practical engineering situations the theory will have certain limitations depending on the degree to which these assumptions can be considered to hold true. The following paragraphs give an indication of when some of the more important assumptions can be taken to be valid and when alternative theories or procedures should be applied.

Assumption: *Stress is proportional to the distance from the axis of zero stress (neutral axis)*, i.e. $\sigma = Ey/R = E\varepsilon$.

Correct for homogeneous beams within the elastic range.

Incorrect (a) for loading conditions outside the elastic range when $\sigma \neq E$,

(b) for composite beams with different materials or properties when 'equivalent sections' must be used; see §4.5

Assumption: *Strain is proportional to the distance from the axis of zero strain*, i.e. $\varepsilon = y/R$.

Correct for initially straight beams or, for engineering purposes, beams with $R/d > 10$ (where d = total depth of section).

Incorrect for initially curved beams for which special theories have been developed or to which correction factors $\sigma = K\,(My/I)$ may be applied.

Assumption: *Neutral axis passes through the centroid of the section.*

Correct for pure bending with no axial load.

Incorrect for combined bending and axial load systems such as eccentric loading. In such cases the loading effects must be separated, stresses arising from each calculated and the results superimposed – see §4.8

Assumption: *Plane cross-sections remain plane.*

Correct (a) for cross-sections at a reasonable distance from points of local loading or stress concentration (usually taken to be at least one-depth of beam),
(b) when change of cross-section with length is gradual,
(c) in the absence of end-condition spurious effects.

These conditions are known as '*St Venant's* principle".

Incorrect (a) for points of local loading;
(b) at positions of stress concentration such as holes, keyways, fillets and other changes in geometry;
(c) in regions of rapid change of cross-section.

In such cases appropriate *stress concentration factors*† must be applied or experimental stress/strain analysis techniques adopted.

Assumption: *The axis of the applied bending moment is coincident with the neutral axis.*

Correct when the axis of bending is a principal axis ($I_{xy} = 0$ see page 432) e.g. on axis of symmetry.

Incorrect for so-called *unsymmetrical bending* cases when the axis of the applied bending moment is not a principal axis.

In such cases the moment should be resolved into components about the principal axes or the other procedures introduced in Chapter 16 applied.

Assumption: *Lateral contraction or expansion is not prevented.*

Correct when the beam can be considered narrow (i.e. width the same order as the depth).

Incorrect for wide beams or plates in which the width may be many times the depth. Special procedures apply for such cases – see Chapter 24.

It should now be evident that care is required in the application of "simple" theory and reference should be made where necessary to more advanced theories.

Examples

Example 4.1

An I-section girder, 200 mm wide by 300 mm deep, with flange and web of thickness 20 mm is used as a simply supported beam over a span of 7 m. The girder carries a distributed load of 5 kN/m and a concentrated load of 20 kN at mid-span. Determine: (a) the second moment of area of the cross-section of the girder, (b) the maximum stress set-up.

† *Stress Concentration Factors*, R. C. Peterson (Wiley & Sons).

Solution

(a) The second moment of area of the cross-section may be found in two ways.

Method 1 – Use of standard forms

For sections with symmetry about the N.A., use can be made of the standard I value for a rectangle about an axis through its centroid, i.e. $bd^3/12$. The section can thus be divided into convenient rectangles for each of which the N.A. passes through the centroid, e.g. in this case, enclosing the girder by a rectangle (Fig. 4.16).

$$I_{\text{girder}} = I_{\text{rectangle}} - I_{\text{shaded portions}}$$

$$= \left[\frac{200 \times 300^3}{12}\right]10^{-12} - 2\left[\frac{90 \times 260^3}{12}\right]10^{-12}$$

$$= (4.5 - 2.64)10^{-4} = \mathbf{1.86 \times 10^{-4}\ m^4}$$

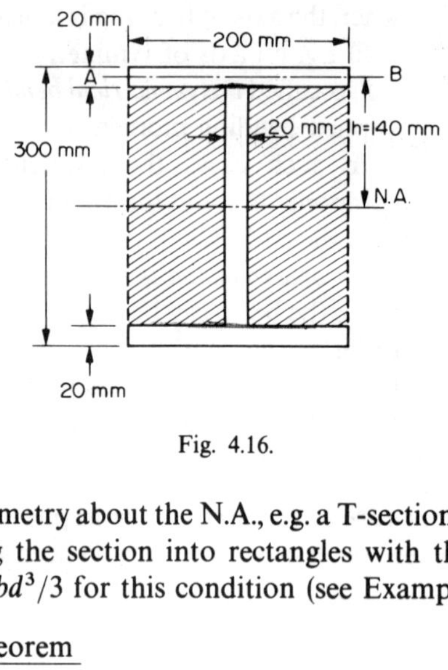

Fig. 4.16.

For sections without symmetry about the N.A., e.g. a T-section, a similar procedure can be adopted, this time dividing the section into rectangles with their edges in the N.A. and applying the standard $I = bd^3/3$ for this condition (see Example 4.2).

Method 2 – Parallel axis theorem

Consider the section divided into three parts – the web and the two flanges.

$$I_{\text{N.A.}} \text{ for the web} = \frac{bd^3}{12} = \left[\frac{20 \times 260^3}{12}\right]10^{-12}$$

$$I \text{ of flange about } AB = \frac{bd^3}{12} = \left[\frac{200 \times 20^3}{12}\right]10^{-12}$$

Therefore using the parallel axis theorem

$$I_{\text{N.A.}} \text{ for flange} = I_{AB} + Ah^2$$

where h is the distance between the N.A. and AB,

$$I_{\text{N.A.}} \text{ for flange} = \left[\frac{200 \times 20^3}{12}\right]10^{-12} + [(200 \times 20)140^2]10^{-12}$$

Therefore total $I_{N.A.}$ of girder

$$= 10^{-12}\left\{\left[\frac{20\times 260^3}{12}\right]+2\left[\frac{200\times 20^3}{12}\right]+200\times 20\times 140^2\right\}$$

$$= 10^{-6}(29.3+0.267+156.8)$$

$$= \mathbf{1.86\times 10^{-4}\ m^4}$$

Both methods thus yield the same value and are equally applicable in most cases. Method 1, however, normally yields the quicker solution.

(b) The maximum stress may be found from the simple bending theory of eqn. (4.4),

i.e. $$\sigma_{max} = \frac{M_{max}\,y_{max}}{I}$$

Now the maximum B.M. for a beam carrying a u.d.l. is at the centre and given by $wL^2/8$. Similarly, the value for the central concentrated load is $WL/4$ also at the centre. Thus, in this case,

$$M_{max} = \frac{WL}{4}+\frac{wL^2}{8} = \left[\frac{20\times 10^3\times 7}{4}\right]+\left[\frac{5\times 10^3\times 7^2}{8}\right]\text{N m}$$

$$= (35.0+30.63)10^3 = 65.63\text{ kN m}$$

$$\therefore \quad \sigma_{max} = \frac{65.63\times 10^3\times 150\times 10^{-3}}{1.9\times 10^{-4}} = \mathbf{51.8\ MN/m^2}$$

The maximum stress in the girder is 52 MN/m^2, this value being compressive on the upper surface and tensile on the lower surface.

Example 4.2

A uniform T-section beam is 100 mm wide and 150 mm deep with a flange thickness of 25 mm and a web thickness of 12 mm. If the limiting bending stresses for the material of the beam are 80 MN/m^2 in compression and 160 MN/m^2 in tension, find the maximum u.d.l. that the beam can carry over a simply supported span of 5 m.

Solution

The second moment of area value I used in the simple bending theory is that about the N.A. Thus, in order to determine the I value of the T-section shown in Fig. 4.17, it is necessary first to position the N.A.

Since this always passes through the centroid of the section we can take moments of area about the base to determine the position of the centroid and hence the N.A.

Thus

$$(100\times 25\times 137.5)10^{-9}+(125\times 12\times 62.5)10^{-9} = 10^{-6}[(100\times 25)+(125\times 12)\bar{y}]$$

$$(343750+93750)10^{-9} = 10^{-6}(2500+1500)\bar{y}$$

$$\bar{y} = \frac{437.5\times 10^{-6}}{4000\times 10^{-6}} = 109.4\times 10^{-3} = 109.4\text{ mm.}$$

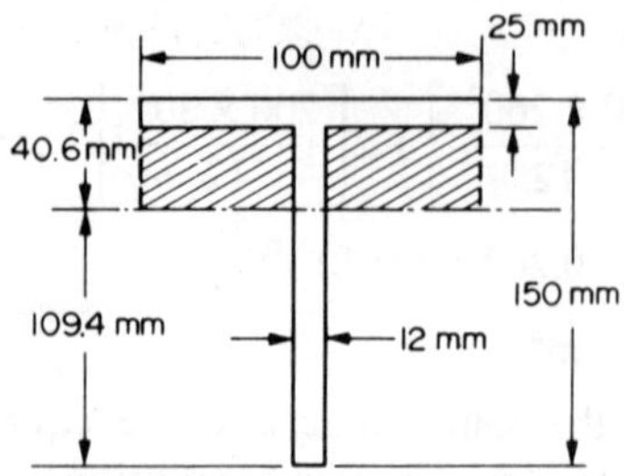

Fig. 4.17.

Thus the N.A. is positioned, as shown, a distance of 109.4 mm above the base. The second moment of area I can now be found as suggested in Example 4.1 by dividing the section into convenient rectangles with their edges in the neutral axis.

$$I = \tfrac{1}{3}[(100 \times 40.6^3) - (88 \times 15.6^3) + (12 \times 109.4^3)]10^{-12}$$

$$= \tfrac{1}{3}(6.69 - 0.33 + 15.71)10^{-6} = 7.36 \times 10^{-6}\ \text{m}^4$$

Now the maximum compressive stress will occur on the upper surface where $y = 40.6$ mm, and, using the limiting compressive stress value quoted,

$$M = \frac{\sigma I}{y} = \frac{80 \times 10^6 \times 7.36 \times 10^{-6}}{40.6 \times 10^{-3}} = 14.5\ \text{kN m}$$

This suggests a maximum allowable B.M. of 14.5 kN m. It is now necessary, however, to check the tensile stress criterion which must apply on the lower surface,

i.e. $$M = \frac{\sigma I}{y} = \frac{160 \times 10^6 \times 7.36 \times 10^{-6}}{109.4 \times 10^{-3}} = 10.76\ \text{kN m}$$

The greatest moment that can therefore be applied to retain stresses within *both* conditions quoted is therefore $M = 10.76$ kN m.

But for a simply supported beam with u.d.l.,

$$M_{max} = \frac{wL^2}{8}$$

$$w = \frac{8M}{L^2} = \frac{8 \times 10.76 \times 10^3}{5^2}$$

$$= \mathbf{3.4\ kN/m}$$

The u.d.l. must be limited to 3.4 kN m.

Example 4.3

A flitched beam consists of two 50 mm × 200 mm wooden beams and a 12 mm × 80 mm steel plate. The plate is placed centrally between the wooden beams and recessed into each so that, when rigidly joined, the three units form a 100 mm × 200 mm section as shown in Fig. 4.18. Determine the moment of resistance of the flitched beam when the maximum bending stress in the timber is 12 MN/m^2. What will then be the maximum bending stress in the steel?

For steel $E = 200$ GN/m^2; for wood $E = 10$ GN/m^2.

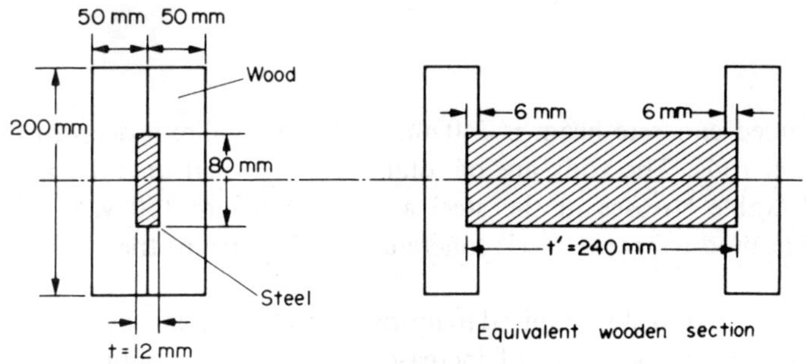

Fig. 4.18.

Solution

The flitched beam may be considered replaced by the equivalent wooden section shown in Fig. 4.18. The thickness t' of the wood equivalent to the steel which it replaces is given by eqn. (4.14),

$$t' = \frac{E}{E'}\,t = \frac{200 \times 10^9}{10 \times 10^9} \times 12 = 240 \text{ mm}$$

Then, for the equivalent section

$$I_{\text{N.A.}} = 2\left[\frac{50 \times 200^3}{12}\right] - 2\left[\frac{6 \times 80^3}{12}\right] + \left[\frac{240 \times 80^3}{12}\right] 10^{-12}$$

$$= (66.67 - 0.51 + 10.2)10^{-6} = 76.36 \times 10^{-6} \text{ m}^4$$

Now the maximum stress in the timber is 12 MN/m^2, and this will occur at $y = 100$ mm; thus, from the bending theory,

$$M = \frac{\sigma I}{y} = \frac{12 \times 10^6 \times 76.36 \times 10^{-6}}{100 \times 10^{-3}} = \mathbf{9.2\ kN\,m}$$

The moment of resistance of the beam, i.e. the bending moment which the beam can withstand within the given limit, is 9.2 kN m.

The maximum stress in the steel with this moment applied is then determined by finding first the maximum stress in the equivalent wood at the same position, i.e. at $y = 40$ mm.

Therefore maximum stress in equivalent wood

$$\sigma'_{\max} = \frac{My}{I} = \frac{9.2 \times 10^3 \times 40 \times 10^{-3}}{76.36 \times 10^{-6}} = 4.82 \times 10^6 \text{ N/m}^2$$

Therefore from eqn. (4.15), the maximum stress in the steel is given by

$$\sigma_{max} = \frac{E}{E'}\sigma'_{max} = \frac{200 \times 10^9}{10 \times 10^9} \times 4.82 \times 10^6$$

$$= 96 \times 10^6 = \mathbf{96\ MN/m^2}$$

Example 4.4

(a) A reinforced concrete beam is 240 mm wide and 450 mm deep to the centre of the reinforcing steel rods. The rods are of total cross-sectional area 1.2×10^{-3} m^2 and the maximum allowable stresses in the steel and concrete are 150 MN/m^2 and 8 MN/m^2 respectively. The modular ratio (steel : concrete) is 16. Determine the moment of resistance of the beam.

(b) If, after installation, it is required to up-rate the service loads by 30% and to replace the above beam with a second beam of increased strength but retaining the same width of 240 mm, determine the new depth and area of steel for tension reinforcement required.

Solution

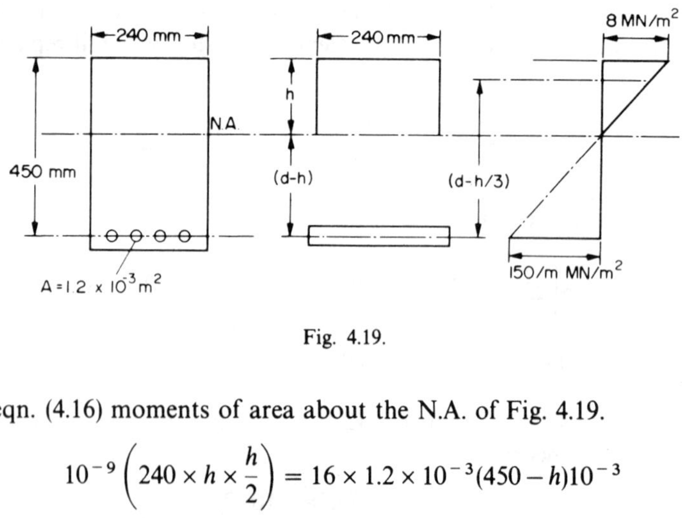

Fig. 4.19.

(a) From eqn. (4.16) moments of area about the N.A. of Fig. 4.19.

$$10^{-9}\left(240 \times h \times \frac{h}{2}\right) = 16 \times 1.2 \times 10^{-3}(450 - h)10^{-3}$$

$$120h^2 = (8640 - 19.2h)10^3$$

$$h^2 + 160h - 72000 = 0$$

From which $$h = 200 \text{ mm}$$

Substituting in eqn. (4.17),

$$\text{moment of resistance (compressive)} = (240 \times 200 \times 10^{-6})\frac{8 \times 10^6}{2}(450 - 66.7)10^{-3}$$

$$= 73.6 \text{ kN m}$$

and from eqn. (4.18)

$$\text{moment of resistance (tensile)} = (16 \times 1.2 \times 10^{-3})\,\frac{150 \times 10^6}{16}\,(450 - 66.7)10^{-3}$$

$$= 69.0 \text{ kN m}$$

Thus the safe moment which the beam can carry within *both* limiting stress values is **69 kN m.**

(b) For this part of the question the dimensions of the new beam are required and it is necessary to assume a *critical* or *economic* section. The position of the N.A. is then determined from eqn. (4.19) by consideration of the proportions of the stress distribution (i.e. assuming that the maximum stresses in the streel and concrete occur together).

Thus from eqn. (4.19)

$$\frac{h}{d} = \frac{1}{1 + \dfrac{t}{mc}} = \frac{1}{1 + \dfrac{150 \times 10^6}{16 \times 8 \times 10^6}} = 0.46$$

From (4.17)

$$M = \frac{bhc}{2}\left(d - \frac{h}{3}\right) = \frac{h}{2d}\left(1 - \frac{h}{3d}\right) c\, bd^2$$

Substituting for $\dfrac{h}{d} = 0.46$ and solving for d gives

$$d = \mathbf{0.49\ m}$$

$\therefore$

$$h = 0.46 \times 0.49 = 0.225 \text{ m}$$

$\therefore$ From (4.20)

$$A = \frac{0.24 \times 0.225 \times 8 \times 10^6}{2 \times 150 \times 10^6}$$

i.e.

$$A = \mathbf{1.44 \times 10^{-3}\ m^2}$$

Example 4.5

(a) A rectangular masonry column has a cross-section 500 mm × 400 mm and is subjected to a vertical compressive load of 100 kN applied at point P shown in Fig. 4.20. Determine the value of the maximum stress produced in the section. (b) Is the section *at any point* subjected to tensile stresses?

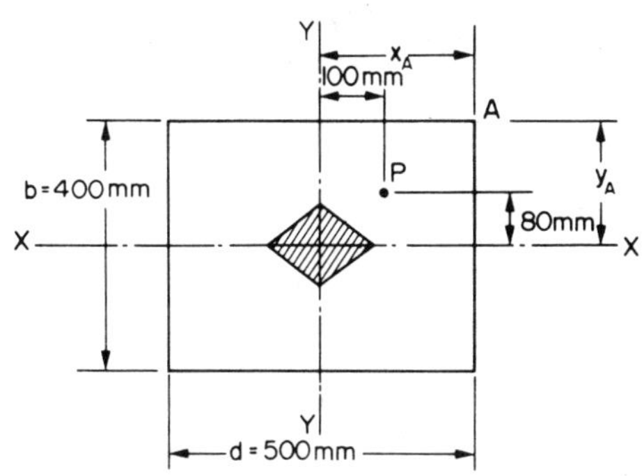

Fig. 4.20.

Solution

In this case the load is eccentric to both the XX and YY axes and bending will therefore take place simultaneously about both axes.

$$\text{Moment about } XX = 100 \times 10^3 \times 80 \times 10^{-3} = 8000 \text{ N m}$$
$$\text{Moment about } YY = 100 \times 10^3 \times 100 \times 10^{-3} = 10000 \text{ N m}$$

Therefore from eqn. (4.26) the maximum stress in the section will be compressive at point A since at this point the compressive effects of bending about both XX and YY add to the direct compressive stress component due to P,

i.e
$$\sigma_{max} = -\left[\frac{P}{A} + \frac{M_{xx}y_A}{I_{xx}} + \frac{M_{yy}x_A}{I_{yy}}\right]$$
$$= -\left[\frac{100 \times 10^3}{500 \times 400 \times 10^{-6}} + \frac{8000 \times 200 \times 10^{-3} \times 12}{(500 \times 400^3)10^{-12}} + \frac{10000 \times 250 \times 10^{-3} \times 12}{(400 \times 500^3)10^{-12}}\right]$$
$$= -(0.5 + 0.6 + 0.6)10^6 = \mathbf{-1.7\ MN/m^2}$$

For the section to contain no tensile stresses, P must be applied within the middle third. Now since $b/3 = 133$ mm and $d/3 = 167$ mm it follows that the maximum possible values of the coordinates x or y for P are $y = \frac{1}{2} \times 133 = 66.5$ mm and $x = \frac{1}{2} \times 167 = 83.5$ mm.

The given position for P lies outside these values so that tensile stresses will certainly exist in the section.

(The full middle-third area is in fact shown in Fig. 4.20 and P is clearly outside this area.)

Example 4.6

The crank of a motor vehicle engine has the section shown in Fig. 4.21 along the line AA. Derive an expression for the stress at any point on this section with the con-rod thrust P

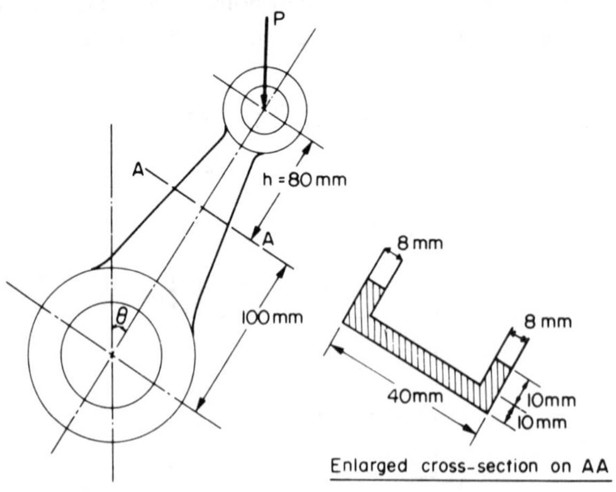

Fig. 4.21.

applied at some angle θ as shown. Hence, if the maximum tensile stress in the section is not to exceed 100 MN/m^2, determine the maximum value of P which can be permitted with $\theta = 60°$. What will be the distribution of stress along the section AA with this value of P applied?

Solution

Assuming that the load P is applied in the plane of the crank the stress at any point along the section AA will be the result of (a) a direct compressive load of magnitude $P\cos\theta$, and (b) a B.M. in the plane of the crank of magnitude $P\sin\theta \times h$; i.e. stress at any point along AA, distance s from the centre-line, is given by eqn. (4.26) as

$$\sigma = \frac{P\cos\theta}{A} \pm \frac{(P\sin\theta \, . \, h}{I_{AA}} \cdot s$$

$$= \frac{P\cos\theta}{A}\left[1 \pm \frac{hs\tan\theta}{k_{AA}^2}\right]$$

where k_{AA} is the radius of gyration of the section AA about its N.A.

Now $$A = [(2 \times 20 \times 8) + (24 \times 10)]10^{-6} = 560 \times 10^{-6}\ \text{m}^2$$

and $$I_{N.A.} = \tfrac{1}{12}[20 \times 40^3 - 10 \times 24^3]\,10^{-12} = 9.51 \times 10^{-8}\ \text{m}^4$$

$\therefore$ $$\sigma = -P\left[\frac{0.5}{560 \times 10^{-6}} \pm \frac{0.866 \times 80 \times 10^{-3} s \times 10^{-3}}{9.51 \times 10^{-8}}\right]$$

$$= -P[0.893 \pm 0.729s]\,10^3\ \text{N/m}^2$$

where s is measured in millimetres,

i.e. $$\text{maximum tensile stress} = P[-0.893 + 0.729 \times 20]10^3\ \text{N/m}^2$$

$$= 13.69P\ \text{kN/m}^2$$

In order that this stress shall not exceed 100 MN/m^2

$$100 \times 10^6 = 13.69 \times P \times 10^3$$

$$P = \mathbf{7.3\ kN}$$

With this value of load applied the direct stress on the section will be

$$-0.893P \times 10^3 = -6.52\ \text{MN/m}^2$$

and the bending stress at each edge

$$\pm 0.729 \times 20 \times 10^3 P = 106.4\ \text{MN/m}^2$$

The stress distribution along AA is then obtained as shown in Fig. 4.22.

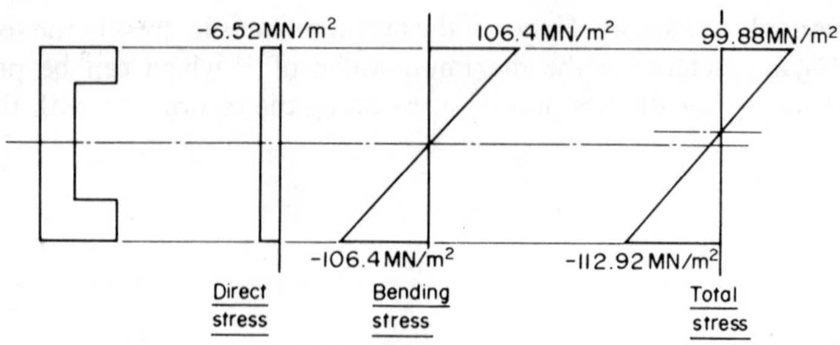

Fig. 4.22.

Problems

4.1 (A). Determine the second moments of area about the axes XX for the sections shown in Fig. 4.23.

[15.69, 7.88, 41.15, 24; all $\times 10^{-6}$ m^4.]

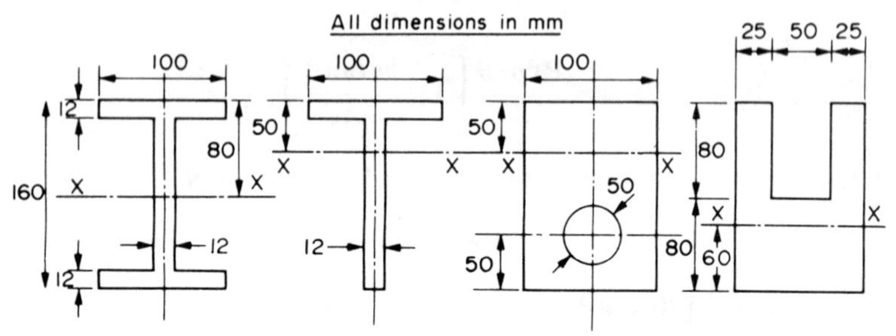

Fig. 4.23.

4.2 (A). A rectangular section beam has a depth equal to twice its width. It is the same material and mass per unit length as an I-section beam 300 mm deep with flanges 25 mm thick and 150 mm wide and a web 12 mm thick. Compare the flexural strengths of the two beams.

[8.59 : 1.]

4.3 (A). A conveyor beam has the cross-section shown in Fig. 4.24 and it is subjected to a bending moment in the plane *YY*. Determine the maximum permissible bending moment which can be applied to the beam (a) for bottom flange in tension, and (b) for bottom flange in compression, if the safe stresses for the material in tension and compression are 30 MN/m^2 and 150 MN/m^2 respectively.

[32.3, 84.8 kN m.]

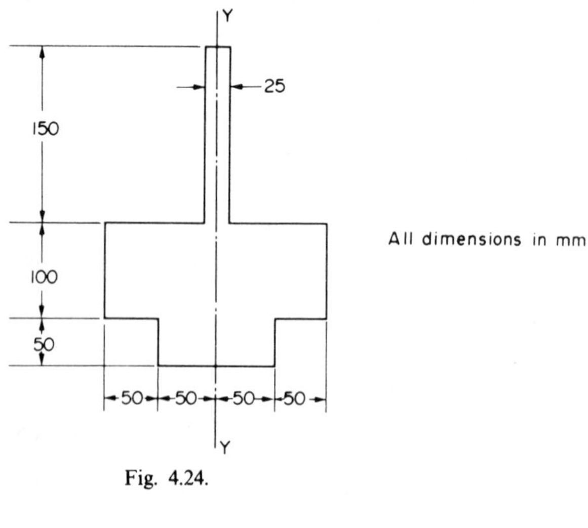

Fig. 4.24.

4.4 (A/B). A horizontal steel girder has a span of 3 m and is built-in at the left-hand end and freely supported at the other end. It carries a uniformly distributed load of 30 kN/m over the whole span, together with a single concentrated load of 20 kN at a point 2 m from the left-hand end. The supporting conditions are such that the reaction at the left-hand end is 65 kN.

(a) Determine the bending moment at the left-hand end and draw the B.M. diagram.
(b) Give the value of the maximum bending moment.
(c) If the girder is 200 mm deep and has a second moment of area of 40×10^{-6} m^4 determine the maximum stress resulting from bending. [I.Mech.E.] [40 kN m; 100 MN/m^2.]

4.5 (A/B). Figure 4.25 represents the cross-section of an extruded alloy member which acts as a simply supported beam with the 75 mm wide flange at the bottom. Determine the moment of resistance of the section if the maximum permissible stresses in tension and compression are respectively 60 MN/m^2 and 45 MN/m^2.

[I.E.I.] [2.62 kN m.]

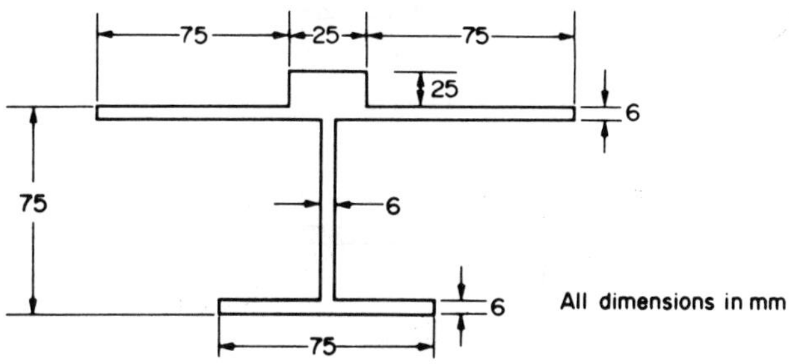

Fig. 4.25.

4.6 (A/B). A trolley consists of a pressed steel section as shown in Fig. 4.26. At each end there are rollers at 350 mm centres.

If the trolley supports a mass of 50 kg evenly distributed over the 350 mm length of the trolley calculate, using the data given in Fig. 4.26, the maximum compressive and tensile stress due to bending in the pressed steel section. State clearly your assumptions. [C.G.] [14.8, 42.6 MN/m^2]

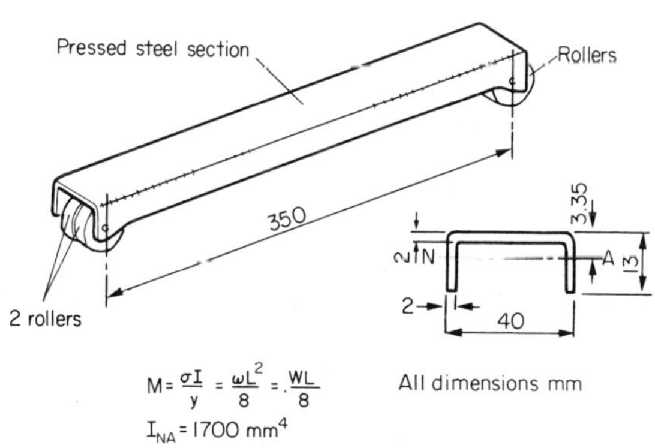

Fig. 4.26.

4.7 (A/B). The channel section of Fig. 4.21 is used as a simply-supported beam over a span of 2.8 m. The channel is used as a guide for a roller of an overhead crane gantry and can be expected to support a maximum load (taken to be a concentrated point load) of 40 kN. At what position of the roller will the bending moment of the channel be a maximum and what will then be the maximum tensile bending stress? If the maximum allowable stress for the material of the beam is 320 MN/m^2 what safety factor exists for the given loading condition.

[Centre, 79.5 MN/m^2, 4]

4.8 (A/B). A 120 × 180 × 15 mm uniform I-section steel girder is used as a cantilever beam carrying a uniformly distributed load ω kN/m over a span of 2.4 m. Determine the maximum value of ω which can be applied before yielding of the outer fibres of the beam cross-section commences. In order to strengthen the girder, steel plates are attached to the outer surfaces of the flanges to double their effective thickness. What width of plate should be added (to the nearest mm) in order to reduce the maximum stress by 30%? The yield stress for the girder material is 320 MN/m^2. [35.5 kN/m, 67 mm]

4.9 (A/B). A 200 mm wide × 300 mm deep timber beam is reinforced by steel plates 200 mm wide × 12 mm deep on the top and bottom surfaces as shown in Fig. 4.27. If the maximum allowable stresses for the steel and timber are 120 MN/m^2 and 8 MN/m^2 respectively, determine the maximum bending moment which the beam can safely carry. For steel $E = 200$ GN/m^2; for timber $E = 10$ GN/m^2. [I.Mech.E.] [103.3 kN m.]

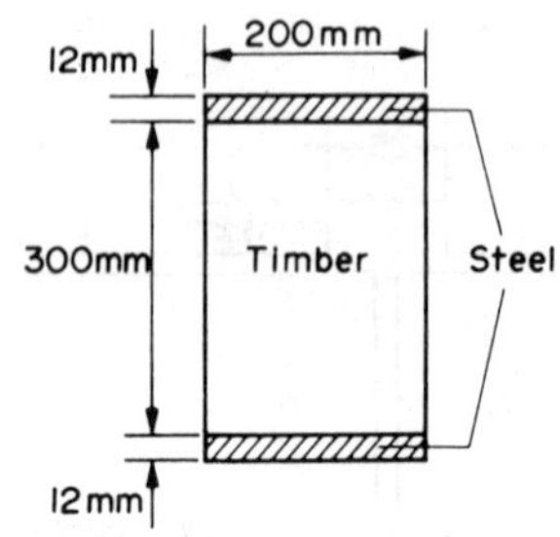

Fig. 4.27.

4.10 (A/B). A composite beam is of the construction shown in Fig. 4.28. Calculate the allowable u.d.l. that the beam can carry over a simply supported span of 7 m if the stresses are limited to 120 MN/m^2 in the steel and 7 MN/m^2 in the timber.

Modular ratio = 20. [1.13 kN/m.]

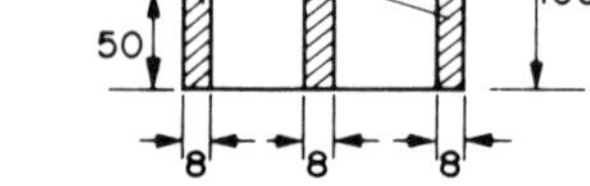

Fig. 4.28.

4.11 (A/B). Two bars, one of steel, the other of aluminium alloy, are each of 75 mm width and are rigidly joined together to form a rectangular bar 75 mm wide and of depth $(t_s + t_A)$, where t_s = thickness of steel bar and t_A = thickness of alloy bar.

Determine the ratio of t_s to t_A in order that the neutral axis of the compound bar is coincident with the junction of the two bars. ($E_s = 210$ GN/m^2; $E_A = 70$ GN/m^2.)

If such a beam is 50 mm deep determine the maximum bending moment the beam can withstand if the maximum stresses in the steel and alloy are limited to 135 MN/m^2 and 37 MN/m^2 respectively. [0.577; 1.47 kN m.]

4.12 (A/B). A brass strip, 50 mm × 12 mm in section, is riveted to a steel strip, 65 mm × 10 mm in section, to form a compound beam of total depth 22 mm, the brass strip being on top and the beam section being symmetrical about the vertical axis. The beam is simply supported on a span of 1.3 m and carries a load of 2 kN at mid-span.

(a) Determine the maximum stresses in each of the materials owing to bending.
(b) Make a diagram showing the distribution of bending stress over the depth of the beam.

Take E for steel = 200 GN/m^2 and E for brass = 100 GN/m^2.

[U.L.] [σ_b = 130 MN/m^2; σ_s = 162.9 MN/m^2.]

4.13 (B). A concrete beam, reinforced in tension only, has a rectangular cross-section, width 200 mm and effective depth to the tensile steel 500 mm, and is required to resist a bending moment of 70 kN m. Assuming a modular ratio of 15, calculate (a) the minimum area of reinforcement required if the stresses in steel and concrete are not to exceed 190 MN/m^2 and 8 MN/m^2 respectively, and (b) the stress in the non-critical material when the bending moment is applied. [E.I.E.] [0.916 × 10^{-3} m^2; 177 MN/m^2.]

4.14 (B). A reinforced concrete beam of rectangular cross-section, b = 200 mm, d (depth to reinforcement) = 300 mm, is reinforced in tension only, the steel ratio, i.e. the ratio of reinforcing steel area to concrete area (neglecting cover), being 1 %. The maximum allowable stresses in concrete and steel are 8 MN/m^2 and 135 MN/m^2 respectively. The modular ratio may be taken as equal to 15. Determine the moment of resistance capable of being developed in the beam. [I.Struct.E.] [20.9 kN m.]

4.15 (B). A rectangular reinforced concrete beam is 200 mm wide and 350 mm deep to reinforcement, the latter consisting of three 20 mm diameter steel rods. If the following stresses are not to be exceeded, calculate: (a) the maximum bending moment which can be sustained, and (b) the steel stress and the maximum concrete stress when the section is subjected to this maximum moment.

Maximum stress in concrete in bending not to exceed 8 MN/m^2.
Maximum steel stress not to exceed 150 MN/m^2.
Modular ratio m = 15. [I.Struct.E.] [38.5 kN m; 138, 8 MN/m^2.]

4.16 (B). A reinforced concrete beam has to carry a bending moment of 100 kN m. The maximum permissible stresses are 8 MN/m^2 and 135 MN/m^2 in the concrete and steel respectively. The beam is to be of rectangular cross-section 300 mm wide. Design a suitable section with "balanced" reinforcement if $E_{steel}/E_{concrete}$ = 12.

[I.Mech.E.] [d = 482.4 mm; A = 1.782 × 10^{-3} m^2.]

CHAPTER 5

SLOPE AND DEFLECTION OF BEAMS

Summary

The following relationships exist between loading, shearing force (S.F.), bending moment (B.M.), slope and deflection of a beam:

$$\text{deflection} = y \quad (\text{or } \delta)$$

$$\text{slope} = i \text{ or } \theta = \frac{dy}{dx}$$

$$\text{bending moment} = M = EI\frac{d^2y}{dx^2}$$

$$\text{shearing force} = Q = EI\frac{d^3y}{dx^3}$$

$$\text{loading} = w = EI\frac{d^4y}{dx^4}$$

In order that the above results should agree mathematically the sign convention illustrated in Fig. 5.4 must be adopted.

Using the above formulae the following standard values for *maximum slopes* and *deflections* of simply supported beams are obtained. (These assume that the beam is uniform, i.e. *EI* is constant throughout the beam.)

MAXIMUM SLOPE AND DEFLECTION OF SIMPLY SUPPORTED BEAMS

Loading condition	Maximum slope	Deflection (y)	Max. deflection (y_{max})
Cantilever with concentrated load W at end	$\frac{WL^2}{2EI}$	$\frac{W}{6EI}[2L^3 - 3L^2x + x^3]$	$\frac{WL^3}{3EI}$
Cantilever with u.d.l. across the complete span	$\frac{wL^3}{6EI}$	$\frac{w}{24EI}[3L^4 - 4L^3x + x^4]$	$\frac{wL^4}{8EI}$
Simply supported beam with concentrated load W at the centre	$\frac{WL^2}{16EI}$	$\frac{Wx}{48EI}[3L^2 - 4x^2]$	$\frac{WL^3}{48EI}$
Simply supported beam with u.d.l. across complete span	$\frac{wL^3}{24EI}$	$\frac{wx}{24EI}[L^3 - 2Lx^2 + x^3]$	$\frac{5wL^4}{384EI}$
Simply supported beam with concentrated load W offset from centre (distance a from one end b from the other)	$0.062\frac{WL^2}{EI}$		$\frac{Wa}{3EIL}\left[\frac{L^2 - a^2}{3}\right]^{3/2}$

Here L is the length of span, EI is known as the flexural rigidity of the member and x for the cantilevers is measured from the free end.

The determination of beam slopes and deflections by simple integration or Macaulay's methods requires a knowledge of certain conditions for various loading systems in order that the constants of integration can be evaluated. They are as follows:

(1) Deflections at supports are assumed zero unless otherwise stated.
(2) Slopes at built-in supports are assumed zero unless otherwise stated.
(3) Slope at the centre of symmetrically loaded and supported beams is zero.
(4) Bending moments at the free ends of a beam (i.e. those not built-in) are zero.

Mohr's theorems for slope and deflection state that if A and B are two points on the deflection curve of a beam and B is a point of zero slope, then

(1) $$\text{slope at } A = \text{area of } \frac{M}{EI} \text{ diagram between } A \text{ and } B$$

For a uniform beam, EI is constant, and the above equation reduces to

$$\text{slope at } A = \frac{1}{EI} \times \text{area of B.M. diagram between } A \text{ and } B$$

N.B.–If B is not a point of zero slope the equation gives the change of slope between A and B.

(2) $$\text{Total deflection of } A \text{ relative to } B = \text{first moment of area of } \frac{M}{EI} \text{ diagram about } A$$

For a uniform beam

$$\text{total deflection of } A \text{ relative to } B = \frac{1}{EI} \times \text{first moment of area of B.M. diagram about } A$$

Again, if B is not a point of zero slope the equation only gives the deflection of A relative to the tangent drawn at B.

Useful quantities for use with uniformly distributed loads are shown in Fig. 5.1.

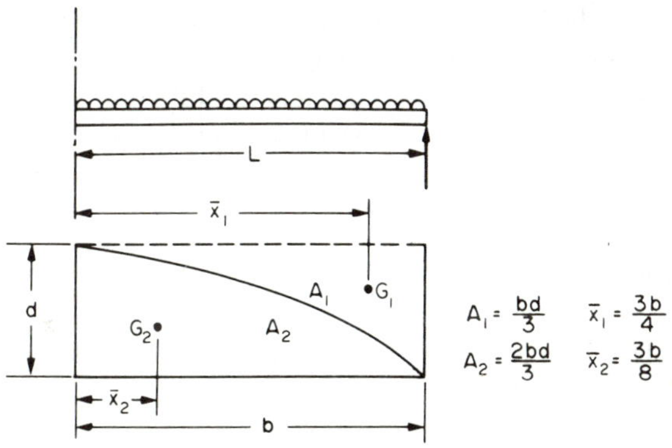

Fig. 5.1.

Both the straightforward integration method and Macaulay's method are based on the relationship $M = EI\dfrac{d^2y}{dx^2}$ (see § 5.2 and § 5.3).

Clapeyron's equations of three moments for continuous beams in its simplest form states that for any portion of a beam on three supports 1, 2 and 3, with spans between of L_1 and L_2, the bending moments at the supports are related by

$$-M_1L_1 - 2M_2(L_1 + L_2) - M_3L_2 = 6\left[\frac{A_1\bar{x}_1}{L_1} + \frac{A_2\bar{x}_2}{L_2}\right]$$

where A_1 is the area of the B.M. diagram, assuming span L_1 simply supported, and $\bar{x}_1$ is the distance of the centroid of this area from the left-hand support. Similarly, A_2 refers to span L_2, with $\bar{x}_2$ the centroid distance from the right-hand support (see Examples 5.6 and 5.7). The following standard results are useful for $\dfrac{6A\bar{x}}{L}$:

(a) Concentrated load W, distance a from the nearest outside support

$$\frac{6A\bar{x}}{L} = \frac{Wa}{L}(L^2 - a^2)$$

(b) Uniformly distributed load w

$$\frac{6A\bar{x}}{L} = \frac{wL^3}{4} \quad \text{(see Example 5.6)}$$

Introduction

In practically all engineering applications limitations are placed upon the performance and behaviour of components and normally they are expected to operate within certain set limits of, for example, stress or deflection. The stress limits are normally set so that the component does not yield or fail under the most severe load conditions which it is likely to meet in service. In certain structural or machine linkage designs, however, maximum stress levels may not be the most severe condition for the component in question. In such cases it is the limitation in the maximum deflection which places the most severe restriction on the operation or design of the component. It is evident, therefore, that methods are required to accurately predict the deflection of members under lateral loads since it is this form of loading which will generally produce the greatest deflections of beams, struts and other structural types of members.

5.1. Relationship between loading, S.F., B.M., slope and deflection

Consider a beam AB which is initially horizontal when unloaded. If this deflects to a new position $A'B'$ under load, the slope at any point C is

$$i = \frac{dy}{dx}$$

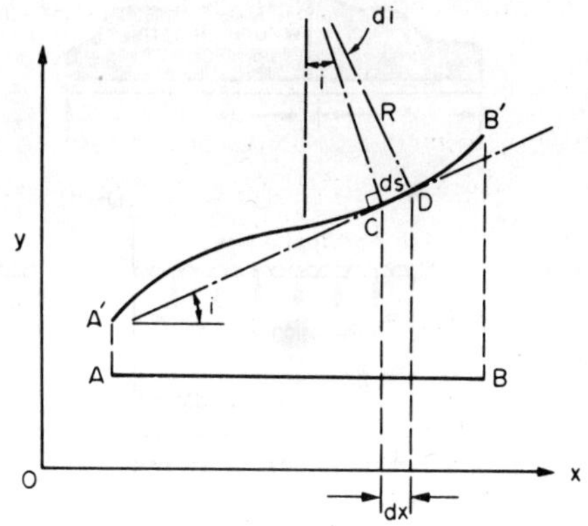

Fig. 5.2. Unloaded beam AB deflected to $A'B'$ under load.

This is usually very small in practice, and for small curvatures

$$ds = dx = R\,di \quad \text{(Fig. 5.2)}$$

$$\therefore \quad \frac{di}{dx} = \frac{1}{R}$$

But

$$i = \frac{dy}{dx}$$

$$\therefore \quad \frac{d^2y}{dx^2} = \frac{1}{R} \tag{5.1}$$

Now from the simple bending theory

$$\frac{M}{I} = \frac{E}{R}$$

$$\therefore \quad \frac{1}{R} = \frac{M}{EI}$$

Therefore substituting in eqn. (5.1)

$$M = EI\frac{d^2y}{dx^2} \tag{5.2}$$

This is the basic differential equation for the deflection of beams.

If the beam is now assumed to carry a distributed loading which varies in intensity over the length of the beam, then a small element of the beam of length dx will be subjected to the loading condition shown in Fig. 5.3. The parts of the beam on either side of the element $EFGH$ carry the externally applied forces, while reactions to these forces are shown on the element itself.

Thus for vertical equilibrium of $EFGH$,

$$Q - w\,dx = Q - dQ$$

$$\therefore \quad dQ = w\,dx$$

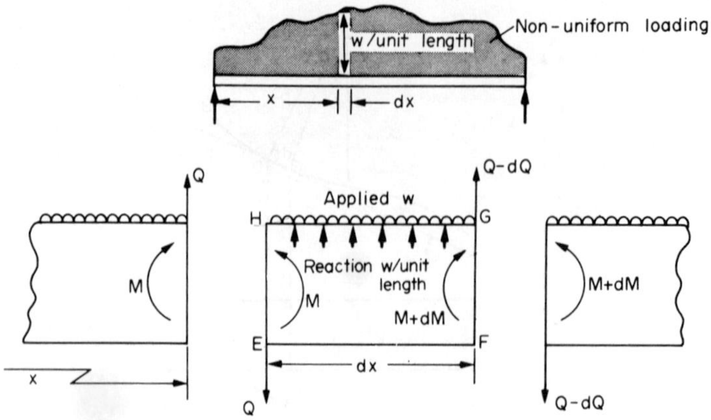

Fig. 5.3. Small element of beam subjected to non-uniform loading (effectively uniform over small length dx).

and integrating,
$$\boldsymbol{Q = \int w dx} \tag{5.3}$$

Also, for equilibrium, moments about any point must be zero.
Therefore taking moments about F,

$$(M + dM) + w dx \frac{dx}{2} = M + Q dx$$

Therefore neglecting the square of small quantities,

$$dM = Q dx$$

and integrating,
$$\boldsymbol{M = \int Q dx} \tag{5.4}$$

The results can then be summarised as follows:

$$\text{deflection} = y$$

$$\text{slope} = \frac{dy}{dx}$$

$$\text{bending moment} = EI\frac{d^2y}{dx^2} \tag{5.4a}$$

$$\text{shear force} = EI\frac{d^3y}{dx^3}$$

$$\text{load distribution} = EI\frac{d^4y}{dx^4}$$

In order that the above results should agree algebraically, i.e. that positive slopes shall have the normal mathematical interpretation of the positive sign and that B.M. and S.F. conventions are consistent with those introduced earlier, it is imperative that the sign convention illustrated in Fig. 5.4 be adopted.

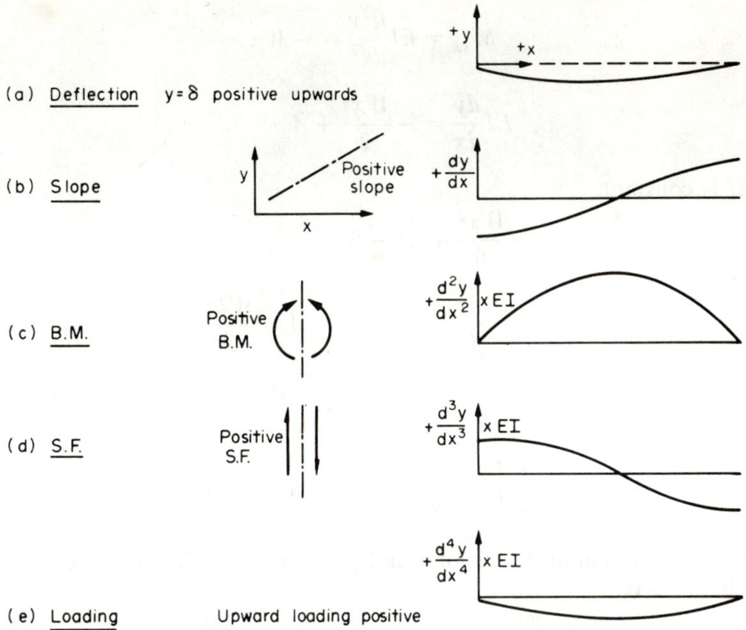

Fig. 5.4. Sign conventions for load, S.F., B.M., slope and deflection.

5.2. Direct integration method

If the value of the B.M. at any point on a beam is known in terms of x, the distance along the beam, and provided that the equation applies along the complete beam, then integration of eqn. (5.4a) will yield slopes and deflections at any point,

i.e.
$$M = EI\frac{d^2y}{dx^2} \quad \text{and} \quad \frac{dy}{dx} = \int \frac{M}{EI}dx + A$$

or
$$y = \int\int\left(\frac{M}{EI}dx\right)dx + Ax + B$$

where A and B are constants of integration evaluated from known conditions of slope and deflection for particular values of x.

(a) *Cantilever with concentrated load at the end* (Fig. 5.5)

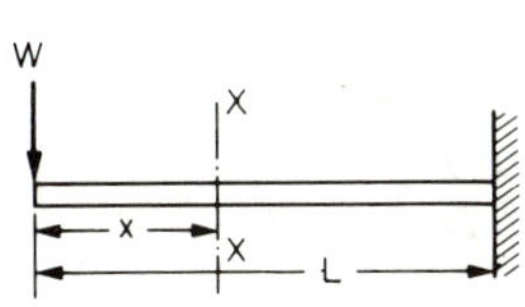

Fig. 5.5.

$$M_{xx} = EI\frac{d^2y}{dx^2} = -Wx$$

$$\therefore \qquad EI\frac{dy}{dx} = -\frac{Wx^2}{2} + A$$

assuming EI is constant.

$$EIy = -\frac{Wx^3}{6} + Ax + B$$

Now when $\quad x = L, \; \frac{dy}{dx} = 0 \quad \therefore \quad A = \frac{WL^2}{2}$

and when $\quad x = L, \; y = 0 \quad \therefore \quad B = \frac{WL^3}{6} - \frac{WL^2}{2}L = -\frac{WL^3}{3}$

$$\therefore \qquad y = \frac{1}{EI}\left[-\frac{Wx^3}{6} + \frac{WL^2x}{2} - \frac{WL^3}{3}\right] \tag{5.5}$$

This gives the deflection at all values of x and produces a maximum value at the tip of the cantilever when $x = 0$,

i.e. $$\textbf{Maximum deflection} = y_{max} = -\frac{WL^3}{3EI} \tag{5.6}$$

The negative sign indicates that deflection is in the negative y direction, i.e. downwards.

Similarly $$\frac{dy}{dx} = \frac{1}{EI}\left[-\frac{Wx^2}{2} + \frac{WL^2}{2}\right] \tag{5.7}$$

and produces a maximum value again when $x = 0$.

$$\textbf{Maximum slope} = \left(\frac{dy}{dx}\right)_{max} = \frac{WL^2}{2EI} \quad \textbf{(positive)} \tag{5.8}$$

(*b*) *Cantilever with uniformly distributed load* (Fig. 5.6)

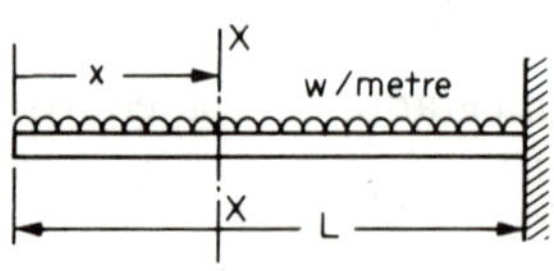

Fig. 5.6.

$$M_{xx} = EI\frac{d^2y}{dx^2} = -\frac{wx^2}{2}$$

$$EI\frac{dy}{dx} = -\frac{wx^3}{6} + A$$

$$EIy = -\frac{wx^4}{24} + Ax + B$$

Again, when $x = L, \quad \dfrac{dy}{dx} = 0 \quad \text{and} \quad A = \dfrac{wL^3}{6}$

$$x = L, \quad y = 0 \quad \text{and} \quad B = \frac{wL^4}{24} - \frac{wL^4}{6} = -\frac{wL^4}{8}$$

$$\therefore \qquad y = \frac{1}{EI}\left[-\frac{wx^4}{24} + \frac{wL^3x}{6} - \frac{wL^4}{8}\right] \tag{5.9}$$

At $x = 0$,
$$y_{max} = -\frac{\mathbf{wL^4}}{\mathbf{8EI}} \quad \textbf{and} \quad \left(\frac{dy}{dx}\right)_{max} = \frac{\mathbf{wL^3}}{\mathbf{6EI}} \tag{5.10}$$

(*c*) *Simply-supported beam with uniformly distributed load* (Fig. 5.7)

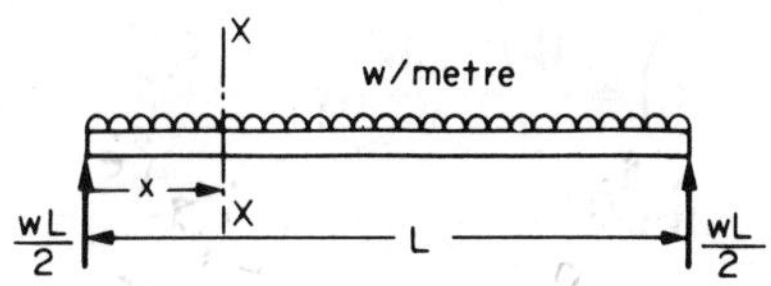

Fig. 5.7.

$$M_{xx} = EI\frac{d^2y}{dx^2} = \frac{wLx}{2} - \frac{wx^2}{2}.$$

$$EI\frac{dy}{dx} = \frac{wLx^2}{4} - \frac{wx^3}{6} + A$$

$$EIy = \frac{wLx^3}{12} - \frac{wx^4}{24} + Ax + B$$

At $\quad x = 0, \quad y = 0 \quad \therefore \quad B = 0$

At $\quad x = L, \quad y = 0 \quad \therefore \quad 0 = \dfrac{wL^4}{12} - \dfrac{wL^4}{24} + AL$

$$\therefore \qquad A = -\frac{wL^3}{24}$$

$$\therefore \qquad y = \frac{1}{EI}\left[\frac{wLx^3}{12} - \frac{wx^4}{24} - \frac{wL^3x}{24}\right] \tag{5.11}$$

In this case the maximum deflection will occur at the centre of the beam where $x = L/2$.

$$\therefore \qquad y_{max} = \frac{1}{EI}\left[\frac{wL}{12}\left(\frac{L^3}{8}\right) - \frac{w}{24}\left(\frac{L^4}{16}\right) - \frac{wL^3}{24}\left(\frac{L}{2}\right)\right]$$

$$= -\frac{\mathbf{5wL^4}}{\mathbf{384EI}} \tag{5.12}$$

Similarly $\quad \left(\dfrac{dy}{dx}\right)_{max} = \pm\dfrac{\mathbf{wL^3}}{\mathbf{24EI}}$ at the ends of the beam. (5.13) (*d*) *Simply supported beam with central concentrated load* (Fig. 5.8)

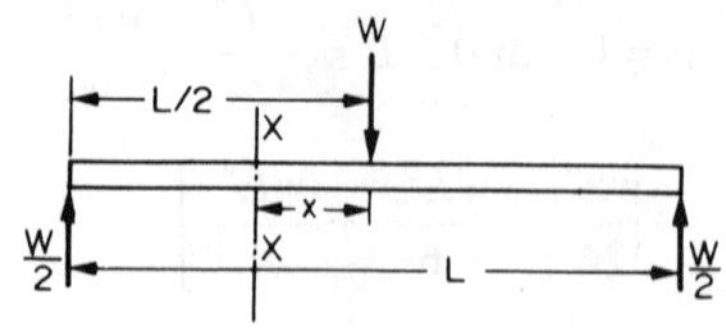

Fig. 5.8.

In order to obtain a single expression for B.M. which will apply across the complete beam in this case it is convenient to take the origin for x at the centre, then:

$$M_{xx} = EI\frac{d^2y}{dx^2} = \frac{W}{2}\left(\frac{L}{2} - x\right) = \frac{WL}{4} - \frac{Wx}{2}$$

$$EI\frac{dy}{dx} = \frac{WL}{4}x - \frac{Wx^2}{4} + A$$

$$EIy = \frac{WLx^2}{8} - \frac{Wx^3}{12} + Ax + B$$

At $\quad x = 0, \quad \frac{dy}{dx} = 0 \quad \therefore \quad A = 0$

$x = \frac{L}{2}, \quad y = 0 \quad \therefore \quad 0 = \frac{WL^3}{32} - \frac{WL^3}{96} + B$

$$\therefore \quad B = -\frac{WL^3}{48}$$

$$\therefore \quad y = \frac{1}{EI}\left[\frac{WLx^2}{8} - \frac{Wx^3}{12} - \frac{WL^3}{48}\right] \tag{5.14}$$

$$\therefore \quad y_{max} = -\frac{WL^3}{48EI} \quad \text{at the centre} \tag{5.15}$$

and

$$\left(\frac{dy}{dx}\right)_{max} = \pm\frac{WL^2}{16EI} \quad \text{at the ends} \tag{5.16}$$

In some cases it is not convenient to commence the integration procedure with the B.M. equation since this may be difficult to obtain. In such cases it is often more convenient to commence with the equation for the loading at the general point XX on the beam. A typical example follows:

(*e*) *Cantilever subjected to non-uniform distributed load* (Fig. 5.9)

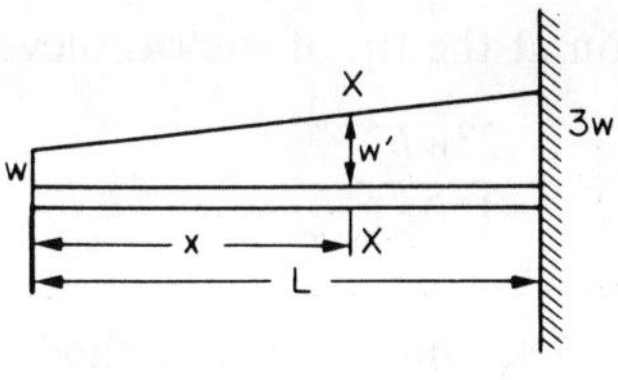

Fig. 5.9.

The loading at section XX is

$$w' = EI\frac{d^4y}{dx^4} = -\left[w + (3w - w)\frac{x}{L}\right] = -w\left(1 + \frac{2x}{L}\right)$$

Integrating,

$$EI\frac{d^3y}{dx^3} = -w\left(x + \frac{x^2}{L}\right) + A \qquad (1)$$

$$EI\frac{d^2y}{dx^2} = -w\left(\frac{x^2}{2} + \frac{x^3}{3L}\right) + Ax + B \qquad (2)$$

$$EI\frac{dy}{dx} = -w\left(\frac{x^3}{6} + \frac{x^4}{12L}\right) + \frac{Ax^2}{2} + Bx + C \qquad (3)$$

$$EIy = -w\left(\frac{x^4}{24} + \frac{x^5}{60L}\right) + \frac{Ax^3}{6} + \frac{Bx^2}{2} + Cx + D \qquad (4)$$

Thus, before the slope or deflection can be evaluated, four constants have to be determined; therefore four conditions are required. They are:

At $x = 0$, S.F. is zero
$\therefore$ from (1) $A = 0$
At $x = 0$, B.M. is zero
$\therefore$ from (2) $B = 0$
At $x = L$, slope $dy/dx = 0$ (slope normally assumed zero at a built-in support)

$\therefore$ from (3)
$$0 = -w\left(\frac{L^3}{6} + \frac{L^3}{12}\right) + C$$

$$C = \frac{wL^3}{4}$$

At $x = L$, $y = 0$

$\therefore$ from (4)
$$0 = -w\left(\frac{L^4}{24} + \frac{L^4}{60}\right) + \frac{wL^4}{4} + D$$

$\therefore$
$$D = -\frac{23wL^4}{120}$$

$$\therefore \quad EIy = -\frac{wx^4}{24} - \frac{wx^5}{60L} + \frac{wL^3x}{4} - \frac{23wL^4}{120}$$

Then, for example, the deflection at the tip of the cantilever, where $x = 0$, is

$$y = -\frac{23wL^4}{120EI}$$

5.3. Macaulay's method

The simple integration method used in the previous examples can only be used when a single expression for B.M. applies along the complete length of the beam. In general this is not the case, and the method has to be adapted to cover all loading conditions.

Consider, therefore, a small portion of a beam in which, at a particular section A, the shearing force is Q and the B.M. is M, as shown in Fig. 5.10. At another section B, distance a along the beam, a concentrated load W is applied which will change the B.M. for points beyond B.

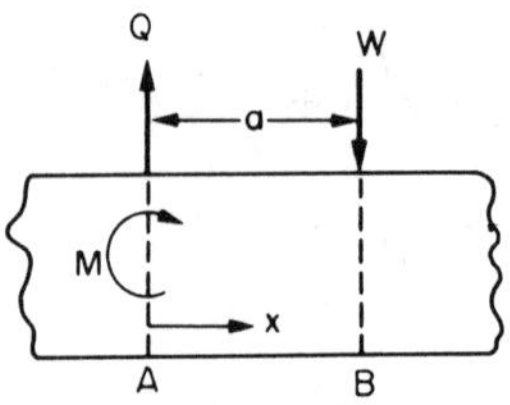

Fig. 5.10.

Between A and B,

$$M = EI\frac{d^2y}{dx^2} = M + Qx \qquad (1)$$

$$\therefore \quad EI\frac{dy}{dx} = Mx + Q\frac{x^2}{2} + C_1 \qquad (2)$$

and

$$EIy = M\frac{x^2}{2} + Q\frac{x^3}{6} + C_1x + C_2 \qquad (3)$$

Beyond B

$$M = EI\frac{d^2y}{dx^2} = M + Qx - W(x-a) \qquad (4)$$

$$\therefore \quad EI\frac{dy}{dx} = Mx + Q\frac{x^2}{2} - W\frac{x^2}{2} + Wax + C_3 \qquad (5)$$

and

$$EIy = M\frac{x^2}{2} + Q\frac{x^3}{6} - W\frac{x^3}{6} + Wa\frac{x^2}{2} + C_3x + C_4 \qquad (6)$$

Now for the same slope at B, equating (2) and (5),

$$Mx + Q\frac{x^2}{2} + C_1 = Mx + Q\frac{x^2}{2} - W\frac{x^2}{2} + Wax + C_3$$

But at $B, x = a$

$$\therefore \quad C_1 = -\frac{Wa^2}{2} + Wa^2 + C_3$$

$$\therefore \quad C_3 = C_1 - \frac{Wa^2}{2}$$

Substituting in (5),

$$EI\frac{dy}{dx} = Mx + Q\frac{x^2}{2} - W\frac{x^2}{2} + Wax + C_1 - \frac{Wa^2}{2}$$

$$\therefore \quad EI\frac{dy}{dx} = Mx + Q\frac{x^2}{2} - \frac{W}{2}(x-a)^2 + C_1 \tag{7}$$

Also, for the same deflection at B equating (3) and (6), with $x = a$

$$\frac{Ma^2}{2} + \frac{Qa^3}{6} + C_1 a + C_2 = \frac{Ma^2}{2} + \frac{Qa^3}{6} - \frac{Wa^3}{6} + \frac{Wa^3}{2} + C_3 a + C_4$$

$$\therefore \quad C_1 a + C_2 = -\frac{Wa^3}{6} + \frac{Wa^3}{2} + C_3 a + C_4$$

$$= -\frac{Wa^3}{6} + \frac{Wa^3}{2} + \left(C_1 - \frac{Wa^2}{2}\right)a + C_4$$

$$\therefore \quad C_4 = C_2 + \frac{Wa^3}{6}$$

Substituting in (6),

$$EIy = M\frac{x^2}{2} + Q\frac{x^3}{6} - W\frac{x^3}{6} + Wa\frac{x^2}{2}\left(C_1 - \frac{Wa^2}{2}\right)x + W\frac{a^3}{6} + C_2$$

$$= M\frac{x^2}{2} + Q\frac{x^3}{6} - W\frac{(x-a)^3}{6} + C_1 x + C_2 \tag{8}$$

Thus, inspecting (4), (7) and (8), we can see that the general method of obtaining slopes and deflections (i.e. integrating the equation for M) will still apply provided that the term $W(x-a)$ is integrated with respect to $(x-a)$ and not x. Thus, when integrated, the term becomes

$$W\frac{(x-a)^2}{2} \quad \text{and} \quad W\frac{(x-a)^3}{6}$$

successively.

In addition, since the term $W(x-a)$ applies only after the discontinuity, i.e. when $x > a$, *it should be considered only when* $x > a$ or when $(x-a)$ *is positive.* For these reasons such terms are conventionally put into square or curly brackets and called *Macaulay terms.*

Thus Macaulay terms must be (a) integrated with respect to themselves and (b) neglected when negative.

For the whole beam, therefore,

$$EI\frac{d^2y}{dx^2} = M + Qx - W[(x-a)]$$

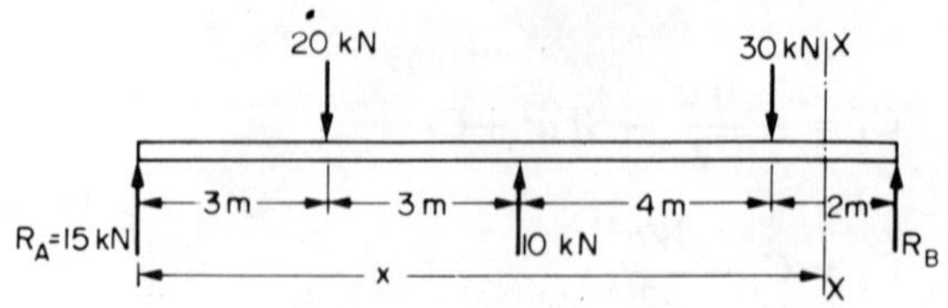

Fig. 5.11.

As an illustration of the procedure consider the beam loaded as shown in Fig. 5.11 for which the central deflection is required. Using the Macaulay method the equation for the B.M. at any general section XX is then given by

$$\text{B.M.}_{XX} = 15x - 20[(x-3)] + 10[(x-6)] - 30[(x-10)]$$

Care is then necessary to ensure that the terms inside the square brackets (Macaulay terms) are treated in the special way noted on the previous page.

Here it must be emphasised that all loads in the right-hand side of the equation are in units of kN (i.e. newtons × 10^3). *In subsequent working, therefore, it is convenient to carry through this factor as a denominator on the left-hand side in order that the expressions are dimensionally correct.*

Integrating,

$$\frac{EI}{10^3}\frac{dy}{dx} = 15\frac{x^2}{2} - 20\left[\frac{(x-3)^2}{2}\right] + 10\left[\frac{(x-6)^2}{2}\right] - 30\left[\frac{(x-10)^2}{2}\right] + A$$

and
$$\frac{EI}{10^3}y = 15\frac{x^3}{6} - 20\left[\frac{(x-3)^3}{6}\right] + 10\left[\frac{(x-6)^3}{6}\right] - 30\left[\frac{(x-10)^3}{6}\right] + Ax + B$$

where A and B are two constants of integration.

Now when $x = 0$, $y = 0$ $\therefore$ $B = 0$

and when $x = 12$, $y = 0$

$$\therefore \quad 0 = \frac{15 \times 12^3}{6} - 20\left[\frac{9^3}{6}\right] + 10\left[\frac{6^3}{6}\right] - 30\left[\frac{2^3}{6}\right] + 12A$$

$$= 4320 - 2430 + 360 - 40 + 12A$$

$$\therefore \quad 12A = -4680 + 2470 = -2210$$

$$\therefore \quad A = -184.2$$

The deflection at any point is given by

$$\frac{EI}{10^3}y = 15\frac{x^3}{6} - 20\left[\frac{(x-3)^3}{6}\right] + 10\left[\frac{(x-6)^3}{6}\right] - 30\left[\frac{(x-10)^3}{6}\right] - 184.2x$$

The deflection at mid-span is thus found by substituting $x = 6$ in the above equation, bearing in mind that the dimensions of the equation are kN m^3.

N.B.–Two of the Macaulay terms then vanish since one becomes zero and the other negative and therefore neglected.

$$\therefore \quad \text{central deflection} = \frac{10^3}{EI}\left[\frac{15 \times 6^3}{6} - \frac{20 \times 3^3}{6} - 184.2 \times 6\right]$$

$$= -\frac{655.2 \times 10^3}{EI}$$

With typical values of $E = 208\ \text{GN/m}^2$ and $I = 82 \times 10^{-6}\ \text{m}^4$

$$\text{central deflection} = 38.4 \times 10^{-3}\ \text{m} = \mathbf{38.4\ mm}$$

5.4. Macaulay's method for u.d.l.s

If a beam carries a uniformly distributed load over the complete span as shown in Fig. 5.12a the B.M. equation is

$$\text{B.M.}_{XX} = EI\frac{d^2y}{dx^2} = R_A x - \frac{wx^2}{2} - W_1[(x-a)] - W_2[(x-b)]$$

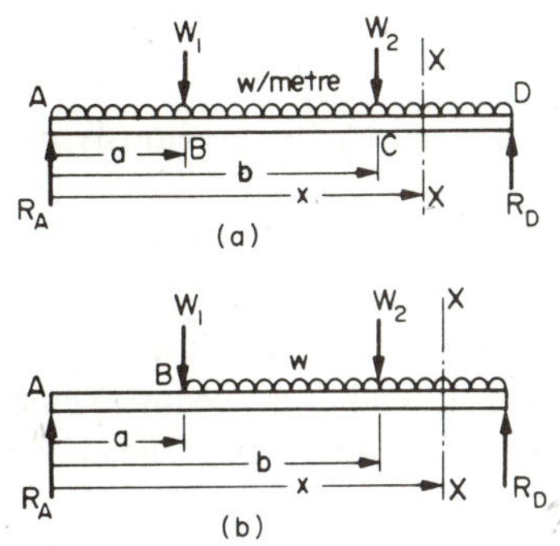

Fig. 5.12.

The u.d.l. term applies across the complete span and does not require the special treatment associated with the Macaulay terms. If, however, the u.d.l. starts at B as shown in Fig. 5.12b the B.M. equation is modified and the u.d.l. term becomes a Macaulay term and is written inside square brackets.

$$\text{B.M.}_{XX} = EI\frac{d^2y}{dx^2} = R_A x - W_1[(x-a)] - w\left[\frac{(x-a)^2}{2}\right] - W_2[(x-b)]$$

Integrating,

$$EI\frac{dy}{dx} = R_A\frac{x^2}{2} - W_1\left[\frac{(x-a)^2}{2}\right] - w\left[\frac{(x-a)^3}{6}\right] - W_2\left[\frac{(x-b)^2}{2}\right] + A$$

$$EIy = R_A\frac{x^3}{6} - W_1\left[\frac{(x-a)^3}{6}\right] - w\left[\frac{(x-a)^4}{24}\right] - W_2\left[\frac{(x-b)^3}{6}\right] + Ax + B$$

Note that Macaulay terms are integrated with respect to, for example, $(x-a)$ and they must be ignored when negative. Substitution of end conditions will then yield the values of the constants A and B in the normal way and hence the required values of slope or deflection.

It must be appreciated, however, that once a term has been entered in the B.M. expression it will apply across the complete beam. The modifications to the procedure required for cases when u.d.l.s. are applied over part of the beam only are introduced in the following theory.

5.5. Macaulay's method for beams with u.d.l. applied over part of the beam

Consider the beam loading case shown in Fig. 5.13a.

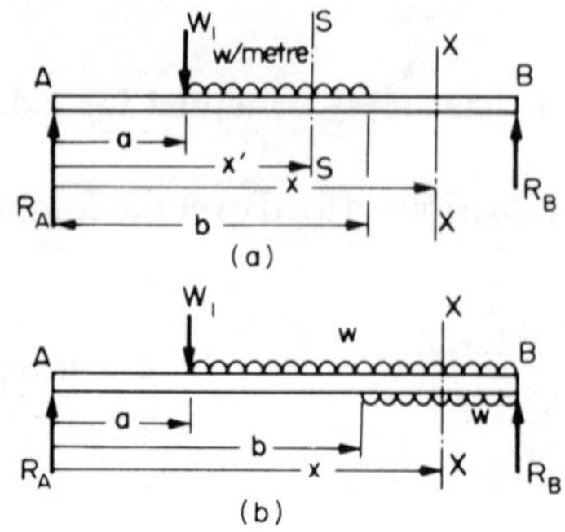

Fig. 5.13.

The B.M. at the section SS is given by the previously introduced procedure as

$$\text{B.M.}_{SS} = R_A x' - W_1[(x'-a)] - W\left[\frac{(x'-a)^2}{2}\right]$$

Having introduced the last (u.d.l.) term, however, it will apply for all values of x' greater than a, i.e. across the rest of the span to the end of the beam. (Remember, Macaulay terms are only neglected when they are negative, e.g. with $x' < a$.) The above equation is *NOT* therefore the correct equation for the load condition shown. The Macaulay method requires that this continuation of the u.d.l. be shown on the loading diagram and the required loading condition can therefore only be achieved by introducing an equal and opposite u.d.l. over the last part of the beam to cancel the unwanted continuation of the initial distributed load. This procedure is shown in Fig. 5.13b.

The correct B.M. equation for any general section XX is then given by

$$\text{B.M.}_{XX} = EI\frac{d^2y}{dx^2} = R_A x - W_1[(x-a)] - w\left[\frac{(x-a)^2}{2}\right] + w\left[\frac{(x-b)^2}{2}\right]$$

This type of approach can be adopted for any beam loading cases in which u.d.l.s are stopped or added to.

A number of examples are shown in Figs. 5.14–17. In each case the required loading system is shown first, followed by the continuation and compensating load system and the resulting B.M. equation.

5.6. Macaulay's method for couple applied at a point

Consider the beam AB shown in Fig. 5.18 with a moment or couple M applied at some point C. Considering the equilibrium of moments about each end in turn produces reactions of

$$R_A = \frac{M}{L} \text{ upwards, } \quad \text{and} \quad R_B = \frac{M}{L} \text{ downwards}$$

These equal and opposite forces then automatically produce the required equilibrium of vertical forces.

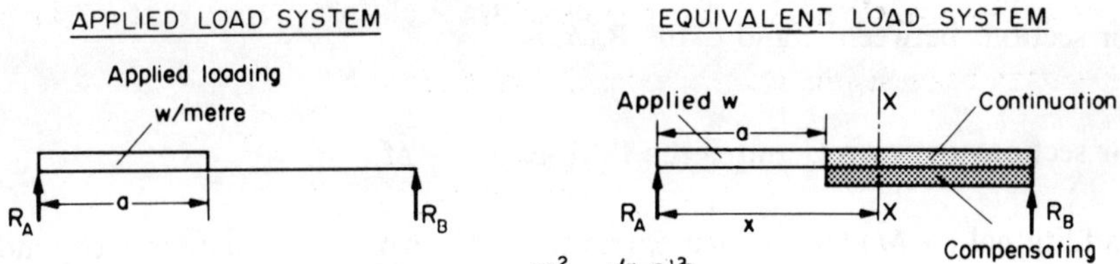

Fig. 5.14. $B.M._{xx} = R_A x - \frac{wx^2}{2} + w\left[\frac{(x-a)^2}{2}\right]$

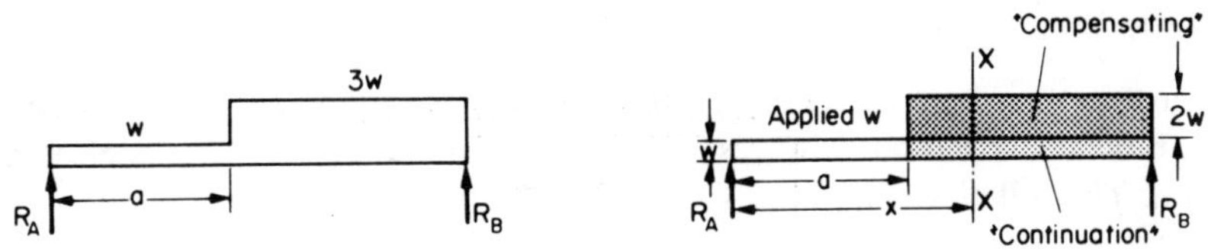

Fig. 5.15. $B.M._{xx} = R_A x - \frac{wx^2}{2} - 2w\left[\frac{(x-a)^2}{2}\right]$

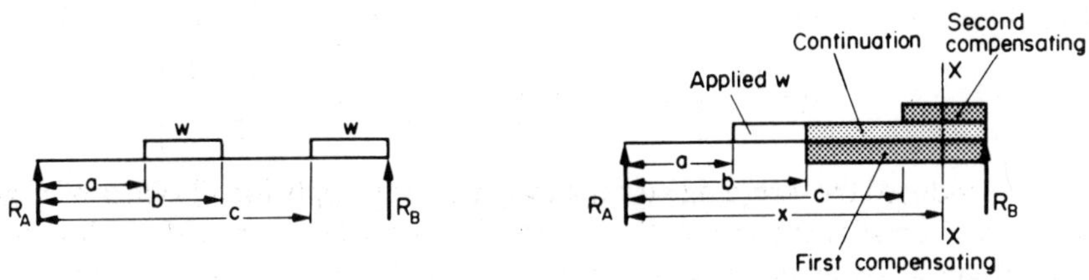

Fig. 5.16. $B.M._{xx} = R_A x - w\left[\frac{(x-a)^2}{2}\right] + w\left[\frac{(x-b)^2}{2}\right] - w\left[\frac{(x-c)^2}{2}\right]$

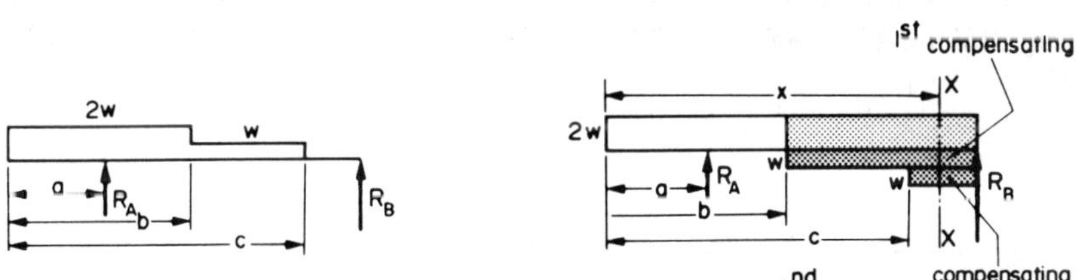

Fig. 5.17. $B.M._{xx} = -2w\frac{x^2}{2} + R_A\left[(x-a)\right] + w\left[\frac{(x-b)^2}{2}\right] + w\left[\frac{(x-c)^2}{2}\right]$

Figs. 5.14, 5.15, 5.16 and 5.17. Typical equivalent load systems for Macaulay method together with appropriate B.M. expressions.

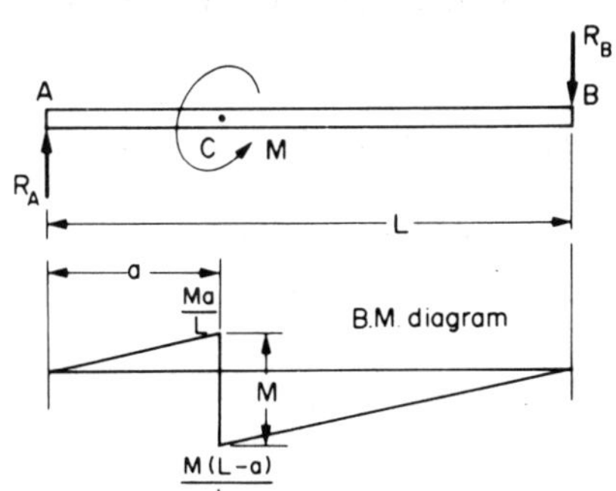

Fig. 5.18. Beam subjected to applied couple or moment *M*.

For sections between A and C the B.M. is $\dfrac{M}{L}x$.

For sections between C and B the B.M. is $\dfrac{Mx}{L} - M$.

The additional $(-M)$ term which enters the B.M. expression for points beyond C can be adequately catered for by the Macaulay method if written in the form

$$M[(x-a)^0]$$

This term can then be treated in precisely the same way as any other Macaulay term, integration being carried out with respect to $(x-a)$ and the term being neglected when x is less than a. The full B.M. equation for the beam is therefore

$$M_{xx} = EI\frac{d^2y}{dx^2} = \frac{Mx}{L} - M[(x-a)^0] \tag{5.17}$$

Then

$$EI\frac{dy}{dx} = \frac{Mx^2}{2L} - M[(x-a)] + A, \quad \text{etc.}$$

5.7. Mohr's "area–moment" method

In applications where the slope or deflection of beams or cantilevers is required at only one position the determination of the complete equations for slope and deflection at all points as obtained by Macaulay's method is rather laborious. In such cases, and in particular where loading systems are relatively simple, the Mohr moment–area method provides a rapid solution.

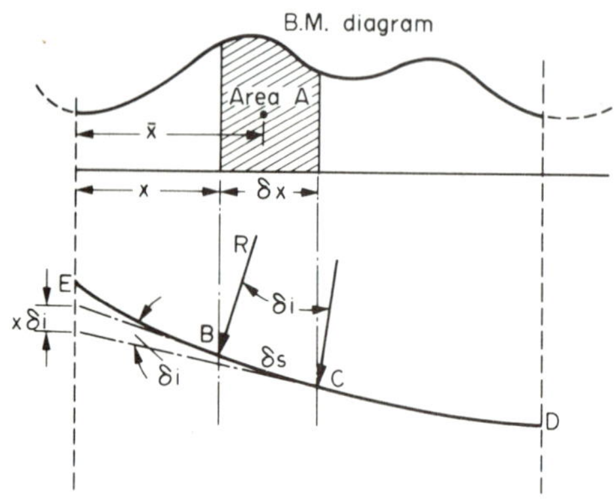

Fig. 5.19.

Figure 5.19 shows the deflected shape of part of a beam ED under the action of a B.M. which varies as shown in the B.M. diagram. Between any two points B and C the B.M. diagram has an area A and centroid distance $\bar{x}$ from E. The tangents at the points B and C give an intercept of $x\delta i$ on the vertical through E, where δi is the angle between the tangents.

Now

$$\delta s = R\delta i$$

and
$$\delta x \simeq \delta s$$
if slopes are small.

$$\therefore \qquad \delta i = \frac{\delta x}{R} = \frac{M}{EI}\delta x$$

$$\therefore \qquad \text{change of slope between } E \text{ and } D = i = \int_E^D \frac{M}{EI}\,dx$$

i.e. *change of slope = area of M/EI diagram between E and D* (5.18)

N.B.–For a uniform beam (EI constant) this equals $\frac{1}{EI}$ × area of B.M. diagram.

Deflection at E resulting from the bending of $BC = x\delta i$

$$\therefore \qquad \text{total deflection resulting from bending of } ED = \int x\delta i$$
$$= \int_E^D \frac{Mx}{EI}\,dx$$

The total deflection of E relative to the tangent at D is equal to the first moment of area of the M/EI diagram about E. (5.19)

Again, if EI is constant this equals $1/EI$ × first moment of area of the B.M. diagram about E.

The theorem is particularly useful when one point on the beam is a point of zero slope since the tangent at this point is then horizontal and deflections relative to the tangent are absolute values of vertical deflections. Thus if D is a point of zero slope the above equations yield the actual slope and deflection at E.

The Mohr area–moment procedure may be summarised in its most useful form as follows: if A and B are two points on the deflection curve of a beam, EI is constant and B is a point of zero slope, then Mohr's theorems state that:

(1) **Slope at $A = 1/EI$ × area of B.M. diagram between A and B.** (5.20)

(2) **Deflection of A relative to $B = 1/EI$ × first moment of area of B.M. diagram between A and B about A.** (5.21)

In many cases of apparently complicated load systems the loading can be separated into a combination of several simple systems which, by the application of the principle of superposition, will produce the same results. This procedure is illustrated in Examples 5.4 and 5.5.

The Mohr method will now be applied to the standard loading cases solved previously by the direct integration procedure.

(*a*) *Cantilever with concentrated load at the end*

In this case B is a point of zero slope and the simplified form of the Mohr theorems stated above can be applied.

$$\text{Slope at } A = \frac{1}{EI}\,[\text{area of B.M. diagram between } A \text{ and } B \text{ (Fig. 5.20)}]$$

$$= \frac{1}{EI}\left[\frac{L}{2}WL\right] = \frac{WL^2}{2EI}$$

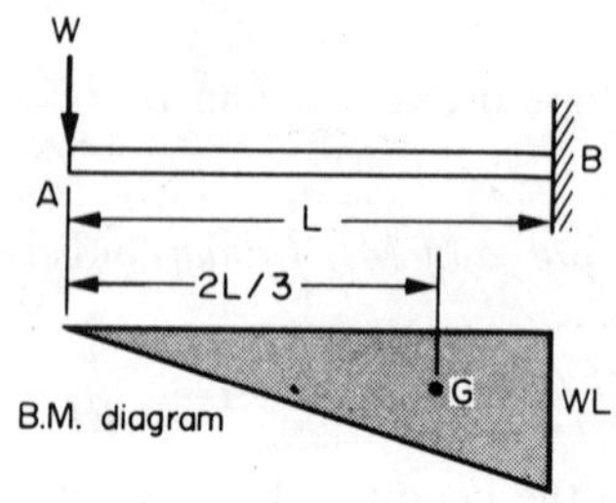

Fig. 5.20.

Deflection at A (relative to B)

$$= \frac{1}{EI}\,[\text{first moment of area of B.M. diagram between } A \text{ and } B \text{ about } A]$$

$$= \frac{1}{EI}\left[\left(\frac{L}{2}WL\right)\frac{2L}{3}\right] = \frac{WL^3}{3EI}$$

(*b*) *Cantilever with u.d.l.*

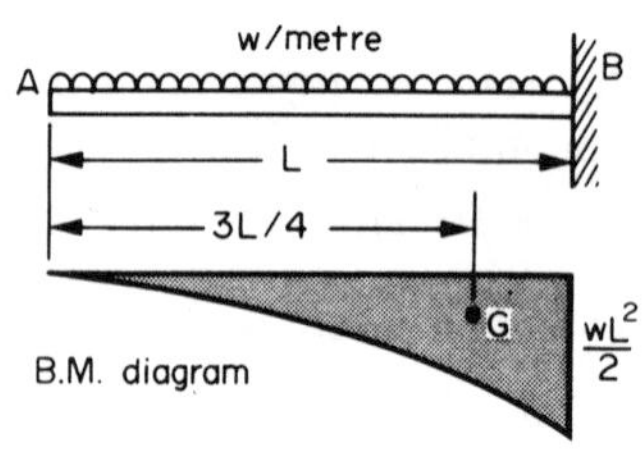

Fig. 5.21.

Again B is a point of zero slope.

$$\therefore \qquad \text{slope at } A = \frac{1}{EI}\,[\text{area of B.M. diagram (Fig. 5.21)}]$$

$$= \frac{1}{EI}\left[\frac{1}{3}L\frac{wL^2}{2}\right]$$

$$= \frac{wL^3}{6EI}$$

$$\text{Deflection at } A = \frac{1}{EI}\,[\text{moment of B.M. diagram about } A]$$

$$= \frac{1}{EI}\left[\left(\frac{1}{3}L\frac{wL^2}{2}\right)\frac{3L}{4}\right] = \frac{wL^4}{8EI}$$

(*c*) *Simply supported beam with u.d.l.*

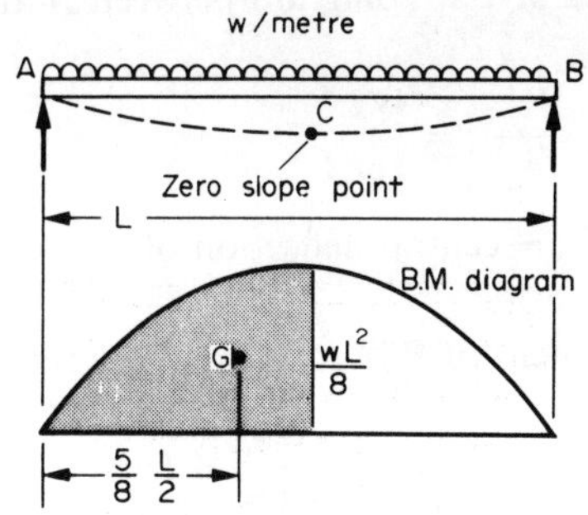

Fig. 5.22.

Here the point of zero slope is at the centre of the beam C. Working relative to C,

$$\text{slope at } A = \frac{1}{EI}\,[\text{area of B.M. diagram between } A \text{ and } C \text{ (Fig. 5.22)}]$$

$$= \frac{1}{EI}\left[\frac{2}{3}\frac{wL^2}{8}\frac{L}{2}\right] = \boldsymbol{\frac{wL^3}{24EI}}$$

Deflection of A relative to C (= central deflection relative to A)

$$= \frac{1}{EI}\,[\text{moment of B.M. diagram between } A \text{ and } C \text{ about } A]$$

$$= \frac{1}{EI}\left[\left(\frac{2}{3}\frac{wL^2}{8}\frac{L}{2}\right)\left(\frac{5L}{16}\right)\right] = \boldsymbol{\frac{5wL^4}{384EI}}$$

(*d*) *Simply supported beam with central concentrated load*

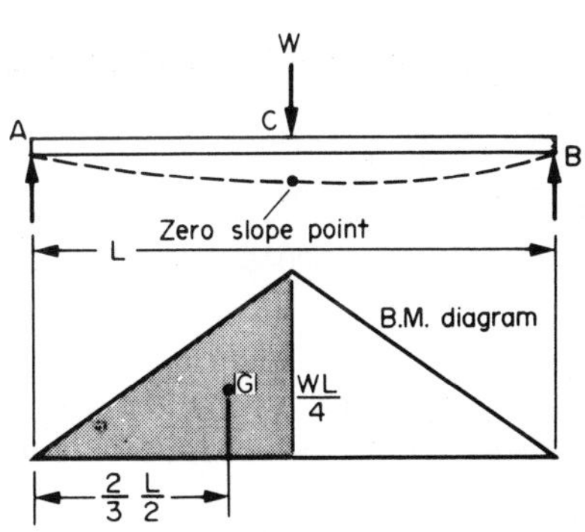

Fig. 5.23.

Again working relative to the zero slope point at the centre C,

$$\text{slope at } A = \frac{1}{EI}[\text{area of B.M. diagram between } A \text{ and } C \text{ (Fig. 5.23)}]$$

$$= \frac{1}{EI}\left[\frac{1}{2}\frac{L}{2}\frac{WL}{4}\right] = \frac{WL^2}{16EI}$$

Deflection of A relative to C (= central deflection of C)

$$= \frac{1}{EI}[\text{moment of B.M. diagram between } A \text{ and } C \text{ about } A]$$

$$= \frac{1}{EI}\left[\left(\frac{1}{2}\frac{L}{2}\frac{WL}{4}\right)\left(\frac{2}{3}\frac{L}{2}\right)\right] = \frac{WL^3}{48EI}$$

5.8. Principle of superposition

The general statement for the principle of superposition asserts that the resultant stress or strain in a system subjected to several forces is the algebraic sum of their effects when applied separately. The principle can be utilised, however, to determine the deflections of beams subjected to complicated loading conditions which, in reality, are merely combinations of a number of simple systems. In addition to the simple standard cases introduced previously, numerous different loading conditions have been solved by various workers and their results may be found in civil or mechanical engineering handbooks or data sheets. Thus, the algebraic sum of the separate deflections caused by a convenient selection of standard loading cases will produce the total deflection of the apparently complex case.

It must be appreciated, however, that the principle of superposition is only valid whilst the beam material remains elastic and for small beam deflections. (Large deflections would produce unacceptable deviation of the lines of action of the loads relative to the beam axis.)

5.9. Energy method

A further, alternative, procedure for calculating deflections of beams or structures is based upon the application of strain energy considerations. This is introduced in detail in Chapter 11 and will not be considered further here.

5.10. Maxwell's theorem of reciprocal displacements

Consider a beam subjected to two loads W_A and W_B at points A and B respectively as shown in Fig. 5.24. Let W_A be gradually applied first, producing a deflection a at A.

$$\text{Work done} = \tfrac{1}{2}W_A a$$

When W_B is applied it will produce a deflection b at B and an additional deflection δ_{ab} at A (the latter occurring in the presence of a now constant load W_A).

$$\text{Extra work done} = \tfrac{1}{2}W_B b + W_A \delta_{ab}$$

$$\therefore \quad \text{total work done} = \tfrac{1}{2}W_A a + \tfrac{1}{2}W_B b + W_A \delta_{ab}$$

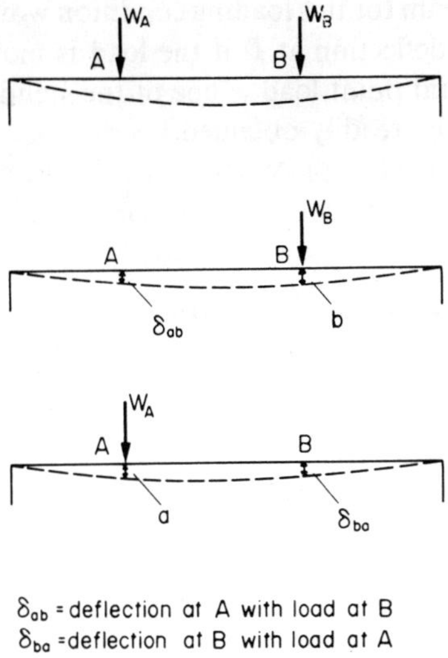

Fig. 5.24. Maxwell's theorem of reciprocal displacements.

Similarly, if the loads were applied in reverse order and the load W_A at A produced an additional deflection δ_{ba} at B, then

$$\text{total work done} = \tfrac{1}{2}W_B b + \tfrac{1}{2}W_A a + W_B \delta_{ba}$$

It should be clear that, regardless of the order in which the loads are applied, the total work done must be the same. Inspection of the above equations thus shows that

$$W_A \delta_{ab} = W_B \delta_{ba}$$

If the two loads are now made equal, then

$$\delta_{ab} = \delta_{ba} \tag{5.22}$$

i.e. *the deflection at A produced by a load at B equals the deflection at B produced by the same load at A.* This is Maxwell's theorem of reciprocal displacements.

As a typical example of the application of this theorem to beams consider the case of a simply supported beam carrying a single concentrated load off-set from the centre (Fig. 5.25).

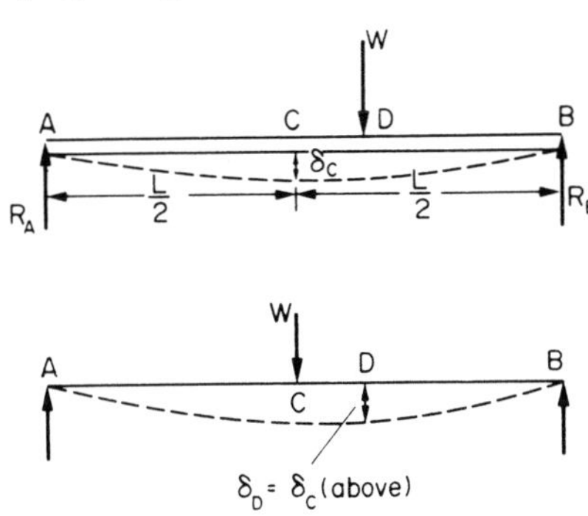

Fig. 5.25.

The central deflection of the beam for this loading condition would be given by the reciprocal displacement theorem as the deflection at D if the load is moved to the centre. Since the deflection equation for a central point load is one of the standard cases treated earlier the required deflection value can be readily obtained.

Maxwell's theorem of reciprocal displacements can also be applied if one or both of the loads are replaced by moments or couples. In this case it can be shown that the theorem is modified to the relevant one of the following forms (a), (b):

(a) *The angle of rotation at A due to a* ***concentrated force*** *at B is numerically equal to the deflection at B due to a* ***couple*** *at A provided that the force and couple are also numerically equal* (Fig. 5.26).

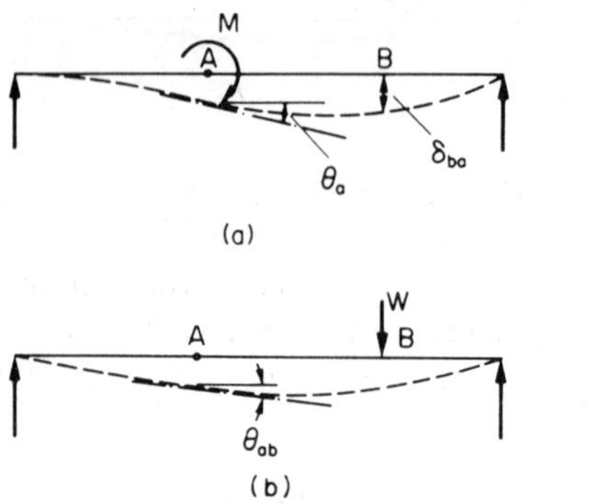

θ_a = slope at A with moment (or load) at A

θ_{ab} = slope at a with load at B

Fig. 5.26.

(b) *The angle of rotation at A due to a* ***couple*** *at B is equal to the rotation at B due to the* ***same couple*** *applied at A* (Fig. 5.27).

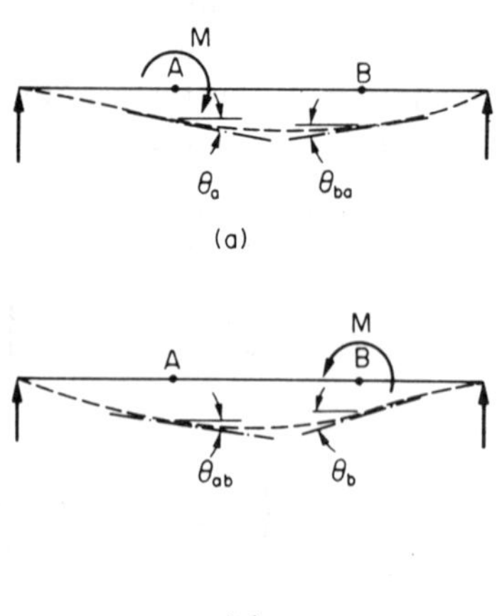

Fig. 5.27.

All three forms of the theorem are quite general in application and are not restricted to beam problems. Any type of component or structure subjected to bending, direct load, shear or torsional deformation may be considered provided always that linear elastic conditions prevail, i.e. Hooke's law applies, and deflections are small enough not to significantly affect the undeformed geometry.

5.11. Continuous beams – Clapeyron's "three-moment" equation

When a beam is supported on more than two supports it is termed *continuous*. In cases such as these it is not possible to determine directly the reactions at the three supports by the normal equations of static equilibrium since there are too many unknowns. An extension of Mohr's area–moment method is therefore used to obtain a relationship between the B.M.s at the supports, from which the reaction values can then be determined and the B.M. and S.F. diagrams drawn.

Consider therefore the beam shown in Fig. 5.28. The areas A_1 and A_2 are the "free" B.M. diagrams, treating the beam as simply supported over two separate spans L_1 and L_2. In general the B.M.s at the three supports will not be zero as this diagram suggests, but will have some values M_1, M_2 and M_3. Thus a *fixing-moment diagram* must be introduced as shown, the actual B.M. diagram then being the algebraic sum of the two diagrams.

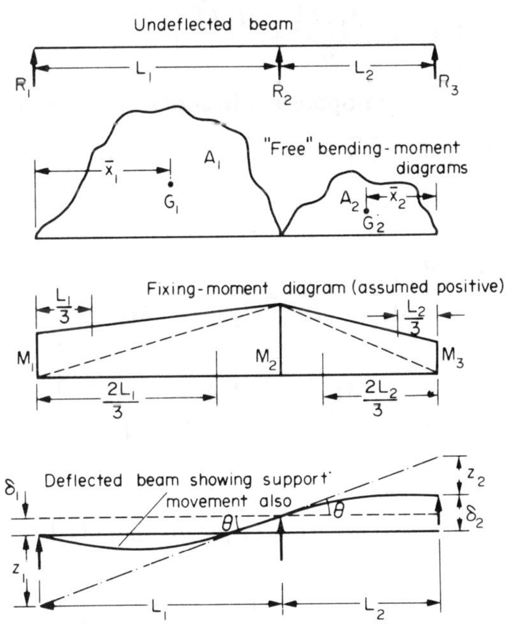

Fig. 5.28. Continuous beam over three supports showing "free" and "fixing" moment diagrams together with the deflected beam form including support movement.

The bottom figure shows the deflected position of the beam, the deflections δ_1 and δ_2 being *relative to the left-hand support*. If a tangent is drawn at the centre support then the intercepts at the end of each span are z_1 and z_2 and θ is the slope of the tangent, and hence the beam, at the centre support.

Now, assuming deflections are small,

$$\theta \text{ (radians)} = \frac{z_1 + \delta_1}{L_1} = \frac{z_2 + \delta_2 - \delta_1}{L_2}$$

$$\therefore \qquad \frac{z_1}{L_1} + \frac{\delta_1}{L_1} = \frac{z_2}{L_2} + \frac{(\delta_2 - \delta_1)}{L_2}$$

But from Mohr's area–moment method,

$$z = \frac{A\bar{x}}{EI}$$

where A is the area of the B.M. diagram over the span to which z refers.

$$\therefore \qquad z_1 = -\frac{1}{EI_1}\left[A_1\bar{x}_1 + \left(\frac{M_1 L_1}{2} \times \frac{L_1}{3}\right) + \left(\frac{M_2 L_1}{2} \times \frac{2L_1}{3}\right)\right]$$

$$= -\frac{1}{EI_1}\left[A_1\bar{x}_1 + \frac{M_1 L_1^2}{6} + \frac{M_2 L_1^2}{3}\right]$$

and

$$z_2 = \frac{1}{EI_2}\left[A_2\bar{x}_2 + \left(\frac{M_3 L_2}{2} \times \frac{L_2}{3}\right) + \left(\frac{M_2 L_2}{2} \times \frac{2L_2}{3}\right)\right]$$

$$= \frac{1}{EI_2}\left[A_2\bar{x}_2 + \frac{M_3 L_2^2}{6} + \frac{M_2 L_2^2}{3}\right]$$

N.B. – Since the intercepts are in opposite directions, they are of opposite sign.

$$\therefore \qquad -\frac{\left[A_1\bar{x}_1 + \frac{M_1 L_1^2}{6} + \frac{M_2 L_1^2}{3}\right]}{EI_1 L_1} + \frac{\delta_1}{L_1} = \frac{\left[A_2\bar{x}_2 + \frac{M_3 L_2^2}{6} + \frac{M_2 L_2^2}{3}\right]}{EI_2 L_2} + \frac{(\delta_2 - \delta_1)}{L_2}$$

$$\therefore \qquad -\frac{A_1\bar{x}_1}{I_1 L_1} - \frac{M_1 L_1}{6I_1} - \frac{M_2 L_1}{3I_1} + \frac{E\delta_1}{L_1} = \frac{A_2\bar{x}_2}{I_2 L_2} + \frac{M_3 L_2}{6I_2} + \frac{M_2 L_2}{3I_2} + \frac{E(\delta_2 - \delta_1)}{L_2}$$

$$\therefore \qquad \boldsymbol{-\frac{M_1 L_1}{I_1} - 2M_2\left[\frac{L_1}{I_1} + \frac{L_2}{I_2}\right] - \frac{M_3 L_2}{I_2} = 6\left[\frac{A_1\bar{x}_1}{I_1 L_1} + \frac{A_2\bar{x}_2}{I_2 L_2}\right] + 6E\left[\frac{(\delta_2 - \delta_1)}{L_2} - \frac{\delta_1}{L_1}\right]} \qquad (5.23)$$

This is the full three-moment equation; it can be greatly simplified **if the beam is uniform,** i.e. $I_1 = I_2 = I$, as follows:

$$-M_1 L_1 - 2M_2[L_1 + L_2] - M_3 L_2 = 6\left[\frac{A_1\bar{x}_1}{L_1} + \frac{A_2\bar{x}_2}{L_2}\right] + 6EI\left[\frac{(\delta_2 - \delta_1)}{L_2} - \frac{\delta_1}{L_1}\right]$$

If the supports are on the same level, i.e. $\delta_1 = \delta_2 = 0$,

$$\boldsymbol{-M_1 L_1 - 2M_2[L_1 + L_2] - M_3 L_2 = 6\left[\frac{A_1\bar{x}_1}{L_1} + \frac{A_2\bar{x}_2}{L_2}\right]} \qquad (5.24)$$

This is the form in which Clapeyron's three-moment equation is normally used.

The following standard results for $\dfrac{6A\bar{x}}{L}$ are very useful:

(1) *Concentrated loads* (Fig. 5.29)

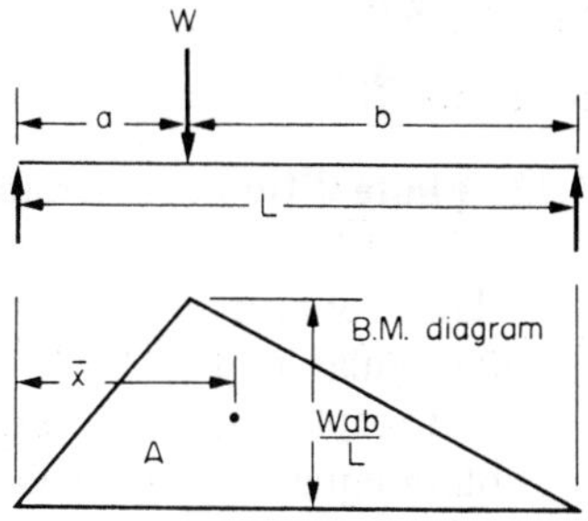

Fig. 5.29.

$$\frac{6A\bar{x}}{L} = \frac{6}{L}\left[\frac{Wab}{L}\times\frac{a}{2}\times\frac{2a}{3}+\frac{Wab}{L}\times\frac{b}{2}\left(a+\frac{b}{3}\right)\right]$$

$$= \frac{6Wab}{L^2}\left[\frac{a^2}{3}+\frac{ab}{2}+\frac{b^2}{6}\right]$$

$$= \frac{Wab}{L^2}[2a^2+3ab+b^2] = \frac{Wab}{L^2}(2a+b)(a+b)$$

$$= \frac{Wab}{L}(2a+b) \qquad (5.25)$$

But $\qquad b = L-a$

$\therefore$
$$\frac{6A\bar{x}}{L} = \frac{Wa}{L}(L-a)(2a+L-a)$$

$$= \frac{Wa}{L}(L^2-a^2) \qquad (5.26)$$

(2) *Uniformly distributed loads* (Fig. 5.30)

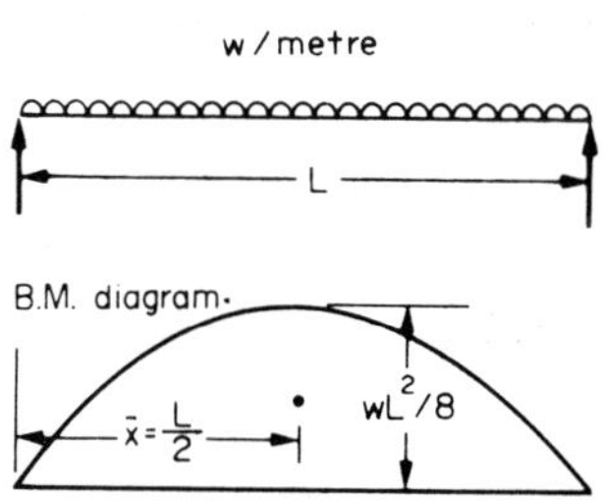

Fig. 5.30.

Here the B.M. diagram is a parabola for which

$$\text{area} = \tfrac{2}{3}\,\text{base} \times \text{height}$$

$$\therefore \qquad \frac{6A\bar{x}}{L} = \frac{6}{L} \times \frac{2}{3} \times L \times \frac{wL^2}{8} \times \frac{L}{2}$$

$$= \frac{wL^3}{4} \qquad (5.27)$$

5.12. Finite difference method

A numerical method for the calculation of beam deflections which is particularly useful for non-prismatic beams or for cases of irregular loading is the so-called *finite difference* method.

The basic principle of the method is to replace the standard differential equation (5.2) by its finite difference approximation, obtain equations for deflections in terms of moments at various points along the beam and solve these simultaneously to yield the required deflection values.

Consider, therefore, Fig. 5.31 which shows part of a deflected beam with the x axis divided into a series of equally spaced intervals. By convention, the ordinates are numbered with respect to the central ordinate B.

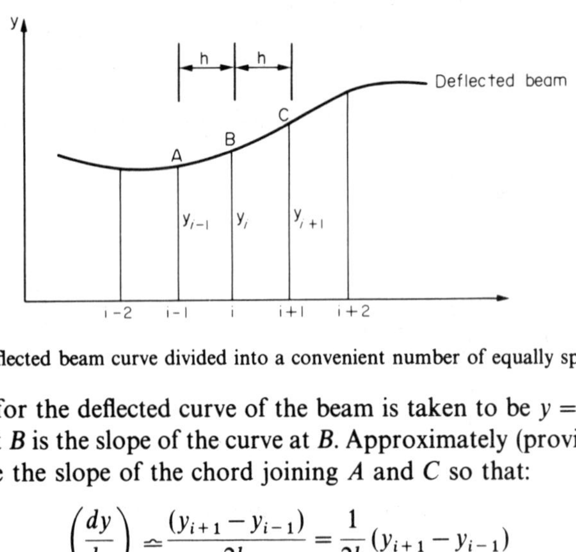

Fig. 5.31. Deflected beam curve divided into a convenient number of equally spaced intervals.

If the equation for the deflected curve of the beam is taken to be $y = f(x)$ then the first derivative dy/dx at B is the slope of the curve at B. Approximately (provided h is small) this can be taken to be the slope of the chord joining A and C so that:

$$\left(\frac{dy}{dx}\right)_i \simeq \frac{(y_{i+1} - y_{i-1})}{2h} = \frac{1}{2h}(y_{i+1} - y_{i-1}) \qquad (5.28)$$

The rate of change of the first derivative, i.e. the rate of change of the slope $\left(= \dfrac{d^2y}{dx^2}\right)$ is given in the same way approximately as the slope to the right of i minus the slope to the left of i divided by the interval between them.

Thus:

$$\left(\frac{d^2y}{dx^2}\right)_i = \frac{\dfrac{(y_{i+1} - y_i)}{h} - \dfrac{(y_i - y_{i-1})}{h}}{h} = \frac{1}{h^2}(y_{i+1} - 2y_i + y_{i-1}) \qquad (5.29)$$

Equations 5.28 and 5.29 are the *finite difference approximations* of the standard beam deflection differential equations and, because they are written in terms of ordinates on either side of the central point i, they are known as *central differences.* Alternative expressions which can be formed to contain only ordinates at, or to the right of i, or ordinates at, or to the left of i are known as forward and backward differences, respectively but these will not be considered here.

Now from eqn. (5.2)

$$M = EI\frac{d^2y}{dx^2}$$

∴ At position i, combining eqn. (5.2) and (5.29).

$$\left(\frac{M}{EI}_i\right) = \frac{1}{h^2}(y_{i+1} - 2y_i + y_{i-1}) \qquad (5.30)$$

A solution for any of the deflection (y) values can then be obtained by applying the finite difference equation at a series of points along the beam and solving the resulting simultaneous equations – see Example 5.8.

The higher the number of points selected the greater the accuracy of solution but the more the number of equations which are required to be solved. The method thus lends itself to computer-assisted evaluation.

In addition to the solution of statically determinate beam problems of the type treated in Example 5.8 the method is also applicable to the analysis of statically indeterminate beams, i.e. those beam loading conditions with unknown (or redundant) quantities such as prop loads or fixing moments – see Example 5.9.

The method is similar in that the bending moment is written in terms of the applied loads and the redundant quantities and equated to the finite difference equation at selected points. Since each redundancy is usually associated with a known (or assumed) condition of slope or deflection, e.g. zero deflection at a propped support, there will always be sufficient equations to allow solution of the unknowns.

The principal advantages of the finite difference method are thus:

(a) that it can be applied to statically determinate or indeterminate beams,
(b) that it can be used for non-prismatic beams,
(c) that it is amenable to computer solutions.

5.13. Deflections due to temperature effects

It has been shown in §2.3 that a uniform temperature increase t on an unconstrained bar of length L will produce an increase in length

$$\Delta L = \alpha L t$$

where α is the coefficient of linear expansion of the material of the bar. Provided that the bar remains unconstrained, i.e. is free to expand, no stresses will result.

Similarly, in the case of a beam supported in such a way that longitudinal expansion can occur freely, no stresses are set up and there will be no tendency for the beam to bend. If, however, the beam is constrained then stresses will result, their values being calculated using the procedure of §2.3 provided that the temperature change is uniform across the whole beam section.

If the temperature is not constant across the beam then, again, stresses and deflections will result and the following procedure must be adopted:

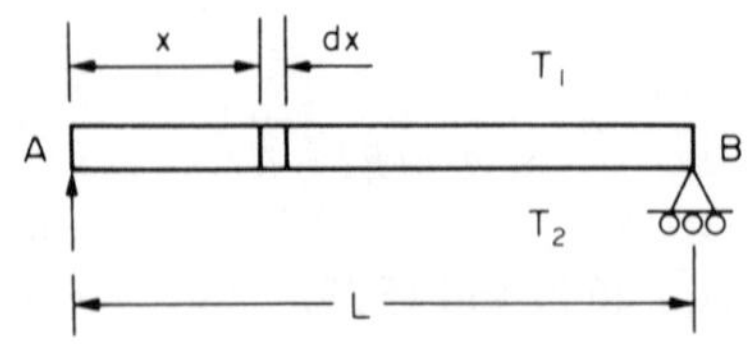

Fig. 5.32(a). Beam initially straight before application of temperature T_1 on the top surface and T_2 on the lower surface. (Beam supported on rollers at B to allow "free" lateral expansion).

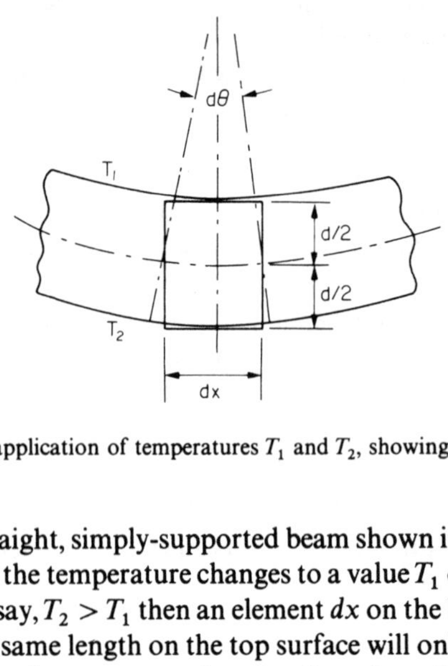

Fig. 5.32(b). Beam after application of temperatures T_1 and T_2, showing distortions of element dx.

Consider the initially straight, simply-supported beam shown in Fig. 5.32(a) with an initial uniform temperature T_0. If the temperature changes to a value T_1 on the upper surface and T_2 on the lower surface with, say, $T_2 > T_1$ then an element dx on the bottom surface will expand to $\alpha(T_2 - T_0).dx$ whilst the same length on the top surface will only expand to $\alpha(T_1 - T_0).dx$. As a result the beam will bend to accommodate the distortion of the element dx, the sides of the element rotating relative to one another by the angle $d\theta$, as shown in Fig. 5.32(b). For a depth of beam d:

$$d.d\theta = \alpha(T_2 - T_0)dx - \alpha(T_1 - T_0)dx$$

or

$$\frac{d\theta}{dx} = \frac{\alpha(T_2 - T_1)}{d} \tag{5.31}$$

The differential equation gives the rate of change of slope of the beam and, since $\theta = dy/dx$,

then

$$\frac{d}{dx}\left(\frac{dy}{dx}\right) = \frac{d^2y}{dx^2} = \frac{\alpha(T_2 - T_1)}{d}$$

Thus the standard differential equation for bending of the beam due to temperature gradient across the beam section is:

$$\frac{d^2y}{dx^2} = \frac{\alpha(T_2 - T_1)}{d} \tag{5.32}$$

This is directly analogous to the standard deflection equation $\frac{d^2y}{dx^2} = \frac{M}{EI}$ so that integration of this equation in exactly the same way as previously for bending moments allows a solution for slopes and deflections produced by the thermal effects.

NB. If the temperature gradient across the beam section is linear, the average temperature $\frac{1}{2}(T_1 + T_2)$ will occur at the mid-height position and, in addition to the bending, the beam will change in overall length by an amount $\alpha L[\frac{1}{2}(T_1 + T_2) - T_0]$ in the absence of any constraint.

Application to cantilevers

Consider the cantilever shown in Fig. 5.33 subjected to temperature T_1 on the top surface and T_2 on the lower surface. In the absence of external loads, and because the cantilever is free to bend, there will be no moment or reaction set up at the built-in end.

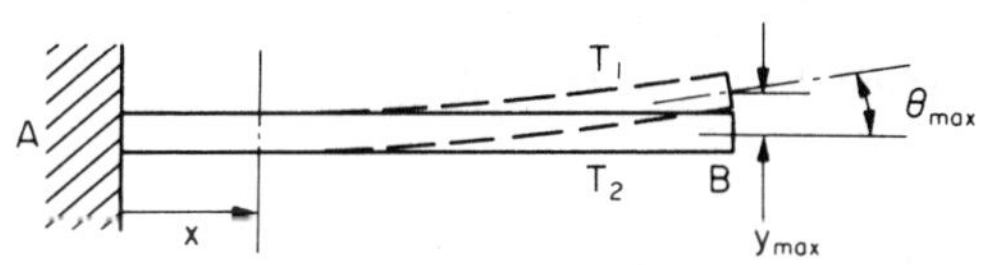

Fig. 5.33. Cantilever with temperature T_1 on the upper surface, T_2 on the lower surface ($T_2 > T_1$).

Applying the differential equation (5.32) we have:

$$\frac{d^2y}{dx^2} = \frac{\alpha(T_2 - T_1)}{d}.$$

Integrating:

$$\frac{dy}{dx} = \frac{\alpha(T_2 - T_1)}{d}x + C_1$$

But at $x = 0$, $\frac{dy}{dx} = 0$, $\therefore$ $C_1 = 0$ and:

$$\frac{dy}{dx} = \frac{\alpha(T_2 - T_1)}{d}x = \theta$$

$\therefore$ The slope at the end of the cantilever is:

$$\theta_{max} = \frac{\alpha(T_2 - T_1)}{d}L. \tag{5.33}$$

Integrating again to find deflections:

$$y = \frac{\alpha(T_2 - T_1)}{d}\frac{x^2}{2} + C_2$$

and, since $y = 0$ at $x = 0$, then $C_2 = 0$, and:

$$y = \frac{\alpha(T_2 - T_1)}{2d}x^2$$

At the end of the cantilever, therefore, the deflection is:

$$y_{max} = \frac{\alpha(T_2 - T_1)}{2d}L^2 \tag{5.34}$$

Application to built-in beams

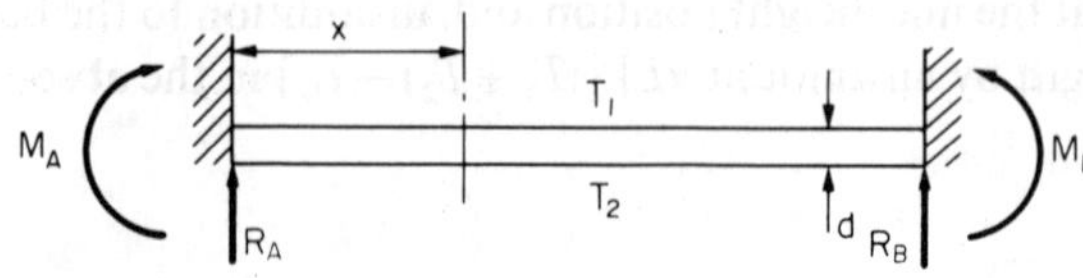

Fig. 5.34. Built-in beam subjected to thermal gradient with temperature T_1 on the upper surface, T_2 on the lower surface.

Consider the built-in beam shown in Fig. 5.34. Using the principle of superposition the differential equation for the beam is given by the combination of the equations for applied bending moment and thermal effects.

$$\text{For bending } EI\frac{d^2y}{dx^2} = M_A + R_A x.$$

$$\text{For thermal effects } \frac{d^2y}{dx^2} = \frac{\alpha(T_2 - T_1)}{d}$$

$$\therefore \quad EI\frac{d^2y}{dx^2} = EI\frac{\alpha(T_2 - T_1)}{d}$$

$\therefore$ The combined differential equation is:

$$EI\frac{d^2y}{dx^2} = M_A + R_A x + EI\frac{\alpha(T_2 - T_1)}{d}.$$

However, in the absence of applied loads and from symmetry of the beam:

$$R_A = R_B = 0,$$

and

$$M_A = M_B = M.$$

$$\therefore \quad EI\frac{d^2y}{dx^2} = M + EI\frac{\alpha(T_2 - T_1)}{d}.$$

Integrating:

$$EI\frac{dy}{dx} = M_x + EI\frac{\alpha(T_2 - T_1)}{d}x + C_1$$

Now at $x = 0, \frac{dy}{dx} = 0 \quad \therefore \quad C_1 = 0,$

and at $x = L \quad \frac{dy}{dx} = 0 \quad \therefore \quad M = -EI\frac{\alpha(T_2 - T_1)}{d}$ (5.35)

Integrating again to find the deflection equation we have:

$$EIy = M.\frac{x^2}{2} + EI\frac{\alpha(T_2 - T_1)}{d}.\frac{x^2}{2} + C_2$$

When $x = 0$, $y = 0$ $\therefore$ $C_2 = 0$,

and, since $M = -EI\dfrac{\alpha(T_2 - T_1)}{d}$ then $y = 0$ for all values of x.

Thus a rather surprising result is obtained whereby the beam will remain horizontal in the presence of a thermal gradient. It will, however, be subject to residual stresses arising from the constraint on overall expansion of the beam under the average temperature $\frac{1}{2}(T_1 + T_2)$. i.e. from §2.3

$$\text{residual stress} = E\alpha[\tfrac{1}{2}(T_1 + T_2)]$$

$$= \tfrac{1}{2}E\alpha(T_1 + T_2). \qquad (5.36)$$

Examples

Example 5.1

(a) A uniform cantilever is 4 m long and carries a concentrated load of 40 kN at a point 3 m from the support. Determine the vertical deflection of the free end of the cantilever if $EI = 65$ MN m^2.

(b) How would this value change if the same total load were applied but uniformly distributed over the portion of the cantilever 3 m from the support?

Solution

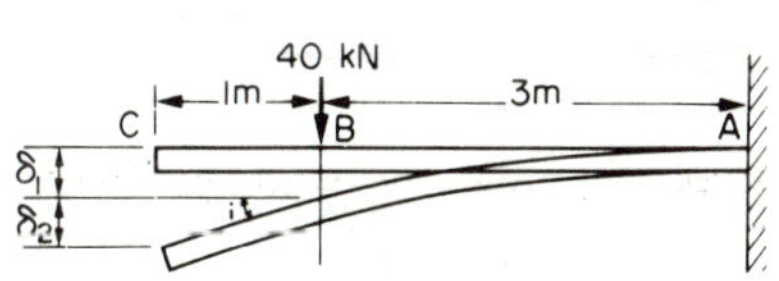

Fig. 5.35.

(a) With the load in the position shown in Fig. 5.35 the cantilever is effectively only 3 m long, the remaining 1 m being unloaded and therefore not bending. Thus, the standard equations for slope and deflections apply between points A and B only.

$$\text{Vertical deflection of } B = -\frac{WL^3}{3EI} = -\frac{40 \times 10^3 \times 3^3}{3 \times 65 \times 10^6} = -5.538 \times 10^{-3}\text{ m} = \delta_1$$

$$\text{Slope at } B = \frac{WL^2}{2EI} = \frac{40 \times 10^3 \times 3^2}{2 \times 65 \times 10^6} = 2.769 \times 10^{-3}\text{ rad} = i$$

Now BC remains straight since it is not subject to bending.

$$\therefore \quad \delta_2 = -iL = -2.769 \times 10^{-3} \times 1 = -2.769 \times 10^{-3}\ \text{m}$$

$$\therefore \quad \text{vertical deflection of } C = \delta_1 + \delta_2 = -(5.538 + 2.769)10^{-3} = \mathbf{-8.31\ mm}$$

The negative sign indicates a deflection in the negative y direction, i.e. downwards.
(b) With the load uniformly distributed,

$$w = \frac{40 \times 10^3}{3} = 13.33 \times 10^3\ \text{N/m}$$

Again using standard equations listed in the summary

$$\delta_1' = -\frac{wL^4}{8EI} = \frac{13.33 \times 10^3 \times 3^4}{8 \times 65 \times 10^6} = -2.076 \times 10^{-3}\ \text{m}$$

$$\text{and slope } i = \frac{wL^3}{6EI} = \frac{13.33 \times 10^3 \times 3^3}{6 \times 65 \times 10^6} = 0.923 \times 10^3\ \text{rad}$$

$$\therefore \quad \delta_2' = -0.923 \times 10^{-3} \times 1 = 0.923 \times 10^{-3}\ \text{m}$$

$$\therefore \quad \text{vertical deflection of } C = \delta_1' + \delta_2' = -(2.076 + 0.923)10^{-3} = \mathbf{-3\ mm}$$

There is thus a considerable (63.9%) reduction in the end deflection when the load is uniformly distributed.

Example 5.2

Determine the slope and deflection under the 50 kN load for the beam loading system shown in Fig. 5.36. Find also the position and magnitude of the maximum deflection. $E = 200\ \text{GN/m}^2$; $I = 83 \times 10^{-6}\ \text{m}^4$.

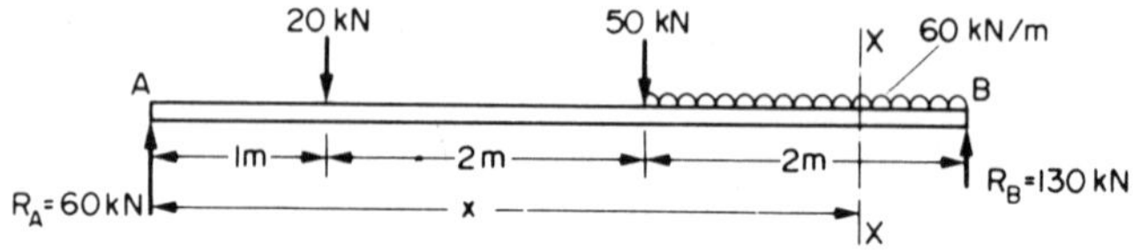

Fig. 5.36.

Solution

Taking moments about either end of the beam gives

$$R_A = 60\ \text{kN} \quad \text{and} \quad R_B = 130\ \text{kN}$$

Applying Macaulay's method,

$$BM_{XX} = \frac{EI}{10^3}\frac{d^2y}{dx^2} = 60x - 20[(x-1)] - 50[(x-3)] - 60\left[\frac{(x-3)^2}{2}\right] \quad (1)$$

The load unit of kilonewton is accounted for by dividing the left-hand side of (1) by 10^3 and the u.d.l. term is obtained by treating the u.d.l. to the left of XX as a concentrated load of $60(x-3)$ acting at its mid-point of $(x-3)/2$ from XX.

Integrating (1),

$$\frac{EI}{10^3}\frac{dy}{dx} = \frac{60x^2}{2} - 20\left[\frac{(x-1)^2}{2}\right] - 50\left[\frac{(x-3)^2}{2}\right] - 60\left[\frac{(x-3)^3}{6}\right] + A \qquad (2)$$

and

$$\frac{EI}{10^3}y = \frac{60x^3}{6} - 20\left[\frac{(x-1)^3}{6}\right] - 50\left[\frac{(x-3)^3}{6}\right] - 60\left[\frac{(x-3)^4}{24}\right] + Ax + B \qquad (3)$$

Now when $x = 0$, $y = 0$ $\therefore B = 0$
when $x = 5$, $y = 0$ $\therefore$ substituting in (3)

$$0 = \frac{60 \times 5^3}{6} - \frac{20 \times 4^3}{6} - \frac{50 \times 2^3}{6} - \frac{60 \times 2^4}{24} + 5A$$

$$0 = 1250 - 213.3 - 66.7 - 40 + 5A$$

$$\therefore \qquad 5A = -930 \qquad A = -186$$

Substituting in (2),

$$\frac{EI}{10^3}\frac{dy}{dx} = \frac{60x^2}{2} - 20\left[\frac{(x-1)^2}{2}\right] - 50\left[\frac{(x-3)^2}{2}\right] - 60\left[\frac{(x-3)^3}{6}\right] - 186$$

$\therefore$ slope at $x = 3$ m (i.e. under the 50 kN load)

$$= \frac{10^3}{EI}\left[\frac{60 \times 3^2}{2} - \frac{20 \times 2^2}{2} - 186\right] = \frac{10^3 \times 44}{200 \times 10^9 \times 83 \times 10^{-6}}$$

$$= \mathbf{0.00265\ rad}$$

And, substituting in (3),

$$\frac{EI}{10^3}y = \frac{60 \times 3^3}{6} - 20\left[\frac{(x-1)^3}{6}\right] - 50\left[\frac{(x-3)^3}{6}\right] - 60\left[\frac{(x-3)^4}{24}\right] - 186x$$

$\therefore$ deflection at $x = 3$ m

$$= \frac{10^3}{EI}\left[\frac{60 \times 3^3}{6} - \frac{20 \times 2^3}{6} - 186 \times 3\right]$$

$$= \frac{10^3}{EI}[270 - 26.67 - 558] = -\frac{10^3 \times 314.7}{200 \times 10^9 \times 83 \times 10^{-6}}$$

$$= -0.01896\,\text{m} = \mathbf{-19\ mm}$$

In order to determine the maximum deflection, its position must first be estimated. In this case, as the slope is positive under the 50 kN load it is reasonable to assume that the maximum deflection point will occur somewhere between the 20 kN and 50 kN loads. For this position, from (2),

$$\frac{EI}{10^3}\frac{dy}{dx} = \frac{60x^2}{2} - 20\frac{(x-1)^2}{2} - 186$$

$$= 30x^2 - 10x^2 + 20x - 10 - 186$$

$$= 20x^2 + 20x - 196$$

But, where the deflection is a maximum, the slope is zero.

$$\therefore \qquad 0 = 20x^2 + 20x - 196$$

$$\therefore \qquad x = \frac{-20 \pm (400 + 15680)^{1/2}}{40} = \frac{-20 \pm 126.8}{40}$$

i.e. $\qquad x = \mathbf{2.67\,m}$

Then, from (3), the maximum deflection is given by

$$\delta_{max} = -\frac{10^3}{EI}\left[\frac{60 \times 2.67^3}{6} - \frac{20 \times 1.67^3}{6} - 186 \times 2.67\right]$$

$$= -\frac{10^3 \times 321.78}{200 \times 10^9 \times 83 \times 10^{-6}} = -0.0194 = -\mathbf{19.4\,mm}$$

In loading situations where this point lies within the portion of a beam covered by a uniformly distributed load the above procedure is cumbersome since it involves the solution of a cubic equation to determine x.

As an alternative procedure it is possible to obtain a reasonable estimate of the position of zero slope, and hence maximum deflection, by sketching the slope diagram, commencing with the slope at either side of the estimated maximum deflection position; slopes will then be respectively positive and negative and the point of zero slope thus may be estimated. Since the slope diagram is generally a curve, the accuracy of the estimate is improved as the points chosen approach the point of maximum deflection.

As an example of this procedure we may re-solve the final part of the question.

Thus, selecting the initial two points as $x = 2$ and $x = 3$,
when $x = 2$,

$$\frac{EI}{10^3}\frac{dy}{dx} = \frac{60 \times 2^2}{2} - \frac{20(1^2)}{2} - 186 = -76$$

when $x = 3$,

$$\frac{EI}{10^3}\frac{dy}{dx} = \frac{60 \times 3^2}{2} - \frac{20(2^2)}{2} - 186 = +44$$

Figure 5.37 then gives a first estimate of the zero slope (maximum deflection) position as $x = 2.63$ on the basis of a straight line between the above-determined values. Recognising the inaccuracy of this assumption, however, it appears reasonable that the required position can

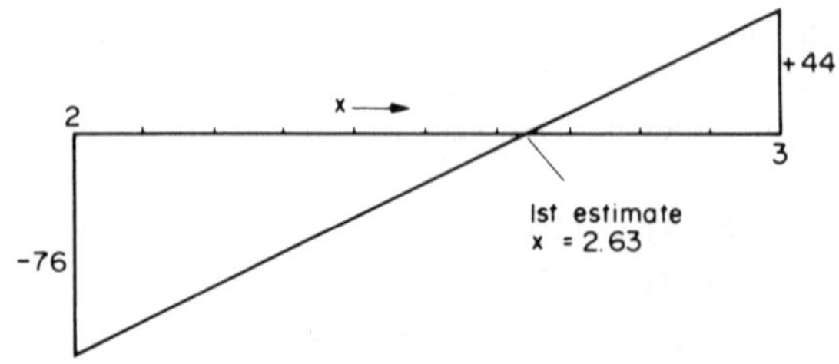

Fig. 5.37.

be more closely estimated as between $x = 2.5$ and $x = 2.7$. Thus, refining the process further, when $x = 2.5$,

$$\frac{EI}{10^3}\frac{dy}{dx} = \frac{60 \times 2.5^2}{2} - \frac{20 \times 1.5^2}{2} - 186 = -21$$

when $x = 2.7$,

$$\frac{EI}{10^3}\frac{dy}{dx} = \frac{60 \times 2.7^2}{2} - \frac{20 \times 1.7^2}{2} - 186 = +3.8$$

Figure 5.38 then gives the improved estimate of

$$x = \mathbf{2.669}$$

which is effectively the same value as that obtained previously.

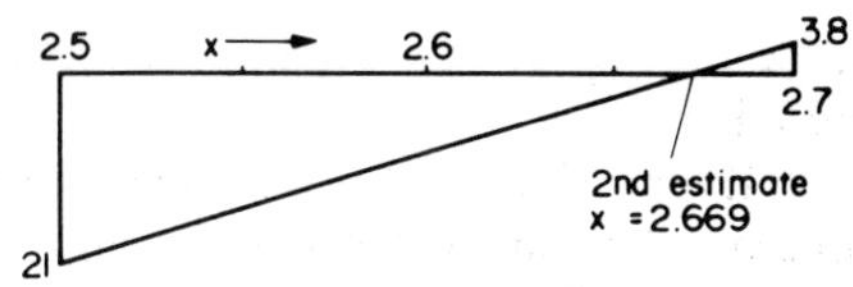

Fig. 5.38.

Example 5.3

Determine the deflection at a point 1 m from the left-hand end of the beam loaded as shown in Fig. 5.39a using Macaulay's method. $EI = 0.65\ \mathrm{MN\,m^2}$.

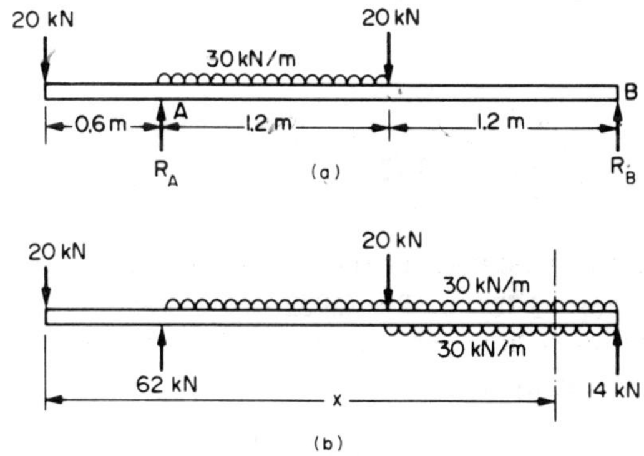

Fig. 5.39.

Solution

Taking moments about B

$$(3 \times 20) + (30 \times 1.2 \times 1.8) + (1.2 \times 20) = 2.4R_A$$

$$\therefore \quad R_A = 62\,\mathrm{kN} \quad \text{and} \quad R_B = 20 + (30 \times 1.2) + 20 - 62 = 14\,\mathrm{kN}$$

Using the modified Macaulay approach for distributed loads over part of a beam introduced in § 5.5 (Fig. 5.39b),

$$M_{xx} = \frac{EI}{10^3}\frac{d^2y}{dx^2} = -20x + 62[(x-0.6)] - 30\left[\frac{(x-0.6)^2}{2}\right] + 30\left[\frac{(x-1.8)^2}{2}\right] - 20[(x-1.8)]$$

$$\frac{EI}{10^3}\frac{dy}{dx} = \frac{-20x^2}{2} + 62\left[\frac{(x-0.6)^2}{2}\right] - 30\left[\frac{(x-0.6)^3}{6}\right] + 30\left[\frac{(x-1.8)^3}{6}\right] - 20\left[\frac{(x-1.8)^2}{2}\right] + A$$

$$\frac{EI}{10^3}y = \frac{-20x^3}{6} + 62\left[\frac{(x-0.6)^3}{6}\right] - 30\left[\frac{(x-0.6)^4}{24}\right] + 30\left[\frac{(x-1.8)^4}{24}\right] - 20\left[\frac{(x-1.8)^3}{6}\right] + Ax + B$$

Now when $x = 0.6$, $y = 0$,

$$\therefore \qquad 0 = -\frac{20 \times 0.6^3}{6} + 0.6A + B$$

$$0.72 = 0.6A + B \tag{1}$$

and when $x = 3$, $y = 0$,

$$\therefore \qquad 0 = -\frac{20 \times 3^3}{6} + \frac{62 \times 2.4^3}{6} - \frac{30 \times 2.4^4}{24} + \frac{30 \times 1.2^4}{24} - \frac{20 \times 1.2^3}{6} + 3A + B$$

$$= -90 + 142.848 - 41.472 + 2.592 - 5.76 + 3A + B$$

$$-8.208 = 3A + B \tag{2}$$

(2) − (1)

$$-8.928 = 2.4A \qquad \therefore \quad A = -3.72$$

Substituting in (1),

$$B = 0.72 - 0.6(-3.72) \qquad B = 2.952$$

Substituting into the Macaulay deflection equation,

$$\frac{EI}{10^3}y = -\frac{20x^3}{6} + 62\left[\frac{(x-0.6)^3}{6}\right] - 30\left[\frac{(x-0.6)^4}{24}\right] + 30\left[\frac{(x-1.8)^4}{24}\right] - 20\left[\frac{(x-1.8)^3}{6}\right] - 3.72x + 2.952$$

At $\qquad x = 1$

$$y = \frac{10^3}{EI}\left[-\frac{20}{6} + \frac{62}{6} \times 0.4^3 - \frac{30 \times 0.4^4}{24} - 3.72 \times 1 + 2.952\right]$$

$$= \frac{10^3}{EI}[-3.33 + 0.661 - 0.032 - 3.72 + 2.952]$$

$$= -\frac{10^3 \times 3.472}{0.65 \times 10^6} = -5.34 \times 10^{-3}\,\text{m} = \mathbf{-5.34\,mm}$$

The beam therefore is deflected *downwards* at the given position.

Example 5.4

Calculate the slope and deflection of the beam loaded as shown in Fig. 5.40 at a point 1.6 m from the left-hand end. $EI = 1.4\,\text{MN}\,\text{m}^2$.

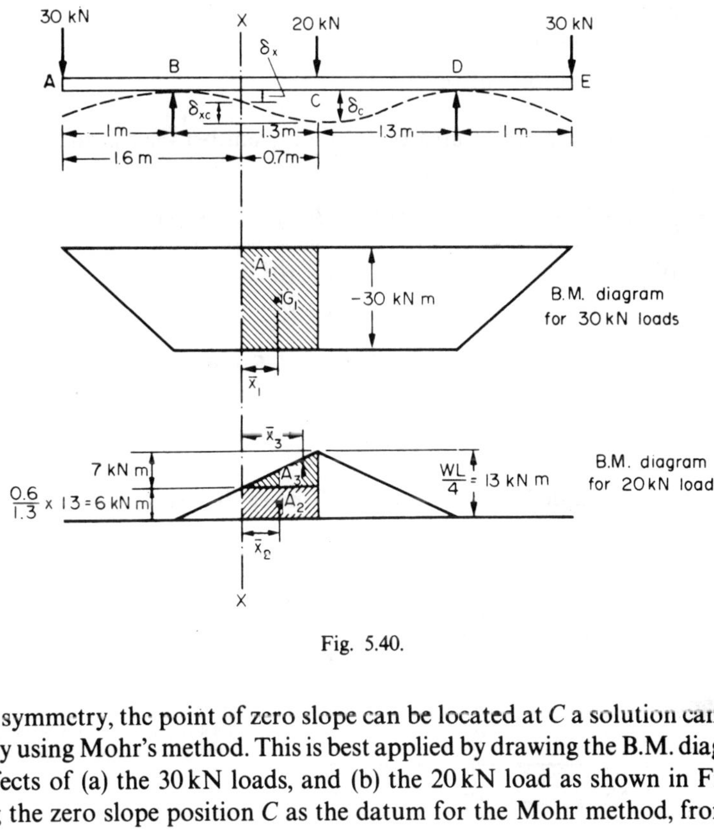

Fig. 5.40.

Solution

Since, by symmetry, the point of zero slope can be located at C a solution can be obtained conveniently using Mohr's method. This is best applied by drawing the B.M. diagrams for the separate effects of (a) the 30 kN loads, and (b) the 20 kN load as shown in Fig. 5.40. Thus, using the zero slope position C as the datum for the Mohr method, from eqn. (5.20)

$$\text{slope at } X = \frac{1}{EI}[\text{area of B.M. diagram between } X \text{ and } C]$$

$$= \frac{10^3}{EI}[(-30 \times 0.7) + (6 \times 0.7) + (\tfrac{1}{2} \times 7 \times 0.7)]$$

$$= \frac{10^3}{EI}[-21 + 4.2 + 2.45] = -\frac{14.35 \times 10^3}{1.4 \times 10^6}$$

$$= -10.25 \times 10^{-3}\,\text{rad}$$

and from eqn. (5.21)

deflection at X relative to the tangent at C

$$= \frac{1}{EI}[\text{first moment of area of B.M. diagram between } X \text{ and } C \text{ about } X]$$

$$\delta_{XC} = \frac{10^3}{EI}[\underset{A_1\bar{x}_1}{(-30 \times 0.7 \times 0.35)} + \underset{A_2\bar{x}_2}{(6 \times 0.7 \times 0.35)} + \underset{A_3\bar{x}_3}{(7 \times 0.7 \times \tfrac{1}{2} \times \tfrac{2}{3} \times 0.7)}]$$

$$= \frac{10^3}{EI}[-7.35 + 1.47 + 1.143] = -\frac{10^3 \times 4.737}{1.4 \times 10^6}$$

$$= -3.38 \times 10^{-3}\,\text{m} = -3.38\,\text{mm}$$

This must now be subtracted from the deflection of C relative to the support B to obtain the actual deflection at X.

Now deflection of C relative to B

$$= \text{deflection of } B \text{ relative to } C$$

$$= \frac{1}{EI}[\text{first moment of area of B.M. diagram between } B \text{ and } C \text{ about } B]$$

$$= \frac{10^3}{EI}[(-30 \times 1.3 \times 0.65) + (13 \times 1.3 \times \tfrac{1}{2} \times 1.3 \times \tfrac{2}{3})]$$

$$= \frac{10^3}{EI}[-25.35 + 7.323] = -\frac{18.027 \times 10^3}{1.4 \times 10^6} = -12.88 \times 10^{-3} = -12.88\,\text{mm}$$

$\therefore$ required deflection of $X = -(12.88 - 3.38) = \mathbf{-9.5\,mm}$

Example 5.5

(a) Find the slope and deflection at the tip of the cantilever shown in Fig. 5.41.

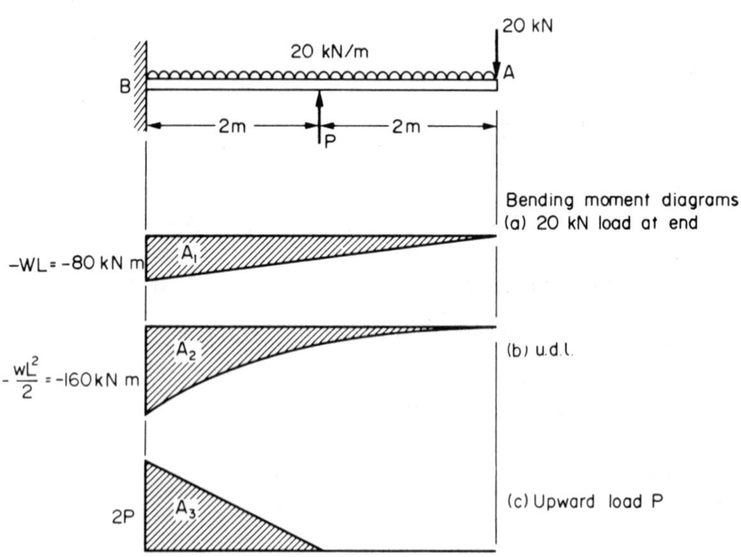

Fig. 5.41.

(b) What load P must be applied upwards at mid-span to reduce the deflection by half? $EI = 20\,\text{MN}\,\text{m}^2$.

Solution

Here again the best approach is to draw separate B.M. diagrams for the concentrated and uniformly distributed loads. Then, since B is a point of zero slope, the Mohr method may be applied.

(a) Slope at $A = \dfrac{1}{EI}$[area of B.M. diagram between A and B]

$$= \frac{1}{EI}[A_1 + A_2] = \frac{10^3}{EI}[\{\tfrac{1}{2} \times 4 \times (-80)\} + \{\tfrac{1}{3} \times 4 \times (-160)\}]$$

$$= \frac{10^3}{EI}[-160 - 213.3] = \frac{373.3 \times 10^3}{20 \times 10^6}$$

$$= \mathbf{18.67 \times 10^{-3}\ rad}$$

Deflection of $A = \dfrac{1}{EI}$[first moment of area of B.M. diagram between A and B about A]

$$= \frac{10^3}{EI}\left[\left(\frac{-80 \times 4 \times \tfrac{2}{3} \times 4}{2}\right) + \left(\frac{-160 \times 4 \times \tfrac{3}{4} \times 4}{3}\right)\right]$$

$$= \frac{-10^3}{EI}[426.6 + 640] = -\frac{1066.6 \times 10^3}{20 \times 10^6} = -53.3 \times 10^{-3}\,\text{m} = \mathbf{-53\,mm}$$

(b) When an extra load P is applied upwards at mid-span its effect on the deflection is required to be $\tfrac{1}{2} \times 53.3 = 26.67\,\text{mm}$. Thus

$$26.67 \times 10^{-3} = \frac{1}{EI}[\text{first moment of area of B.M. diagram for } P \text{ about } A]$$

$$= \frac{10^3}{EI}[\tfrac{1}{2} \times 2P \times 2(2 + \tfrac{2}{3} \times 2)]$$

$$\therefore \qquad P = \frac{26.67 \times 20 \times 10^6}{10^3 \times 6.66} = \mathbf{80 \times 10^3\ N}$$

The required load at mid-span is 80 kN.

Example 5.6

The uniform beam of Fig. 5.42 carries the loads indicated. Determine the B.M. at B and hence draw the S.F. and B.M. diagrams for the beam.

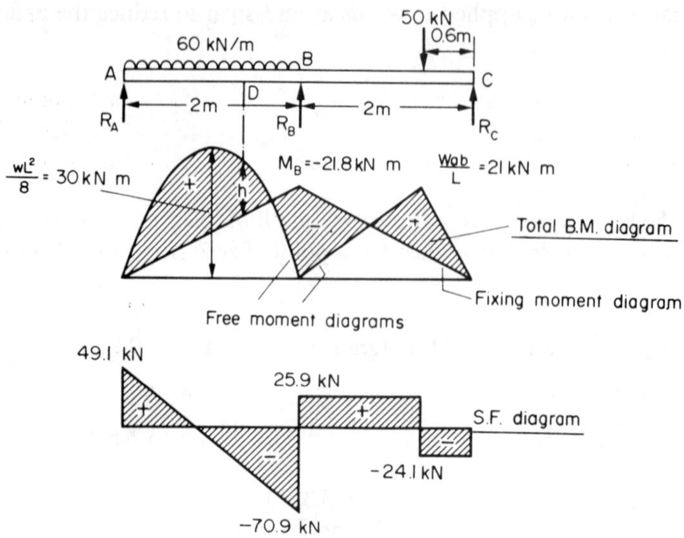

Fig. 5.42.

Solution

Applying the three-moment equation (5.24) to the beam we have,

$$-M_A L_1 - 2M_B(L_1 + L_2) - M_C L_2 = \frac{wL_1^3}{4} + \frac{Wa}{L_2}(L_2^2 - a^2)$$

(Note that the dimension a is always to the "outside" support of the particular span carrying the concentrated load.)

Now with A and C simply supported

$$M_A = M_C = 0$$

$$\therefore \quad -2M_B(2+2) = \frac{60 \times 10^3 \times 2^3}{4} + \frac{50 \times 10^3 \times 0.6}{2}(2^2 - 0.6^2)$$

$$-8M_B = (120 + 54.6)10^3 = 174.6 \times 10^3$$

$$M_B = -\mathbf{21.8\ kN\ m}$$

With the normal B.M. sign convention the B.M. at B is therefore -21.8 kN m.

Taking moments about B (forces to left),

$$2R_A - (60 \times 10^3 \times 2 \times 1) = -21.8 \times 10^3$$

$$R_A = \tfrac{1}{2}(-21.8 + 120)10^3 = \mathbf{49.1\ kN}$$

Taking moments about B (forces to right),

$$2R_C - (50 \times 10^3 \times 1.4) = -21.8 \times 10^3$$

$$R_C = \tfrac{1}{2}(-21.8 + 70) = \mathbf{24.1\ kN}$$

and, since the total load $= R_A + R_B + R_C = 50 + (60 \times 2) = 170\,\text{kN}$

$$\therefore \quad R_B = 170 - 49.1 - 24.1 = \mathbf{96.8\,kN}$$

The B.M. and S.F. diagrams are then as shown in Fig. 5.42. The fixing moment diagram can be directly subtracted from the free moment diagrams since M_B is negative. The final B.M. diagram is then as shown shaded, values at any particular section being measured from the fixing moment line as datum,

e.g. B.M. at $D = +h$ (to scale)

Example 5.7

A beam *ABCDE* is continuous over four supports and carries the loads shown in Fig. 5.43. Determine the values of the fixing moment at each support and hence draw the S.F. and B.M. diagrams for the beam.

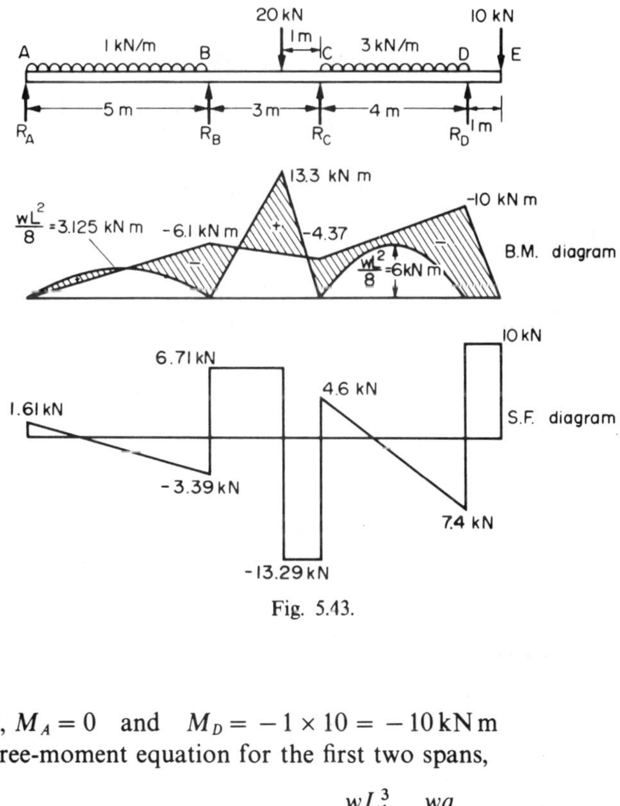

Fig. 5.43.

Solution

By inspection, $M_A = 0$ and $M_D = -1 \times 10 = -10\,\text{kN m}$

Applying the three-moment equation for the first two spans,

$$-M_A L_1 - 2M_B(L_1 + L_2) - M_C L_2 = \frac{wL_1^3}{4} + \frac{wa}{L_2}(L_2^2 - a^2)$$

$$0 - 2M_B(5+3) - 3M_C = \left[\frac{1 \times 5^3}{4} + \frac{20 \times 1}{3}(3^2 - 1^2)\right]10^3$$

$$-16M_B - 3M_C = (31.25 + 53.33)10^3$$

$$-16M_B - 3M_C = 84.58 \times 10^3 \tag{1}$$

and, for the second and third spans,

$$-M_BL_2 - 2M_C(L_2+L_3) - M_DL_3 = \frac{wa}{L_2}(L_2^2 - a^2) + \frac{wL_3^3}{4}$$

$$-3M_B - 2M_C(3+4) - (-10\times10^3)4 = \left[\frac{20\times2}{3}(3^2-2^2) + \frac{(3\times4^3)}{4}\right]10^3$$

$$-3M_B - 14M_C + (40\times10^3) = (66.67+48)10^3$$

$$-3M_B - 14M_C = 74.67\times10^3 \qquad (2)$$

$$(2)\times16/3 \qquad -16M_B - 74.67M_C = 398.24\times10^3 \qquad (3)$$

$$(3)-(1) \qquad -71.67M_C = 313.66\times10^3$$

$$M_C = \mathbf{-4.37\times10^3\ Nm}$$

Substituting in (1),

$$-16M_B - 3(-4.37\times10^3) = 84.58\times10^3$$

$$M_B = -\frac{(84.58-13.11)10^3}{16}$$

$$= \mathbf{-4.47\ kN\ m}$$

Moments about B (to left),

$$R_A\times5 - \left(\frac{1\times10^3}{2}\times5^2\right) = -4.47\times10^3$$

$$5R_A = (-4.47+12.5)10^3$$

$$R_A = \mathbf{1.61\ kN}$$

Moments about C (to left),

$$R_A\times8 - (1\times10^3\times5\times5.5) + (R_B\times3) - (20\times10^3\times1) = -4.37\times10^3$$

$$3R_B = -4.37\times10^3 + 27.5\times10^3 + 20\times10^3 - 8\times1.61\times10^3$$

$$3R_B = 30.3\times10^3$$

$$R_B = \mathbf{10.1\ kN}$$

Moments about C (to right),

$$(-10\times10^3\times5) + 4R_D - (3\times10^3\times4\times2) = -4.37\times10^3$$

$$4R_D = (-4.37+50+24)10^3$$

$$R_D = \mathbf{17.4\ kN}$$

Then, since

$$R_A + R_B + R_C + R_D = 47\ \text{kN}$$

$$1.61 + 10.1 + R_C + 17.4 = 47$$

$$R_C = \mathbf{17.9\ kN}$$

This value should then be checked by taking moments to the right of B,

$$(-10\times10^3\times8)+7R_D+3R_C-(3\times10^3\times4\times5)-(20\times10^3\times2) = -4.47\times10^3$$

$$3R_C = (-4.47+40+60+80-121.8)10^3 = 53.73\times10^3$$

$$R_C = \mathbf{17.9\,kN}$$

The S.F. and B.M. diagrams for the beam are shown in Fig. 5.43.

Example 5.8

Using the finite difference method, determine the central deflection of a simply-supported beam carrying a uniformly distributed load ω over its complete span. The beam can be assumed to have constant flexural rigidity EI throughout.

Solution

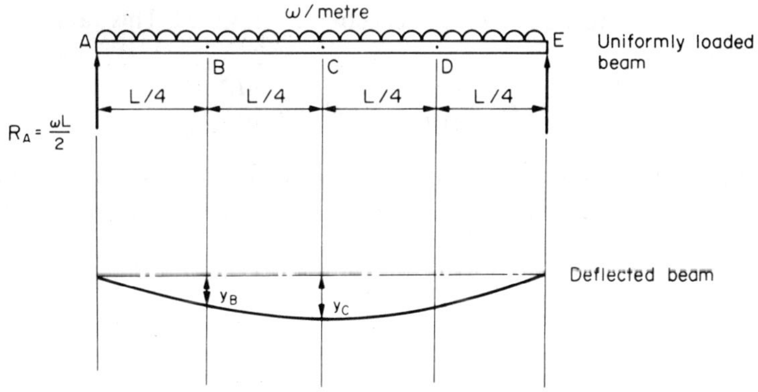

Fig. 5.44.

As a simple demonstration of the finite difference approach, assume that the beam is divided into only four equal segments (thus reducing the accuracy of the solution from that which could be achieved with a greater number of segments).

Then, B.M. at B $= \dfrac{\omega L}{2}\times\dfrac{L}{4}-\dfrac{\omega L}{4}\cdot\dfrac{L}{8}=\dfrac{3\omega L^2}{32}=M_B$

but, from eqn. (5.30):

$$\frac{M_B}{EI}=\frac{1}{h^2}(y_{i+1}-2y_i+y_{i-1})$$

$$\therefore \qquad \frac{I}{EI}\left(\frac{3\omega L^2}{32}\right)=\frac{1}{(L/4)^2}(y_c-2y_B+y_A)$$

and, since $y_A = 0$,

$$\frac{3\omega L^2}{512\,EI}=y_c-2y_B. \qquad (1)$$

Similarly $$\text{B.M. at } C = \frac{\omega L}{2}\cdot\frac{L}{2} - \frac{\omega L}{2}\cdot\frac{L}{4} = \frac{\omega L^2}{8} = M_c.$$

and, from eqn. (5.30)

$$\frac{1}{EI}\left(\frac{\omega L^2}{8}\right) = \frac{1}{(L/4)^2}(y_B - 2y_c + y_D)$$

Now, from symmetry, $y_D = y_B$

$$\therefore \qquad \frac{\omega L^4}{128EI} = 2y_B - 2y_c \qquad (2)$$

Adding eqns. (1) and (2);

$$-y_c = \frac{\omega L^4}{128EI} + \frac{3\omega L^4}{512EI}$$

$$\therefore \qquad y_c = \frac{-7\omega L^4}{512EI} = -0.0137\frac{\omega L^4}{EI}$$

the negative sign indicating a downwards deflection as expected. This value compares with the "exact" value of:

$$y_c = \frac{5\omega L^4}{384EI} = -0.01302\frac{\omega L^4}{EI}$$

a difference of about 5 %. As stated earlier, this comparison could be improved by selecting more segments but, nevertheless, it is remarkably accurate for the very small number of segments chosen.

Example 5.9

The statically indeterminate propped cantilever shown in Fig. 5.45 is propped at B and carries a central load W. It can be assumed to have a constant flexural rigidity EI throughout.

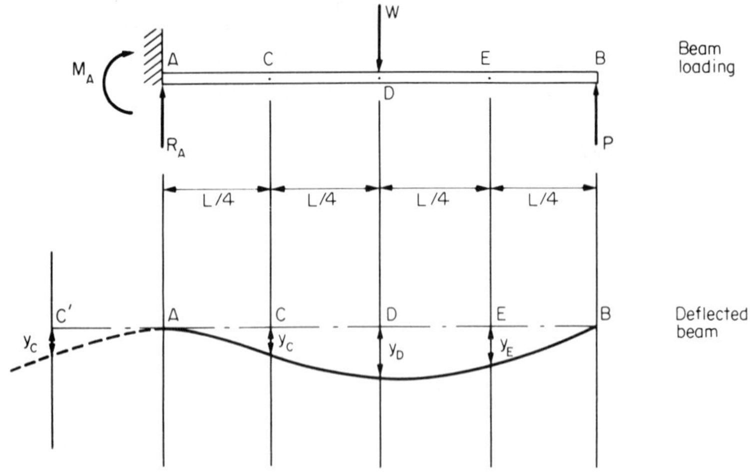

Fig. 5.45.

Determine, using a finite difference approach, the values of the reaction at the prop and the central deflection.

Solution

Whilst at first sight, perhaps, there appears to be a number of redundancies in the cantilever loading condition, in fact the problem reduces to that of a single redundancy, say the unknown prop load P, since with a knowledge of P the other "unknowns" M_A and R_A can be evaluated easily.

Thus, again for simplicity, consider the beam divided into four equal segments giving three unknown deflections y_C, y_D and y_E (assuming zero deflection at the prop B) and one redundancy. Four equations are thus required for solution and these may be obtained by applying the difference equation at four selected points on the beam:
From eqn. (5.30)

$$\text{B.M. at } E = M_E = \frac{PL}{4} = \frac{EI}{(L/4)^2}(y_B - 2y_E + y_D)$$

but $y_B = 0$

$$y_D - 2y_E = \frac{PL^3}{64EI} \qquad (1)$$

$$\text{B.M. at } D = M_D = \frac{PL}{2} = \frac{EI}{(L/4)^2}(y_E - 2y_D + y_C)$$

$\therefore$

$$y_E - 2y_D + y_C = \frac{PL^3}{32EI} \qquad (2)$$

$$\text{B.M. at } C = M_C = P.\frac{3L}{4} - \frac{WL}{4} = \frac{EI}{(L/4)^2}(y_A - 2y_C + y_D)$$

But $y_A = 0$

$\therefore$

$$y_D - 2y_C = \frac{3PL^3}{64} - \frac{WL^3}{64} \qquad (3)$$

At point A it is necessary to introduce the mirror image of the beam giving point C' to the left of A with a deflection $y'_C = y_C$ in order to produce the fourth equation.
Then:

$$\text{B.M. at } A = M_A = PL - \frac{WL}{2} = \frac{EI}{(L/4)^2}(y'_C - 2y_A + y_C)$$

and again since $y_A = 0$

$$y_C = \frac{PL^3}{EI\,32} - \frac{WL^3}{EI\,64} \qquad (4)$$

Solving equations (1) to (4) simultaneously gives the required prop load:

$$P = \frac{7W}{22} = \mathbf{0.318\,W},$$

and the central deflection:

$$y_D = -\frac{17WL^3}{1408EI} = -\mathbf{0.0121}\frac{\mathbf{WL^3}}{\mathbf{EI}}$$

Problems

5.1 (A/B). A beam of length 10 m is symmetrically placed on two supports 7 m apart. The loading is 15 kN/m between the supports and 20 kN at each end. What is the central deflection of the beam? $E = 210\,\text{GN/m}^2$; $I = 200 \times 10^{-6}\,\text{m}^4$. [6.8 mm.]

5.2 (A/B). Derive the expression for the maximum deflection of a simply supported beam of negligible weight carrying a point load at its mid-span position. The distance between the supports is L, the second moment of area of the cross-section is I and the modulus of elasticity of the beam material is E.

The maximum deflection of such a simply supported beam of length 3 m is 4.3 mm when carrying a load of 200 kN at its mid-span position. What would be the deflection at the free end of a cantilever of the same material, length and cross-section if it carries a load of 100 kN at a point 1.3 m from the free end? [13.4 mm.]

5.3 (A/B). A horizontal beam, simply supported at its ends, carries a load which varies uniformly from 15 kN/m at one end to 60 kN/m at the other. Estimate the central deflection if the span is 7 m, the section 450 mm deep and the maximum bending stress 100 MN/m². $E = 210\,\text{GN/m}^2$. [U.L.] [21.9 mm.]

5.4 (A/B). A beam AB, 8 m long, is freely supported at its ends and carries loads of 30 kN and 50 kN at points 1 m and 5 m respectively from A. Find the position and magnitude of the maximum deflection. $E = 210\,\text{GN/m}^2$; $I = 200 \times 10^{-6}\,\text{m}^4$. [14.4 mm.]

5.5 (A/B). A beam 7 m long is simply supported at its ends and loaded as follows: 120 kN at 1 m from one end A, 20 kN at 4 m from A and 60 kN at 5 m from A. Calculate the position and magnitude of the maximum deflection. The second moment of area of the beam section is $400 \times 10^{-6}\,\text{m}^4$ and E for the beam material is 210 GN/m². [9.8 mm at 3.474 m.]

5.6 (B). A beam $ABCD$, 6 m long, is simply-supported at the right-hand end D and at a point B 1 m from the left-hand end A. It carries a vertical load of 10 kN at A, a second concentrated load of 20 kN at C, 3 m from D, and a uniformly distributed load of 10 kN/m between C and D. Determine the position and magnitude of the maximum deflection if $E = 208\,\text{GN/m}^2$ and $I = 35 \times 10^{-6}\,\text{m}^4$. [3.553 m from A, 11.95 mm.]

5.7 (B). A 3 m long cantilever ABC is built-in at A, partially supported at B, 2 m from A, with a force of 10 kN and carries a vertical load of 20 kN at C. A uniformly distributed load of 5 kN/m is also applied between A and B. Determine a) the values of the vertical reaction and built-in moment at A and b) the deflection of the free end C of the cantilever.

Develop an expression for the slope of the beam at any position and hence plot a slope diagram. $E = 208\,\text{GN/m}^2$ and $I = 24 \times 10^{-6}\,\text{m}^4$. [20 kN, 50 kN m, − 15 mm.]

5.8 (B). Develop a general expression for the slope of the beam of question 5.6 and hence plot a slope diagram for the beam. Use the slope diagram to confirm the answer given in question 5.6 for the position of the maximum deflection of the beam.

5.9 (B). What would be the effect on the end deflection for question 5.7, if the built-in end A were replaced by a simple support at the same position and point B becomes a full simple support position (i.e. the force at B is no longer 10 kN). What general observation can you make about the effect of built-in constraints on the stiffness of beams? [5.7 mm.]

5.10 (B). A beam AB is simply supported at A and B over a span of 3 m. It carries loads of 50 kN and 40 kN at 0.6 m and 2 m respectively from A, together with a uniformly distributed load of 60 kN/m between the 50 kN and 40 kN concentrated loads. If the cross-section of the beam is such that $I = 60 \times 10^{-6}\,\text{m}^4$ determine the value of the deflection of the beam under the 50 kN load. $E = 210\,\text{GN/m}^2$. Sketch the S.F. and B.M. diagrams for the beam. [3.7 mm.]

5.11 (B). Obtain the relationship between the B.M., S.F., and intensity of loading of a laterally loaded beam. A simply supported beam of span L carries a distributed load of intensity kx^2/L^2 where x is measured from one support towards the other. Find:

(a) the location and magnitude of the greatest bending moment;
(b) the support reactions. [U.Birm.] [$0.63L$, $0.0393kL^2$, $kL/12$, $kL/4$.]

5.12 (B). A uniform beam 4 m long is simply supported at its ends, where couples are applied, each 3 kN m in magnitude but opposite in sense. If $E = 210\,\text{GN/m}^2$ and $I = 90 \times 10^{-6}\,\text{m}^4$ determine the magnitude of the deflection at mid-span.

What load must be applied at mid-span to reduce the deflection by half? [0.317 mm, 2.25 kN.]

5.13 (B). A 500 mm × 175 mm steel beam of length 8 m is supported at the left-hand end and at a point 1.6 m from the right-hand end. The beam carries a uniformly distributed load of 12 kN/m on its whole length, an additional uniformly distributed load of 18 kN/m on the length between the supports and a point load of 30 kN at the right-hand end. Determine the slope and deflection of the beam at the section midway between the supports and also at the right-hand end. EI for the beam is $1.5 \times 10^8\,\text{Nm}^2$. [U.L.] [$1.13 \times 10^{-4}$, 3.29 mm, 9.7×10^{-4}, 1.71 mm.]

5.14 (B). A cantilever, 2.6 m long, carrying a uniformly distributed load w along the entire length, is propped at its free end to the level of the fixed end. If the load on the prop is then 30 kN, calculate the value of w. Determine also the slope of the beam at the support. If any formula for deflection is used it must first be proved. $E = 210\,\text{GN/m}^2$; $I = 4 \times 10^{-6}\,\text{m}^4$. [U.E.I.] [30.8 kN/m, 0.014 rad.]

5.15 (B). A beam ABC of total length L is simply supported at one end A and at some point B along its length. It carries a uniformly distributed load of w per unit length over its whole length. Find the optimum position of B so that the greatest bending moment in the beam is as low as possible. [U.Birm.] [$L/2$.]

5.16 (B). A beam AB, of constant section, depth 400 mm and $I_{max} = 250 \times 10^{-6}\,\text{m}^4$, is hinged at A and simply supported on a non-yielding support at C. The beam is subjected to the given loading (Fig. 5.46). For this loading determine (a) the vertical deflection of B; (b) the slope of the tangent to the bent centre line at C. $E = 80\,\text{GN/m}^2$. [I.Struct.E.] [27.3 mm, 0.0147 rad.]

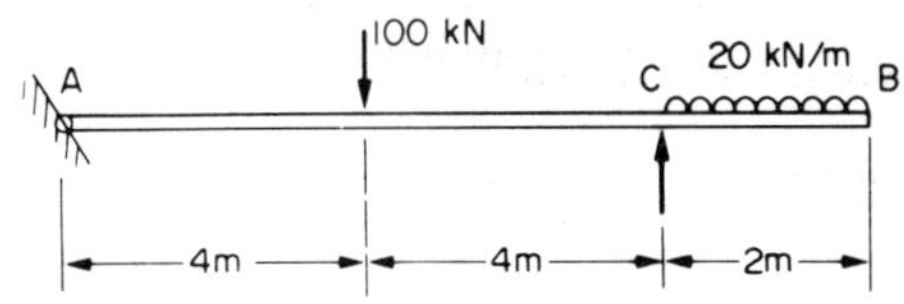

Fig. 5.46.

5.17 (B). A simply supported beam AB is 7 m long and carries a uniformly distributed load of 30 kN/m run. A couple is applied to the beam at a point C, 2.5 m from the left-hand end, A, the couple being clockwise in sense and of magnitude 70 kNm. Calculate the slope and deflection of the beam at a point D, 2 m from the left-hand end. Take $EI = 5 \times 10^7\,\text{Nm}^2$. [E.M.E.U.] [$5.78 \times 10^{-3}$ rad, 16.5 mm.]

5.18 (B). A uniform horizontal beam ABC is 0.75 m long and is simply supported at A and B, 0.5 m apart, by supports which can resist upward or downward forces. A vertical load of 50 N is applied at the free end C, which produces a deflection of 5 mm at the centre of span AB. Determine the position and magnitude of the maximum deflection in the span AB, and the magnitude of the deflection at C. [E.I.E.] [5.12 mm (upwards), 20.1 mm.]

5.19 (B). A continuous beam ABC rests on supports at A, B and C. The portion AB is 2 m long and carries a central concentrated load of 40 kN, and BC is 3 m long with a u.d.l. of 60 kN/m on the complete length. Draw the S.F. and B.M. diagrams for the beam. [−3.25, 148.75, 74.5 kN (Reactions); $M_B = -46.5$ kN m.]

5.20 (B). State Clapeyron's theorem of three moments. A continuous beam $ABCD$ is constructed of built-up sections whose effective flexural rigidity EI is constant throughout its length. Bay lengths are $AB = 1$ m, $BC = 5$ m, $CD = 4$ m. The beam is simply supported at B, C and D, and carries point loads of 20 kN and 60 kN at A and midway between C and D respectively, and a distributed load of 30 kN/m over BC. Determine the bending moments and vertical reactions at the supports and sketch the B.M. and S.F. diagrams.
[U.Birm.] [−20, −66.5, 0 kN m; 85.7, 130.93, 13.37 kN.]

5.21 (B). A continuous beam $ABCD$ is simply supported over three spans $AB = 1$ m, $BC = 2$ m and $CD = 2$ m. The first span carries a central load of 20 kN and the third span a uniformly distributed load of 30 kN/m. The central span remains unloaded. Calculate the bending moments at B and C and draw the S.F. and B.M. diagrams. The supports remain at the same level when the beam is loaded.
[1.36, −7.84 kN m; 11.36, 4.03, 38.52, 26.08 kN (Reactions).]

5.22 (B). A beam, simply supported at its ends, carries a load which increases uniformly from 15 kN/m at the left-hand end to 100 kN/m at the right-hand end. If the beam is 5 m long find the equation for the rate of loading and, using this, the deflection of the beam at mid-span if $E = 200\,\text{GN/m}^2$ and $I = 600 \times 10^{-6}\,\text{m}^4$.
[$w = -(15 + 85x/L)$; 3.9 mm.]

5.23 (B). A beam 5 m long is firmly fixed horizontally at one end and simply supported at the other by a prop. The beam carries a uniformly distributed load of 30 kN/m run over its whole length together with a concentrated load of 60 kN at a point 3 m from the fixed end. Determine:
(a) the load carried by the prop if the prop remains at the same level as the end support;
(b) the position of the point of maximum deflection. [B.P.] [82.16 kN; 2.075 m.]

5.24 (B/C). A continuous beam $ABCDE$ rests on five simple supports A, B, C, D and E. Spans AB and BC carry a u.d.l. of 60 kN/m and are respectively 2 m and 3 m long. CD is 2.5 m long and carries a concentrated load of 50 kN at 1.5 m from C. DE is 3 m long and carries a concentrated load of 50 kN at the centre and a u.d.l. of 30 kN/m. Draw the B.M. and S.F. diagrams for the beam.
[Fixing moments: 0, −44.91, −25.1, −38.95, 0 kN m. Reactions: 37.55, 179.1, 97.83, 118.5, 57.02 kN.]

CHAPTER 6

BUILT-IN BEAMS

Summary

The maximum bending moments and maximum deflections for built-in beams with standard loading cases are as follows:

MAXIMUM B.M. AND DEFLECTION FOR BUILT-IN BEAMS

Loading case	Maximum B.M.	Maximum deflection
Central concentrated load W	$\dfrac{WL}{8}$	$\dfrac{WL^3}{192EI}$
Uniformly distributed load w/metre (total load W)	$\dfrac{wL^2}{12} = \dfrac{WL}{12}$	$\dfrac{wL^4}{384EI} = \dfrac{WL^3}{384EI}$
Concentrated load W not at mid-span	$\dfrac{Wab^2}{L^2}$ or $\dfrac{Wa^2b}{L^2}$	$\dfrac{2Wa^3b^2}{3EI(L+2a)^2}$ at $x = \dfrac{2aL}{(L+2a)}$ $\left(\text{where } a < \dfrac{L}{2}\right)$ $= \dfrac{Wa^3b^3}{3EIL^3}$ under load
Distributed load w' varying in intensity between $x = x_1$ and $x = x_2$	$M_A = -\displaystyle\int_{x_1}^{x_2} \frac{w'(L-x)^2}{L^2}dx$ $M_B = -\displaystyle\int_{x_1}^{x_2} \frac{w'(L-x)x^2}{L^2}dx$	

Effect of movement of supports

If one end B of an initially horizontal built-in beam AB moves through a distance δ relative to end A, end moments are set up of value

$$M_A = -M_B = \frac{6EI\delta}{L^2}$$

and the reactions at each support are

$$R_A = -R_B = \frac{12EI\delta}{L^3}$$

Thus, in most practical situations where loaded beams sink at the supports the above values represent *changes* in fixing moment and reaction values, their directions being indicated in Fig. 6.6.

Introduction

When both ends of a beam are rigidly fixed the beam is said to be *built-in*, *encastred* or *encastré*. Such beams are normally treated by a modified form of Mohr's area–moment method or by Macaulay's method.

Built-in beams are assumed to have zero slope at each end, so that the total change of slope along the span is zero. Thus, from Mohr's first theorem,

$$\text{area of } \frac{M}{EI} \text{ diagram across the span} = 0$$

or, if the beam is uniform, EI is constant, and

$$\textbf{area of B.M. diagram} = \mathbf{0} \qquad (6.1)$$

Similarly, if both ends are level the deflection of one end relative to the other is zero. Therefore, from Mohr's second theorem:

$$\text{first moment of area of } \frac{M}{EI} \text{ diagram about one end} = 0$$

and, if EI is constant,

$$\textbf{first moment of area of B.M. diagram about one end} = \mathbf{0} \qquad (6.2)$$

To make use of these equations it is convenient to break down the B.M. diagram for the built-in beam into two parts:

(a) that resulting from the loading, assuming simply supported ends, and known as the *free-moment diagram*;
(b) that resulting from the end moments or fixing moments which must be applied at the ends to keep the slopes zero and termed the *fixing-moment diagram.*

6.1. Built-in beam carrying central concentrated load

Consider the centrally loaded built-in beam of Fig. 6.1. A_a is the area of the free-moment diagram and A_b that of the fixing-moment diagram.

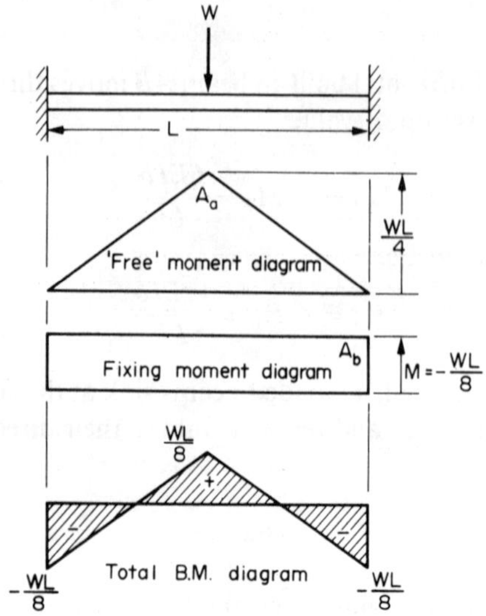

Fig. 6.1.

By symmetry the fixing moments are equal at both ends. Now from eqn. (6.1)

$$A_a + A_b = 0$$

$$\therefore \quad \tfrac{1}{2} \times L \times \frac{WL}{4} = -ML$$

$$\therefore \quad \boldsymbol{M = \frac{WL}{8}} \qquad (6.3)$$

The B.M. diagram is therefore as shown in Fig. 6.1, the maximum B.M. occurring at both the ends and the centre.

Applying Mohr's second theorem for the deflection at mid-span,

$$\delta = \left[\begin{array}{c}\text{first moment of area of B.M. diagram between centre and} \\ \text{one end about the centre}\end{array}\right] \times \frac{1}{EI}$$

$$= \frac{1}{EI}\left[\frac{1}{2}\left(\frac{1}{2} \times \frac{WL}{4} \times L\right)\left(\frac{1}{3} \times \frac{L}{2}\right) + \left(\frac{ML}{2} \times \frac{L}{4}\right)\right]$$

$$= \frac{1}{EI}\left[\frac{WL^3}{96} + \frac{ML^2}{8}\right] = \frac{1}{EI}\left[\frac{WL^3}{96} - \frac{WL^3}{64}\right]$$

$$= -\boldsymbol{\frac{WL^3}{192EI}} \quad \text{(i.e. downward deflection)} \qquad (6.4)$$

6.2. Built-in beam carrying uniformly distributed load across the span

Consider now the uniformly loaded beam of Fig. 6.2.

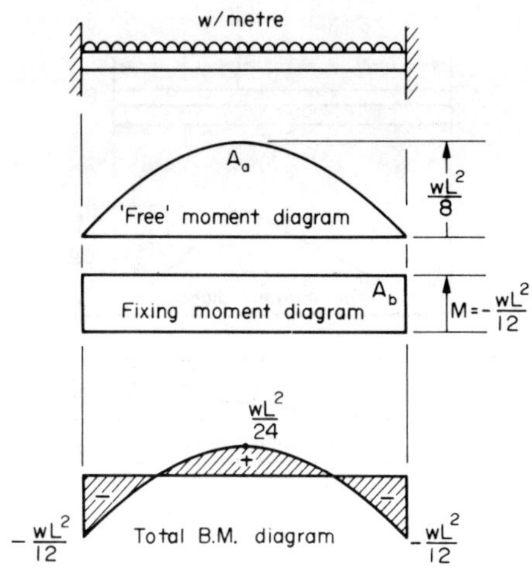

Fig. 6.2.

Again, for zero change of slope along the span,

$$A_a + A_b = 0$$

$$\therefore \quad \frac{2}{3} \times \frac{wL^2}{8} \times L = -ML$$

$$\therefore \quad \boldsymbol{M = \frac{wL^2}{12}} \tag{6.5}$$

The deflection at the centre is again given by Mohr's second theorem as the moment of one-half of the B.M. diagram about the centre.

$$\therefore \quad \delta = \left[\left(\frac{2}{3} \times \frac{wL^2}{8} \times \frac{L}{2}\right)\left(\frac{3}{8} \times \frac{L}{2}\right) + \left(\frac{ML}{2} \times \frac{L}{4}\right)\right]\frac{1}{EI}$$

$$= \frac{1}{EI}\left[\frac{3wL^4}{384} + \frac{ML^2}{8}\right] = \frac{1}{EI}\left[\frac{3wL^4}{384} - \frac{wL^4}{96}\right]$$

$$= -\boldsymbol{\frac{wL^4}{384EI}} \tag{6.6}$$

The negative sign again indicates a downwards deflection.

6.3. Built-in beam carrying concentrated load offset from the centre

Consider the loaded beam of Fig. 6.3.

Since the slope at both ends is zero the change of slope across the span is zero, i.e. the total area between A and B of the B.M. diagram is zero (Mohr's theorem).

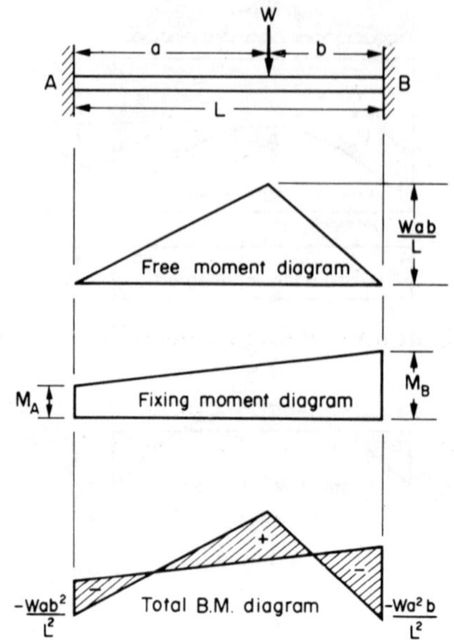

Fig. 6.3.

$$\therefore \quad \left(\frac{1}{2}\times\frac{Wab}{L}\times L\right)+\tfrac{1}{2}(M_A+M_B)L=0$$

$$M_A+M_B=-\frac{Wab}{L} \tag{1}$$

Also the deflection of A relative to B is zero; therefore the moment of the B.M. diagram between A and B about A is zero.

$$\therefore \quad \left[\frac{1}{2}\times\frac{Wab}{L}\times a\right]\frac{2a}{3}+\left[\frac{1}{2}\times\frac{Wab}{L}\times b\right]\left(a+\frac{b}{3}\right)+\left(\tfrac{1}{2}M_A L\times\frac{L}{3}\right)+\left(\tfrac{1}{2}M_B L\times\frac{2L}{3}\right)=0$$

$$\frac{L^2}{6}(M_A+2M_B)+\frac{Wa^3b}{3L}+\frac{Wab^2}{3L}\left(a+\frac{b}{3}\right)=0$$

$$M_A+2M_B=-\frac{Wab}{L^3}[2a^2+3ab+b^2] \tag{2}$$

Subtracting (1),

$$M_B=-\frac{Wab}{L^3}[2a^2+3ab+b^2-L^2]$$

but $L=a+b$,

$$\therefore \quad M_B=-\frac{Wab}{L^3}[2a^2+3ab+b^2-a^2-2ab-b^2]$$

$$=-\frac{Wab}{L^3}[a^2+ab]=-\frac{Wa^2bL}{L^3}$$

$$=-\frac{\mathbf{Wa^2b}}{\mathbf{L^2}} \tag{6.7}$$

Substituting in (1),

$$M_A = -\frac{Wab}{L} + \frac{Wa^2b}{L^2}$$

$$= -\frac{Wab(a+b)}{L^2} + \frac{Wa^2b}{L^2}$$

$$= -\frac{Wab^2}{L^2} \qquad (6.8)$$

6.4. Built-in beam carrying a non-uniform distributed load

Let w' be the distributed load varying in intensity along the beam as shown in Fig. 6.4. On a short length dx at a distance x from A there is a load of $w'dx$. Contribution of this load to M_A

$$= -\frac{Wab^2}{L^2} \quad (\text{where } W = w'dx)$$

$$= -\frac{w'dx \times x(L-x)^2}{L^2}$$

$$\therefore \qquad \text{total } M_A = -\int_0^L \frac{w'x(L-x)^2 dx}{L^2} \qquad (6.9)$$

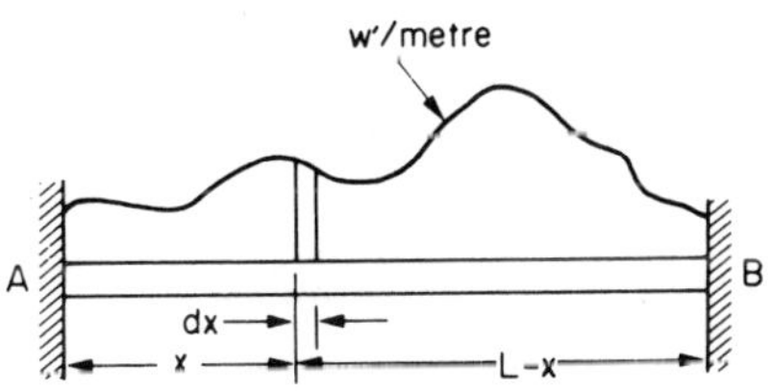

Fig. 6.4. Built-in (*encastré*) beam carrying non-uniform distributed load.

Similarly,

$$M_B = -\int_0^L \frac{w'(L-x)x^2}{L^2} dx \qquad (6.10)$$

If the distributed load is across only part of the span the limits of integration must be changed to take account of this: i.e. for a distributed load w' applied between $x = x_1$ and $x = x_2$ and varying in intensity,

$$M_A = -\int_{x_1}^{x_2} \frac{w'x(L-x)^2}{L^2} dx \qquad (6.11)$$

$$M_B = -\int_{x_1}^{x_2} \frac{w'(L-x)x^2}{L^2} dx \qquad (6.12)$$

6.5. Advantages and disadvantages of built-in beams

Provided that perfect end fixing can be achieved, built-in beams carry smaller maximum B.M.s (and hence are subjected to smaller maximum stresses) and have smaller deflections than the corresponding simply supported beams with the same loads applied; in other words built-in beams are stronger and stiffer. Although this would seem to imply that built-in beams should be used whenever possible, in fact this is not the case in practice. The principal reasons are as follows:

(1) The need for high accuracy in aligning the supports and fixing the ends during erection increases the cost.
(2) Small subsidence of either support can set up large stresses.
(3) Changes of temperature can also set up large stresses.
(4) The end fixings are normally sensitive to vibrations and fluctuations in B.M.s, as in applications introducing rolling loads (e.g. bridges, etc.).

These disadvantages can be reduced, however, if hinged joints are used at points on the beam where the B.M. is zero, i.e. at *points of inflexion or contraflexure*. The beam is then effectively a central beam supported on two end cantilevers, and for this reason the construction is sometimes termed the *double-cantilever* construction. The beam is then free to adjust to changes in level of the supports and changes in temperature (Fig. 6.5).

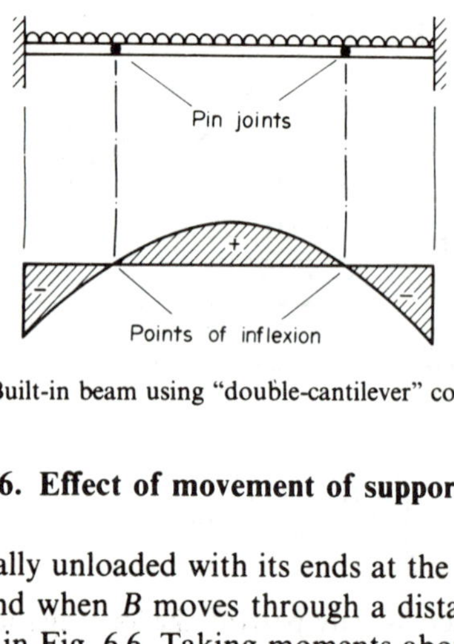

Fig. 6.5. Built-in beam using "double-cantilever" construction.

6.6. Effect of movement of supports

Consider a beam AB initially unloaded with its ends at the same level. If the slope is to remain horizontal at each end when B moves through a distance δ relative to end A, the moments must be as shown in Fig. 6.6. Taking moments about B

$$R_A \times L = M_A + M_B$$

and, by symmetry,
$$M_A = M_B = M$$

$$\therefore \qquad R_A = \frac{2M}{L}$$

Similarly,
$$R_B = \frac{2M}{L}$$

in the direction shown.

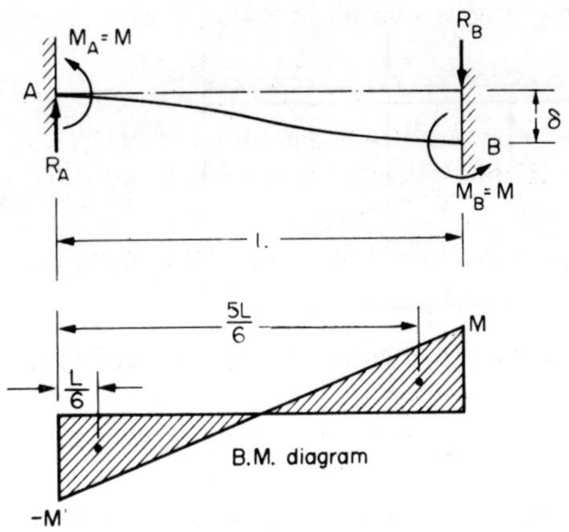

Fig. 6.6. Effect of support movement on B.M.s.

Now from Mohr's second theorem the deflection of A relative to B is equal to the first moment of area of the B.M. diagram about $A \times 1/EI$.

$$\therefore \quad \delta = \left[\left(-\tfrac{1}{2}M \times \frac{L}{2}\right)\frac{L}{6} + \left(\tfrac{1}{2}M \times \frac{L}{2}\right)\frac{5L}{6}\right]\frac{1}{EI}$$

$$= \frac{ML^2}{24EI}(-1+5) = \frac{ML^2}{6EI} \qquad (6.13)$$

$$\therefore \quad M = \frac{6EI\delta}{L^2} \quad \text{and} \quad R_A = R_B = \frac{12EI\delta}{L^3} \qquad (6.14)$$

in the directions shown in Fig. 6.6.

These values will also represent the *changes* in the fixing moments and end reactions for a beam under load when one end sinks relative to the other.

Examples

Example 6.1

An encastre beam has a span of 3 m and carries the loading system shown in Fig. 6.7. Draw the B.M. diagram for the beam and hence determine the maximum bending stress set up. The beam can be assumed to be uniform, with $I = 42 \times 10^{-6}\ \text{m}^4$ and with an overall depth of 200 mm.

Solution

Using the *principle of superposition* the loading system can be reduced to the three cases for which the B.M. diagrams have been drawn, together with the fixing moment diagram, in Fig. 6.7.

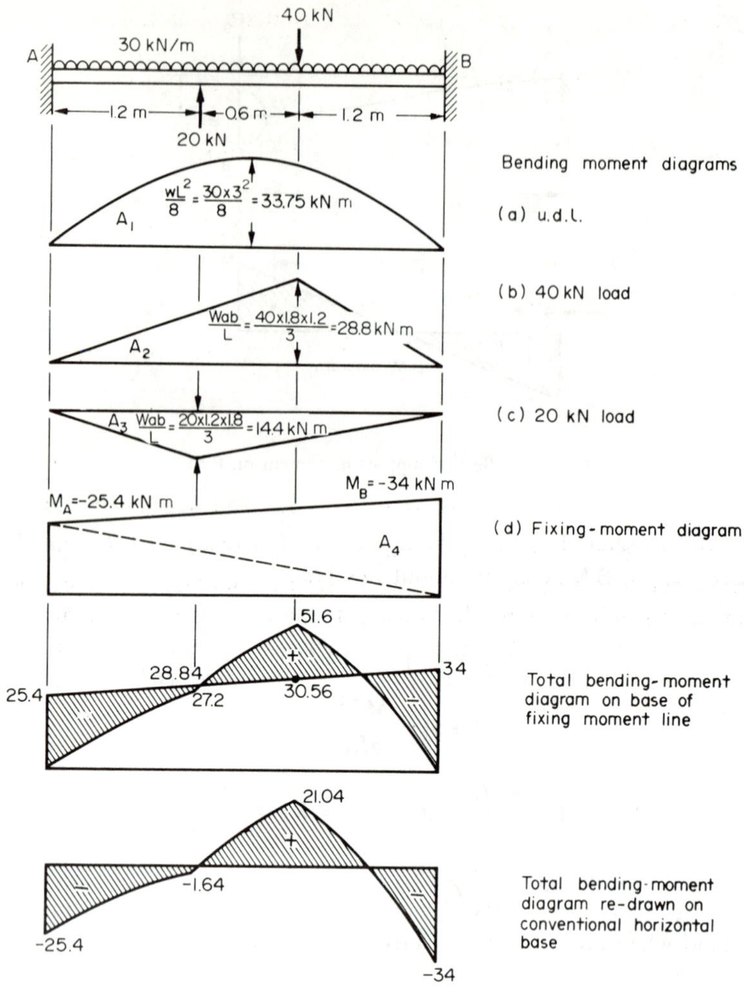

Fig. 6.7. Illustration of the application of the "principle of superposition" to Mohr's area–moment method of solution.

Now from eqn. (6.1)

$$A_1 + A_2 + A_4 = A_3$$

$$(\tfrac{2}{3} \times 33.75 \times 10^3 \times 3) + (\tfrac{1}{2} \times 28.8 \times 10^3 \times 3) + [\tfrac{1}{2}(M_A + M_B)3] = (\tfrac{1}{2} \times 14.4 \times 10^3 \times 3)$$

$$67.5 \times 10^3 + 43.2 \times 10^3 + 1.5(M_A + M_B) = 21.6 \times 10^3$$

$$M_A + M_B = -59.4 \times 10^3 \qquad (1)$$

Also, from eqn. (6.2), taking moments of area about A,

$$A_1\bar{x}_1 + A_2\bar{x}_2 + A_4\bar{x}_4 = A_3\bar{x}_3$$

and, dividing areas A_2 and A_4 into the convenient triangles shown,

$$(67.5\times10^3\times1.5)+(\tfrac{1}{2}\times28.8\times10^3\times1.8)\frac{2\times1.8}{3}+(\tfrac{1}{2}\times28.8\times10^3\times1.2)(1.8+\tfrac{1}{3}\times1.2)$$

$$+(\tfrac{1}{2}M_A\times3\times\tfrac{1}{3}\times3)+(\tfrac{1}{2}M_B\times3\times\tfrac{2}{3}\times3)=(\tfrac{1}{2}\times14.4\times10^3\times1.2)\tfrac{2}{3}\times1.2$$

$$+(\tfrac{1}{2}\times14.4\times10^3\times1.8)\left(1.2+\frac{1.8}{3}\right)$$

$$(101.25+31.1+38.0)10^3+1.5M_A+3M_B=(6.92+23.3)10^3$$

$$1.5\,M_A+3M_B=-140\times10^3$$

$\therefore$

$$M_A+2M_B=-93.4\times10^3 \qquad (2)$$

(2) − (1),

$$M_B=-34\times10^3\,\text{N m}=-34\,\text{kN m}$$

and from (1),

$$M_A=-25.4\times10^3\,\text{N m}=-25.4\,\text{kN m}$$

The fixing moments are therefore negative and not positive as assumed in Fig. 6.7. The total B.M. diagram is then found by combining all the separate loading diagrams and the fixing moment diagram to produce the result shown in Fig. 6.7. It will be seen that the maximum B.M. occurs at the built-in end B and has a value of 34 kN m. This will therefore be the position of the maximum bending stress also, the value being determined from the simple bending theory

$$\sigma_{max}=\frac{My}{I}=\frac{34\times10^3\times100\times10^{-3}}{42\times10^{-6}}$$

$$=81\times10^6=\mathbf{81\ MN/m^2}$$

Example 6.2

A built-in beam, 4 m long, carries combined uniformly distributed and concentrated loads as shown in Fig. 6.8. Determine the end reactions, the fixing moments at the built-in supports and the magnitude of the deflection under the 40 kN load. Take $EI = 14\,\text{MN m}^2$.

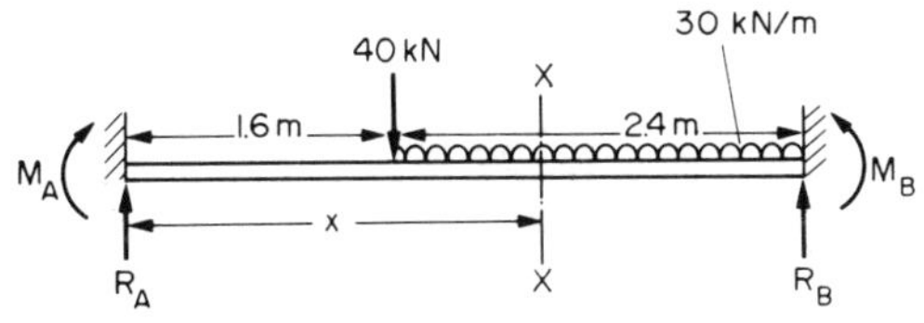

Fig. 6.8.

Solution

Using Macaulay's method (see page 106)

$$M_{xx}=\frac{EI}{10^3}\frac{d^2y}{dx^2}=M_A+R_Ax-40[x-1.6]-30\left[\frac{(x-1.6)^2}{2}\right]$$

Note that the unit of load of kilonewton is conveniently accounted for by dividing EI by 10^3. It can then be assumed in further calculation that R_A is in kN and M_A in kN m.

Integrating,

$$\frac{EI}{10^3}\frac{dy}{dx} = M_A x + R_A \frac{x^2}{2} - \frac{40}{2}[(x-1.6)^2] - \frac{30}{6}[(x-1.6)^3] + A$$

and

$$\frac{EI}{10^3} y = M_A \frac{x^2}{2} + R_A \frac{x^3}{6} - \frac{40}{6}[(x-1.6)^3] - \frac{30}{24}[(x-1.6)^4] + Ax + B$$

Now, when $x = 0$, $y = 0$ $\quad \therefore \quad B = 0$

and when $x = 0$, $\dfrac{dy}{dx} = 0 \quad \therefore \quad A = 0$

When $x = 4$, $y = 0$

$$0 = M_A \times \frac{4^2}{2} + R_A \times \frac{4^3}{6} - \frac{40}{6}(2.4)^3 - \frac{30}{24}(2.4)^4$$

$$0 = 8M_A + 10.67\,R_A - 92.16 - 41.47$$

$$133.6 = 8M_A + 10.67\,R_A \qquad (1)$$

When $x = 4$, $\dfrac{dy}{dx} = 0$

$\therefore$

$$0 = 4M_A + \frac{4^2}{2} R_A - \frac{40}{2}(2.4)^2 - \frac{30}{6}(2.4)^3$$

$$0 = 4M_A + 8R_A - 115.2 - 69.12$$

$$184.32 = 4M_A + 8R_A \qquad (2)$$

Multiply (2) × 2,

$$368.64 = 8M_A + 16R_A \qquad (3)$$

(3) − (1),

$$235.04 = 5.33\,R_A$$

$$R_A = \frac{235.04}{5.33} = \mathbf{44.1\,kN}$$

Now $\quad R_A + R_B = 40 + (2.4 \times 30) = 112\,\text{kN}$

$\therefore \quad R_B = 112 - 44.1 = \mathbf{67.9\,kN}$

Substituting in (2),

$$4M_A + 352.8 = 184.32$$

$\therefore \quad M_A = \frac{1}{4}(184.32 - 352.8) = \mathbf{-42.12\,kN\,m}$

i.e. M_A is in the opposite direction to that assumed in Fig. 6.8.

Taking moments about A,

$$M_B + 4R_B - (40 \times 1.6) - (30 \times 2.4 \times 2.8) - (-42.12) = 0$$

$$\therefore \quad M_B = -(67.9 \times 4) + 64 + 201.6 - 42.12 = \mathbf{-48.12\,kN\,m}$$

i.e. again in the opposite direction to that assumed in Fig. 6.8.

(Alternatively, and more conveniently, this value could have been obtained by substitution into the original Macaulay expression with $x = 4$, which is, in effect, taking moments about B. The need to take additional moments about A is then overcome.)

Substituting into the Macaulay deflection expression,

$$\frac{EI}{10^3}y = -42.1\frac{x^2}{2} + \frac{44.1x^3}{6} - \frac{20}{3}[x-1.6]^3 - \tfrac{5}{4}[x-1.6]^4$$

Thus, under the 40 kN load, where $x = 1.6$ (and neglecting negative Macaulay terms),

$$y = \frac{10^3}{EI}\left[\frac{-(42.1 \times 2.56)}{2} + \frac{(44.1 \times 4.1)}{6} - 0 - 0\right]$$

$$= -\frac{23.75 \times 10^3}{14 \times 10^6} = -1.7 \times 10^{-3}\,\mathrm{m}$$

$$= \mathbf{-1.7\,mm}$$

The negative sign as usual indicates a deflection downwards.

Example 6.3

Determine the fixing moment at the left-hand end of the beam shown in Fig. 6.9 when the load varies linearly from 30 kN/m to 60 kN/m along the span of 4 m.

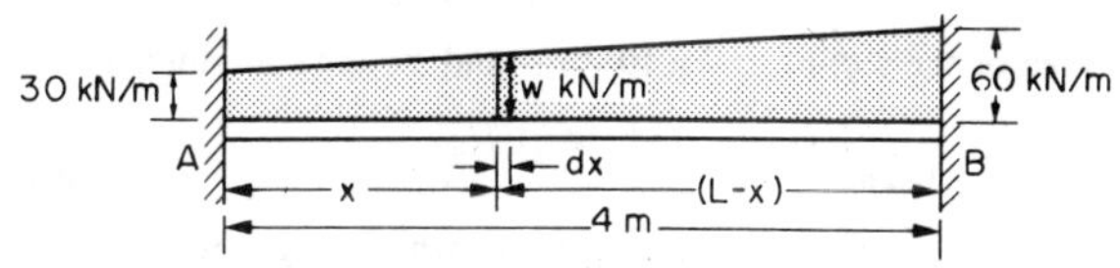

Fig. 6.9.

Solution

From §6.4

$$M_A = -\int_0^L \frac{w'x(L-x)^2}{L^2}\,dx$$

Now $$w' = \left(30 + \frac{30x}{4}\right)10^3 = (30 + 7.5x)10^3\ \mathrm{N/m}$$

$$\therefore \qquad M_A = -\int_0^4 \frac{(30+7.5x)10^3(4-x)^2}{4^2}\,x\,dx$$

$$= -\frac{10^3}{16}\int_0^4 (30+7.5x)(16-8x+x^2)x\,dx$$

$$= -\frac{10^3}{16}\int_0^4 (480x-240x^2+30x^3+120x^2-60x^3+7.5x^4)\,dx$$

$$= -\frac{10^3}{16}\int_0^4 (480x-120x^2-30x^3+7.5x^4)\,dx$$

$$= -\frac{10^3}{16}\left[\frac{480x^2}{2}-\frac{120x^3}{3}-\frac{30x^4}{4}+\frac{7.5x^5}{5}\right]_0^4$$

$$= -\frac{10^3}{16}[240\times 16-40\times 64-30\times 64+2.5\times 1024]$$

$$= \mathbf{-120\times 10^3\ N\,m}$$

The required moment at A is thus 120 kN m in the opposite direction to that shown in Fig. 6.8.

Problems

6.1 (A/B). A straight beam $ABCD$ is rigidly built-in at A and D and carries point loads of 5 kN at B and C.

$$AB = BC = CD = 1.8\ \text{m}$$

If the second moment of area of the section is $7\times 10^{-6}\ \text{m}^4$ and Young's modulus is 210 GN/m², calculate:

(a) the end moments;
(b) the central deflection of the beam. [U.Birm.] [-6 kN m; 4.13 mm.]

6.2 (A/B). A beam of uniform section with rigidly fixed ends which are at the same level has an effective span of 10 m. It carries loads of 30 kN and 50 kN at 3 m and 6 m respectively from the left-hand end. Find the vertical reactions and the fixing moments at each end of the beam. Determine the bending moments at the two points of loading and sketch, approximately to scale, the B.M. diagram for the beam.
[41.12, 38.88 kN; -92, -90.9, 31.26, 64.62 kN m.]

6.3 (A/B). A beam of uniform section and of 7 m span is "fixed" horizontally at the same level at each end. It carries a concentrated load of 100 kN at 4 m from the left-hand end. Neglecting the weight of the beam and working from first principles, find the position and magnitude of the maximum deflection if $E = 210\ \text{GN/m}^2$ and $I = 190\times 10^{-6}\ \text{m}^4$. [3.73 from l.h. end; 4.28 mm.]

6.4 (A/B). A uniform beam, built-in at each end, is divided into four equal parts and has equal point loads, each W, placed at the centre of each portion. Find the deflection at the centre of this beam and prove that it equals the deflection at the centre of the same beam when carrying an equal total load uniformly distributed along the entire length.

[U.C.L.I.] $\left[\dfrac{WL^3}{96EI}.\right]$

6.5 (A/B). A horizontal beam of I-section, rigidly built-in at the ends and 7 m long, carries a total uniformly distributed load of 90 kN as well as a concentrated central load of 30 kN. If the bending stress is limited to 90 MN/m^2 and the deflection must not exceed 2.5 mm, find the depth of section required. Prove the deflection formulae if used, or work from first principles. $E = 210$ GN/m^2. [U.L.C.I.] [583 mm.]

6.6 (A/B). A beam of uniform section is built-in at each end so as to have a clear span of 7 m. It carries a uniformly distributed load of 20 kN/m on the left-hand half of the beam, together with a 120 kN load at 5 m from the left-hand end. Find the reactions and the fixing moments at the ends and draw a B.M. diagram for the beam, inserting the principal values. [U.L.] [−105.4, −148 kN; 80.7, 109.3 kN m.]

6.7 (A/B). A steel beam of 10 m span is built-in at both ends and carries two point loads, each of 90 kN, at points 2.6 m from the ends of the beam. The middle 4.8 m has a section for which the second moment of area is 300×10^{-6} m^4 and the 2.6 m lengths at either end have a section for which the second moment of area is 400×10^{-6} m^4. Find the fixing moments at the ends and calculate the deflection at mid-span. Take $E = 210$ GN/m^2 and neglect the weight of the beam. [U.L.] [$M_A = M_B = 173.2$ kN m; 8.1 mm.]

6.8 (B.) A loaded horizontal beam has its ends securely built-in; the clear span is 8 m and $I = 90 \times 10^{-6}$ m^4. As a result of subsidence one end moves vertically through 12 mm. Determine the changes in the fixing moments and reactions. For the beam material $E = 210$ GN/m^2. [21.26 kN m; 5.32 kN.]

CHAPTER 7

SHEAR STRESS DISTRIBUTION

Summary

The *shear stress* in a beam at any transverse cross-section in its length, and at a point a vertical distance y from the neutral axis, *resulting from bending* is given by

$$\tau = \frac{QA\bar{y}}{Ib} \quad \text{or} \quad \tau = \frac{Q}{Ib}\int y\,dA$$

where Q is the applied vertical shear force at that section; A is the area of cross-section "above" y, i.e. the area between y and the outside of the section, which may be above or below the neutral axis (N.A.); $\bar{y}$ is the distance of the centroid of area A from the N.A.; I is the second moment of area of the complete cross-section; and b is the breadth of the section at position y.

For **rectangular sections,**

$$\tau = \frac{6Q}{bd^3}\left[\frac{d^2}{4} - y^2\right] \quad \text{with} \quad \tau_{max} = \frac{3Q}{2bd} \quad \text{when} \quad y = 0$$

For **I-section beams** the *vertical shear* in the web is given by

$$\tau = \frac{Q}{2I}\left[\frac{h^2}{4} - y^2\right] + \frac{Qb}{2It}\left[\frac{d^2}{4} - \frac{h^2}{4}\right]$$

with a *maximum* value of

$$\tau_{max} = \frac{Qh^2}{8I} + \frac{Qb}{2It}\left[\frac{d^2}{4} - \frac{h^2}{4}\right]$$

The *maximum* value of the *horizontal shear* in the flanges is

$$\tau_{max} = \frac{Qb}{4I}(d - t_1)$$

For **circular sections**

$$\tau = \frac{4Q}{3\pi R^2}\left[1 - \left(\frac{y}{R}\right)^2\right]$$

with a *maximum* value of

$$\tau_{max} = \frac{4Q}{3\pi R^2}$$

The *shear centre* of a section is that point, in or outside the section, through which load must be applied to produce zero twist of the section. Should a section have two axes of symmetry, the point where they cross is automatically the shear centre.

The shear centre of a **channel section** is given by

$$e = \frac{k^2h^2t}{4I}$$

Introduction

If a horizontal beam is subjected to vertical loads a shearing force (S.F.) diagram can be constructed as described in Chapter 3 to determine the value of the vertical S.F. at any section. This force tends to produce relative sliding between adjacent vertical sections of the beam, and it will be shown in Chapter 13, §13.2, that it is always accompanied by *complementary shears* which in this case will be horizontal. Since the concept of complementary shear is sometimes found difficult to appreciate, the following explanation is offered.

Consider the case of two rectangular-sectioned beams lying one on top of the other and supported on simple supports as shown in Fig. 7.1. If some form of vertical loading is applied the beams will bend as shown in Fig. 7.2, i.e. if there is negligible friction between the mating surfaces of the beams each beam will bend independently of the other and as a result the lower surface of the top beam will slide relative to the upper surface of the lower beam.

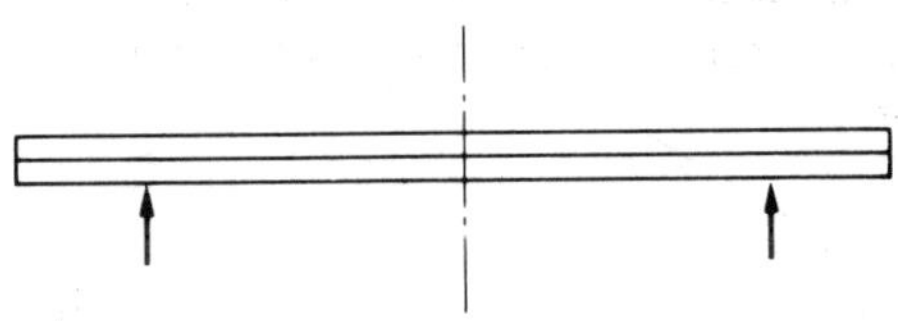

Fig. 7.1. Two beams (unconnected) on simple supports prior to loading.

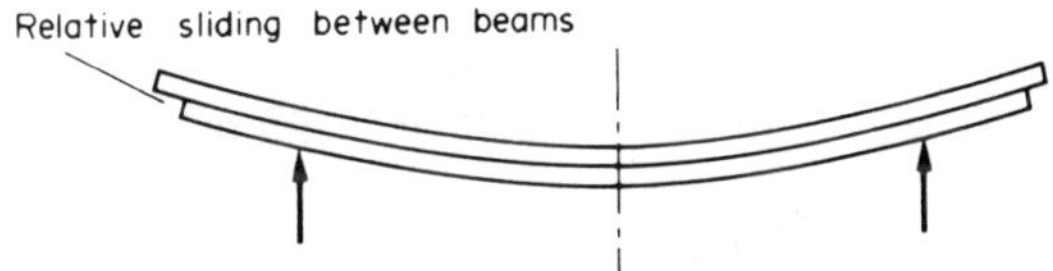

Fig. 7.2. Illustration of the presence of shear (relative sliding) between adjacent planes of a beam in bending.

If, therefore, the two beams are replaced by a single solid bar of depth equal to the combined depths of the initial two beams, then there must be some internal system of forces, and hence stresses, set up within the beam to prevent the above-mentioned sliding at the central fibres as bending takes place. Since the normal bending theory indicates that direct stresses due to bending are zero at the centre of a rectangular section beam, the prevention of sliding can only be achieved by horizontal shear stresses set up by the bending action.

Now on any element it will be shown in §13.2 that applied shears are always accompanied by complementary shears of equal value but opposite rotational sense on the perpendicular faces. Thus the *horizontal shears* due to bending are always associated with *complementary vertical shears* of equal value. For an element at either the top or bottom surface, however, there can be no vertical shears if the surface is "free" or unloaded and hence the horizontal shear is also zero. It is evident, therefore, that, for beams in bending, shear stresses are set up both vertically and horizontally varying from some as yet undetermined value at the centre to zero at the top and bottom surfaces.

The method of determination of the remainder of the shear stress distribution across beam sections is considered in detail below.

7.1. Distribution of shear stress due to bending

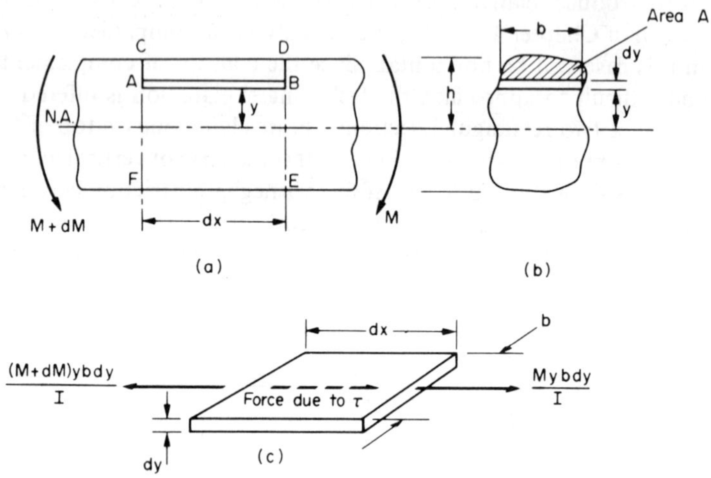

Fig. 7.3.

Consider the portion of a beam of length dx, as shown in Fig. 7.3a, and an element AB distance y from the N.A. Under any loading system the B.M. across the beam will change from M at B to $(M + dM)$ at A. Now as a result of bending,

$$\text{longitudinal stress } \sigma = \frac{My}{I}$$

$$\text{longitudinal stress at } A = \frac{(M + dM)y}{I}$$

and $$\text{longitudinal stress at } B = \frac{My}{I}$$

$\therefore$ $$\text{longitudinal force on the element at } A = \sigma A = \frac{(M + dM)y}{I} \times bdy$$

and $$\text{longitudinal force on the element at } B = \frac{My}{I} \times bdy$$

The force system on the element is therefore as shown in Fig. 7.3c with a net out-of-balance force to the left

$$= \frac{(M+dM)y}{I} bdy - \frac{My}{I} bdy$$

$$= \frac{dM}{I} ybdy$$

Therefore total out-of-balance force from all sections above height y

$$= \int_y^h \frac{dM}{I} ybdy$$

For equilibrium, this force is resisted by a shear force set up on the section of length dx and breadth b, as shown in Fig. 7.4.

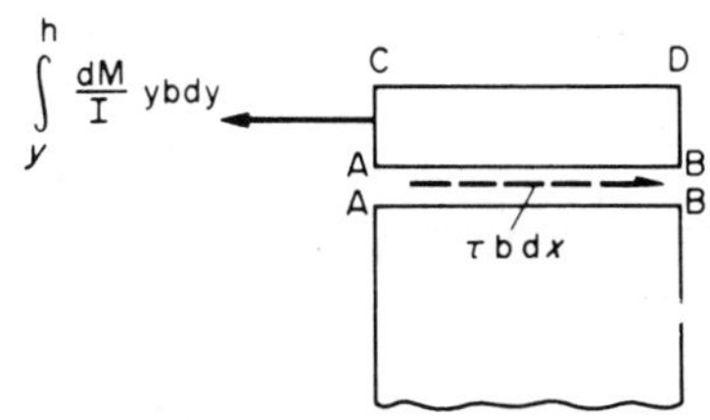

Fig. 7.4.

Thus if the shear stress is τ, then

$$\tau bdx = \frac{dM}{I} \int_y^h ybdy \tag{7.1}$$

But $\int_y^h ybdy$ = first moment of area of shaded portion of Fig. 7.3b about the N.A.

$= A\bar{y}$

where A is the area of shaded portion and $\bar{y}$ the distance of its centroid from the N.A.

Also $\frac{dM}{dx}$ = rate of change of the B.M.

= S.F. Q at the section

$$\therefore \qquad \tau = \frac{QA\bar{y}}{Ib} \tag{7.2}$$

or, alternatively,

$$\tau = \frac{Q}{Ib} \int_y^h ydA \quad \text{where} \quad dA = bdy \tag{7.3}$$

7.2. Application to rectangular sections

Consider now the rectangular-sectioned beam of Fig. 7.5 subjected at a given transverse cross-section to a S.F. Q.

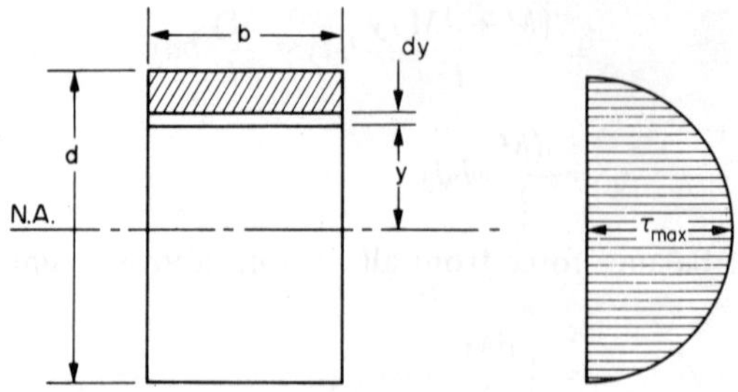

Fig. 7.5. Shear stress distribution due to bending of a rectangular section beam.

$$\tau = \frac{QA\bar{y}}{Ib}$$

$$= \frac{Q}{Ib} \times b\left(\frac{d}{2} - y\right) \times \left[y + \frac{\left(\frac{d}{2} - y\right)}{2}\right]$$

$$= \frac{Q \times 12}{bd^3 \times b} \times b\left(\frac{d}{2} - y\right)\frac{\left(\frac{d}{2} + y\right)}{2}$$

$$= \frac{6Q}{bd^3}\left[\frac{d^2}{4} - y^2\right] \quad \text{(i.e. } a \text{ } parabola\text{)} \tag{7.4}$$

Now $$\tau_{max} = \frac{6Q}{bd^3} \times \frac{d^2}{4} = \frac{3Q}{2bd} \quad \text{when } y = 0 \tag{7.5}$$

and $$\text{average } \tau = \frac{Q}{bd}$$

$\therefore$ $$\tau_{max} = \tfrac{3}{2} \times \tau_{average} \tag{7.6}$$

7.3. Application to I-section beams

Consider the I-section beam shown in Fig. 7.6.

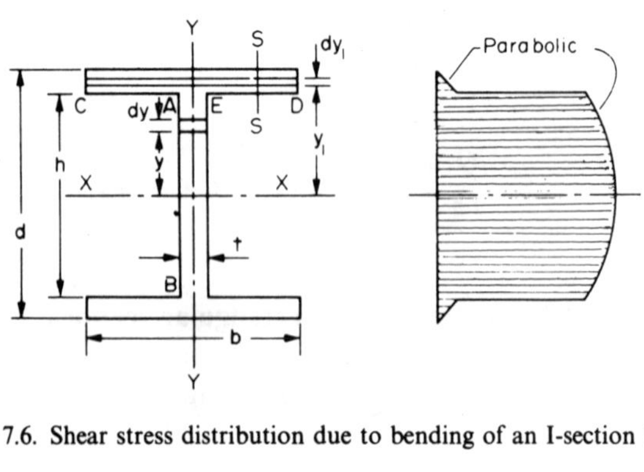

Fig. 7.6. Shear stress distribution due to bending of an I-section beam.

7.3.1. Vertical shear in the web

The distribution of shear stress due to bending at any point in a given transverse cross-section is given, in general, by eqn. (7.3)

$$\tau = \frac{Q}{Ib}\int_{y}^{d/2} y\,dA$$

In the case of the I-beam, however, the width of the section is not constant so that the quantity dA will be different in the web and the flange. Equation (7.3) must therefore be modified to

$$\tau = \frac{Q}{It}\int_{y}^{h/2} ty\,dy + \frac{Q}{It}\int_{h/2}^{d/2} by_1\,dy_1$$

$$= \frac{Q}{2I}\left[\frac{h^2}{4} - y^2\right] + \frac{Qb}{2It}\left[\frac{d^2}{4} - \frac{h^2}{4}\right]$$

As for the rectangular section, the first term produces a parabolic stress distribution. The second term is a constant and equal to the value of the shear stress at the top and bottom of the web, where $y = h/2$,

i.e.
$$\tau_A = \tau_B = \frac{Qb}{2It}\left[\frac{d^2}{4} - \frac{h^2}{4}\right] \qquad (7.7)$$

The *maximum shear* occurs at the N.A., where $y = 0$,

$$\tau_{max} = \frac{Qh^2}{8I} + \frac{Qb}{2It}\left[\frac{d^2}{4} - \frac{h^2}{4}\right] \qquad (7.8)$$

7.3.2. Vertical shear in the flanges

(a) Along the central section YY

The vertical shear in the flange where the width of the section is b is again given by eqn. (7.3) as

$$\tau = \frac{Q}{Ib}\int_{y_1}^{d/2} y_1\,dA$$

$$= \frac{Q}{Ib}\int_{y_1}^{d/2} y_1 b\,dy_1 = \frac{Q}{I}\left[\frac{d^2}{8} - \frac{y_1^2}{2}\right] \qquad (7.9)$$

The *maximum* value is that at the bottom of the flange when $y_1 = h/2$,

$$\tau_{max} = \frac{Q}{I}\left[\frac{d^2}{8} - \frac{h^2}{8}\right] = \frac{Q}{8I}[d^2 - h^2] \qquad (7.10)$$

this value being considerably smaller than that obtained at the top of the web.

At the outside of the flanges, where $y_1 = d/2$, the vertical shear (and the complementary horizontal shear) are zero. At intermediate points the distribution is again parabolic producing the total stress distribution indicated in Fig. 7.6. As a close approximation, however, the distribution across the flanges is often taken to be linear since its effect is minimal compared with the values in the web.

(b) Along any other section SS, removed from the web

At the general section SS in the flange the shear stress at both the upper and lower edges must be zero. The distribution across the thickness of the flange is then the same as that for a rectangular section of the same dimensions.

The discrepancy between the values of shear across the free surfaces CA and ED and those at the web-flange junction indicate that the distribution of shear at the junction of the web and flange follows a more complicated relationship which cannot be investigated by the elementary analysis used here. Advanced elasticity theory must be applied to obtain a correct solution, but the values obtained above are normally perfectly adequate for general design work particularly in view of the following comments.

As stated above, the vertical shear stress in the flanges is very small in comparison with that in the web and is often neglected. Thus, in girder design, it is normally assumed that the web carries all the vertical shear. Additionally, the thickness of the web t is often very small in comparison with b such that eqns. (7.7) and (7.8) are nearly equal. The distribution of shear across the web in such cases is then taken to be uniform and equal to the total shear force Q divided by the cross-sectional area (th) of the web alone.

7.3.3. Horizontal shear in the flanges

The proof of §7.1 considered the equilibrium of an element in a vertical section of a component similar to element A of Fig. 7.9. Consider now a similar element B in the horizontal flange of the channel section (or I section) shown in Fig. 7.7.

The element has dimensions dz, t and dx comparable directly to the element previously treated of dy, b and dx. The proof of §7.1 can be applied in precisely the same way to this flange element giving an out-of-balance force on the element, from Fig. 7.9(b),

$$= \frac{(M+dM)}{I} y \, . \, tdz - \frac{My \, . \, tdz}{I}$$

$$= \frac{dM}{I} y \, . \, tdz$$

with a total out-of-balance force for the sections between z and L

$$= \int_z^L \frac{dM}{I} y \, . \, tdz.$$

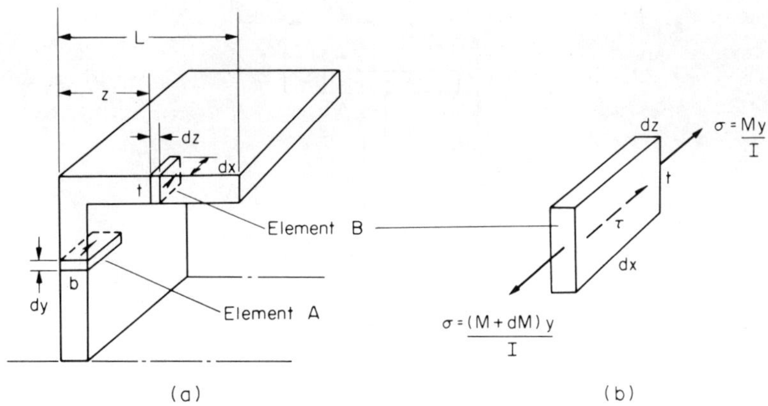

Fig. 7.7. Horizontal shear in flanges.

This force being reacted by the shear on the element shown in Fig. 7.9(c),

$$= \tau t dx$$

$\therefore$
$$\tau t dx = \int_z^L \frac{dM}{I} \cdot y\, t dz$$

and
$$\tau = \frac{dM}{dx} \cdot \frac{1}{It} \int_z^L t dz \cdot y$$

But
$$\int_z^L t dz \,.\, y = A\bar{y} \quad \text{and} \quad \frac{dM}{dx} = Q.$$

$\therefore$
$$\tau = \frac{QA\bar{y}}{It} \qquad (7.11)$$

Thus the same form of expression is obtained to that of eqn (7.2) but with the breadth b of the web replaced by thickness t of the flange: I and y still refer to the N.A. and A is the area of the flange 'beyond' the point being considered.

Thus the horizontal shear stress distribution in the flanges of the I section of Fig. 7.8 can

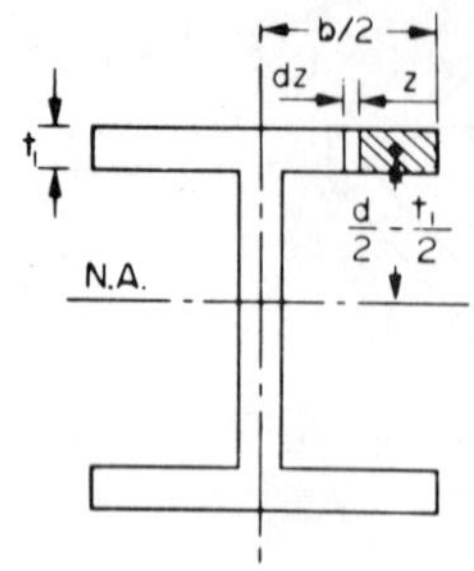

Fig. 7.8.

now be obtained from eqn. (7.11):

$$\tau = \frac{QA\bar{y}}{It}$$

with
$$\bar{y} = \frac{d}{2} - \frac{t_1}{2} = \tfrac{1}{2}(d - t_1)$$

$$A = t_1 dz$$

$$t = t_1$$

Thus
$$\tau = \frac{Q}{It_1}\int_0^{b/2} \tfrac{1}{2}(d - t_1)t_1 dz = \frac{Q}{2I}(d - t_1)\left[z\right]_0^{b/2}$$

The distribution is therefore linear from zero at the free ends of the flange to a maximum value of

$$\boldsymbol{\tau_{max} = \frac{Qb}{4I}(d - t_1)} \quad \text{at the centre} \tag{7.12}$$

7.4. Application to circular sections

In this case it is convenient to use the alternative form of eqn. (7.2), namely (7.1),

$$\tau b dx = \frac{dM}{I}\int_z^h by dy$$

$$\tau = \frac{Q}{Ib}\int_z^h by dy$$

Consider now the element of thickness dz and breadth b shown in Fig. 7.9.

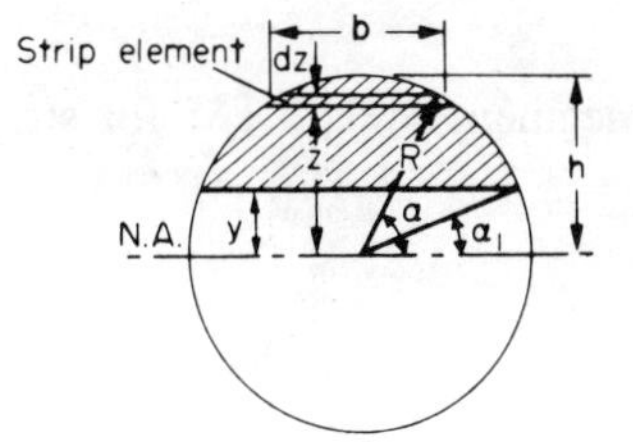

Fig. 7.9.

Now $b = 2R\cos\alpha$, $y = z = R\sin\alpha$ and $dz = R\cos\alpha d\alpha$ and, at section distance y from the N.A., $b = 2R\cos\alpha_1$,

$$\therefore \quad \tau = \frac{Q}{I \times 2R\cos\alpha_1}\int_{\alpha_1}^{\pi/2} 2R\cos\alpha\, R\sin\alpha\, R\cos\alpha\, d\alpha$$

$$= \frac{Q \times 4}{2R\cos\alpha_1 \times \pi R^4}\int_{\alpha_1}^{\pi/2} 2R^3\cos^2\alpha\sin\alpha\, d\alpha \quad \text{since } I = \frac{\pi R^4}{4}$$

$$= \frac{4Q}{\pi R^2\cos\alpha_1}\left[-\frac{\cos^3\alpha}{3}\right]_{\alpha_1}^{\pi/2}$$

$$- \frac{4Q}{3\pi R^2\cos\alpha_1}\left[\cos^3\alpha_1\right]_{\pi/2}^{\alpha_1} - \frac{4Q\cos^2\alpha_1}{3\pi R^2}$$

$$= \frac{4Q}{3\pi R^2}[1-\sin^2\alpha_1] = \frac{4Q}{3\pi R^2}\left[1-\left(\frac{y}{R}\right)^2\right] \tag{7.13}$$

i.e. *a parabola* with its maximum value at $y = 0$.

Thus $$\tau_{max} = \frac{4Q}{3\pi R^2}$$

Now $$\text{mean stress} = \frac{Q}{\pi R^2}$$

$$\therefore \quad \frac{\textbf{maximum shear stress}}{\textbf{mean shear stress}} = \frac{\dfrac{4Q}{3\pi R^2}}{\dfrac{Q}{\pi R^2}} = \frac{4}{3} \tag{7.14}$$

Alternative procedure

Using eqn. (7.2), namely $\tau = \dfrac{QA\bar{y}}{Ib}$, and referring to Fig. 7.9,

$$\frac{b}{2} = (R^2 - z^2)^{1/2} = R\cos\alpha \quad \text{and} \quad \sin\alpha = \frac{z}{R}$$

$$A\bar{y} \text{ for the shaded segment} = \int_{R\sin\alpha}^{R} A\bar{y} \quad \text{for strip element}$$

$$= \int_{R\sin\alpha}^{R} b\,dz\,z$$

$$= 2\int_{R\sin\alpha}^{R} (R^2 - z^2)^{1/2} z\,dz$$

$$= \tfrac{2}{3}\left[(R^2 - z^2)^{3/2}\right]_R^{R\sin\alpha}$$

$$= \tfrac{2}{3}[R^2(1 - \sin^2\alpha)]^{3/2}$$

$$= \tfrac{2}{3}R^3(\cos^2\alpha)^{3/2} = \tfrac{2}{3}R^3\cos^3\alpha$$

$$\therefore \qquad \tau = \frac{QA\bar{y}}{Ib} = \frac{Q \times \frac{2}{3}R^3\cos^3\alpha}{\dfrac{\pi R^4}{4} \times 2R\cos\alpha}$$

since
$$I = \frac{\pi R^4}{4}$$

$$\therefore \qquad \tau = \frac{4Q}{3\pi R^2}\cos^2\alpha = \frac{4Q}{3\pi R^2}[1 - \sin^2\alpha]$$

$$= \frac{4Q}{3\pi R^2}\left[1 - \left(\frac{y}{R}\right)^2\right] \qquad (7.13) \text{ bis.}$$

7.5. Limitation of shear stress distribution theory

There are certain practical situations where eq. (7.2) leads to an incomplete solution and it is necessary to consider other conditions, such as equilibrium at a free surface, before a valid solution is obtained. Take, for example, the case of the bending of a bar or beam having a circular cross-section as shown in Fig. 7.10(a).

The shear stress distribution across the section owing to bending is given by eqn. (7.2) as:

$$\tau = \frac{Q}{3I}\left[\left(\frac{d}{2}\right)^2 - y^2\right] = \tau_{yz} \qquad (7.15)$$

At some horizontal section AB, therefore, the shear stresses will be as indicated in Fig. 7.10, and will be equal along AB, with a value given by eqn. (7.15).

Now, for an element at A, for example, there should be no component of stress normal to the surface since it is unloaded and equilibrium would not result. The vertical shear given by the equation above, however, clearly would have a component normal to the surface. A valid solution can only be obtained therefore if a secondary shear stress τ_{xz} is set up in the z direction which, together with τ_{yz}, produces a resultant shear stress tangential to the free surface—see Fig. 7.10(b).

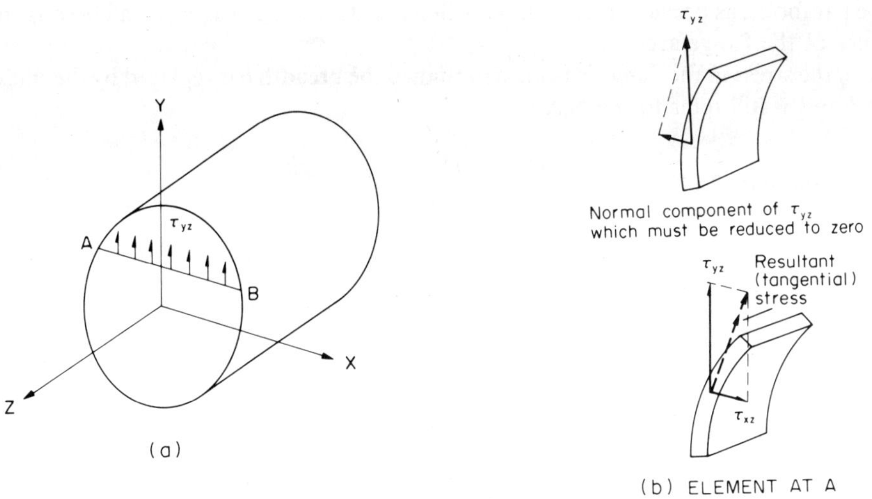

Fig. 7.10.

Solutions for the value of τ_{xz} and its effect on τ_{yz} are beyond the scope of this text† but the principal outlined indicates a limitation of the shear stress distribution theory which should be appreciated.

7.6. Shear centre

Consider the channel section of Fig. 7.11 under the action of a shearing force Q at a distance e from the centre of the web. The shearing stress at any point on the cross-section of the channel is then given by the equation $\tau = \dfrac{QA\bar{y}}{Ib}$. The distribution in the rectangular web

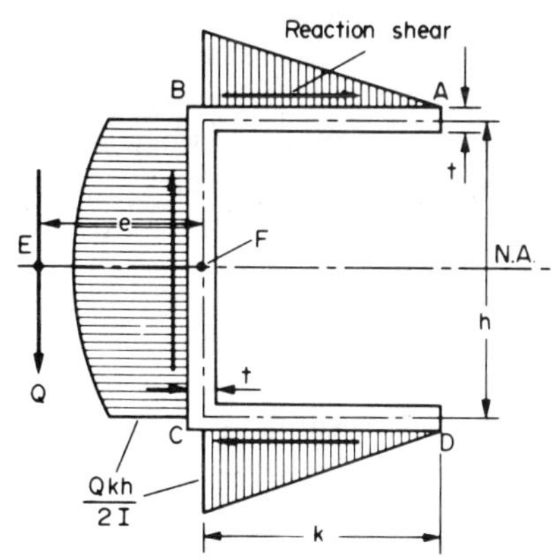

Fig. 7.11.

† I. S. Sokdnikoff, *Mathematical Theory of Elasticity*, 2nd edition (McGraw Hill, New York, 1956).

will be parabolic, as previously found, but will not reduce to zero at each end because of the presence of the flange areas.

When the stress in the flange is being determined the breadth b is replaced by the thickness t, but I and $\bar{y}$ still refer to the N.A.

Thus from eqn. 7.11, $$\tau_B = \frac{QA\bar{y}}{It} = \frac{Q \times kt}{It} \times \frac{h}{2} = \frac{Qkh}{2I}$$

and $$\tau_A = 0 \quad \text{since area beyond } A = 0$$

Between A and B the distribution is linear, since τ is directly proportional to the distance along AB (Q, t, h and I all being constant). An exactly similar distribution will be obtained for CD.

The stresses in the flanges give rise to forces represented by

$$\text{average stress} \times \text{area} = \tfrac{1}{2} \times \frac{Qkh}{2I} \times kt = \frac{Qk^2ht}{4I}$$

These produce a torque about F which must equal the applied torque, stresses in the web producing forces which have no moment about F.

Equating torques about F for equilibrium:

$$Q \times e = \frac{Qk^2ht}{4I} \times h$$

$$\therefore \qquad \boldsymbol{e = \frac{k^2h^2t}{4I}} \qquad (7.16)$$

Thus if a force acts on the axis of symmetry, distance e from the centre of the web, there will be no tendency for the section to twist since moments will be balanced. The point E is then termed the *shear centre* of the section.

The shear centre of a section is therefore defined as that point through which load must be applied for zero twist of the section. With loads applied at the shear centre of beam sections, stresses will be produced due to pure bending, and evaluation of the stresses produced will be much easier than would be the case if torsion were also present. The requirement for load to be applied through the shear centre of a section is a particular condition used throughout Chapter 16 (Volume 2) on the bending of unsymmetrical sections.

It should be noted that if a section has two axes of symmetry the point where they cross is automatically the shear centre.

Examples

Example 7.1

At a given position on a beam of uniform I-section the beam is subjected to a shear force of 100 kN. Plot a curve to show the variation of shear stress across the section and hence determine the ratio of the maximum shear stress to the mean shear stress.

Solution

Consider the I-section shown in Fig. 7.12. By symmetry, the centroid of the section is at

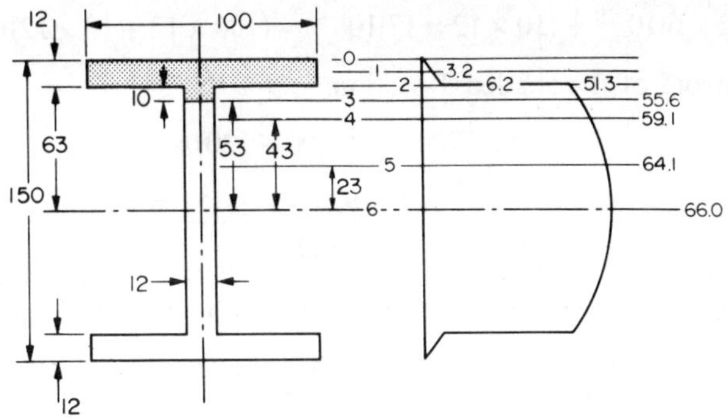

(a) Beam cross-section (all dimensions in mm)

(b) Shear stress distribution (MN/m^2)

Fig. 7.12.

mid-height and the neutral axis passes through this position. The second moment of area of the section is then given by

$$I = \frac{100 \times 150^3 \times 10^{-12}}{12} - \frac{88 \times 126^3 \times 10^{-12}}{12}$$

$$= (28.125 - 14.67)10^{-6} = 13.46 \times 10^{-6}\ \mathrm{m}^4$$

The distribution of shear stress across the section is

$$\tau = \frac{QA\bar{y}}{Ib} = \frac{100 \times 10^3 A\bar{y}}{13.46 \times 10^{-6} b} = 7.43 \times 10^9 \frac{A\bar{y}}{b}$$

The solution of this equation is then best completed in tabular form as shown below. In this case, because of symmetry, only sections above the N.A. need be considered since a similar distribution will be obtained below the N.A.

It should be noted that two values of shear stress are required at section 2 to take account of the change in breadth at this section. The values of A and $\bar{y}$ for sections 3, 4, 5 and 6 are those of a T-section beam and may be found as shown for section 3 (shaded area of Fig. 7.12a).

Section	$A \times 10^{-6}$ (m^2)	$\bar{y} \times 10^{-3}$ (m)	$b \times 10^{-3}$ (m)	$\tau = \frac{QA\bar{y}}{Ib}$ (MN/m^2)
0	0	–	–	–
1	$100 \times 6 = 600$	72	100	3.2
2	$100 \times 12 = 1200$	69	100	6.2
2	1200	69	12	51.3
3	1320	68	12	55.6
4	1440	66.3	12	59.1
5	1680	61.6	12	64.1
6	1956	54.5	12	66.0

For section 3:

Taking moments about the top edge (Fig. 7.12a),

$$(100 \times 12 \times 6)10^{-9} + (10 \times 12 \times 17)10^{-9} = (100 \times 12 + 10 \times 12)h \times 10^{-9}$$

where h is the centroid of the shaded T-section,

$$7200 + 2040 = (1200 + 120)\,h$$

$$h = \frac{9240}{1320} = 7\,\text{mm}$$

$$\therefore \qquad \bar{y}_3 = 75 - 7 = 68\,\text{mm}$$

The distribution of shear stress due to bending is then shown in Fig. 7.12b, giving a maximum shear stress of $\boldsymbol{\tau_{max} = 66\,\text{MN/m}^2}$.

Now the mean shear stress across the section is:

$$\tau_{\text{mean}} = \frac{\text{shear force}}{\text{area}} = \frac{100 \times 10^3}{3.912 \times 10^{-3}}$$

$$= 25.6\,\text{MN/m}^2$$

$$\frac{\text{max. shear stress}}{\text{mean shear stress}} = \frac{66}{25.6} = \mathbf{2.58}$$

Example 7.2

At a certain section a beam has the cross-section shown in Fig. 7.13. The beam is simply supported at its ends and carries a central concentrated load of 500 kN together with a load of 300 kN/m uniformly distributed across the complete span of 3 m. Draw the shear stress distribution diagram for a section 1 m from the left-hand support.

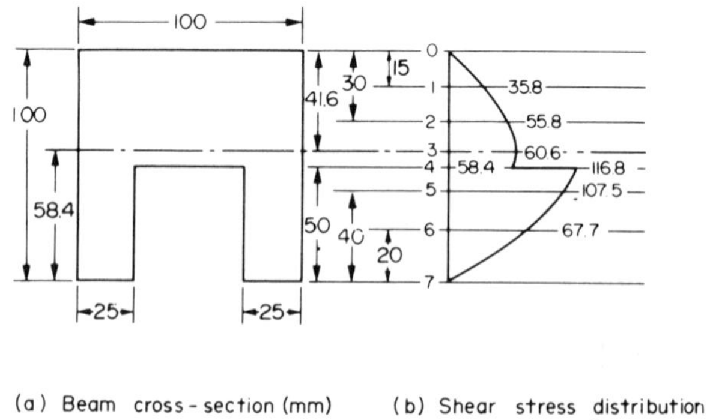

Fig. 7.13.

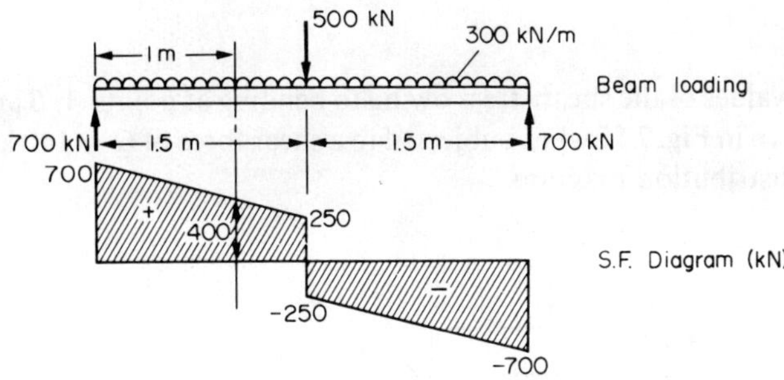

Fig. 7.14.

Solution

From the S.F. diagram for the beam (Fig. 7.14) it is evident that the S.F. at the section 1 m from the left-hand support is 400 kN,

i.e. $$Q = 400\,\text{kN}$$

To find the position of the N.A. of the beam section of Fig. 7.13(a) take moments of area about the base.

$$(100 \times 100 \times 50)10^{-9} - (50 \times 50 \times 25)10^{-9} = (100 \times 100 - 50 \times 50)\,\bar{y} \times 10^{-9}$$

$$500000 - 62500 = (10000 - 2500)\,\bar{y}$$

$$\bar{y} = \frac{437500}{7500} = 58.4\,\text{mm}$$

Then $$I_{\text{N.A.}} = \left[\frac{100 \times 41.6^3}{3} + 2\left(\frac{25 \times 58.4^3}{3}\right) + \left(\frac{50 \times 8.4^3}{3}\right)\right]10^{-12}$$

$$= (2.41 + 3.3 + 0.0099)10^{-6} = 5.72 \times 10^{-6}\,\text{m}^4$$

$$\therefore \qquad \tau = \frac{QA\bar{y}}{Ib} = \frac{400 \times 10^3}{5.72 \times 10^{-6}}\frac{A\bar{y}}{b} = 7 \times 10^{10}\frac{A\bar{y}}{b}$$

Section	$A \times 10^{-6}$ (m^2)	$\bar{y} \times 10^{-3}$ (m)	$b \times 10^{-3}$ (m)	$\tau = 7 \times 10^4 \frac{A\bar{y}}{b}$ (MN/m^2)
0	0	–	–	0
1	1500	34.1	100	35.8
2	3000	26.6	100	55.8
3	4160	20.8	100	60.6
4	2500	33.4	100	58.4
4	2500	33.4	50	116.8
5	2000	38.4	50	107.5
6	1000	48.4	50	67.7
7	0	–	–	0

The shear stress distribution across the section is shown in Fig. 7.13b.

Example 7.3

Determine the values of the shear stress owing to bending at points A, B and C in the beam cross-section shown in Fig. 7.15 when subjected to a shear force of $Q = 140$ kN. Hence sketch the shear stress distribution diagram.

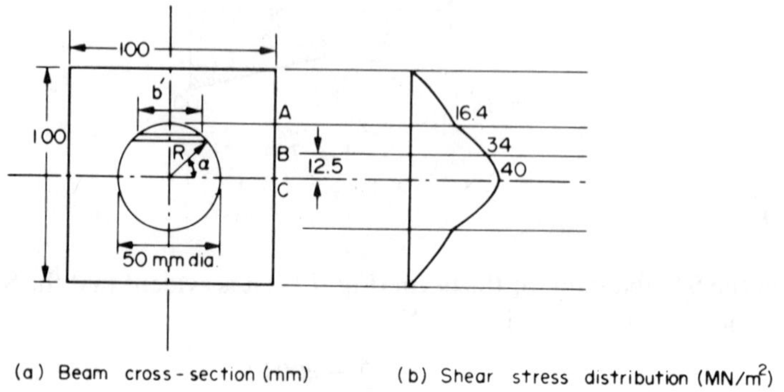

Fig. 7.15.

Solution

By symmetry the centroid will be at the centre of the section and

$$I_{N.A.} = \frac{100 \times 100^3}{12} \times 10^{-12} - \frac{\pi \times 50^4}{64} \times 10^{-12}$$

$$= (8.33 - 0.31)10^{-6} = 8.02 \times 10^{-6}\,\text{m}^4$$

At A:

$$\tau_A = \frac{QA\bar{y}}{Ib} = \frac{140 \times 10^3 \times (100 \times 25 \times 37.5)10^{-9}}{8.02 \times 10^{-6} \times 100 \times 10^{-3}} = \mathbf{16.4\,MN/m^2}$$

At B:

Here the required $A\bar{y}$ is obtained by subtracting $A\bar{y}$ for the portion of the circle above B from that of the rectangle above B. Now for the circle

$$A\bar{y} = \int_{12.5}^{25} b'y\,dy \quad \text{(Fig. 7.15a)}$$

But, when $y = 12.5$, $\quad \sin\alpha = \dfrac{12.5}{25} = \tfrac{1}{2} \quad \therefore \quad \alpha = \pi/6$

and when $y = 25$, $\quad \sin\alpha = \dfrac{25}{25} = 1 \quad \therefore \quad \alpha = \pi/2$

Also $\quad b' = 2R\cos\alpha, \quad y = R\sin\alpha \quad \text{and} \quad dy = R\cos\alpha\,d\alpha$

$$\therefore \text{ for circle portion, } A\bar{y} = \int_{\pi/6}^{\pi/2} 2R\cos\alpha\, R\sin\alpha\, R\cos\alpha\, d\alpha$$

$$= \int_{\pi/6}^{\pi/2} 2R^3 \cos^2\alpha \sin\alpha\, d\alpha$$

$$= 2R^3\left[-\frac{\cos^3\alpha}{3}\right]_{\pi/6}^{\pi/2}$$

$$= \frac{2 \times 25^3 \times 10^{-9}}{3}\left[\left(\frac{\sqrt{3}}{2}\right)^3\right] = 6.75 \times 10^{-6}\,\text{m}^3$$

$\therefore$ for complete section above B

$$A\bar{y} = 100 \times 37.5 \times 31.25 \times 10^{-9} - 6.75 \times 10^{-6}$$

$$= 110.25 \times 10^{-6}\,\text{m}^3$$

and $$b' = 2R\cos\pi/6 = 2 \times 25 \times 10^{-3} \times \sqrt{3}/2 = 43.3 \times 10^{-3}\,\text{m}$$

$\therefore$ $$b = (100 - 43.3)10^{-3} = 56.7 \times 10^{-3}\,\text{m}$$

$\therefore$ $$\tau_B = \frac{QA\bar{y}}{Ib} = \frac{140 \times 10^3 \times 110.25 \times 10^{-6}}{8.02 \times 10^{-6} \times 56.7 \times 10^{-3}} = \mathbf{34\,MN/m^2}$$

At C:

$$A\bar{y} \text{ for semicircle} = 2R^3\left[-\frac{\cos^3\alpha}{3}\right]_0^{\pi/2}$$

$$= \frac{2 \times 25^3 \times 10^{-9}}{3}[-(0-1)] = 10.41 \times 10^{-6}\,\text{m}^3$$

$A\bar{y}$ for section above C

$$= (100 \times 50 \times 25)10^{-9} - 10.41 \times 10^{-6} = 114.59 \times 10^{-6}\,\text{m}^3$$

and $$b = (100 - 50)10^{-3} = 50 \times 10^{-3}\,\text{m}$$

$\therefore$ $$\tau_c = \frac{140 \times 10^3 \times 114.59 \times 10^{-6}}{8.02 \times 10^{-6} \times 50 \times 10^{-3}} = \mathbf{40\,MN/m^2}$$

The total shear stress distribution across the section is then sketched in Fig. 7.15*b*.

Example 7.4

A beam having the cross-section shown in Fig. 7.16 is constructed from material having a constant thickness of 1.3 mm. Through what point must vertical loads be applied in order that there shall be no twisting of the section? Sketch the shear stress distribution.

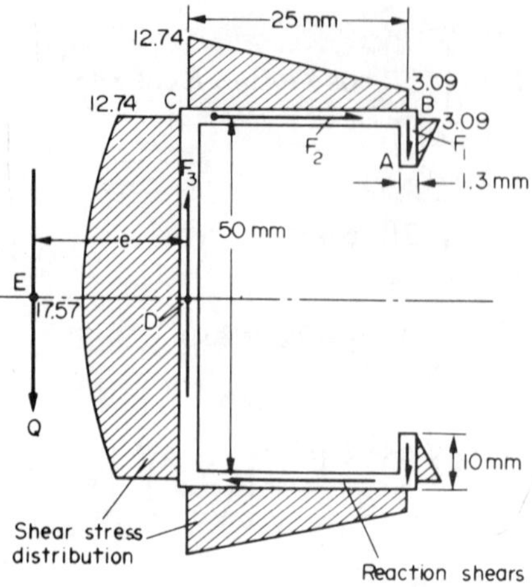

Fig. 7.16.

Solution

Let a load of QN be applied through the point E, distance e from the centre of the web.

$$I_{N.A.} = \left[\frac{1.3 \times 50^3}{12} + 2\left(\frac{25 \times 1.3^3}{12} + 25 \times 1.3 \times 25^2\right) + 2\left(\frac{1.3 \times 10^3}{12} + 1.3 \times 10 \times 20^2\right)\right] \times 10^{-12}$$

$$= [1.354 + 2(0.00046 + 2.03) + 2(0.011 + 0.52)]10^{-8}$$

$$= 6.48 \times 10^{-8}\,\text{m}^4$$

Shear stress $$\tau_A = \frac{QA\bar{y}}{Ib} = \frac{Q \times 0}{Ib} = 0$$

$$\tau_B = \frac{Q \times (10 \times 1.3 \times 20)10^{-9}}{6.48 \times 10^{-8} \times 1.3 \times 10^{-3}} = 3.09Q\,\text{kN/m}^2$$

$$\tau_C = 3.09Q + \frac{Q(25 \times 1.3 \times 25)10^{-9}}{6.48 \times 10^{-8} \times 1.3 \times 10^{-3}}$$

$$= 3.09Q + 9.65Q = 12.74Q\,\text{kN/m}^2$$

$$\tau_D = 12.74Q + \frac{Q(25 \times 1.3 \times 12.5)10^{-9}}{6.48 \times 10^{-8} \times 1.3 \times 10^{-3}}$$

$$= 12.74Q + 4.83Q = 17.57Q\,\text{kN/m}^2$$

The shear stress distribution is then sketched in Fig. 7.16. It should be noted that whilst the distribution is linear along BC it is not strictly so along AB. For ease of calculation of the shear centre, however, it is usually assumed to be linear since the contribution of this region to moments about D is small (the shear centre is the required point through which load must be applied to produce zero twist of the section).

Thus taking moments of *forces* about D for equilibrium,

$$Q \times e \times 10^{-3} = 2F_1 \times 25 \times 10^{-3} + 2F_2 \times 25 \times 10^{-3}$$

$$= 50 \times 10^{-3}\left[\tfrac{1}{2} \times 3.09\,Q \times 10^3 \times (10 \times 1.3 \times 10^{-6}) + 10^3\frac{(3.09Q + 12.74Q)}{2}(25 \times 1.3 \times 10^{-6})\right]$$

$$= 13.866Q \times 10^{-3}$$

$$e = \mathbf{13.87\ mm}$$

Thus, loads must be applied through the point E, 13.87 mm to the left of the web centre-line for zero twist of the section.

Problems

7.1 (A/B). A uniform I-section beam has flanges 150 mm wide by 8 mm thick and a web 180 mm wide and 8 mm thick. At a certain section there is a shearing force of 120 kN. Draw a diagram to illustrate the distribution of shear stress across the section as a result of bending. What is the maximum shear stress? [86.7 MN/m².]

7.2 (A/B). A girder has a uniform T cross-section with flange 250 mm × 50 mm and web 220 mm × 50 mm. At a certain section of the girder there is a shear force of 360 kN.

Plot neatly to scale the shear–stress distribution across the section, stating the values:

(a) where the web and the flange of the section meet;
(b) at the neutral axis. [B.P.] [7.47, 37.4, 39.6 MN/m².]

7.3 (A/B). A beam having an inverted T cross-section has an overall width of 150 mm and overall depth of 200 mm. The thickness of the crosspiece is 50 mm and of the vertical web 25 mm. At a certain section along the beam the vertical shear force is found to be 120 kN. Draw neatly to scale, using 20 mm spacing except where closer intervals are required, a shear–stress distribution diagram across this section. If the mean stress is calculated over the whole of the cross-sectional area, determine the ratio of the maximum shear stress to the mean shear stress. [B.P.] [3.37.]

7.4 (A/B). The channel section shown in Fig. 7.17 is simply supported over a span of 5 m and carries a uniformly distributed load of 15 kN/m run over its whole length. Sketch the shearing-stress distribution diagram at the point of maximum shearing force and mark important values. Determine the ratio of the maximum shearing stress to the average shearing stress. [B.P.] [3, 9.2, 9.3 MN/m²; 2.42.]

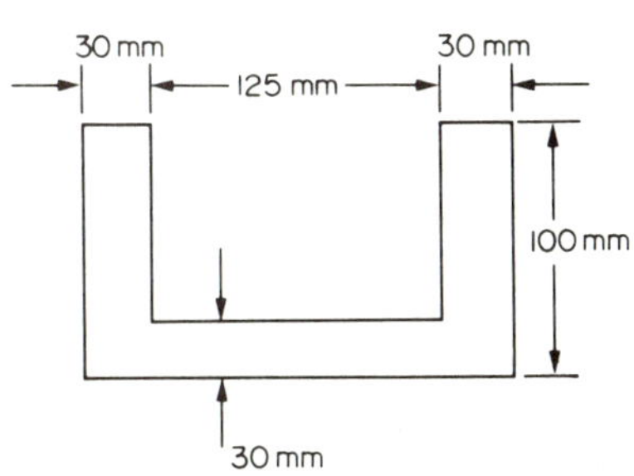

Fig. 7.17.

7.5 (A/B). Fig. 7.18 shows the cross-section of a beam which carries a shear force of 20 kN. Plot a graph to scale which shows the distribution of shear stress due to bending across the cross-section.

[I.Mech.E.] [21.7, 5.2, 5.23 MN/m².]

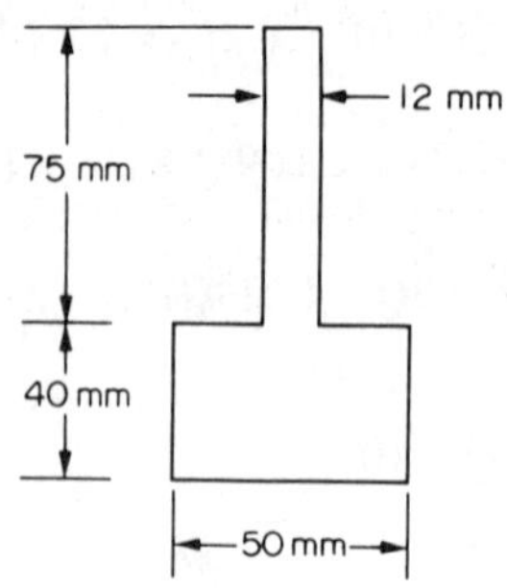

Fig. 7.18.

7.6 (B). Show that the difference between the maximum and mean shear stress in the web of an I-section beam is $\frac{Qh^2}{24I}$ where Q is the shear force on the cross-section, h is the depth of the web and I is the second moment of area of the cross-section about the neutral axis of bending. Assume the I-section to be built of rectangular sections, the flanges having width B and thickness t and the web a thickness b. Fillet radii are to be ignored. [I.Mech.E.]

7.7 (B). Deduce an expression for the shearing stress at any point in a section of a beam owing to the shearing force at that section. State the assumptions made.

A simply supported beam carries a central load W. The cross-section of the beam is rectangular of depth d. At what distance from the neutral axis will the shearing stress be equal to the mean shearing stress on the section?

[U.L.C.I.] [$d/\sqrt{12}$.]

7.8 (B). A steel bar rolled to the section shown in Fig. 7.19 is subjected to a shearing force of 200 kN applied in the direction *YY*. Making the usual assumptions, determine the average shearing stress at the sections A, B, C and D, and find the ratio of the maximum to the mean shearing stress in the section. Draw to scale a diagram to show the variation of the average shearing stress across the section.

[U.L.] [*Clue*: treat as equivalent section similar to that of Example 7.3.] [7.2, 12.3, 33.6, 43.8 MN/m², 3.93.]

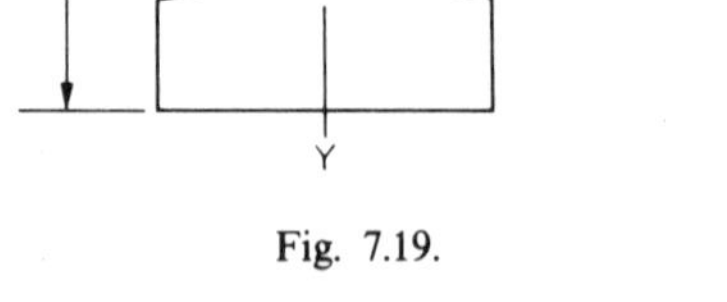

Fig. 7.19.

7.9 (C). Using customary notation, show that the shear stress over the cross-section of a loaded beam is given by $\tau = \frac{QA\bar{y}}{Ib}$.

The cross-section of a beam is an isosceles triangle of base B and height H, the base being arranged in a horizontal plane. Find the shear stress at the neutral axis owing to a shear force Q acting on the cross-section and express it in terms of the mean shear stress.

(The second moment of area of a triangle about its base is $BH^3/12$.) [U.L.C.I.] $\left[\frac{8Q}{3BH}; \frac{4}{3}\tau_{mean}.\right]$

7.10 (C). A hollow steel cylinder, of 200 mm external and 100 mm internal diameter, acting as a beam, is subjected to a shearing force $Q = 10$ kN perpendicular to the axis. Determine the mean shearing stress across the section and, making the usual assumptions, the average shearing stress at the neutral axis and at sections 25, 50 and 75 mm from the neutral axis as fractions of the mean value.

Draw a diagram to show the variation of average shearing stress across the section of the cylinder.

[U.L.] [0.425 MN/m^2; 1.87, 1.65, 0.8, 0.47 MN/m^2.]

7.11 (C). A hexagonal-cross-section bar is used as a beam with its greatest dimension vertical and simply supported at its ends. The beam carries a central load of 60 kN. Draw a stress distribution diagram for a section of the beam at quarter span. All sides of the bar have a length of 25 mm.

($I_{N.A.}$ for triangle $= bh^3/36$ where b = base and h = height.)

[0, 9.2, 14.8, 25.9 MN/m^2 at 12.5 mm intervals above and below the N.A.]

CHAPTER 8

TORSION

Summary

For a *solid or hollow shaft* of uniform circular cross-section throughout its length, the theory of pure torsion states that

$$\frac{T}{J} = \frac{\tau}{R} = \frac{G\theta}{L}$$

where T is the applied external torque, constant over length L;

J is the polar second moment of area of shaft cross-section

$$= \frac{\pi D^4}{32} \text{ for a solid shaft and } \frac{\pi(D^4 - d^4)}{32} \text{ for a hollow shaft;}$$

D is the outside diameter; R is the outside radius;

d is the inside diameter;

τ is the shear stress at radius R and is the maximum value for both solid and hollow shafts;

G is the modulus of rigidity (shear modulus); and

θ is the angle of twist in *radians* on a length L.

For *very thin-walled hollow shafts*

$J = 2\pi r^3 t$, where r is the mean radius of the shaft wall and t is the thickness.

Shear stress and shear strain are related to the angle of twist thus:

$$\tau = \frac{G\theta}{L} R = G\gamma$$

Strain energy in torsion is given by

$$U = \frac{T^2 L}{2GJ} = \frac{GJ\theta^2}{2L} \left(= \frac{\tau^2}{4G} \times \text{volume for } \textit{solid shafts} \right)$$

For a circular shaft subjected to *combined bending and torsion* the *equivalent bending moment* is

$$M_e = \tfrac{1}{2}[M + \sqrt{(M^2 + T^2)}]$$

and the *equivalent torque* is

$$T_e = \tfrac{1}{2}\sqrt{(M^2 + T^2)}$$

where M and T are the applied bending moment and torque respectively.

The *power transmitted* by a shaft carrying torque T at ω rad/s $= T\omega$.

8.1. Simple torsion theory

When a uniform circular shaft is subjected to a torque it can be shown that every section of the shaft is subjected to a state of pure shear (Fig. 8.1), the moment of resistance developed by the shear stresses being everywhere equal to the magnitude, and opposite in sense, to the applied torque. For the purposes of deriving a simple theory to describe the behaviour of shafts subjected to torque it is necessary to make the following basic assumptions:

(1) The material is homogeneous, i.e. of uniform elastic properties throughout.
(2) The material is elastic, following Hooke's law with shear stress proportional to shear strain.
(3) The stress does not exceed the elastic limit or limit of proportionality.
(4) Circular sections remain circular.
(5) Cross-sections remain plane. (This is certainly not the case with the torsion of non-circular sections.)
(6) Cross-sections rotate as if rigid, i.e. every diameter rotates through the same angle.

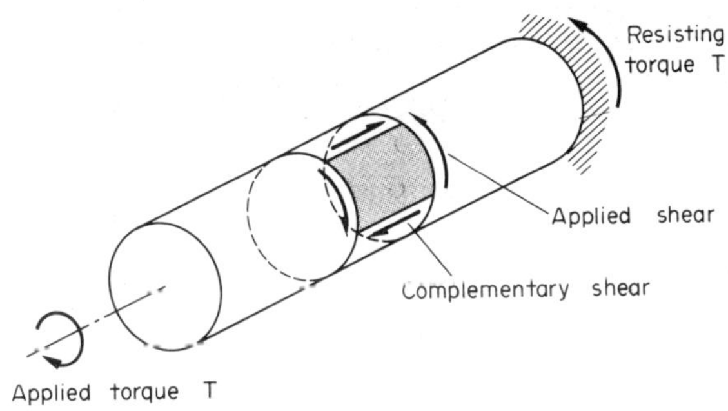

Fig. 8.1. Shear system set up on an element in the surface of a shaft subjected to torsion.

Practical tests carried out on circular shafts have shown that the theory developed below on the basis of these assumptions shows excellent correlation with experimental results.

(a) *Angle of twist*

Consider now the solid circular shaft of radius R subjected to a torque T at one end, the other end being fixed (Fig. 8.2). Under the action of this torque a radial line at the free end of the shaft twists through an angle θ, point A moves to B, and AB subtends an angle γ at the fixed end. This is then the angle of distortion of the shaft, i.e. *the shear strain.*

Since angle in radians = arc ÷ radius

$$\text{arc } AB = R\theta = L\gamma$$

$$\therefore \quad \gamma = R\theta/L \tag{8.1}$$

From the definition of rigidity modulus

$$G = \frac{\text{shear stress } \tau}{\text{shear strain } \gamma}$$

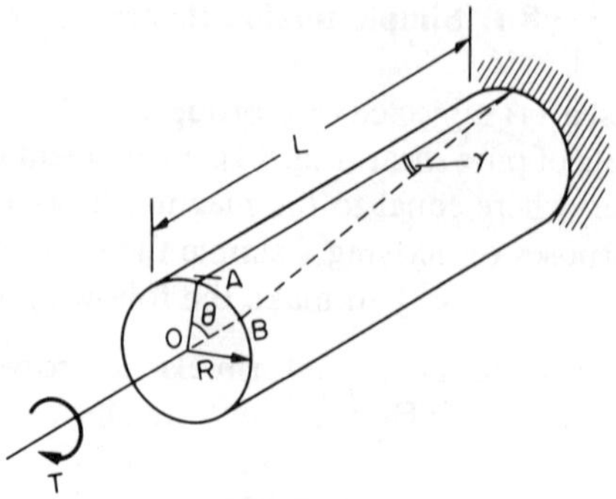

Fig. 8.2.

$$\therefore \qquad \gamma = \frac{\tau}{G} \tag{8.2}$$

where τ is the shear stress set up at radius R.

Therefore equating eqns. (8.1) and (8.2),

$$\frac{R\theta}{L} = \frac{\tau}{G}$$

$$\frac{\tau}{R} = \frac{G\theta}{L}\left(= \frac{\tau'}{r}\right) \tag{8.3}$$

where τ' is the shear stress at any other radius r.

(b) Stresses

Let the cross-section of the shaft be considered as divided into elements of radius r and thickness dr as shown in Fig. 8.3 each subjected to a shear stress τ'.

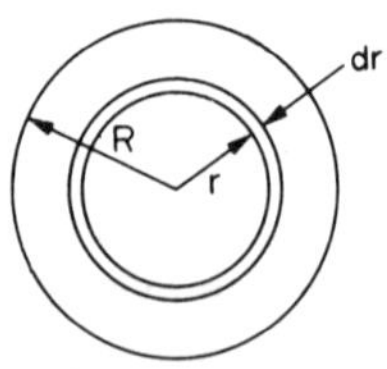

Fig. 8.3. Shaft cross-section.

The force set up on each element

$$= \text{stress} \times \text{area}$$

$$= \tau' \times 2\pi r\,dr \text{ (approximately)}$$

This force will produce a moment about the centre axis of the shaft, providing a contribution to the torque

$$= (\tau' \times 2\pi r\, dr) \times r$$

$$= 2\pi\tau' r^2\, dr$$

The total torque on the section T will then be the sum of all such contributions across the section,

i.e.

$$T = \int_0^R 2\pi\tau' r^2\, dr$$

Now the shear stress τ' will vary with the radius r and must therefore be replaced in terms of r before the integral is evaluated.

From eqn. (8.3)

$$\tau' = \frac{G\theta}{L} r$$

$$\therefore \quad T = \int_0^R 2\pi \frac{G\theta}{L} r^3\, dr$$

$$= \frac{G\theta}{L} \int_0^R 2\pi r^3\, dr$$

The integral $\int_0^R 2\pi r^3\, dr$ is called the *polar second moment of area J*, and may be evaluated as a standard form for solid and hollow shafts as shown in §8.2 below.

$$\therefore \quad T = \frac{G\theta}{L} J$$

or

$$\frac{T}{J} = \frac{G\theta}{L} \tag{8.4}$$

Combining eqns. (8.3) and (8.4) produces the so-called simple theory of torsion:

$$\frac{T}{J} = \frac{\tau}{R} = \frac{G\theta}{L} \tag{8.5}$$

8.2. Polar second moment of area

As stated above the polar second moment of area J is defined as

$$J = \int_0^R 2\pi r^3\, dr$$

For a solid shaft,

$$J = 2\pi\left[\frac{r^4}{4}\right]_0^R$$

$$= \frac{2\pi R^4}{4} \quad \text{or} \quad \frac{\pi D^4}{32} \tag{8.6}$$

For a hollow shaft of internal radius r,

$$J = 2\pi\int_r^R r^3\,dr = 2\pi\left[\frac{r^4}{4}\right]_r^R$$

$$= \frac{\pi}{2}(R^4 - r^4) \quad \text{or} \quad \frac{\pi}{32}(D^4 - d^4) \tag{8.7}$$

For thin-walled hollow shafts the values of D and d may be nearly equal, and in such cases there can be considerable errors in using the above equation involving the difference of two large quantities of similar value. It is therefore convenient to obtain an alternative form of expression for the polar moment of area.

Now

$$J = \int_0^R 2\pi r^3\,dr = \Sigma(2\pi r\,dr)r^2$$

$$= \Sigma\, Ar^2$$

where $A(= 2\pi r\,dr)$ is the area of each small element of Fig. 8.3, i.e. J is the sum of the Ar^2 terms for all elements.

If a thin hollow cylinder is therefore considered as just one of these small elements with its wall thickness $t = dr$, then

$$J = Ar^2 = (2\pi r\,t)r^2$$

$$= 2\pi r^3 t \text{ (approximately)} \tag{8.8}$$

8.3. Shear stress and shear strain in shafts

The shear stresses which are developed in a shaft subjected to pure torsion are indicated in Fig. 8.1, their values being given by the simple torsion theory as

$$\tau = \frac{G\theta}{L}R$$

Now from the definition of the shear or rigidity modulus G,

$$\tau = G\gamma$$

It therefore follows that the two equations may be combined to relate the shear stress and strain in the shaft to the angle of twist per unit length, thus

$$\tau = \frac{G\theta}{L}R = G\gamma \tag{8.9}$$

or, in terms of some internal radius r,

$$\tau' = \frac{G\theta}{L} r = G\gamma \tag{8.10}$$

These equations indicate that the shear stress and shear strain vary linearly with radius and have their maximum value at the outside radius (Fig. 8.4). The applied shear stresses in the plane of the cross-section are accompanied by complementary stresses of equal value on longitudinal planes as indicated in Figs. 8.1 and 8.4. The significance of these longitudinal shears to material failure is discussed further in §8.10.

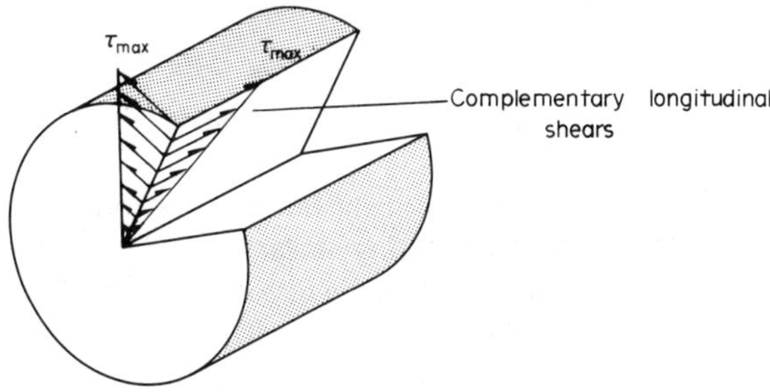

Fig. 8.4. Complementary longitudinal shear stress in a shaft subjected to torsion.

8.4. Section modulus

It is sometimes convenient to re-write part of the torsion theory formula to obtain the maximum shear stress in shafts as follows:

$$\frac{T}{J} = \frac{\tau}{R}$$

$$\therefore \qquad \tau = \frac{TR}{J}$$

With R the outside radius of the shaft the above equation yields the greatest value possible for τ (Fig. 8.4),

i.e. $$\tau_{max} = \frac{TR}{J}$$

$$\therefore \qquad \tau_{max} = \frac{T}{Z} \tag{8.11}$$

where $Z = J/R$ is termed the *polar section modulus*. It will be seen from the preceding section that:

for solid shafts, $$Z = \frac{\pi D^3}{16} \tag{8.12}$$

and for hollow shafts, $$Z = \frac{\pi(D^4 - d^4)}{16D} \tag{8.13}$$

8.5. Torsional rigidity

The angle of twist per unit length of shafts is given by the torsion theory as

$$\frac{\theta}{L} = \frac{T}{GJ}$$

The quantity GJ is termed the *torsional rigidity* of the shaft and is thus given by

$$GJ = \frac{T}{\theta/L} \tag{8.14}$$

i.e. the torsional rigidity is the torque divided by the angle of twist (in radians) per unit length.

8.6. Torsion of hollow shafts

It has been shown above that the maximum shear stress in a solid shaft is developed in the outer surface, values at other radii decreasing linearly to zero at the centre. It is clear, therefore, that if there is to be some limit set on the maximum allowable working stress in the shaft material then only the outer surface of the shaft will reach this limit. The material within the shaft will work at a lower stress and, particularly near the centre, will not contribute as much to the torque-carrying capacity of the shaft. In applications where weight reduction is of prime importance as in the aerospace industry, for instance, it is often found advisable to use hollow shafts.

The relevant formulae for hollow shafts have been introduced in §8.2 and will not be repeated here. As an example of the increased torque-to-weight ratio possible with hollow shafts, however, it should be noted for a hollow shaft with an inside diameter half the outside diameter that the maximum stress increases by 6% over that for a solid shaft of the same outside diameter whilst the weight reduction achieved is approximately 25%.

8.7. Torsion of thin-walled tubes

The torsion of thin-walled tubes of circular and non-circular cross-section will be treated fully in § §20.4 and 20.6.

8.8. Composite shafts – series connection

If two or more shafts of different material, diameter or basic form are connected together in such a way that each carries the same torque, then the shafts are said to be connected in series and the composite shaft so produced is therefore termed *series-connected* (Fig. 8.5) (see Example 8.3). In such cases the composite shaft strength is treated by considering each component shaft separately, applying the torsion theory to each in turn; the composite shaft will therefore be as weak as its weakest component. If relative dimensions of the various parts are required then a solution is usually effected by equating the torques in each shaft, e.g. for two shafts in series

$$T = \frac{G_1 J_1 \theta_1}{L_1} = \frac{G_2 J_2 \theta_2}{L_2} \tag{8.15}$$

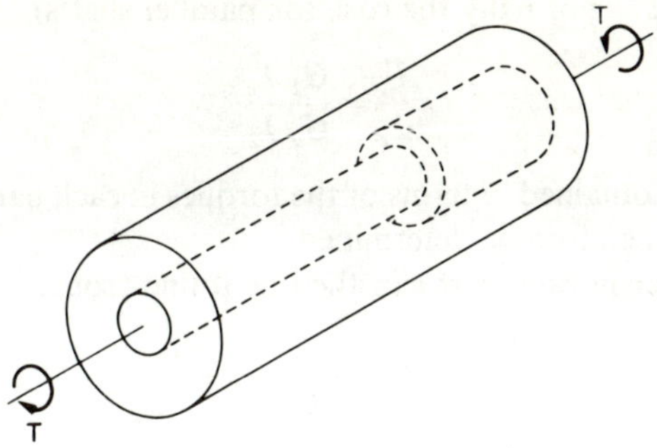

Fig. 8.5. "Series-connected" shaft – common torque.

In some applications it is convenient to ensure that the angles of twist in each shaft are equal, i.e. $\theta_1 = \theta_2$, so that for similar materials in each shaft

$$\frac{J_1}{L_1} = \frac{J_2}{L_2}$$

or

$$\frac{L_1}{L_2} = \frac{J_1}{J_2} \tag{8.16}$$

8.9. Composite shafts – parallel connection

If two or more materials are rigidly fixed together such that the applied torque is shared between them then the composite shaft so formed is said to be *connected in parallel* (Fig. 8.6).

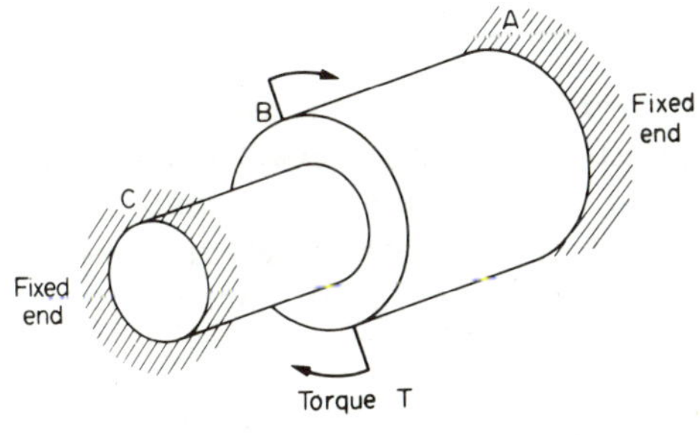

Fig. 8.6. "Parallel-connected" shaft – shared torque.

For parallel connection,

$$\text{total torque } T = T_1 + T_2 \tag{8.17}$$

In this case the angles of twist of each portion are equal and

$$\frac{T_1 L_1}{G_1 J_1} = \frac{T_2 L_2}{G_2 J_2} \tag{8.18}$$

i.e. for equal lengths (as is normally the case for parallel shafts)

$$\frac{T_1}{T_2} = \frac{G_1 J_1}{G_2 J_2} \tag{8.19}$$

Thus two equations are obtained in terms of the torques in each part of the composite shaft and these torques can therefore be determined.

The maximum stresses in each part can then be found from

$$\tau_1 = \frac{T_1 R_1}{J_1} \quad \text{and} \quad \tau_2 = \frac{T_2 R_2}{J_2}$$

8.10. Principal stresses

It will be shown in §13.2 that a state of pure shear as produced by the torsion of shafts is equivalent to a system of biaxial direct stresses, one stress tensile, one compressive, of equal value and at 45° to the shaft axis as shown in Fig. 8.7; these are then the principal stresses.

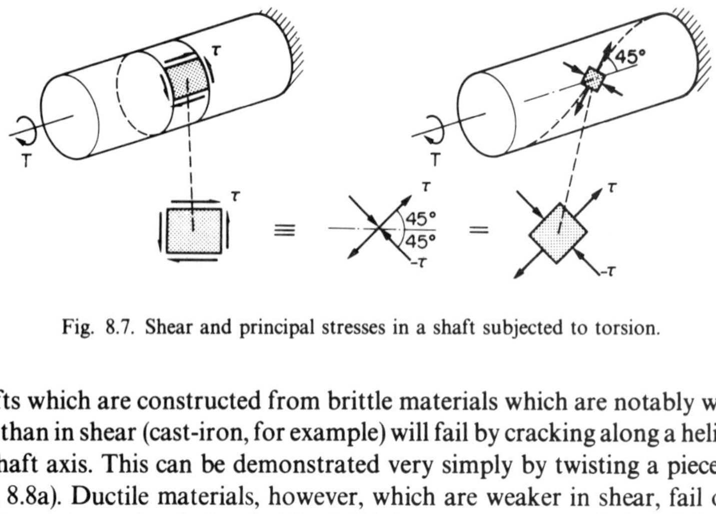

Fig. 8.7. Shear and principal stresses in a shaft subjected to torsion.

Thus shafts which are constructed from brittle materials which are notably weaker under direct stress than in shear (cast-iron, for example) will fail by cracking along a helix inclined at 45° to the shaft axis. This can be demonstrated very simply by twisting a piece of chalk to failure (Fig. 8.8a). Ductile materials, however, which are weaker in shear, fail on the shear planes at right angles to the shaft axis (Fig. 8.8b). In some cases, e.g. timber, failure occurs by cracking along the shear planes parallel to the shaft axis owing to the nature of the material with fibres generally parallel to the axis producing a weakness in shear longitudinally rather than transversely. The complementary shears of Fig. 8.4 then assume greater significance.

8.11. Strain energy in torsion

It will be shown in §11.4 that the strain energy stored in a solid circular bar or shaft subjected to a torque T is given by the alternative expressions

$$U = \tfrac{1}{2}T\theta = \frac{T^2 L}{2GJ} = \frac{GJ\theta^2}{2L} = \frac{\tau^2}{4G} \times \text{volume} \tag{8.20}$$

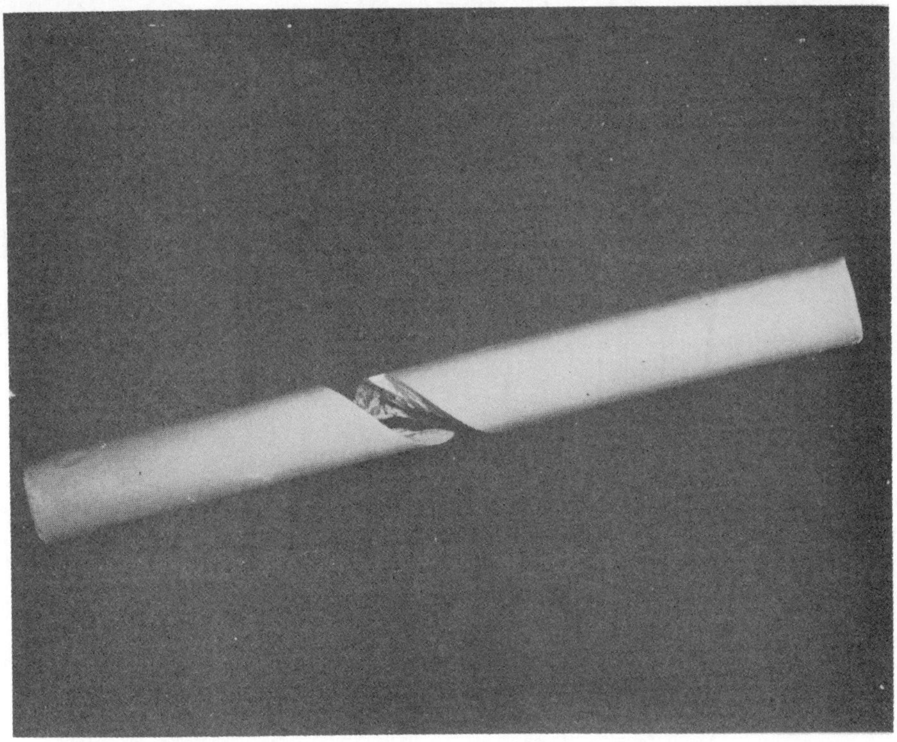

Fig. 8.8a. Typical failure of a brittle material (chalk) in torsion. Failure occurs on a 45° helix owing to the action of the direct tensile stresses produced at 45° by the applied torque.

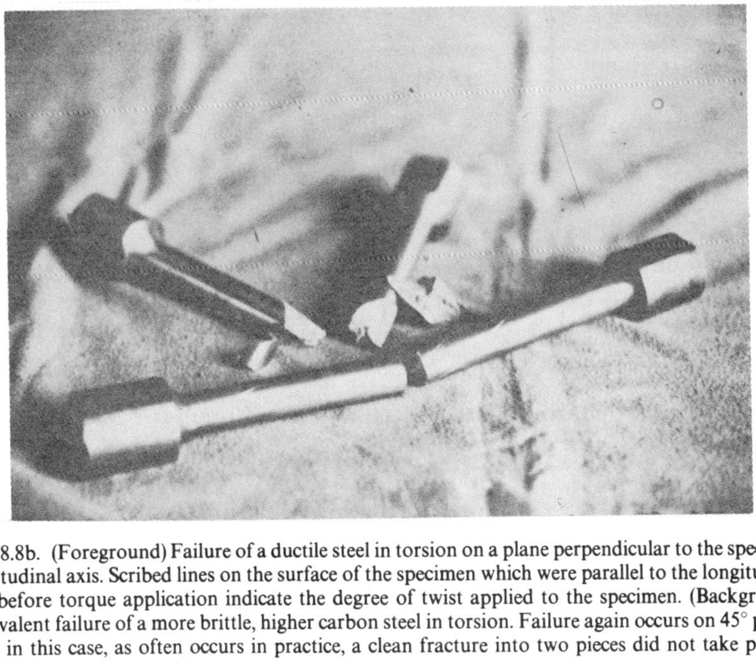

Fig. 8.8b. (Foreground) Failure of a ductile steel in torsion on a plane perpendicular to the specimen longitudinal axis. Scribed lines on the surface of the specimen which were parallel to the longitudinal axis before torque application indicate the degree of twist applied to the specimen. (Background) Equivalent failure of a more brittle, higher carbon steel in torsion. Failure again occurs on 45° planes but in this case, as often occurs in practice, a clean fracture into two pieces did not take place.

8.12. Variation of data along shaft length – torsion of tapered shafts

This section illustrates the procedure which may be adopted when any of the quantities normally used in the torsion equations vary along the length of the shaft. Provided the variation is known in terms of x, the distance along the shaft, then a solution can be obtained.

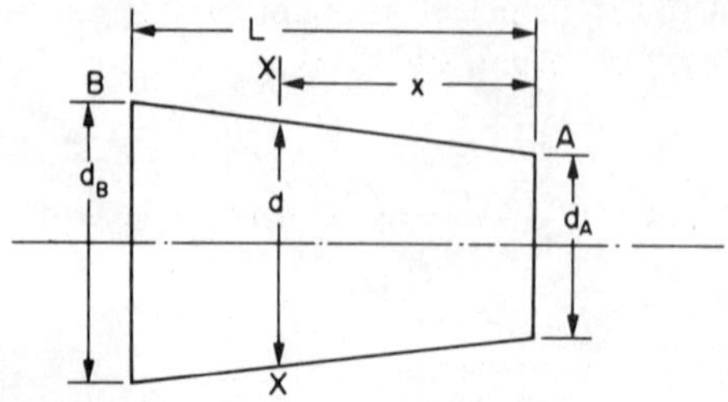

Fig. 8.9. Torsion of a tapered shaft.

Consider, therefore, the tapered shaft shown in Fig. 8.9 with its diameter changing linearly from d_A to d_B over a length L. The diameter at any section x from end A is then given by

$$d = d_A + (d_B - d_A)\frac{x}{L}$$

Provided that the angle of the taper is not too great, the simple torsion theory may be applied to an element at section XX in order to determine the angle of twist of the shaft, i.e. for the element shown,

$$\frac{G d\theta}{dx} = \frac{T}{J_{XX}}$$

Therefore the total angle of twist of the shaft is given by

$$\theta = \int_0^L \frac{T}{GJ_{XX}}\,dx$$

Now
$$J_{XX} = \frac{\pi d^4}{32} = \frac{\pi}{32}\left[d_A + (d_B - d_A)\frac{x}{L}\right]^4$$

Substituting and integrating,

$$\theta = \frac{32TL}{3\pi G}\left[\frac{1}{d_A^3} - \frac{1}{d_B^3}\right]\left[\frac{1}{d_B} - \frac{1}{d_A}\right] = \frac{32TL}{3\pi G}\left[\frac{d_A^2 + d_A d_B + d_B^2}{d_A^3 d_B^3}\right]$$

When $d_A = d_B = d$ this reduces to $\theta = \dfrac{32TL}{\pi G d^4}$ the standard result for a parallel shaft.

8.13. Power transmitted by shafts

If a shaft carries a torque T Newton metres and rotates at ω rad/s it will do work at the rate of

$$T\omega \text{ Nm/s (or joule/s).}$$

Now the rate at which a system works is defined as its power, the basic unit of power being the Watt (1 Watt = 1 Nm/s).
Thus, the power transmitted by the shaft:

$$= T\omega \text{ Watts.}$$

Since the Watt is a very small unit of power in engineering terms use is normally made of S.I. multiples, i.e. kilowatts (kW) or megawatts (MW).

8.14. Combined stress systems – combined bending and torsion

In most practical transmission situations shafts which carry torque are also subjected to bending, if only by virtue of the self-weight of the gears they carry. Many other practical applications occur where bending and torsion arise simultaneously so that this type of loading represents one of the major sources of complex stress situations.

In the case of shafts, bending gives rise to tensile stress on one surface and compressive stress on the opposite surface whilst torsion gives rise to pure shear throughout the shaft. An element on the tensile surface will thus be subjected to the stress system indicated in Fig. 8.10 and eqn. (13.11) or the Mohr circle procedure of §13.6 can be used to obtain the principal stresses present.

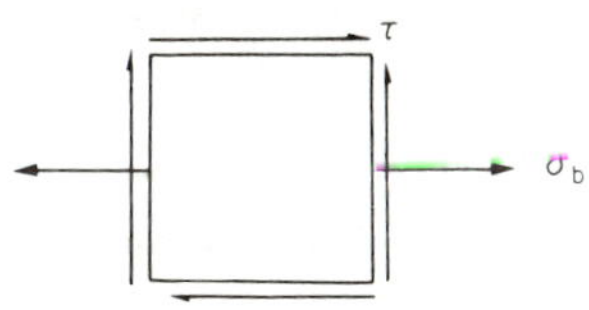

Fig. 8.10. Stress system on the surface of a shaft subjected to torque and bending.

Alternatively, the shaft can be considered to be subjected to *equivalent torques* or *equivalent bending moments* as described below.

8.15. Combined bending and torsion – equivalent bending moment

For shafts subjected to the simultaneous application of a bending moment M and torque T the *principal stresses* set up in the shaft can be shown to be equal to those produced by an *equivalent bending moment*, of a certain value M_e acting alone.

From the simple bending theory the maximum direct stresses set up at the outside surface of the shaft owing to the bending moment M are given by

$$\sigma = \frac{My_{\max}}{I} = \frac{MD}{2I}$$

Similarly, from the torsion theory, the maximum shear stress in the surface of the shaft is given by

$$\tau = \frac{TR}{J} = \frac{TD}{2J}$$

But for a circular shaft $J = 2I$,

$$\therefore \qquad \tau = \frac{TD}{4I}$$

The principal stresses for this system can now be obtained by applying the formula derived in § 13.4,

i.e.
$$\sigma_1 \text{ or } \sigma_2 = \tfrac{1}{2}(\sigma_x + \sigma_y) \pm \tfrac{1}{2}\sqrt{[(\sigma_x - \sigma_y)^2 + 4\tau^2]}$$

and, with $\sigma_y = 0$, the maximum principal stress σ_1 is given by

$$\sigma_1 = \frac{1}{2}\left(\frac{MD}{2I}\right) + \frac{1}{2}\sqrt{\left[\left(\frac{MD}{2I}\right)^2 + 4\left(\frac{TD}{4I}\right)^2\right]}$$

$$= \frac{1}{2}\left(\frac{D}{2I}\right)[M + \sqrt{(M^2 + T^2)}]$$

Now if M_e is the bending moment which, acting alone, will produce the same maximum stress, then

$$\sigma_1 = \frac{M_e y_{max}}{I} = \frac{M_e D}{2I}$$

$$\therefore \qquad \frac{M_e D}{2I} = \frac{1}{2}\left(\frac{D}{2I}\right)[M + \sqrt{(M^2 + T^2)}]$$

i.e. the equivalent bending moment is given by

$$\boldsymbol{M_e = \tfrac{1}{2}[M + \sqrt{(M^2 + T^2)}]} \qquad (8.21)$$

and it will produce the same maximum direct stress as the combined bending and torsion effects.

8.16. Combined bending and torsion – equivalent torque

Again considering shafts subjected to the simultaneous application of a bending moment M and a torque T the *maximum shear stress* set up in the shaft may be determined by the application of an *equivalent torque* of value T_e acting alone.

From the preceding section the principal stresses in the shaft are given by

$$\sigma_1 = \frac{1}{2}\left(\frac{D}{2I}\right)[M + \sqrt{(M^2 + T^2)}] = \frac{1}{2}\left(\frac{D}{J}\right)[M + \sqrt{(M^2 + T^2)}]$$

and
$$\sigma_2 = \frac{1}{2}\left(\frac{D}{2I}\right)[M - \sqrt{(M^2 + T^2)}] = \frac{1}{2}\left(\frac{D}{J}\right)[M - \sqrt{(M^2 + T^2)}]$$

Now the maximum shear stress is given by eqn. (13.12)

$$\tau_{max} = \tfrac{1}{2}(\sigma_1 - \sigma_2) = \frac{1}{2}\left(\frac{D}{J}\right)\sqrt{(M^2 + T^2)}$$

But, from the torsion theory, the equivalent torque T_e will set up a maximum shear stress of

$$\tau_{max} = \frac{T_e D}{2J}$$

Thus if these maximum shear stresses are to be equal,

$$T_e = \sqrt{(M^2 + T^2)} \tag{8.22}$$

It must be remembered that the equivalent moment M_e and equivalent torque T_e are merely convenient devices to obtain the maximum principal direct stress or maximum shear stress, respectively, under the combined stress system. They should not be used for other purposes such as the calculation of power transmitted by the shaft; this depends solely on the torque T carried by the shaft (not on T_e).

8.17. Combined bending, torsion and direct thrust

Additional stresses arising from the action of direct thrusts on shafts may be taken into account by adding the direct stress due to the thrust σ_d to that of the direct stress due to bending σ_b taking due account of sign. The complex stress system resulting on any element in the shaft is then as shown in Fig. 8.11 and may be solved to determine the principal stresses using Mohr's stress circle method of solution described in § 13.6.

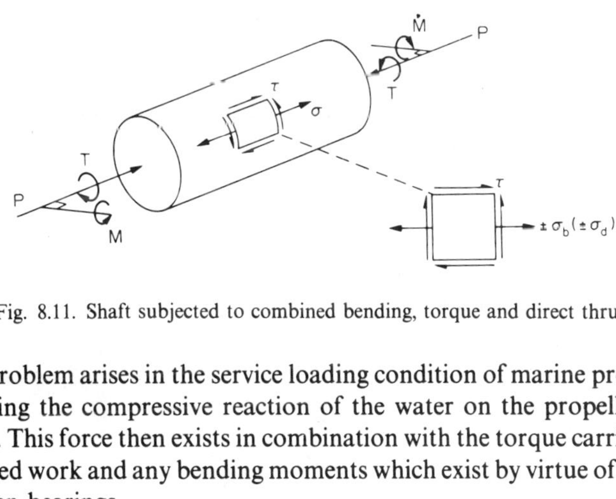

Fig. 8.11. Shaft subjected to combined bending, torque and direct thrust.

This type of problem arises in the service loading condition of marine propeller shafts, the direct thrust being the compressive reaction of the water on the propeller as the craft is pushed forward. This force then exists in combination with the torque carried by the shaft in doing the required work and any bending moments which exist by virtue of the self-weight of the shaft between bearings.

The compressive stress σ_d arising from the propeller reaction is thus superimposed on the bending stresses; on the compressive bending surface it will be additive to σ_b whilst on the "tensile" surface it will effectively reduce the value of σ_b, see Fig. 8.11.

8.18. Combined bending, torque and internal pressure

In the case of pressurised cylinders, direct stresses will be introduced in two perpendicular directions. These have been introduced in Chapters 9 and 10 as the radial and circumferential stresses σ_r and σ_H. If the cylinder also carries a torque then shear stresses will be introduced, their value being calculated from the simple torsion theory of § 8.3. The stress system on an element will thus become that shown in Fig. 8.12.

If bending is present it will generally be on the x axis and will result in a modification to the value of σ_x. If the element is taken on the tensile surface of the cylinder then the bending stress σ_b will add to the value of σ_H, if on the compressive surface it must be subtracted from σ_H.

Once again a solution to such problems can be effected either by application of eqn. (13.11) or by a Mohr circle approach.

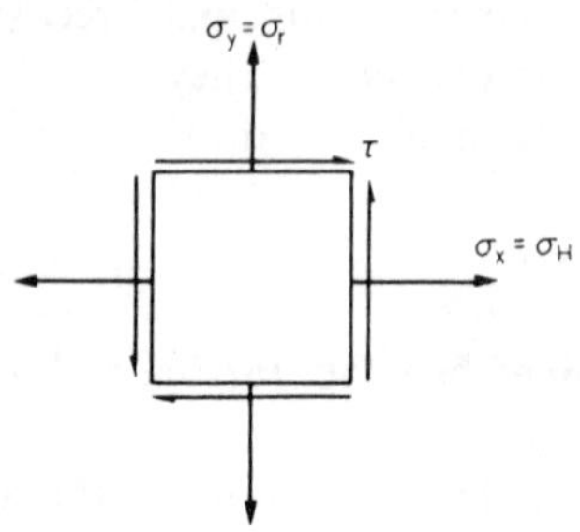

Fig. 8.12. Stress system under combined torque and internal pressure.

Examples

Example 8.1

(a) A solid shaft, 100 mm diameter, transmits 75 kW at 150 rev/min. Determine the value of the maximum shear stress set up in the shaft and the angle of twist per metre of the shaft length if $G = 80\ \text{GN/m}^2$.

(b) If the shaft were now bored in order to reduce weight to produce a tube of 100 mm outside diameter and 60 mm inside diameter, what torque could be carried if the same maximum shear stress is not to be exceeded? What is the percentage increase in power/weight ratio effected by this modification?

Solution

(a)
$$\text{Power} = T\omega \quad \therefore \quad \text{torque } T = \frac{\text{power}}{\omega}$$

$$\therefore \quad T = \frac{75 \times 10^3}{150 \times 2\pi/60} = 4.77\ \text{kN m}$$

From the torsion theory

$$\frac{T}{J} = \frac{\tau}{R} \quad \text{and} \quad J = \frac{\pi}{32} \times 100^4 \times 10^{-12} = 9.82 \times 10^{-6}\ \text{m}^4$$

$$\therefore \quad \tau_{max} = \frac{TR_{max}}{J} = \frac{4.77 \times 10^3 \times 50 \times 10^{-3}}{9.82 \times 10^{-6}} = \mathbf{24.3\ MN/m^2}$$

Also from the torsion theory

$$\theta = \frac{TL}{GJ} = \frac{4.77 \times 10^3 \times 1}{80 \times 10^9 \times 9.82 \times 10^{-6}} = 6.07 \times 10^{-3} \text{ rad/m}$$

$$= 6.07 \times 10^{-3} \times \frac{360}{2\pi} = \mathbf{0.348\ degrees/m}$$

(b) When the shaft is bored, the polar moment of area J is modified thus:

$$J = \frac{\pi}{32}(D^4 - d^4) = \frac{\pi}{32}(100^4 - 60^4)10^{-12} = 8.545 \times 10^{-6}\ \text{m}^4$$

The torque carried by the modified shaft is then given by

$$T = \frac{\tau J}{R} = \frac{24.3 \times 10^6 \times 8.545 \times 10^{-6}}{50 \times 10^{-3}} = \mathbf{4.15 \times 10^3\ Nm}$$

Now, weight/metre of original shaft

$$= \frac{\pi}{4}(100)^2 \times 10^{-6} \times 1 \times \rho g = 7.854 \times 10^{-3}\ \rho g$$

where ρ is the density of the shaft material.

Also, weight/metre of modified shaft $= \frac{\pi}{4}(100^2 - 60^2)10^{-6} \times 1 \times \rho g$

$$= 5.027 \times 10^{-3}\ \rho g$$

Power/weight ratio for original shaft $= \dfrac{T\omega}{\text{weight/metre}}$

$$= \frac{4.77 \times 10^3\ \omega}{7.854 \times 10^{-3}\ \rho g} = 6.073 \times 10^5 \frac{\omega}{\rho g}$$

Power/weight ratio for modified shaft

$$= \frac{4.15 \times 10^3\ \omega}{5.027 \times 10^{-3}\ \rho g} = 8.255 \times 10^5 \frac{\omega}{\rho g}$$

Therefore percentage increase in power/weight ratio

$$= \frac{(8.255 - 6.073)}{6.073} \times 100 = \mathbf{36\%}$$

Example 8.2

Determine the dimensions of a hollow shaft with a diameter ratio of 3:4 which is to transmit 60 kW at 200 rev/min. The maximum shear stress in the shaft is limited to 70 MN/m^2 and the angle of twist to 3.8° in a length of 4 m.

For the shaft material $G = 80$ GN/m^2.

Solution

The two limiting conditions stated in the question, namely maximum shear stress and angle of twist, will each lead to different values for the required diameter. The larger shaft must then be chosen as the one for which neither condition is exceeded.

Maximum shear stress condition

Since power $= T\omega$ and $\omega = 200 \times \dfrac{2\pi}{60} = 20.94$ rad/s

then
$$T = \frac{60 \times 10^3}{20.94} = 2.86 \times 10^3 \text{ Nm}$$

From the torsion theory
$$J = \frac{TR}{\tau}$$

$$\therefore \qquad \frac{\pi}{32}(D^4 - d^4) = \frac{2.86 \times 10^3 \times D}{70 \times 10^6 \times 2}$$

But $d/D = 0.75$

$$\therefore \qquad \frac{\pi}{32} D^4 (1 - 0.75^4) = 20.43 \times 10^{-6} D$$

$$D^3 = \frac{20.43 \times 10^{-6}}{0.0671} = 304.4 \times 10^{-6}$$

$$\therefore \qquad D = 0.0673 \text{ m} = 67.3 \text{ mm}$$

and
$$d = 50.5 \text{ mm}$$

Angle of twist condition

Again from the torsion theory
$$J = \frac{TL}{G\theta}$$

$$\frac{\pi}{32}(D^4 - d^4) = \frac{2.86 \times 10^3 \times 4 \times 360}{80 \times 10^9 \times 3.8 \times 2\pi}$$

$$\frac{\pi}{32} D^4 (1 - 0.75^4) = 2.156 \times 10^{-6}$$

$$D^4 = \frac{2.156 \times 10^{-6}}{0.0671} = 32.12 \times 10^{-6}$$

$$D = 0.0753 \text{ m} = 75.3 \text{ mm}$$

and
$$d = 56.5 \text{ mm}$$

Thus the dimensions required for the shaft to satisfy both conditions are **outer diameter 75.3 mm; inner diameter 56.5 mm.**

Example 8.3

(a) A steel transmission shaft is 510 mm long and 50 mm external diameter. For part of its length it is bored to a diameter of 25 mm and for the rest to 38 mm diameter. Find the maximum power that may be transmitted at a speed of 210 rev/min if the shear stress is not to exceed 70 MN/m^2.

(b) If the angle of twist in the length of 25 mm bore is equal to that in the length of 38 mm bore, find the length bored to the latter diameter.

Solution

(a) This is, in effect, a question on *shafts in series* since each part is subjected to the same torque.

From the torsion theory

$$T = \frac{\tau J}{R}$$

and as the maximum stress and the radius at which it occurs (the outside radius) are the same for both shafts the torque allowable for a known value of shear stress is dependent only on the value of J. This will be least where the internal diameter is greatest since

$$J = \frac{\pi}{32}(D^4 - d^4)$$

$$\therefore \qquad \text{least value of } J = \frac{\pi}{32}(50^4 - 38^4)10^{-12} = 0.41 \times 10^{-6}\,\text{m}^4$$

Therefore maximum allowable torque if the shear stress is not to exceed 70 MN/m^2 (at 25 mm radius) is given by

$$T = \frac{70 \times 10^6 \times 0.41 \times 10^{-6}}{25 \times 10^{-3}} = 1.15 \times 10^3\,\text{Nm}$$

$$\text{Maximum power} = T\omega = 1.15 \times 10^3 \times 210 \times \frac{2\pi}{60}$$

$$= 25.2 \times 10^3 = \mathbf{25.2\,kW}$$

(b) Let suffix 1 refer to the 38 mm diameter bore portion and suffix 2 to the other part. Now for shafts in series, eqn. (8.16) applies,

i.e.
$$\frac{J_1}{L_1} = \frac{J_2}{L_2}$$

$$\therefore \quad \frac{L_2}{L_1} = \frac{J_2}{J_1} = \frac{\frac{\pi}{32}(50^4 - 25^4)10^{-12}}{\frac{\pi}{32}(50^4 - 38^4)10^{-12}} = 1.43$$

$$\therefore \quad L_2 = 1.43\,L_1$$

But
$$L_1 + L_2 = 510\,\text{mm}$$

$$\therefore \quad L_1(1 + 1.43) = 510$$

$$L_1 = \frac{510}{2.43} = \mathbf{210\,mm}$$

Example 8.4

A circular bar *ABC*, 3 m long, is rigidly fixed at its ends *A* and *C*. The portion *AB* is 1.8 m long and of 50 mm diameter and *BC* is 1.2 m long and of 25 mm diameter. If a twisting moment of 680 N m is applied at *B*, determine the values of the resisting moments at *A* and *C* and the maximum stress in each section of the shaft. What will be the angle of twist of each portion?

For the material of the shaft $G = 80\ \text{GN/m}^2$.

Solution

In this case the two portions of the *shaft* are *in parallel* and the applied torque is shared between them. Let suffix 1 refer to portion *AB* and suffix 2 to portion *BC*.

Since the angles of twist in each portion are equal and *G* is common to both sections,

then
$$\frac{T_1 L_1}{J_1} = \frac{T_2 L_2}{J_2}$$

$$\therefore \quad T_1 = \frac{J_1}{J_2} \times \frac{L_2}{L_1} \times T_2 = \frac{\frac{\pi}{32} \times 50^4}{\frac{\pi}{32} \times 25^4} \times \frac{1.2}{1.8} \times T_2$$

$$= \frac{16 \times 1.2}{1.8} T_2 = 10.67 T_2$$

Total torque $= T_1 + T_2 = T_2(10.67 + 1) = 680$

$$\therefore \quad T_2 = \frac{680}{11.67} = \mathbf{58.3\,N\,m}$$

and
$$T_1 = \mathbf{621.7\,N\,m}$$

For portion *AB*,

$$\tau_{\max} = \frac{T_1 R_1}{J_1} = \frac{621.7 \times 25 \times 10^{-3}}{\frac{\pi}{32} \times 50^4 \times 10^{-12}} = \mathbf{25.33 \times 10^6\ N/m^2}$$

For portion *BC*,

$$\tau_{max} = \frac{T_2 R_2}{J_2} = \frac{58.3 \times 12.5 \times 10^{-3}}{\frac{\pi}{32} \times 25^4 \times 10^{-12}} = \mathbf{19.0 \times 10^6\ N/m^2}$$

$$\text{Angle of twist for each portion} = \frac{T_1 L_1}{J_1 G}$$

$$= \frac{621.7 \times 1.8}{\frac{\pi}{32} \times 50^4 \times 10^{-12} \times 80 \times 10^9} = 0.0228\ \text{rad} = \mathbf{1.3\ degrees}$$

Problems

8.1 (A). A solid steel shaft *A* of 50 mm diameter rotates at 250 rev/min. Find the greatest power that can be transmitted for a limiting shearing stress of 60 MN/m^2 in the steel.

It is proposed to replace *A* by a hollow shaft *B*, of the same external diameter but with a limiting shearing stress of 75 MN/m^2. Determine the internal diameter of *B* to transmit the same power at the same speed. [38.6 kW, 33.4 mm.]

8.2 (A). Calculate the dimensions of a hollow steel shaft which is required to transmit 750 kW at a speed of 400 rev/min if the maximum torque exceeds the mean by 20% and the greatest intensity of shear stress is limited to 75 MN/m^2. The internal diameter of the shaft is to be 80% of the external diameter. (The mean torque is that derived from the horsepower equation.) [135.2, 108.2 mm.]

8.3 (A). A steel shaft 3 m long is transmitting 1 MW at 240 rev/min. The working conditions to be satisfied by the shaft are:

(a) that the shaft must not twist more than 0.02 radian on a length of 10 diameters;
(b) that the working stress must not exceed 60 MN/m^2.

If the modulus of rigidity of steel is 80 GN/m^2 what is

(i) the diameter of the shaft required;
(ii) the actual working stress;
(iii) the angle of twist of the 3 m length? [B.P.] [150 mm; 60 MN/m^2; 0.030 rad.]

8.4 (A). A hollow shaft has to transmit 6 MW at 150 rev/min. The maximum allowable stress is not to exceed 60 MN/m^2 nor the angle of twist 0.3° per metre length of shafting. If the outside diameter of the shaft is 300 mm find the minimum thickness of the hollow shaft to satisfy the above conditions. $G = 80$ GN/m^2. [61.5 mm.]

8.5 (A). A flanged coupling having six bolts placed at a pitch circle diameter of 180 mm connects two lengths of solid steel shafting of the same diameter. The shaft is required to transmit 80 kW at 240 rev/min. Assuming the allowable intensities of shearing stresses in the shaft and bolts are 75 MN/m^2 and 55 MN/m^2 respectively, and the maximum torque is 1.4 times the mean torque, calculate:

(a) the diameter of the shaft;
(b) the diameter of the bolts. [B.P.] [67.2, 13.8 mm.]

8.6 (A). A hollow low carbon steel shaft is subjected to a torque of 0.25 MN m. If the ratio of internal to external diameter is 1 to 3 and the shear stress due to torque has to be limited to 70 MN/m^2 determine the required diameters and the angle of twist in degrees per metre length of shaft.
$G = 80$ GN/m^2. [I.Struct.E.] [264, 88 mm; 0.38°.]

8.7 (A). Describe how you would carry out a torsion test on a low carbon steel specimen and how, from data taken, you would find the modulus of rigidity and yield stress in shear of the steel. Discuss the nature of the torque–twist curve and compare it with the shear stress–shear strain relationship. [U.Birm.]

8.8 (A/B). Opposing axial torques are applied at the ends of a straight bar *ABCD*. Each of the parts *AB*, *BC* and *CD* is 500 mm long and has a hollow circular cross-section, the inside and outside diameters being, respectively, *AB* 25 mm and 60 mm, *BC* 25 mm and 70 mm, *CD* 40 mm and 70 mm. The modulus of rigidity of the material is 80 GN/m^2 throughout. Calculate:

(a) the maximum torque which can be applied if the maximum shear stress is not to exceed 75 MN/m^2;
(b) the maximum torque if the twist of *D* relative to *A* is not to exceed 2°. [E.I.E.] [3.085 kN m, 3.25 kN m.]

8.9 (A/B). A solid steel shaft of 200 mm diameter transmits 5 MW at 500 rev/min. It is proposed to alter the horsepower to 7 MW and the speed to 440 rev/min and to replace the solid shaft by a hollow shaft made of the same type of steel but having only 80 % of the weight of the solid shaft. The length of both shafts is the same and the hollow shaft is to have the same maximum shear stress as the solid shaft. Find:
(a) the ratio between the torque per unit angle of twist per metre for the two shafts;
(b) the external and internal diameters for the hollow shaft. [I.Mech.E.] [2.085; 261, 190 mm.]

8.10 (A/B). A shaft *ABC* rotates at 600 rev/min and is driven through a coupling at the end *A*. At *B* a pulley takes off two-thirds of the power, the remainder being absorbed at *C*. The part *AB* is 1.3 m long and of 100 mm diameter; *BC* is 1.7 m long and of 75 mm diameter. The maximum shear stress set up in *BC* is 40 MN/m^2. Determine the maximum stress in *AB* and the power transmitted by it, and calculate the total angle of twist in the length *AC*. Take $G = 80$ GN/m^2. [I.Mech.E.] [16.9 MN/m^2; 208 kW; 1.61°.]

8.11 (A/B). A composite shaft consists of a steel rod of 75 mm diameter surrounded by a closely fitting brass tube firmly fixed to it. Find the outside diameter of the tube such that when a torque is applied to the composite shaft it will be shared equally by the two materials.
$G_S = 80$ GN/m^2; $G_B = 40$ GN/m^2.
If the torque is 16 kN m, calculate the maximum shearing stress in each material and the angle of twist on a length of 4 m. [U.L.] [98.7 mm; 96.6, 63.5 MN/m^2; 7.38°.]

8.12 (A/B). A circular bar 4 m long with an external radius of 25 mm is solid over half its length and bored to an internal radius of 12 mm over the other half. If a torque of 120 N m is applied at the centre of the shaft, the two ends being fixed, determine the maximum shear stress set up in the surface of the shaft and the work done by the torque in producing this stress. [2.51 MN/m^2; 0.151 N m.]

8.13 (A/B). The shaft of Problem 8.12 is now fixed at one end only and the torque applied at the free end. How will the values of maximum shear stress and work done change? [5.16 MN/m^2; 0.603 N m.]

8.14 (B). Calculate the minimum diameter of a solid shaft which is required to transmit 70 kW at 600 rev/min if the shear stress is not to exceed 75 MN/m^2. If a bending moment of 300 N m is now applied to the shaft find the speed at which the shaft must be driven in order to transmit the same horsepower for the same value of maximum shear stress. [630 rev/min.]

8.15 (B). A solid shaft of 75 mm diameter and 4 m span supports a flywheel of weight 2.5 kN at a point 1.8 m from one support. Determine the maximum direct stress produced in the surface of the shaft when it transmits 35 kW at 200 rev/min. [65.9 MN/m^2.]

8.16 (B). The shaft of Problem 12.15 is now subjected to an axial compressive end load of 80 kN, the other conditions remaining unchanged. What will be the magnitudes of the maximum principal stress in the shaft? [84 MN/m^2.]

8.17 (B). A horizontal shaft of 75 mm diameter projects from a bearing, and in addition to the torque transmitted the shaft carries a vertical load of 8 kN at 300 mm from the bearing. If the safe stress for the material, as determined in a simple tension test, is 135 MN/m^2 find the safe torque to which the shaft may be subjected using as the criterion (a) the maximum shearing stress, (b) the maximum strain energy per unit volume. Poisson's ratio $\nu = 0.29$. [U.L.] [5.05, 8.3 kN m.]

8.18 (B). A pulley subjected to vertical belt drive develops 10 kW at 240 rev/min, the belt tension ratio being 0.4. The pulley is fixed to the end of a length of overhead shafting which is supported in two self-aligning bearings, the centre line of the pulley overhanging the centre line of the left-hand bearing by 150 mm. If the pulley is of 250 mm diameter and weight 270 N, neglecting the weight of the shafting, find the minimum shaft diameter required if the maximum allowable stress intensity at a point on the top surface of the shaft at the centre line of the left-hand bearing is not to exceed 90 MN/m^2 direct or 40 MN/m^2 shear. [50.5 mm.]

8.19 (B). A hollow steel shaft of 100 mm external diameter and 50 mm internal diameter transmits 0.6 MW at 500 rev/min and is subjected to an end thrust of 45 kN. Find what bending moment may safely be applied if the greater principal stress is not to exceed 90 MN/m^2. What will then be the value of the smaller principal stress? [City U.] [3.6 kN m; -43.1 MN/m^2.]

8.20 (B). A solid circular shaft is subjected to an axial torque T and to a bending moment M. If $M = kT$, determine in terms of k the ratio of the maximum principal stress to the maximum shear stress. Find the power transmitted by a 50 mm diameter shaft, at a speed of 300 rev/min when $k = 0.4$ and the maximum shear stress is 75 MN/m^2. [I.Mech.] [$1 + k/\sqrt{(k^2+1)}$; 57.6 kW.]

8.21 (B). (a) A solid circular steel shaft is subjected to a bending moment of 10 kN m and is required to transmit a maximum power of 550 kW at 420 rev/min. Assuming the shaft to be simply supported at each end and neglecting the shaft weight, determine the ratio of the maximum principal stress to the maximum shear stress induced in the shaft material.

(b) A 300 mm external diameter and 200 mm internal diameter hollow steel shaft operates under the following conditions:

power transmitted = 2280 kW; maximum torque = 1.2 × mean torque; maximum bending moment = 11 kN m; maximum end thrust = 66 kN; maximum principal compressive stress = 40 MN/m^2.

Determine the maximum safe speed of rotation for the shaft. [1.625 : 1; 169 rev/min.]

8.22 (C). A uniform solid shaft of circular cross-section will drive the propeller of a ship. It will therefore necessarily be subject simultaneously to a thrust load and a torque. The magnitude of the thrust can be related to the magnitude of the torque by the simple relationship $N = KT$, where N denotes the magnitude of the thrust, T that of the torque and K is a constant. There will also be some bending moment on the shaft. Assuming that the design requirement is that the maximum shearing stress in the material shall nowhere exceed a certain value, denoted by τ, show that the maximum bending moment that can be allowed is given by the expression

$$\text{bending moment, } M = \left[\left(\frac{\tau\pi^2 r^6}{4T^2} - 1\right)^{1/2} - \frac{Kr}{4}\right]T$$

where r denotes the radius of the shaft cross-section. [City U.]

CHAPTER 9

THIN CYLINDERS AND SHELLS

Summary

The stresses set up in the walls of a *thin cylinder* owing to an internal pressure p are:

$$\text{circumferential or hoop stress } \sigma_H = \frac{pd}{2t}$$

$$\text{longitudinal or axial stress } \sigma_L = \frac{pd}{4t}$$

where d is the internal diameter and t is the wall thickness of the cylinder.

Then:

$$\text{longitudinal strain } \varepsilon_L = \frac{1}{E}[\sigma_L - \nu\sigma_H]$$

$$\text{hoop strain } \varepsilon_H = \frac{1}{E}[\sigma_H - \nu\sigma_L]$$

$$\text{change of internal volume of cylinder under pressure} = \frac{pd}{4tE}[5-4\nu]\,V$$

$$\text{change of volume of contained liquid under pressure} = \frac{pV}{K}$$

where K is the bulk modulus of the liquid.

For *thin rotating cylinders* of mean radius R the tensile hoop stress set up when rotating at ω rad/s is given by $\sigma_H = \rho\omega^2R^2.$

For *thin spheres*:

$$\text{circumferential or hoop stress } \sigma_H = \frac{pd}{4t}$$

$$\text{change of volume under pressure} = \frac{3pd}{4tE}[1-\nu]\,V$$

Effects of end plates and joints – add "joint efficiency factor" η to denominator of stress equations above.

9.1. Thin cylinders under internal pressure

When a thin-walled cylinder is subjected to internal pressure, three mutually perpendicular principal stresses will be set up in the cylinder material, namely the *circumferential* or *hoop* stress, the *radial* stress and the *longitudinal* stress. Provided that the ratio of thickness to inside diameter of the cylinder is less than 1/20, it is reasonably accurate to assume that the hoop and longitudinal stresses are constant across the wall thickness and that the magnitude of the radial stress set up is so small in comparison with the hoop and longitudinal stresses that it can be neglected. This is obviously an approximation since, in practice, it will vary from zero at the outside surface to a value equal to the internal pressure at the inside surface. For the purpose of the initial derivation of stress formulae it is also assumed that the ends of the cylinder and any riveted joints present have no effect on the stresses produced; in practice they will have an effect and this will be discussed later (§9.6).

9.1.1. Hoop or circumferential stress

This is the stress which is set up in resisting the bursting effect of the applied pressure and can be most conveniently treated by considering the equilibrium of half of the cylinder as shown in Fig. 9.1.

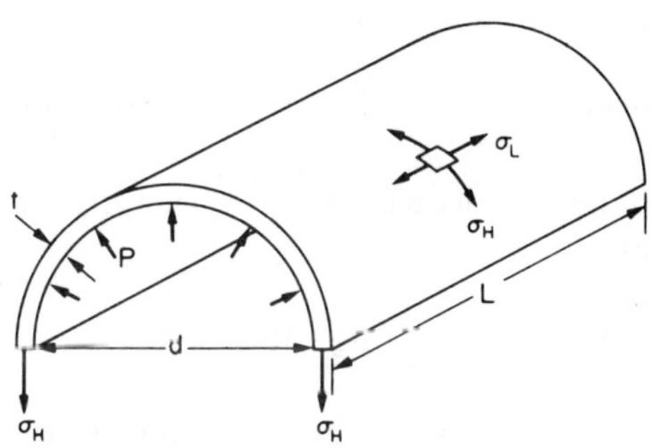

Fig. 9.1. Half of a thin cylinder subjected to internal pressure showing the hoop and longitudinal stresses acting on any element in the cylinder surface.

Total force on half-cylinder owing to internal pressure $= p \times$ projected area $= p \times dL$

Total resisting force owing to hoop stress σ_H set up in the cylinder walls

$$= 2\sigma_H \times Lt$$

$$\therefore \qquad 2\sigma_H Lt = pdL$$

$$\therefore \qquad \textbf{circumferential or hoop stress } \sigma_H = \frac{pd}{2t} \qquad (9.1)$$

9.1.2. Longitudinal stress

Consider now the cylinder shown in Fig. 9.2.

Total force on the end of the cylinder owing to internal pressure

$$= \text{pressure} \times \text{area} = p \times \frac{\pi d^2}{4}$$

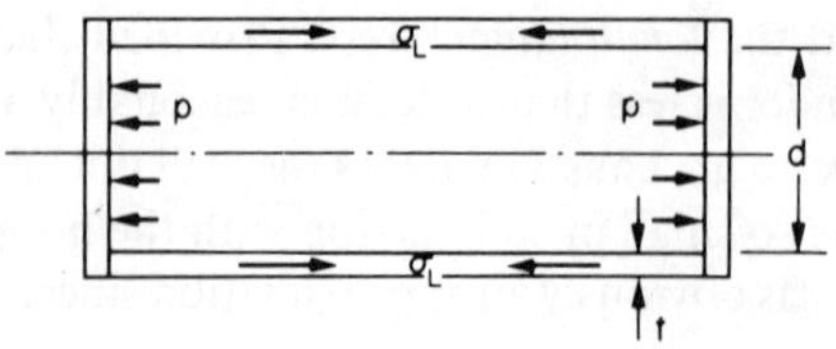

Fig. 9.2. Cross-section of a thin cylinder.

Area of metal resisting this force = πdt (approximately)

$$\therefore \qquad \text{stress set up} = \frac{\text{force}}{\text{area}} = p \times \frac{\pi d^2/4}{\pi dt} = \frac{pd}{4t}$$

i.e. $$\textbf{longitudinal stress } \sigma_L = \frac{pd}{4t} \qquad (9.2)$$

9.1.3. Changes in dimensions

(a) Change in length

The change in length of the cylinder may be determined from the longitudinal strain, i.e. neglecting the radial stress.

$$\text{Longitudinal strain} = \frac{1}{E}[\sigma_L - \nu\sigma_H]$$

and $$\text{change in length} = \text{longitudinal strain} \times \text{original length}$$

$$= \frac{1}{E}[\sigma_L - \nu\sigma_H]L$$

$$= \frac{pd}{4tE}[1 - 2\nu]L \qquad (9.3)$$

(b) Change in diameter

As above, the change in diameter may be determined from the strain on a diameter, i.e. the *diametral* strain.

$$\text{Diametral strain} = \frac{\text{change in diameter}}{\text{original diameter}}$$

Now the change in diameter may be found from a consideration of the circumferential change. The stress acting around a circumference is the hoop or circumferential stress σ_H giving rise to the circumferential strain ε_H.

$$\text{Change in circumference} = \text{strain} \times \text{original circumference}$$

$$= \varepsilon_H \times \pi d$$

$$\text{New circumference} = \pi d + \pi d\varepsilon_H = \pi d\,(1+\varepsilon_H)$$

But this is the circumference of a circle of diameter $d\,(1+\varepsilon_H)$

$$\therefore \quad \text{New diameter} = d\,(1+\varepsilon_H)$$

$$\therefore \quad \text{Change in diameter} = d\varepsilon_H$$

$$\text{Diametral strain } \varepsilon_D = \frac{d\varepsilon_H}{d} = \varepsilon_H$$

i.e. **the diametral strain equals the hoop or circumferential strain** (9.4)

Thus

$$\text{change in diameter} = d\varepsilon_H = \frac{d}{E}[\sigma_H - \nu\sigma_L]$$

$$= \frac{pd^2}{4tE}[2-\nu] \qquad (9.5)$$

(*c*) *Change in internal volume*

$$\text{Change in volume} = \text{volumetric strain} \times \text{original volume}$$

From the work of § 14.5, page 364.

$$\text{volumetric strain} = \text{sum of three mutually perpendicular direct strains}$$

$$= \varepsilon_L + 2\varepsilon_D$$

$$= \frac{1}{E}[\sigma_L - \nu\sigma_H] + \frac{2}{E}[\sigma_H - \nu\sigma_L]$$

$$= \frac{1}{E}[\sigma_L + 2\sigma_H - \nu(\sigma_H + 2\sigma_L)]$$

$$= \frac{pd}{4tE}[1+4-\nu(2+2)]$$

$$= \frac{pd}{4tE}[5-4\nu]$$

Therefore with original internal volume V

$$\textbf{change in internal volume} = \frac{pd}{4tE}[5-4\nu]\,V \qquad (9.6)$$

9.2. Thin rotating ring or cylinder

Consider a thin ring or cylinder as shown in Fig. 9.3 subjected to a radial pressure p caused by the centrifugal effect of its own mass when rotating. The centrifugal effect on a unit length

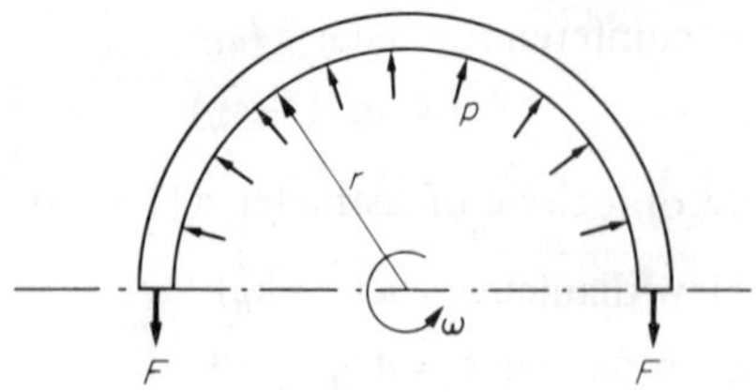

Fig. 9.3. Rotating thin ring or cylinder.

of the circumference is:

$$p = m\omega^2 r$$

Thus, considering the equilibrium of half the ring shown in the figure:

$$2F = p \times 2r$$
$$F = pr$$

where F is the hoop tension set up owing to rotation.

The cylinder wall is assumed to be so thin that the centrifugal effect can be taken to be constant across the wall thickness.

$$\therefore \quad F = pr = m\omega^2 r^2$$

This tension is transmitted through the complete circumference and therefore is restricted by the complete cross-sectional area.

$$\therefore \quad \text{hoop stress} = \frac{F}{A} = \frac{m\omega^2 r^2}{A}$$

where A is the cross-sectional area of the ring.

Now with unit length assumed, m/A is the mass of the ring material per unit volume, i.e. the density ρ.

$$\therefore \quad \textbf{hoop stress} = \boldsymbol{\rho\omega^2 r^2} \qquad (9.7)$$

9.3. Thin spherical shell under internal pressure

Because of the symmetry of the sphere the stresses set up owing to internal pressure will be two mutually perpendicular hoop or circumferential stresses of equal value and a radial stress. As with thin cylinders having thickness to diameter ratios less than 1:20, the radial stress is assumed negligible in comparison with the values of hoop stress set up. The stress system is therefore one of equal biaxial hoop stresses.

Consider, therefore, the equilibrium of the half-sphere shown in Fig. 9.4.

Force on half-sphere owing to internal pressure

$$= \text{pressure} \times \text{projected area}$$

$$= p \times \frac{\pi d^2}{4}$$

$$\text{Resisting force} = \sigma_H \times \pi d t \quad \text{(approximately)}$$

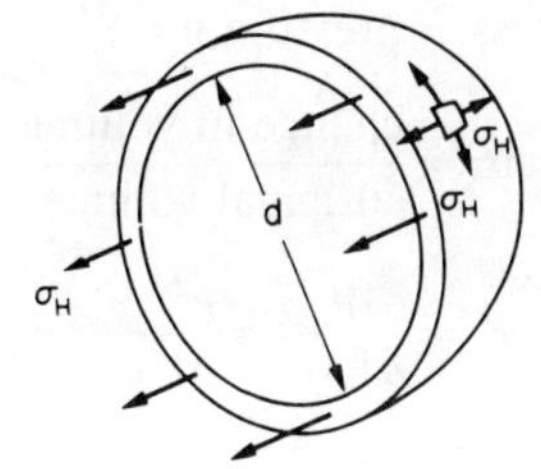

Fig. 9.4. Half of a thin sphere subjected to internal pressure showing uniform hoop stresses acting on a surface element.

$$\therefore \qquad p \times \frac{\pi d^2}{4} = \sigma_H \times \pi d t$$

or
$$\sigma_H = \frac{pd}{4t}$$

i.e. $$\textbf{circumferential or hoop stress} = \frac{pd}{4t} \qquad (9.8)$$

9.3.1. Change in internal volume

As for the cylinder,

change in volume = original volume × volumetric strain

but

$$\begin{aligned} \text{volumetric strain} &= \text{sum of three mutually perpendicular strains (in this case all equal)} \\ &= 3\varepsilon_D = 3\varepsilon_H \\ &= \frac{3}{E}[\sigma_H - \nu\sigma_H] \\ &= \frac{3pd}{4tE}[1-\nu] \end{aligned}$$

$$\therefore \qquad \textbf{change in internal volume} = \frac{3pd}{4tE}[1-\nu]V \qquad (9.9)$$

9.4. Vessels subjected to fluid pressure

If a fluid is used as the pressurisation medium the fluid itself will change in volume as pressure is increased and this must be taken into account when calculating the amount of fluid which must be pumped into the cylinder in order to raise the pressure by a specified amount, the cylinder being initially full of fluid at atmospheric pressure.

Now the *bulk modulus* of a fluid is defined as follows:

$$\text{bulk modulus } K = \frac{\text{volumetric stress}}{\text{volumetric strain}}$$

where, in this case, volumetric stress = pressure p

and volumetric strain $= \dfrac{\text{change in volume}}{\text{original volume}} = \dfrac{\delta V}{V}$

$$\therefore \qquad K = \frac{p}{\delta V/V} = \frac{pV}{\delta V}$$

i.e. **change in volume of fluid under pressure** $= \dfrac{pV}{K}$ (9.10)

The extra fluid required to raise the pressure must, therefore, take up this volume together with the increase in internal volume of the cylinder itself.

∴ extra fluid required to raise *cylinder* pressure by p

$$= \frac{pd}{4tE}[5-4\nu]V + \frac{pV}{K} \qquad (9.11)$$

Similarly, for *spheres*, the extra fluid required is

$$= \frac{3pd}{4tE}[1-\nu]V + \frac{pV}{K} \qquad (9.12)$$

9.5. Cylindrical vessel with hemispherical ends

Consider now the vessel shown in Fig. 9.5 in which the wall thickness of the cylindrical and hemispherical portions may be different (this is sometimes necessary since the hoop stress in the cylinder is twice that in a sphere of the same radius and wall thickness). For the purpose of the calculation the internal diameter of both portions is assumed equal.
From the preceding sections the following formulae are known to apply:

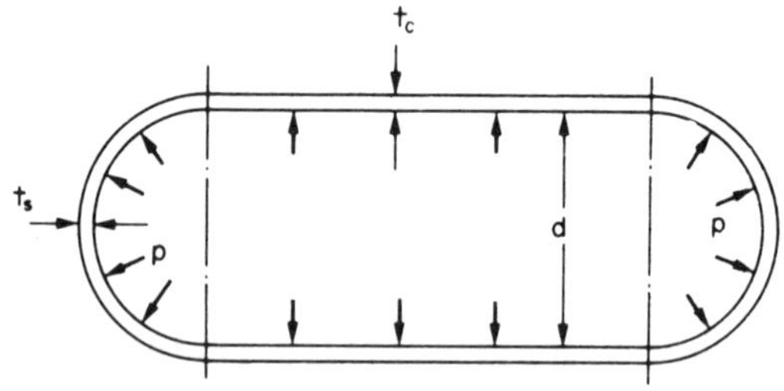

Fig. 9.5. Cross-section of a thin cylinder with hemispherical ends.

(a) *For the cylindrical portion*

$$\text{hoop or circumferential stress} = \sigma_{H_c} = \frac{pd}{2t_c}$$

$$\text{longitudinal stress} = \sigma_{L_c} = \frac{pd}{4t_c}$$

$$\therefore \quad \text{hoop or circumferential strain} = \frac{1}{E}[\sigma_{H_c} - \nu\sigma_{L_c}]$$

$$= \frac{pd}{4t_c E}[2 - \nu]$$

(b) *For the hemispherical ends*

$$\text{hoop stress} = \sigma_{H_s} = \frac{pd}{4t_s}$$

$$\therefore \quad \text{hoop strain} = \frac{1}{E}[\sigma_{H_s} - \nu\sigma_{H_s}]$$

$$= \frac{pd}{4t_s E}[1 - \nu]$$

Thus equating the two strains in order that there shall be no distortion of the junction,

$$\frac{pd}{4t_c E}[2 - \nu] = \frac{pd}{4t_s E}[1 - \nu]$$

i.e.

$$\frac{t_s}{t_c} = \frac{(1 - \nu)}{(2 - \nu)} \qquad (9.13)$$

With the normally accepted value of Poisson's ratio for general steel work of 0.3, the thickness ratio becomes

$$\frac{t_s}{t_c} = \frac{0.7}{1.7}$$

i.e. the thickness of the cylinder walls must be approximately 2.4 times that of the hemispherical ends for no distortion of the junction to occur. In these circumstances, because of the reduced wall thickness of the ends, the maximum stress will occur in the ends. For *equal maximum stresses* in the two portions the thickness of the cylinder walls must be twice that in the ends but some distortion at the junction will then occur.

9.6. Effects of end plates and joints

The preceding sections have all assumed uniform material properties throughout the components and have neglected the effects of endplates and joints which are necessary requirements for their production. In general, the strength of the components will be reduced by the presence of, for example, riveted joints, and this should be taken into account by the introduction of a *joint efficiency factor* η into the equations previously derived.

Thus, for *thin cylinders*:

$$\text{hoop stress} = \frac{pd}{2t\eta_L}$$

where η_L is the efficiency of the longitudinal joints,

$$\text{longitudinal stress} = \frac{pd}{4t\eta_C}$$

where η_C is the efficiency of the circumferential joints.

For *thin spheres*:

$$\text{hoop stress} = \frac{pd}{4t\eta}$$

Normally the joint efficiency is stated in percentage form and this must be converted into equivalent decimal form before substitution into the above equations.

9.7. Wire-wound thin cylinders

In order to increase the ability of thin cylinders to withstand high internal pressures without excessive increases in wall thickness, and hence weight and associated material cost, they are sometimes wound with high tensile steel tape or wire under tension. This subjects the cylinder to an initial hoop, compressive, stress which must be overcome by the stresses owing to internal pressure before the material is subjected to tension. There then remains at this stage the normal pressure capacity of the cylinder before the maximum allowable stress in the cylinder is exceeded.

It is normally required to determine the tension necessary in the tape during winding in order to ensure that the maximum hoop stress in the cylinder will not exceed a certain value when the internal pressure is applied.

Consider, therefore, the half-cylinder of Fig. 9.6, where σ_H denotes the hoop stress in the cylinder walls and σ_t the stress in the rectangular-sectioned tape. Let conditions before pressure is applied be denoted by suffix 1 and after pressure is applied by suffix 2.

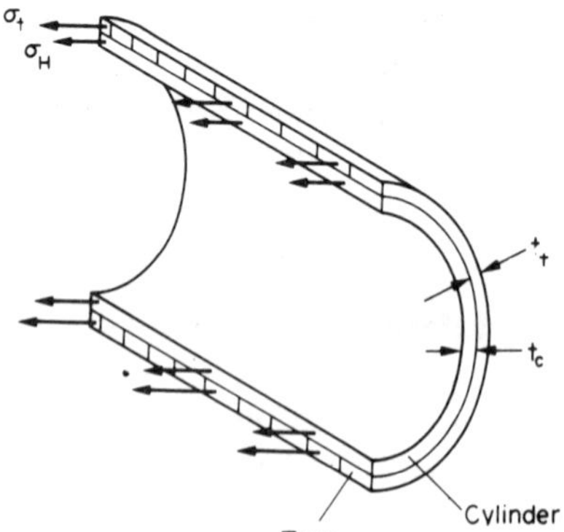

Fig. 9.6. Section of a thin cylinder with an external layer of tape wound on with a tension.

Now force owing to tape $= \sigma_{t_1} \times \text{area}$

$$= \sigma_{t_1} \times 2Lt_t$$

resistive force in the cylinder material $= \sigma_{H_1} \times 2Lt_c$

i.e. for equilibrium

$$\sigma_{t_1} \times 2Lt_t = \sigma_{H_1} \times 2Lt_c$$

or

$$\sigma_{t_1} \times t_t = \sigma_{H_1} \times t_c$$

so that the *compressive* hoop stress set up in the cylinder walls after winding and before pressurisation is given by

$$\sigma_{H_1} = \sigma_{t_1} \times \frac{t_t}{t_c} \text{ (compressive)} \qquad (9.14)$$

This equation will be modified if wire of circular cross-section is used for the winding process in preference to rectangular-sectioned tape. The area carrying the stress σ_{t_1} will then be $2na$ where a is the cross-sectional area of the wire and n is the number of turns along the cylinder length.

After pressure has been applied another force is introduced

$$= \text{pressure} \times \text{projected area} = pdL$$

Again, equating forces for equilibrium of the half-cylinder,

$$pdL = (\sigma_{H_2} \times 2Lt_c) + (\sigma_{t_2} \times 2Lt_t) \qquad (9.15)$$

where σ_{H_2} is the hoop stress in the cylinder after pressurisation and σ_{t_2} is the final stress in the tape after pressurisation.

Since the limiting value of σ_{H_2} is known for any given internal pressure p, this equation yields the value of σ_{t_2}.

Now the change in strain on the outside surface of the cylinder must equal that on the inside surface of the tape if they are to remain in contact.

$$\text{Change in strain in the tape} = \frac{\sigma_{t_2} - \sigma_{t_1}}{E_t}$$

where E_t is Young's modulus of the tape.

In the absence of any internal pressure originally there will be no longitudinal stress or strain so that the original strain in the cylinder walls is given by σ_{H_1}/E_c, where E_c is Young's modulus of the cylinder material. When pressurised, however, the cylinder will be subjected to a longitudinal strain so that the final strain in the cylinder walls is given by

$$\frac{1}{E_c}[\sigma_{H_2} - \nu\sigma_L] = \frac{1}{E_c}\left[\sigma_{H_2} - \nu\frac{pd}{4t_c}\right]$$

$\therefore$ change in strain on the cylinder $= \frac{1}{E_c}\left[\sigma_{H_2} - \nu\frac{pd}{4t_c} - \sigma_{H_1}\right]$

$$\therefore \quad \frac{1}{E_t}[\sigma_{t_2} - \sigma_{t_1}] = \frac{1}{E_c}\left[\sigma_{H_2} - \nu\frac{pd}{4t_c} - \sigma_{H_1}\right]$$

Thus with σ_{H_1} obtained in terms of σ_{t_1} from eqn. (9.14), p and σ_{H_2} known, and σ_{t_2} found from eqn. (9.15) the only unknown σ_{t_1} can be determined.

Examples

Example 9.1

A thin cylinder 75 mm internal diameter, 250 mm long with walls 2.5 mm thick is subjected to an internal pressure of 7 MN/m^2. Determine the change in internal diameter and the change in length.

If, in addition to the internal pressure, the cylinder is subjected to a torque of 200 N m, find the magnitude and nature of the principal stresses set up in the cylinder. $E = 200\ GN/m^2$. $\nu = 0.3$.

Solution

(a) From eqn. (9.5), change in diameter $= \dfrac{pd^2}{4tE}(2-\nu)$

$$= \frac{7\times 10^6 \times 75^2 \times 10^{-6}}{4\times 2.5\times 10^{-3}\times 200\times 10^9}(2-0.3)$$

$$= 33.4\times 10^{-6}\ \text{m}$$

$$= \mathbf{33.4\ \mu m}$$

(b) From eqn. (9.3), change in length $= \dfrac{pdL}{4tE}(1-2\nu)$

$$= \frac{7\times 10^6\times 75\times 10^{-3}\times 250\times 10^{-3}}{4\times 2.5\times 10^{-3}\times 200\times 10^9}(1-0.6)$$

$$= \mathbf{26.2\ \mu m}$$

(c) Hoop stress $\sigma_H = \dfrac{pd}{2t} = \dfrac{7\times 10^6\times 75\times 10^{-3}}{2\times 2.5\times 10^{-3}}$

$$= \mathbf{105\ MN/m^2}$$

Longitudinal stress $\sigma_L = \dfrac{pd}{4t} = \dfrac{7\times 10^6\times 75\times 10^{-3}}{2\times 2.5\times 10^{-3}}$

$$= \mathbf{52.5\ MN/m^2}$$

In addition to these stresses a shear stress τ is set up.

From the torsion theory,

$$\frac{T}{J} = \frac{\tau}{R} \qquad \therefore \quad \tau = \frac{TR}{J}$$

Now $$J = \frac{\pi}{32}\frac{(80^4 - 75^4)}{10^{12}} = \frac{\pi}{32}\frac{(41-31.6)}{10^6} = 0.92\times 10^{-6}\ \text{m}^4$$

Then $$\text{shear stress } \tau = \frac{200\times 20\times 10^{-3}}{0.92\times 10^{-6}} = 4.34\ \text{MN/m}^2.$$

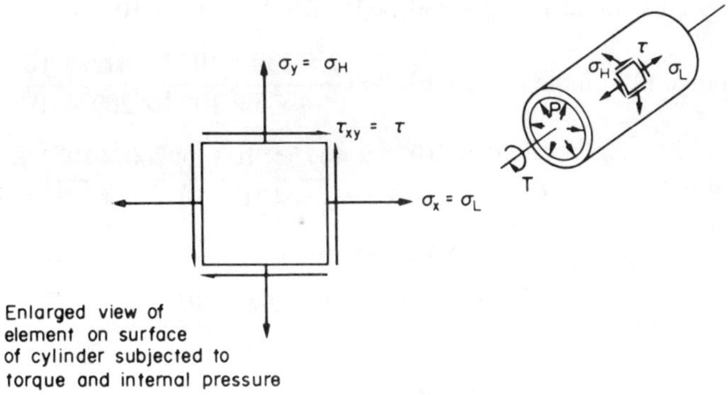

Fig. 9.7. Enlarged view of the stresses acting on an element in the surface of a thin cylinder subjected to torque and internal pressure.

The stress system then acting on any element of the cylinder surface is as shown in Fig. 9.7. The principal stresses are then given by eqn. (13.11),

$$\sigma_1 \text{ and } \sigma_2 = \tfrac{1}{2}(\sigma_x + \sigma_y) \pm \tfrac{1}{2}\sqrt{[(\sigma_x - \sigma_y)^2 + 4\tau_{xy}{}^2]}$$

$$= \tfrac{1}{2}(105 + 52.5) \pm \tfrac{1}{2}\sqrt{[(105 - 52.5)^2 + 4(4.34)^2]}$$

$$= \tfrac{1}{2} \times 157.5 \pm \tfrac{1}{2}\sqrt{(2760 + 75.3)}$$

$$= 78.75 \pm 26.6$$

Then $\sigma_1 = 105.35 \text{ MN/m}^2$ and $\sigma_2 = 52.15 \text{ MN/m}^2$

The principal stresses are

105.4 MN/m² and **52.2 MN/m²** both tensile.

Example 9.2

A cylinder has an internal diameter of 230 mm, has walls 5 mm thick and is 1 m long. It is found to change in internal volume by 12.0×10^{-6} m³ when filled with a liquid at a pressure p. If $E = 200 \text{ GN/m}^2$ and $\nu = 0.25$, and assuming rigid end plates, determine:

(a) the values of hoop and longitudinal stresses;
(b) the modifications to these values if joint efficiencies of 45% (hoop) and 85% (longitudinal) are assumed;
(c) the necessary change in pressure p to produce a further increase in internal volume of 15%. The liquid may be assumed incompressible.

Solution

(a) From eqn. (9.6)

$$\text{change in internal volume} = \frac{pd}{4tE}(5 - 4\nu)V$$

original volume $V = \frac{\pi}{4} \times 230^2 \times 10^{-6} \times 1 = 41.6 \times 10^{-3}\,\text{m}^3$

Then change in volume $= 12 \times 10^{-6} = \dfrac{p \times 230 \times 10^{-3} \times 41.6 \times 10^{-3}}{4 \times 5 \times 10^{-3} \times 200 \times 10^9}(5-1)$

Thus
$$p = \frac{12 \times 10^{-6} \times 4 \times 5 \times 10^{-3} \times 200 \times 10^9}{230 \times 10^{-3} \times 41.6 \times 10^{-3} \times 4}$$
$$= \mathbf{1.25\,MN/m^2}$$

Hence,
$$\text{hoop stress} = \frac{pd}{2t} = \frac{1.25 \times 10^6 \times 230 \times 10^{-3}}{2 \times 5 \times 10^{-3}}$$
$$= \mathbf{28.8\,MN/m^2}$$
$$\text{longitudinal stress} = \frac{pd}{4t} = \mathbf{14.4\,MN/m^2}$$

(b) Hoop stress, acting on the longitudinal joints (§9.6)

$$= \frac{pd}{2t\eta_L} = \frac{1.25 \times 10^6 \times 230 \times 10^{-3}}{2 \times 5 \times 10^{-3} \times 0.85}$$
$$= \mathbf{33.9\,MN/m^2}$$

Longitudinal stress (acting on the circumferential joints)

$$= \frac{pd}{4t\eta_c} = \frac{1.25 \times 10^6 \times 230 \times 10^{-3}}{4 \times 5 \times 10^{-3} \times 0.45}$$
$$= \mathbf{32\,MN/m^2}$$

(c) Since the change in volume is directly proportional to the pressure, the necessary 15% increase in volume is achieved by increasing the pressure also by 15%.

$$\text{Necessary increase in } p = 0.15 \times 1.25 \times 10^6$$
$$= \mathbf{1.86\,MN/m^2}$$

Example 9.3

(a) A sphere, 1 m internal diameter and 6 mm wall thickness, is to be pressure-tested for safety purposes with water as the pressure medium. Assuming that the sphere is initially filled with water at atmospheric pressure, what extra volume of water is required to be pumped in to produce a pressure of 3 MN/m^2 gauge? For water, $K = 2.1$ GN/m^2.

(b) The sphere is now placed in service and filled with gas until there is a volume change of 72×10^{-6} m^3. Determine the pressure exerted by the gas on the walls of the sphere.

(c) To what value can the gas pressure be increased before failure occurs according to the maximum principal stress theory of elastic failure?

For the material of the sphere $E = 200$ GN/m^2, $\nu = 0.3$ and the yield stress σ_y in simple tension $= 280$ MN/m^2.

Solution

(a) Bulk modulus $K = \dfrac{\text{volumetric stress}}{\text{volumetric strain}}$

Now volumetric stress = pressure $p = 3\ \text{MN/m}^2$

and volumetric strain = change in volume ÷ original volume

i.e.
$$K = \frac{p}{\delta V/V}$$

$\therefore$ change in volume of water $= \dfrac{pV}{K} = \dfrac{3 \times 10^6}{2.1 \times 10^9} \times \dfrac{4\pi}{3}(0.5)^3$

$$= \mathbf{0.748 \times 10^{-3}\ m^3}$$

(b) From eqn. (9.9) the change in volume is given by

$$\delta V = \frac{3pd}{4tE}(1-\nu)V$$

$$\therefore \quad 72 \times 10^{-6} = \frac{3p \times 1 \times \frac{4}{3}\pi(0.5)^3(1-0.3)}{4 \times 6 \times 10^{-3} \times 200 \times 10^9}$$

$$\therefore \quad p = \frac{72 \times 10^{-6} \times 4 \times 6 \times 200 \times 10^6 \times 3}{3 \times 4\pi(0.5)^3 \times 0.7}$$

$$= 314 \times 10^3\ \text{N/m}^2 = \mathbf{314\ kN/m^2}$$

(c) The maximum stress set up in the sphere will be the hoop stress,

i.e.
$$\sigma_1 = \sigma_H = \frac{pd}{4t}$$

Now, according to the maximum principal stress theory (see §15.2) failure will occur when the maximum principal stress equals the value of the yield stress of a specimen subjected to simple tension,

i.e. when
$$\sigma_1 = \sigma_y = 280\ \text{MN/m}^2$$

Thus
$$280 \times 10^6 = \frac{pd}{4t}$$

$$p = \frac{280 \times 10^6 \times 4 \times 6 \times 10^{-3}}{1}$$

$$= 6.72 \times 10^6\ \text{N/m}^2 = \mathbf{6.7\ MN/m^2}$$

The sphere would therefore yield at a pressure of $6.7\ \text{MN/m}^2$.

Example 9.4

A closed thin copper cylinder of 150 mm internal diameter having a wall thickness of 4 mm is closely wound with a single layer of steel tape having a thickness of 1.5 mm, the tape being wound on when the cylinder has no internal pressure. Estimate the tensile stress in the steel tape when it is being wound to ensure that when the cylinder is subjected to an internal pressure of 3.5 MN/m^2 the tensile hoop stress in the cylinder will not exceed 35 MN/m^2. For copper, Poisson's ratio $\nu = 0.3$ and $E = 100\,\text{GN/m}^2$; for steel, $E = 200\,\text{GN/m}^2$.

Solution

Let σ_t be the stress in the tape and let conditions before pressure is applied be denoted by suffix 1 and after pressure is applied by suffix 2.

Consider the half-cylinder shown (before pressure is applied) in Fig. 9.6 (see page 206):

$$\text{force owing to tension in tape} = \sigma_{t_1} \times \text{area}$$
$$= \sigma_{t_1} \times 1.5 \times 10^{-3} \times L \times 2$$

$$\text{resistive force in the material of cylinder wall} = \sigma_{H_1} \times 4 \times 10^{-3} \times L \times 2$$

$$\therefore \quad 2\sigma_{H_1} \times 4 \times 10^{-3} \times L = 2\sigma_{t_1} \times 1.5 \times 10^{-3} \times L$$

$$\therefore \quad \sigma_{H_1} = \frac{1.5}{4}\sigma_{t_1} = 0.375\,\sigma_{t_1} \text{(compressive)} \qquad (1)$$

After pressure is applied another force is introduced

$$= \text{pressure} \times \text{projected area}$$
$$= p(dL)$$

Equating forces now acting on the half-cylinder,

$$pdL = (\sigma_{H_2} \times 2 \times 4 \times 10^{-3} \times L) + (\sigma_{t_2} \times 2 \times 1.5 \times 10^{-3} \times L)$$

but

$$p = 3.5 \times 10^6\ \text{N/m}^2 \quad \text{and} \quad \sigma_{H_2} = 35 \times 10^6\ \text{N/m}^2$$

$$\therefore \quad 3.5 \times 10^6 \times 150 \times 10^{-3} L = (35 \times 10^6 \times 2 \times 4 \times 10^{-3} L) + (\sigma_{t_2} \times 2 \times 1.5 \times 10^{-3} \times L)$$

$$\therefore \quad 525 \times 10^6 = 280 \times 10^6 + 3\sigma_{t_2}$$

$$\therefore \quad \sigma_{t_2} = \frac{(525 - 280)}{3} 10^6$$

$$\sigma_{t_2} = 82\ \text{MN/m}^2$$

The change in strain on the outside of the cylinder and on the inside of the tape must be equal:

$$\text{change in strain in tape} = \frac{\sigma_{t_2} - \sigma_{t_1}}{E_s}$$

$$\text{original strain in cylinder walls} = \frac{\sigma_{H_1}}{E_c}$$

(Since there is no pressure in the cylinder in the original condition there will be no longitudinal stress.)

Final strain in cylinder (after pressurising)

$$= \frac{\sigma_{H_2}}{E_c} - \frac{\nu\sigma_L}{E_c}$$

$$= \frac{1}{Ec}\left(\sigma_{H_2} - \frac{\nu pd}{4t}\right)$$

Then change in strain in cylinder

$$= \frac{1}{E_c}\left(\sigma_{H_2} - \frac{\nu pd}{4t} - \sigma_{H_1}\right)$$

Then $$\frac{1}{E_s}(\sigma_{t_2} - \sigma_{t_1}) = \frac{1}{E_c}\left(\sigma_{H_2} - \frac{\nu pd}{4t} - \sigma_{H_1}\right)$$

Substituting for σ_{H_1} from eqn. (1)

$$\frac{82 \times 10^6 - \sigma_{t_1}}{200 \times 10^9} = \frac{1}{100 \times 10^9}\left[35 \times 10^6 - \frac{0.3 \times 3.5 \times 10^6 \times 154 \times 10^{-3}}{4 \times 4 \times 10^{-3}} - 0.375\sigma_{t_1}\right]$$

$$82 \times 10^6 - \sigma_{t_1} = 2(35 \times 10^6 - 10.1 \times 10^6 - 0.375\sigma_{t_1})$$

$$= 49.8 \times 10^6 - 0.75\sigma_{t_1}$$

Then $$1.75\sigma_{t_1} = (82.0 - 49.8)10^6$$

$$\sigma_{t_1} = \frac{32.2 \times 10^6}{1.75}$$

$$= \mathbf{18.4\ MN/m^2}$$

Problems

9.1 (A). Determine the hoop and longitudinal stresses set up in a thin boiler shell of circular cross-section, 5 m long and of 1.3 m internal diameter when the internal pressure reaches a value of 2.4 bar (240 kN/m^2). What will then be its change in diameter? The wall thickness of the boiler is 25 mm. $E = 210\,GN/m^2$; $\nu = 0.3$.
[6.24, 3.12 MN/m^2; 0.033 mm.]

9.2 (A). Determine the change in volume of a thin cylinder of original volume $65.5 \times 10^{-3}\,m^3$ and length 1.3 m if its wall thickness is 6 mm and the internal pressure 14 bar (1.4 MN/m^2). For the cylinder material $E = 210\,GN/m^2$; $\nu = 0.3$.
[$17.5 \times 10^{-6}\,m^3$.]

9.3 (A). What must be the wall thickness of a thin spherical vessel of diameter 1 m if it is to withstand an internal pressure of 70 bar (7 MN/m^2) and the hoop stresses are limited to 270 MN/m^2?
[12.96 mm.]

9.4 (A/B). A steel cylinder 1 m long, of 150 mm internal diameter and plate thickness 5 mm, is subjected to an internal pressure of 70 bar (7 MN/m^2); the increase in volume owing to the pressure is $16.8 \times 10^{-6}\,m^3$. Find the values of Poisson's ratio and the modulus of rigidity. Assume $E = 210\,GN/m^2$. [U.L.] [0.299; 80.8 GN/m^2.]

9.5 (B). Define bulk modulus K, and show that the decrease in volume of a fluid under pressure p is pV/K. Hence derive a formula to find the extra fluid which must be pumped into a thin cylinder to raise its pressure by an amount p.

How much fluid is required to raise the pressure in a thin cylinder of length 3 m, internal diameter 0.7 m, and wall thickness 12 mm by 0.7 bar (70 kN/m^2)? $E = 210\,GN/m^2$ and $\nu = 0.3$ for the material of the cylinder and $K = 2.1\,GN/m^2$ for the fluid.
[$5.981 \times 10^{-3}\,m^3$.]

9.6 (B). A spherical vessel of 1.7 m diameter is made from 12 mm thick plate, and it is to be subjected to a hydraulic test. Determine the additional volume of water which it is necessary to pump into the vessel, when the vessel is initially just filled with water, in order to raise the pressure to the proof pressure of 116 bar (11.6 MN/m^2). The bulk modulus of water is 2.9 GN/m^2. For the material of the vessel, $E = 200\,GN/m^2$, $\nu = 0.3$.
[$26.14 \times 10^{-3}\,m^3$.]

9.7 (B). A thin-walled steel cylinder is subjected to an internal fluid pressure of 21 bar (2.1 MN/m^2). The boiler is of 1 m inside diameter and 3 m long and has a wall thickness of 30 mm. Calculate the hoop and longitudinal stresses present in the cylinder and determine what torque may be applied to the cylinder if the principal stress is limited to 150 MN/m^2. [35, 17.5 MN/m^2; 6 MN m.]

9.8 (B). A thin cylinder of 300 mm internal diameter and 12 mm thickness is subjected to an internal pressure p while the ends are subjected to an external pressure of $\frac{1}{2}p$. Determine the value of p at which elastic failure will occur according to (a) the maximum shear stress theory, and (b) the maximum shear strain energy theory, if the limit of proportionality of the material in simple tension is 270 MN/m^2. What will be the volumetric strain at this pressure? $E = 210\,GN/m^2$; $\nu = 0.3$ [21.6, 23.6 MN/m^2, 2.289×10^{-3}, 2.5×10^{-3}.]

9.9 (C). A brass pipe has an internal diameter of 400 mm and a metal thickness of 6 mm. A single layer of high-tensile wire of diameter 3 mm is wound closely round it at a tension of 500 N. Find (a) the stress in the pipe when there is no internal pressure; (b) the maximum permissible internal pressure in the pipe if the working tensile stress in the brass is 60 MN/m^2; (c) the stress in the steel wire under condition (b). Treat the pipe as a thin cylinder and neglect longitudinal stresses and strains. $E_S = 200\,GN/m^2$; $E_B = 100\,GN/m^2$.

[U.L.] [27.8, 3.04 MN/m^2; 104.8 MN/m^2.]

9.10 (B). A cylindrical vessel of 1 m diameter and 3 m long is made of steel 12 mm thick and filled with water at 16°C. The temperature is then raised to 50°C. Find the stresses induced in the material of the vessel given that over this range of temperature water increases 0.006 per unit volume. (Bulk modulus of water = 2.9 GN/m^2; E for steel = 210 GN/m^2 and $\nu = 0.3$.) Neglect the expansion of the steel owing to temperature rise.

[663, 331.5 MN/m^2.]

9.11 (C). A 3 m long aluminium-alloy tube, of 150 mm outside diameter and 5 mm wall thickness, is closely wound with a single layer of 2.5 mm diameter steel wire at a tension of 400 N. It is then subjected to an internal pressure of 70 bar (7 MN/m^2).

(a) Find the stress in the tube before the pressure is applied.

(b) Find the final stress in the tube.

$E_A = 70\,GN/m^2$; $\nu_A = 0.28$; $E_S = 200\,GN/m^2$. [−32, 20.5 MN/m^2.]

9.12 (B). (a) Derive the equations for the circumferential and longitudinal stresses in a thin cylindrical shell.

(b) A thin cylinder of 300 mm internal diameter, 3 m long and made from 3 mm thick metal, has its ends blanked off. Working from first principles, except that you may use the equations derived above, find the change in capacity of this cylinder when an internal fluid pressure of 20 bar is applied. $E = 200\,GN/m^2$; $\nu = 0.3$. [$201 \times 10^{-6}\,m^3$.]

9.13 (A/B). Show that the tensile hoop stress set up in a thin rotating ring or cylinder is given by:

$$\sigma_H = \rho\omega^2 r^2.$$

Hence determine the maximum angular velocity at which the disc can be rotated if the hoop stress is limited to 20 MN/m^2. The ring has a mean diameter of 260 mm. [3800 rev/min.]

CHAPTER 10

THICK CYLINDERS

Summary

The *hoop and radial stresses* at any point in the wall cross-section of a thick cylinder at radius r are given by the Lamé equations:

$$\text{hoop stress } \sigma_H = A + \frac{B}{r^2}$$

$$\text{radial stress } \sigma_r = A - \frac{B}{r^2}$$

With internal and external pressures P_1 and P_2 and internal and external radii R_1 and R_2 respectively, the *longitudinal stress* in a cylinder with closed ends is

$$\sigma_L = \frac{P_1 R_1^2 - P_2 R_2^2}{(R_2^2 - R_1^2)} = \text{Lamé constant } A$$

Changes in dimensions of the cylinder may then be determined from the following strain formulae:

$$\text{circumferential or hoop strain} = \text{diametral strain}$$

$$= \frac{\sigma_H}{E} - \nu\frac{\sigma_r}{E} - \nu\frac{\sigma_L}{E}$$

$$\text{longitudinal strain} = \frac{\sigma_L}{E} - \nu\frac{\sigma_r}{E} - \nu\frac{\sigma_H}{E}$$

For *compound tubes* the resultant hoop stress is the algebraic sum of the hoop stresses resulting from shrinkage and the hoop stresses resulting from internal and external pressures.

For *force and shrink fits* of cylinders made of *different materials*, the total interference or shrinkage allowance (on radius) is

$$[\varepsilon_{H_o} - \varepsilon_{H_i}]r$$

where ε_{H_o} and ε_{H_i} are the hoop strains existing in the outer and inner cylinders respectively at the common radius r. For cylinders of the *same material* this equation reduces to

$$\frac{r}{E}[\sigma_{H_o} - \sigma_{H_i}]$$

For a *hub or sleeve shrunk on a solid shaft* the shaft is subjected to constant hoop and radial stresses, each equal to the pressure set up at the junction. The hub or sleeve is then treated as a thick cylinder subjected to this internal pressure.

Wire-wound thick cylinders

If the internal and external radii of the cylinder are R_1 and R_2 respectively and it is wound with wire until its external radius becomes R_3, the radial and hoop stresses *in the wire* at any radius r between the radii R_2 and R_3 are found from:

$$\text{radial stress} = \left(\frac{r^2 - R_1^2}{2r^2}\right) T \log_e \left(\frac{R_3^2 - R_1^2}{r^2 - R_1^2}\right)$$

$$\text{hoop stress} = T\left\{1 - \left(\frac{r^2 + R_1^2}{2r^2}\right) \log_e \left(\frac{R_3^2 - R_1^2}{r^2 - R_1^2}\right)\right\}$$

where T is the constant tension stress in the wire.

The hoop and radial stresses *in the cylinder* can then be determined by considering the cylinder to be subjected to an external pressure equal to the value of the radial stress above when $r = R_2$.

When an additional internal pressure is applied the final stresses will be the algebraic sum of those resulting from the internal pressure and those resulting from the wire winding.

Plastic yielding of thick cylinders

For initial yield, the internal pressure P_1 is given by:

$$P_1 = \frac{\sigma_y}{2R_2^2}[R_2^2 - R_1^2]$$

For yielding to a radius R_p,

$$P_1 = \sigma_y \left[\log_e \frac{R_1}{R_p} - \frac{1}{2R_2}(R_2^2 - R_p^2)\right]$$

and for complete collapse,

$$P_1 = \sigma_y \left[\log_e \frac{R_1}{R_2}\right]$$

10.1. Difference in treatment between thin and thick cylinders – basic assumptions

The theoretical treatment of thin cylinders assumes that the hoop stress is constant across the thickness of the cylinder wall (Fig. 10.1), and also that there is no pressure gradient across the wall. Neither of these assumptions can be used for thick cylinders for which the variation of hoop and radial stresses is shown in Fig. 10.2, their values being given by the *Lamé equations*:

$$\sigma_H = A + \frac{B}{r^2} \quad \text{and} \quad \sigma_r = A - \frac{B}{r^2}$$

Development of the theory for thick cylinders is concerned with sections remote from the

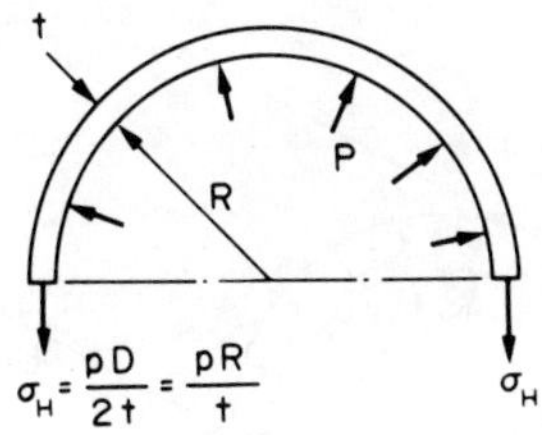

Fig. 10.1. Thin cylinder subjected to internal pressure.

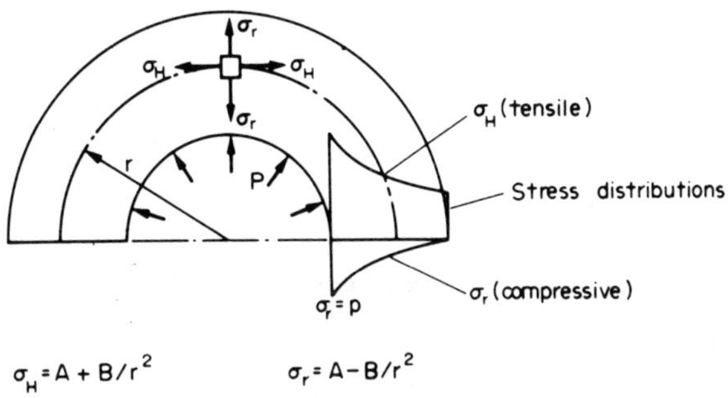

Fig. 10.2. Thick cylinder subjected to internal pressure.

ends since distribution of the stresses around the joints makes analysis at the ends particularly complex. For central sections the applied pressure system which is normally applied to thick cylinders is symmetrical, and all points on an annular element of the cylinder wall will be displaced by the same amount, this amount depending on the radius of the element. Consequently there can be no shearing stress set up on transverse planes and stresses on such planes are therefore principal stresses (see page 331). Similarly, since the radial shape of the cylinder is maintained there are no shears on radial or tangential planes, and again stresses on such planes are principal stresses. Thus, consideration of any element in the wall of a thick cylinder involves, in general, consideration of a mutually prependicular, tri-axial, principal stress system, the three stresses being termed **radial**, **hoop** (tangential or circumferential) and **longitudinal** (axial) stresses.

10.2. Development of the Lamé theory

Consider the thick cylinder shown in Fig. 10.3. The stresses acting on an element of unit length at radius r are as shown in Fig. 10.4, the radial stress increasing from σ_r to $\sigma_r + d\sigma_r$ over the element thickness dr (*all stresses are assumed tensile*),

For radial equilibrium of the element:

$$(\sigma_r + d\sigma_r)(r + dr)\,d\theta \times 1 - \sigma_r \times r d\theta \times 1 = 2\sigma_H \times dr \times 1 \times \sin\frac{d\theta}{2}$$

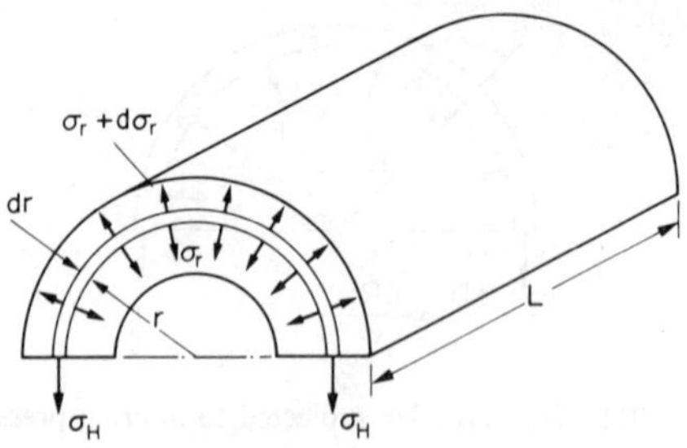

Fig. 10.3.

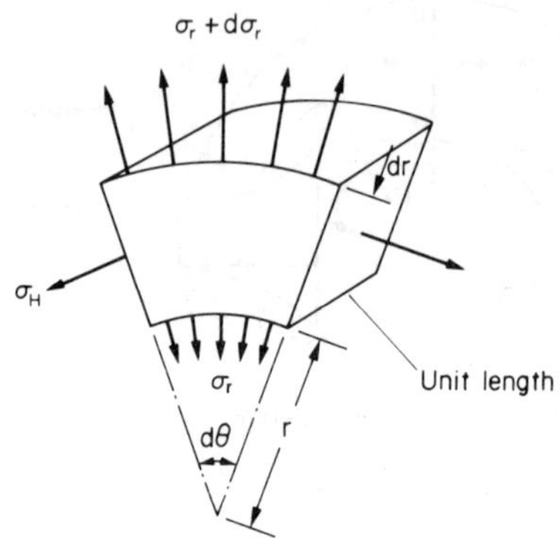

Fig. 10.4.

For small angles:

$$\sin\frac{d\theta}{2} \simeq \frac{d\theta}{2} \text{ radian}$$

Therefore, neglecting second-order small quantities,

$$r d\sigma_r + \sigma_r dr = \sigma_H dr$$

$\therefore$
$$\sigma_r + r\frac{d\sigma_r}{dr} = \sigma_H$$

or
$$\boldsymbol{\sigma_H - \sigma_r = r\frac{d\sigma_r}{dr}} \tag{10.1}$$

Assuming now that plane sections remain plane, i.e. the longitudinal strain ε_L is constant across the wall of the cylinder,

then
$$\varepsilon_L = \frac{1}{E}[\sigma_L - \nu\sigma_r - \nu\sigma_H]$$

$$= \frac{1}{E}[\sigma_L - \nu(\sigma_r + \sigma_H)] = \text{constant}$$

It is also assumed that the longitudinal stress σ_L is constant across the cylinder walls at points remote from the ends.

$\therefore$
$$\sigma_r + \sigma_H = \text{constant} = 2A \text{ (say)} \tag{10.2}$$

Substituting in (10.1) for σ_H,

$$2A - \sigma_r - \sigma_r = r\frac{d\sigma_r}{dr}$$

Multiplying through by r and rearranging,

$$2\sigma_r r + r^2\frac{d\sigma_r}{dr} - 2Ar = 0$$

i.e.

$$\frac{d}{dr}(\sigma_r r^2 - Ar^2) = 0$$

Therefore, integrating, $\sigma_r r^2 - Ar^2 = \text{constant} = -B$ (say)

$$\therefore \qquad \sigma_r = A - \frac{B}{r^2} \tag{10.3}$$

and from eqn. (10.2)

$$\sigma_H = A + \frac{B}{r^2} \tag{10.4}$$

The above equations yield the radial and hoop stresses at any radius r in terms of constants A and B. For any pressure condition there will always be two known conditions of stress (usually radial stress) which enable the constants to be determined and the required stresses evaluated.

10.3. Thick cylinder – internal pressure only

Consider now the thick cylinder shown in Fig. 10.5 subjected to an internal pressure P, *the external pressure being zero.*

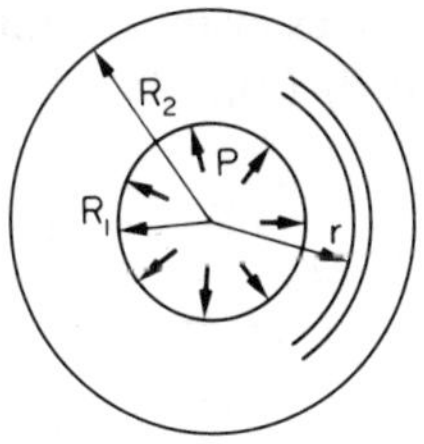

Fig. 10.5. Cylinder cross-section.

The two known conditions of stress which enable the Lamé constants A and B to be determined are:

$$\text{At } r = R_1 \quad \sigma_r = -P \quad \text{and} \quad \text{at } r = R_2 \quad \sigma_r = 0$$

N.B. –*The internal pressure is considered as a* ***negative*** *radial stress* since it will produce a radial compression (i.e. thinning) of the cylinder walls and the normal stress convention takes compression as negative.

Substituting the above conditions in eqn. (10.3),

$$-P = A - \frac{B}{R_1^2}$$

$$0 = A - \frac{B}{R_2^2}$$

i.e.

$$A = \frac{PR_1^2}{(R_2^2 - R_1^2)} \quad \text{and} \quad B = \frac{PR_1^2 R_2^2}{(R_2^2 - R_1^2)}$$

$\therefore$

$$\text{radial stress } \sigma_r = A - \frac{B}{r^2}$$

$$= \frac{PR_1^2}{(R_2^2 - R_1^2)}\left[1 - \frac{R_2^2}{r^2}\right]$$

$$= \frac{PR_1^2}{(R_2^2 - R_1^2)}\left[\frac{r^2 - R_2^2}{r^2}\right] = -P\left[\frac{(R_2/r)^2 - 1}{k^2 - 1}\right] \quad (10.5)$$

where k is the diameter ratio $D_2/D_1 = R_2/R_1$

and

$$\text{hoop stress } \sigma_H = \frac{PR_1^2}{(R_2^2 - R_1^2)}\left[1 + \frac{R_2^2}{r^2}\right]$$

$$= \frac{PR_1^2}{(R_2^2 - R_1^2)}\left[\frac{r^2 + R_2^2}{r^2}\right] = P\left[\frac{(R_2/r)^2 + 1}{k^2 - 1}\right] \quad (10.6)$$

These equations yield the stress distributions indicated in Fig. 10.2 with maximum values of both σ_r and σ_H at the inside radius.

10.4. Longitudinal stress

Consider now the cross-section of a thick cylinder with closed ends subjected to an internal pressure P_1 and an external pressure P_2 (Fig. 10.6).

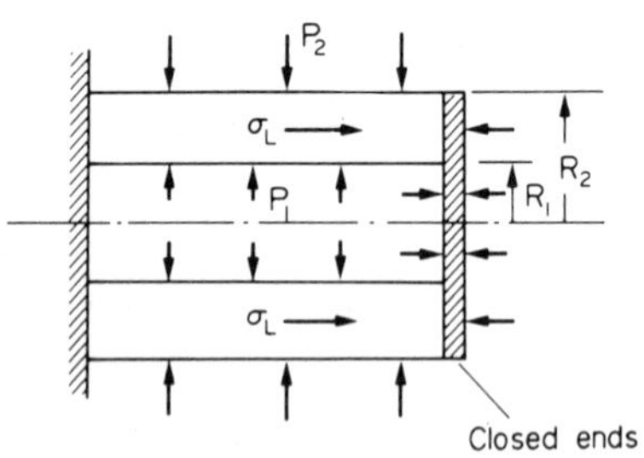

Fig. 10.6. Cylinder longitudinal section.

For horizontal equilibrium:

$$P_1 \times \pi R_1^2 - P_2 \times \pi R_2^2 = \sigma_L \times \pi(R_2^2 - R_1^2)$$

where σ_L is the longitudinal stress set up in the cylinder walls,

$$\therefore \qquad \text{longitudinal stress } \sigma_L = \frac{P_1 R_1^2 - P_2 R_2^2}{R_2^2 - R_1^2} \tag{10.7}$$

i.e. a constant.

It can be shown that the constant has the same value as the constant A of the Lame equations. This can be verified for the "internal pressure only" case of §10.3 by substituting $P_2 = 0$ in eqn. (10.7) above.

For combined internal and external pressures, the relationship $\sigma_L = A$ *also applies.*

10.5. Maximum shear stress

It has been stated in §10.1 that the stresses on an element at any point in the cylinder wall are principal stresses.

It follows, therefore, that the maximum shear stress at any point will be given by eqn. (13.12) as

$$\tau_{max} = \frac{\sigma_1 - \sigma_3}{2}$$

i.e. *half the difference between the greatest and least principal stresses.*

Therefore, in the case of the thick cylinder, normally,

$$\tau_{max} = \frac{\sigma_H - \sigma_r}{2}$$

since σ_H is normally tensile, whilst σ_r is compressive and both exceed σ_l in magnitude.

$$\therefore \qquad \tau_{max} = \frac{1}{2}\left[\left(A + \frac{B}{r^2}\right) - \left(A - \frac{B}{r^2}\right)\right]$$

$$\tau_{max} = \frac{B}{r^2} \tag{10.8}$$

The greatest value of τ_{max} *thus normally occurs at the inside radius where* $r = R_1$.

10.6. Change of cylinder dimensions

(*a*) *Change of diameter*

It has been shown in §9.3 that the diametral strain on a cylinder equals the hoop or circumferential strain.

Therefore change of diameter = diametral strain × original diameter

= circumferential strain × original diameter

With the principal stress system of hoop, radial and longitudinal stresses, all assumed tensile, the circumferential strain is given by

$$\varepsilon_H = \frac{1}{E}[\sigma_H - \nu\sigma_r - \nu\sigma_L]$$

Thus the change of diameter at any radius r of the cylinder is given by

$$\Delta D = \frac{2r}{E}[\sigma_H - \nu\sigma_r - \nu\sigma_L] \tag{10.9}$$

(b) *Change of length*

Similarly, the change of length of the cylinder is given by

$$\Delta L = \frac{L}{E}[\sigma_L - \sigma\nu_r - \nu\sigma_H] \tag{10.10}$$

10.7. Comparison with thin cylinder theory

In order to determine the limits of D/t ratio within which it is safe to use the simple thin cylinder theory, it is necessary to compare the values of stress given by both thin and thick cylinder theory for given pressures and D/t values. Since the maximum hoop stress is normally the limiting factor, it is this stress which will be considered.

From thin cylinder theory:

$$\sigma_H = P\frac{D}{2t}$$

i.e.

$$\frac{\sigma_H}{P} = \frac{K}{2} \quad \text{where } K = D/t$$

For thick cylinders, from eqn. (10.6),

$$\sigma_H = \frac{PR_1^2}{(R_2^2 - R_1^2)}\left[1 + \frac{R_2^2}{r^2}\right]$$

i.e.

$$\sigma_{H_{max}} = P\frac{(R_1^2 + R_2^2)}{(R_2^2 - R_1^2)} \quad \text{at } r = R_1 \tag{10.11}$$

Now, substituting for $R_2 = R_1 + t$ and $D = 2R_1$,

$$\sigma_{H_{max}} = \left[\frac{\frac{1}{2}D^2 + t(D+t)}{t(D+t)}\right]P$$

$$= \left[\frac{D^2}{2t^2(D/t+1)} + 1\right]P$$

i.e.

$$\frac{\sigma_{H_{max}}}{P} = \frac{K^2}{2(K+1)} + 1 \tag{10.12}$$

Thus for various D/t ratios the stress values from the two theories may be plotted and compared; this is shown in Fig. 10.7.

Also indicated in Fig. 10.7 is the percentage error involved in using the thin cylinder theory. It will be seen that the error will be held within 5% if D/t ratios in excess of 15 are used.

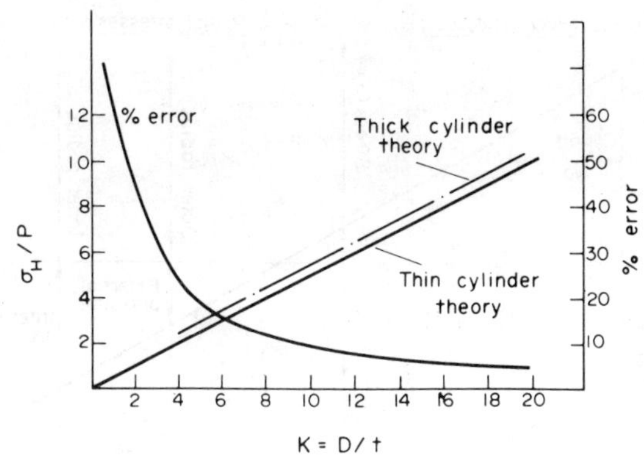

Fig. 10.7. Comparison of thin and thick cylinder theories for various diameter/thickness ratios.

However, if D is taken as the *mean* diameter for calculation of the thin cylinder values instead of the inside diameter as used here, the percentage error reduces from 5% to approximately 0.25% at $D/t = 15$.

10.8. Graphical treatment – Lamé line

The Lamé equations when plotted on stress and $1/r^2$ axes produce straight lines, as shown in Fig. 10.8.

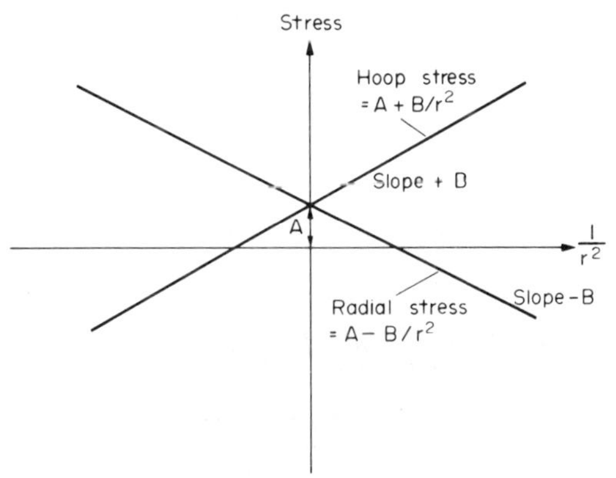

Fig. 10.8. Graphical representation of Lamé equations – Lamé line.

Both lines have exactly the same intercept A and the same magnitude of slope B, the only difference being the sign of their slopes. The two are therefore combined by plotting hoop stress values to the left of the σ axis (again against $1/r^2$) instead of to the right to give the single line shown in Fig. 10.9. In most questions one value of σ_r and one value of σ_H, or alternatively two values of σ_r, are given. In both cases the single line can then be drawn.

When a thick cylinder is subjected to external pressure only, the radial stress at the inside radius is zero and the graph becomes the straight line shown in Fig. 10.10.

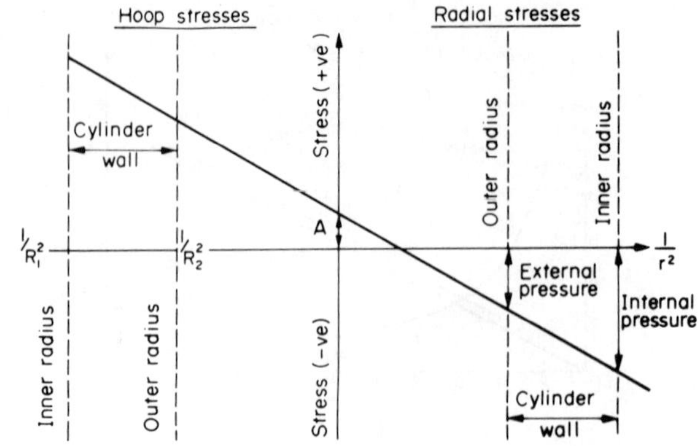

Fig. 10.9. Lamé line solution for cylinder with internal and external pressures.

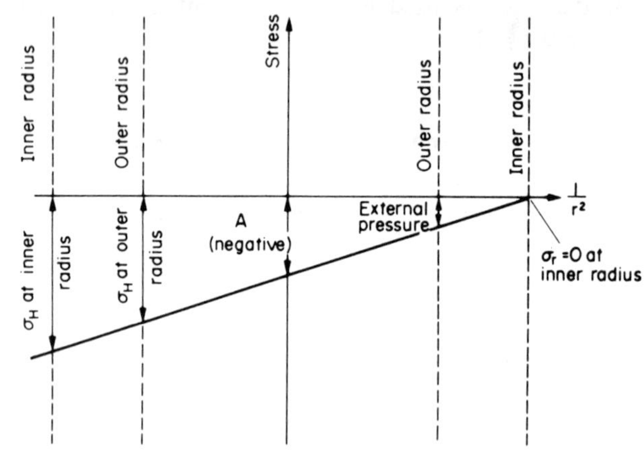

Fig. 10.10. Lamé line solution for cylinder subjected to external pressure only.

N.B.–From §10.4 the value of the *longitudinal stress* σ_L is given by the *intercept A on the σ axis.*

It is not sufficient simply to read off stress values from the axes since this can introduce appreciable errors. Accurate values must be obtained from proportions of the figure using similar triangles.

10.9. Compound cylinders

From the sketch of the stress distributions in Fig. 10.2 it is evident that there is a large variation in hoop stress across the wall of a cylinder subjected to internal pressure. The material of the cylinder is not therefore used to its best advantage. To obtain a more uniform hoop stress distribution, cylinders are often built up by shrinking one tube on to the outside of another. When the outer tube contracts on cooling the inner tube is brought into a state of compression. The outer tube will conversely be brought into a state of tension. If this compound cylinder is now subjected to internal pressure the resultant hoop stresses will be the algebraic sum of those resulting from internal pressure and those resulting from shrinkage as drawn in Fig. 10.11; thus a much smaller total fluctuation of hoop stress is obtained. A similar effect is obtained if a cylinder is wound with wire or steel tape under tension (see §10.19).

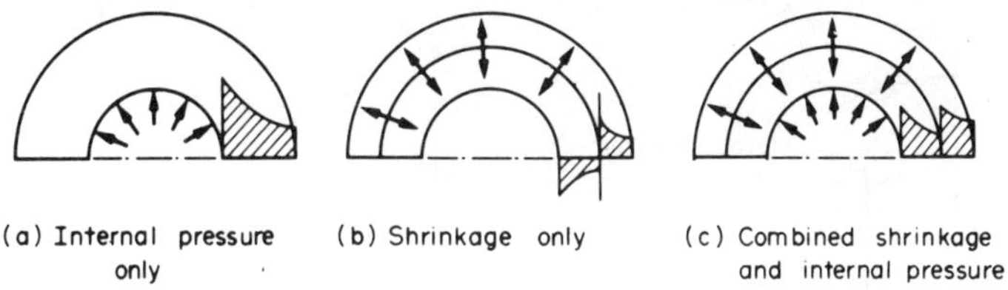

Fig. 10.11. Compound cylinders – combined internal pressure and shrinkage effects.

(*a*) *Same materials*

The method of solution for compound cylinders constructed from similar materials is to break the problem down into three separate effects:

(a) shrinkage pressure only on the inside cylinder;
(b) shrinkage pressure only on the outside cylinder;
(c) internal pressure only on the complete cylinder (Fig. 10.12).

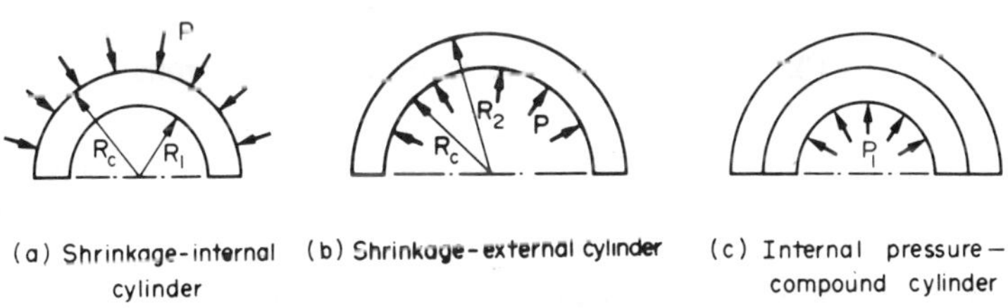

Fig. 10.12. Method of solution for compound cylinders.

For each of the resulting load conditions there are two known values of radial stress which enable the Lamé constants to be determined in each case:

i.e. condition (a) *shrinkage – internal cylinder:*

At $r = R_1$, $\sigma_r = 0$

At $r = R_c$, $\sigma_r = -p$ (compressive since it tends to reduce the wall thickness)

condition (b) *shrinkage – external cylinder:*

At $r = R_2$, $\sigma_r = 0$

At $r = R_c$, $\sigma_r = -p$

condition (c) *internal pressure – compound cylinder:*

At $r = R_2$, $\sigma_r = 0$

At $r = R_1$, $\sigma_r = -P_1$

Thus for each condition the hoop and radial stresses at any radius can be evaluated and the principle of superposition applied, i.e. the various stresses are then combined algebraically to produce the stresses in the compound cylinder subjected to both shrinkage and internal pressure. In practice this means that the compound cylinder is able to withstand greater internal pressures before failure occurs or, alternatively, that a thinner compound cylinder (with the associated reduction in material cost) may be used to withstand the same internal pressure as the single thick cylinder it replaces.

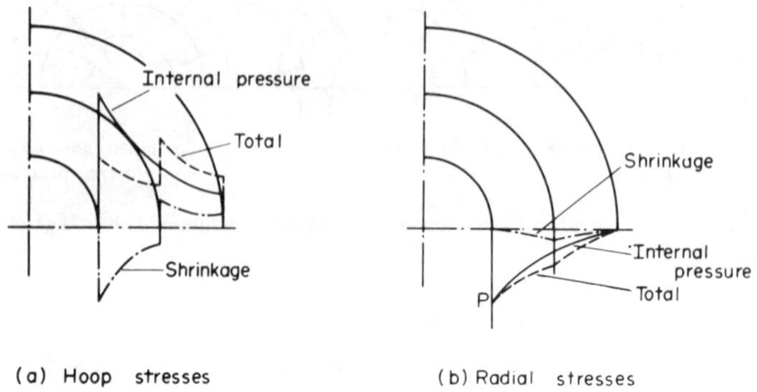

Fig. 10.13. Distribution of hoop and radial stresses through the walls of a compound cylinder.

(*b*) *Different materials*

(See §10.14.)

10.10. Compound cylinders – graphical treatment

The graphical, or Lamé line, procedure introduced in §10.8 can be used for solution of compound cylinder problems. The vertical lines representing the boundaries of the cylinder walls may be drawn at their appropriate $1/r^2$ values, and the solution for condition (c) of Fig. 10.12 may be carried out as before, producing a single line across both cylinder sections (Fig. 10.14a).

The graphical representation of the effect of shrinkage does not produce a single line, however, and the effect on each cylinder must therefore be determined by projection of known lines on the radial side of the graph to the respective cylinder on the hoop stress side, i.e. conditions (a) and (b) of Fig. 10.12 must be treated separately as indeed they are in the analytical approach. The resulting graph will then appear as in Fig. 10.14b.

The total effect of combined shrinkage and internal pressure is then given, as before, by the algebraic combination of the separate effects, i.e. the graphs must be added together, taking due account of sign to produce the graph of Fig. 10.14c. In practice this is the only graph which need be constructed, all effects being considered on the single set of axes. Again, *all values should be calculated from proportions of the figure*, i.e. by the use of similar triangles.

10.11. Shrinkage or interference allowance

In the design of compound cylinders it is important to relate the difference in diameter of the mating cylinders to the stresses this will produce. This difference in diameter at the

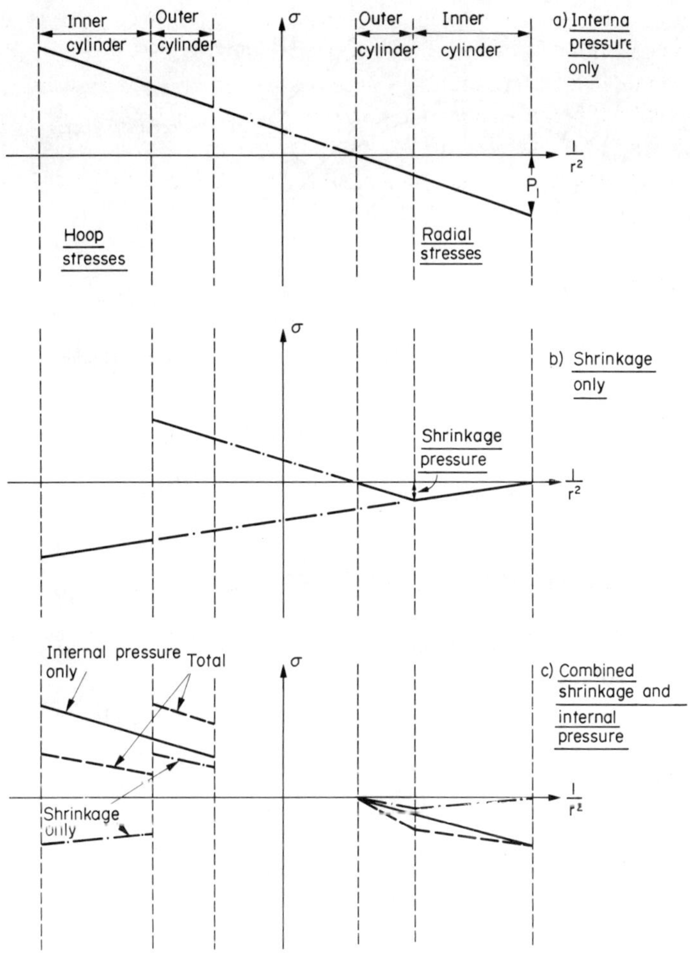

Fig. 10.14. Graphical (Lamé line) solution for compound cylinders.

"common" surface is normally termed the *shrinkage* or *interference allowance* whether the compound cylinder is formed by a shrinking or a force fit procedure respectively. Normally, however, the shrinking process is used, the outer cylinder being heated until it will freely slide over the inner cylinder thus exerting the required junction or shrinkage pressure on cooling.

Consider, therefore, the compound cylinder shown in Fig. 10.15, *the material of the two cylinders not necessarily being the same.*

Let the pressure set up at the junction of two cylinders owing to the force or shrink fit be p. Let the hoop stresses set up at the junction on the inner and outer tubes resulting from the pressure p be σ_{H_i} (compressive) and σ_{H_o} (tensile) respectively.

Then, if

$$\delta_o = \text{radial shift of outer cylinder}$$

and

$$\delta_i = \text{radial shift of inner cylinder (as shown in Fig. 10.15)}$$

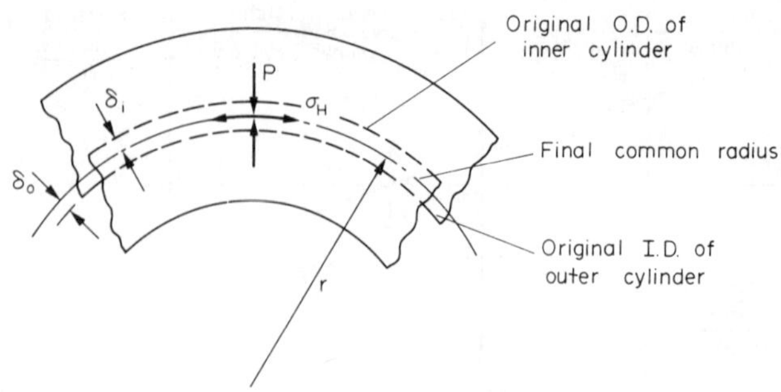

Fig. 10.15. Interference or shrinkage allowance for compound cylinders – total interference = $\delta_o + \delta_i$.

since circumferential strain = diametral strain

$$\text{circumferential strain at radius } r \text{ on outer cylinder} = \frac{2\delta_o}{2r} = \frac{\delta_o}{r} = \varepsilon_{H_o}$$

$$\text{circumferential strain at radius } r \text{ on inner cylinder} = \frac{2\delta_i}{2r} = \frac{\delta_i}{r} = -\varepsilon_{H_i}$$

(negative since it is a *decrease* in diameter).

$$\begin{aligned}\text{Total interference or shrinkage} &= \delta_o + \delta_i\\ &= r\varepsilon_{H_o} + r(-\varepsilon_{H_i})\\ &= (\varepsilon_{H_o} - \varepsilon_{H_i})r\end{aligned}$$

Now assuming open ends, i.e. $\sigma_L = 0$,

$$\varepsilon_{H_o} = \frac{\sigma_{H_o}}{E_1} - \frac{\nu_1}{E_1}(-p) \qquad \text{since } \sigma_{r_o} = -p$$

and

$$\varepsilon_{H_i} = \frac{\sigma_{H_i}}{E_2} - \frac{\nu_2}{E_2}(-p) \qquad \text{since } \sigma_{r_i} = -p$$

where E_1 and ν_1, E_2 and ν_2 are the elastic modulus and Poisson's ratio of the two tubes respectively.

Therefore total interference or shrinkage allowance (based on radius)

$$= \left[\frac{1}{E_1}(\sigma_{H_o} + \nu_1 p) - \frac{1}{E_2}(\sigma_{H_i} + \nu_2 p)\right] r \qquad (10.13)$$

where r is the initial nominal radius of the mating surfaces.

N.B. σ_{H_i}, being compressive, will change the negative sign to a positive one when its value is substituted. Shrinkage allowances *based on diameter* will be *twice* this value, i.e. replacing radius r by diameter d.

Generally, however, the tubes are of the same material.

$$\therefore \qquad E_1 = E_2 = E \quad \text{and} \quad \nu_1 = \nu_2 = \nu$$

$$\therefore \qquad \textbf{Shrinkage allowance} = \frac{r}{E}(\sigma_{H_o} - \sigma_{H_i}) \qquad (10.14)$$

The values of σ_{H_o} and σ_{H_i} may be determined graphically or analytically in terms of the shrinkage pressure p which can then be evaluated for any known shrinkage or interference allowance. Other stress values throughout the cylinder can then be determined as described previously.

10.12. Hub on solid shaft

The Lamé equations give

$$\sigma_H = A + \frac{B}{r^2} \quad \text{and} \quad \sigma_r = A - \frac{B}{r^2}$$

Since σ_H and σ_r cannot be infinite when $r = 0$, i.e. at the centre of the solid shaft, it follows that B must be zero since this is the only solution which can yield finite values for the stresses.

From the above equations, therefore, it follows that $\sigma_H = \sigma_r = A$ for all values of r.

Now at the outer surface of the shaft

$$\sigma_r = -p \qquad \text{the shrinkage pressure.}$$

Therefore the hoop and radial stresses throughout a solid shaft are everywhere equal to the shrinkage or interference pressure and both are compressive. The maximum shear stress $= \frac{1}{2}(\sigma_1 - \sigma_2)$ is thus zero throughout the shaft.

10.13. Force fits

It has been stated that compound cylinders may be formed by *shrinking* or by *force-fit* techniques. In the latter case the interference allowance is small enough to allow the outer cylinder to be pressed over the inner cylinder with a large axial force.

If the interference pressure set up at the common surface is p, the normal force N between the mating cylinders is then

$$N = p \times 2\pi r L$$

where L is the axial length of the contact surfaces.

The friction force F between the cylinders which has to be overcome by the applied force is thus

$$F = \mu N$$

where μ is the coefficient of friction between the contact surfaces.

$$\therefore \qquad F = \mu(p \times 2\pi r L)$$

$$= \mathbf{2\pi\mu p r L} \qquad (10.15)$$

With a knowledge of the magnitude of the applied force required the value of p may be determined.

Alternatively, for a known interference between the cylinders the procedure of § 10.11 may be carried out to determine the value of p which will be produced and hence the force F which will be required to carry out the press-fit operation.

10.14. Compound cylinder – different materials

The value of the shrinkage or interference allowance for compound cylinders constructed from cylinders of different materials is given by eqn. (10.13). The value of the shrinkage pressure set up owing to a known amount of interference can then be calculated as with the standard compound cylinder treatment, each component cylinder being considered separately subject to the shrinkage pressure.

Having constructed the compound cylinder, however, the treatment is different for the analysis of stresses owing to applied internal and/or external pressures. Previously the compound cylinder has been treated as a single thick cylinder and, e.g., a single Lamé line drawn across both cylinder walls for solution. In the case of cylinders of different materials, however, each component cylinder must be considered separately as with the shrinkage effects. Thus, for a known internal pressure P_1 which sets up a common junction pressure p, the Lamé line solution takes the form shown in Fig. 10.16.

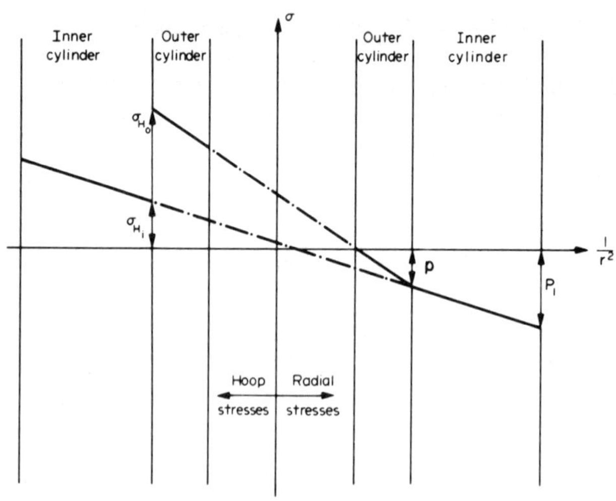

Fig. 10.16. Graphical solution for compound tubes of different materials.

For a full solution of problems of this type it is often necessary to make use of the equality of *diametral* strains at the common junction surface, i.e. to realise that for the cylinders to maintain contact with each other the diametral strains must be equal at the common surface.

Now diametral strain = circumferential strain

$$= \frac{1}{E}[\sigma_H - \nu\sigma_r - \nu\sigma_L]$$

Therefore at the common surface, ignoring longitudinal strains and stresses,

$$\frac{1}{E}[\sigma_{H_o} - \nu_o\sigma_r] = \frac{1}{E_i}[\sigma_{H_i} - \nu_i\sigma_r] \qquad (10.16)$$

where E_o and ν_o = Young's modulus and Poisson's ratio of outer cylinder,
E_i and ν_i = Young's modulus and Poisson's ratio of inner cylinder,
$\sigma_r = -p$ = radial stress at common surface,
and σ_{H_o} and σ_{H_i} = (as before) the hoop stresses at the common surface for the outer and inner cylinders respectively.

10.15. Uniform heating of compound cylinders of different materials

When an initially unstressed compound cylinder constructed from two tubes of the same material is heated uniformly, all parts of the cylinder will expand at the same rate, this rate depending on the value of the coefficient of expansion for the cylinder material.

If the two tubes are of different materials, however, each will attempt to expand at different rates and *differential thermal stresses* will be set up as described in §2.3. The method of treatment for such compound cylinders is therefore similar to that used for compound bars in the section noted.

Consider, therefore, two tubes of different material as shown in Fig. 10.17. Here it is convenient, for simplicity of treatment, to take as an example steel and brass for the two materials since the coefficients of expansion for these materials are known, the value for brass being greater than that for steel. Thus if the inner tube is of brass, as the temperature rises the brass will attempt to expand at a faster rate than the outer steel tube, the "free" expansions being indicated in Fig. 10.17a. In practice, however, when the tubes are joined as a compound cylinder, the steel will restrict the expansion of the brass and, conversely, the brass will force the steel to expand beyond its "free" expansion position. As a result a compromise situation is reached as shown in Fig. 10.17b, both tubes being effectively compressed radially (i.e. on their thickness) through the amounts shown. An effective increase p_t in "shrinkage" pressure is thus introduced.

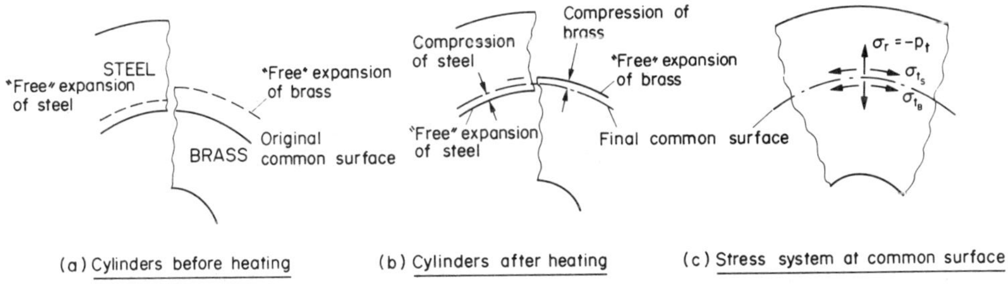

Fig. 10.17. Uniform heating of compound cylinders constructed from tubes of different materials – in this case, steel and brass.

p_t is the radial pressure introduced at the common interface by virtue of the differential thermal expansions.

Therefore, as for the compound bar treatment of §2.3:

compression of steel + compression of brass = difference in "free" lengths

$$\varepsilon_s d + \varepsilon_B d = (\alpha_B - \alpha_s) t d$$

$$= (\alpha_B - \alpha_s)(T_2 - T_1) d \qquad (10.17)$$

where d is the initial nominal diameter of the mating surfaces, α_B and α_s are the coefficients of linear expansion for the brass and steel respectively, $t = T_2 - T_1$ is the temperature change, and ε_B and ε_s are the diametral strains in the two materials.

Alternatively, using a treatment similar to that used in the derivation of the thick cylinder "shrinkage fit" expressions

$$\Delta d = (\varepsilon_{H_o} - \varepsilon_{H_i})d = (\alpha_B - \alpha_s)td$$

ε_{H_i} being compressive, then producing an identical expression to that obtained above.

Now since diametral strain = circumferential or hoop strain

$$\varepsilon_s = \frac{1}{E_s}[\sigma_{t_s} - \nu_s \sigma_r]$$

$$\varepsilon_B = \frac{1}{E_B}[\sigma_{t_B} - \nu_B \sigma_r]$$

σ_{t_s} and σ_{t_B} being the hoop stresses set up at the common interface surfaces in the steel and brass respectively *due to the differential thermal expansion* and σ_r the effective increase in radial stress at the common junction surface caused by the same effect, i.e. $\sigma_r = -p_t$.

However, any radial pressure at the common interface will produce hoop tension in the outer cylinder but hoop compression in the inner cylinder. The expression for ε_B obtained above will thus always be negative when the appropriate stress values have been inserted. Since eqn. (10.17) deals with **magnitudes** of displacements only, it follows that a negative sign must be introduced to the value of ε_B before it can be substituted into eqn. (10.17).

Substituting for ε_S and ε_B in eqn. (10.17) with $\sigma_r = -p_t$

$$\frac{1}{E_S}[\sigma_{t_s} + \nu_s p_t] - \frac{1}{E_B}[\sigma_{t_B} + \nu_B p_t] = [\alpha_B - \alpha_S][T_2 - T_1] \qquad (10.18)$$

The values of σ_{t_s} and σ_{t_B} are found in terms of the radial stress p_t at the junction surfaces by calculation or by graphical means as shown in Fig. 10.18.

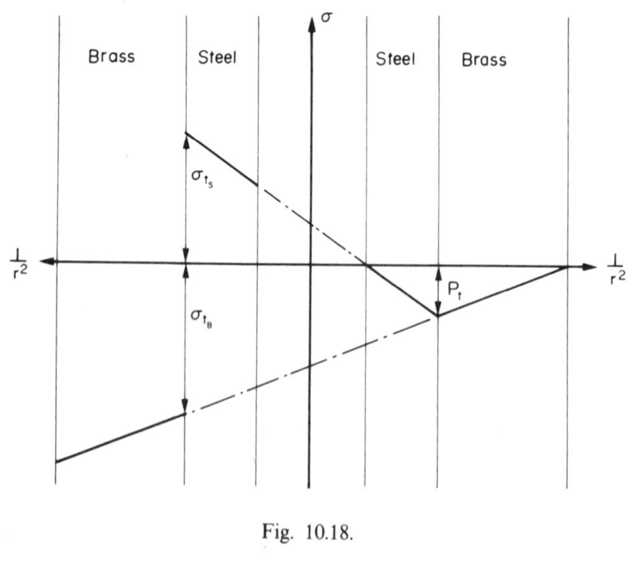

Fig. 10.18.

Substitution in eqn. (10.18) then yields the value of the "unknown" p_t and hence the other resulting stresses.

10.16. Failure theories – yield criteria

For thick cylinder design the Tresca (maximum shear stress) criterion is normally used for *ductile materials* (see Chapter 15), i.e. the maximum shear stress in the cylinder wall is equated to the maximum shear stress at yield in simple tension,

$$\tau_{max} = \sigma_y/2$$

Now the maximum shear stress is at the inside radius (§10.5) and is given by

$$\tau_{max} = \frac{\sigma_H - \sigma_r}{2}$$

Therefore, for cylinder failure

$$\frac{\sigma_y}{2} = \frac{\sigma_H - \sigma_r}{2}$$

i.e.

$$\sigma_y = \sigma_H - \sigma_r$$

Here, σ_H and σ_r are the hoop and radial stresses at the inside radius and σ_y is the allowable yield stress of the material taking into account any safety factors which may be introduced by the company concerned.

For brittle materials such as cast iron the Rankine (maximum principal stress) theory is used. In this case failure is deemed to occur when

$$\sigma_y = \sigma_{H_{max}}.$$

10.17. Plastic yielding – "auto-frettage"

It has been shown that the most highly stressed part of a thick cylinder is at the inside radius. It follows, therefore, that if the internal pressure is increased sufficiently, yielding of the cylinder material will take place at this position. Fortunately the condition is not too serious at this stage since there remains a considerable bulk of elastic material surrounding the yielded area which contains the resulting strains within reasonable limits. As the pressure is increased further, however, plastic penetration takes place deeper and deeper into the cylinder wall and eventually the whole cylinder will yield.

If the pressure is such that plastic penetration occurs only partly into the cylinder wall, on release of that pressure the elastic outer zone attempts to return to its original dimensions but is prevented from doing so by the permanent deformation or "set" of the yielded material. The result is that the elastic material is held in a state of residual tension whilst the inside is brought into residual compression. This process is termed *auto-frettage* and it has the same effect as shrinking one tube over another without the necessary complications of the shrinking procedure, i.e. on subsequent loading cycles the cylinder is able to withstand a higher internal pressure since the compressive residual stress at the inside surface has to be overcome before this region begins to experience tensile stresses. For this reason gun barrels and other pressure vessels are often pre-stressed in this way prior to service.

A full theoretical treatment of the auto-frettage process is introduced in Chapter 18 together with associated plastic collapse theory.

10.18. Wire-wound thick cylinders

Consider a thick cylinder with inner and outer radii R_1 and R_2 respectively, wound with wire under tension until its external radius becomes R_3. The resulting hoop and radial stresses developed in the cylinder will depend upon the way in which the tension T in the wire varies. The simplest case occurs when the tension in the wire is held constant throughout the winding process, and the solution for this condition will be introduced here. Solution for more complicated tension conditions will be found in more advanced texts and are not deemed appropriate for this volume. The method of solution, however, is similar.

(a) *Stresses in the wire*

Let the combined tube and wire be considered as a thick cylinder. The tension in the wire produces an "effective" external pressure on the tube and hence a compressive hoop stress.

Now for a thick cylinder subjected to an *external* pressure P the hoop and radial stresses are given by

$$\sigma_H = -\frac{PR_2^2}{(R_2^2 - R_1^2)}\left[\frac{r^2 + R_1^2}{r^2}\right]$$

and

$$\sigma_r = -\frac{PR_2^2}{(R_2^2 - R_1^2)}\left[\frac{r^2 - R_1^2}{r^2}\right]$$

i.e.

$$\sigma_H = \sigma_r\left[\frac{r^2 + R_1^2}{r^2 - R_1^2}\right]$$

If the initial tensile stress in the wire is T the final tensile hoop stress in the winding at any radius r is less than T by an amount equal to the compressive hoop stress set up by the effective "external" pressure caused by the winding,

i.e. final hoop stress in the winding at radius $r = T - \sigma_r\left[\frac{(r^2 + R_1^2)}{(r^2 - R_1^2)}\right]$ (10.19)

Using the same analysis outlined in §10.2,

$$\sigma_H = \sigma_r + r\frac{d\sigma_r}{dr}$$

$$\therefore \quad \sigma_r + r\frac{d\sigma_r}{dr} = T - \sigma_r\left[\frac{(r^2 + R_1^2)}{(r^2 - R_1^2)}\right]$$

$$\therefore \quad r\frac{d\sigma_r}{dr} = T - \sigma_r\left[\frac{r^2 + R_1^2 - r^2 + R_1^2}{(r^2 - R_1^2)}\right]$$

$$= T - \left[\frac{2R_1^2\sigma_r}{r^2 - R_1^2}\right]$$

Multiplying through by $\dfrac{r}{(r^2-R_1^2)}$ and rearranging,

$$\frac{r^2}{(r^2-R_1^2)}\frac{d\sigma_r}{dr}+2\frac{R_1^2 r}{(r^2-R_1^2)^2}\sigma_r=\frac{Tr}{(r^2-R_1^2)}$$

$$\therefore \qquad \frac{d}{dr}\left[\frac{r^2}{(r^2-R_1^2)}\sigma_r\right]=\frac{Tr}{(r^2-R_1^2)}$$

$$\frac{r^2}{(r^2-R_1^2)}\sigma_r=\frac{T}{2}\log_e(r^2-R_1^2)+A \qquad (10.20)$$

But $\sigma_r = 0$ when $r = R_3$,

$$\therefore \qquad 0=\frac{T}{2}\log_e(R_3^2-R_1^2)+A$$

$$A=-\frac{T}{2}\log_e(R_3^2-R_1^2)$$

Therefore substituting in eqn. (10.20),

$$\frac{r^2}{(r^2-R_1^2)}\sigma_r=\frac{T}{2}\log_e\frac{(r^2-R_1^2)}{(R_3^2-R_1^2)}$$

$$\therefore \qquad \sigma_r=\frac{(r^2-R_1^2)}{2r^2}T\log_e\frac{(r^2-R_1^2)}{(R_3^2-R_1^2)}$$

$$=-\frac{(r^2-R_1^2)}{2r^2}T\log_e\frac{(R_3^2-R_1^2)}{(r^2-R_1^2)} \qquad (10.21)$$

From eqn. (10.19),

$$\sigma_H=T-\sigma_r\left[\frac{r^2+R_1^2}{r^2-R_1^2}\right]$$

Therefore since the sign of σ_r has been taken into account in setting up eqn. (10.19)

$$\sigma_H=T\left[1-\frac{(r^2+R_1^2)}{2r^2}\log_e\frac{(R_3^2-R_1^2)}{(r^2-R_1^2)}\right] \qquad (10.22)$$

Thus eqns. (10.21) and (10.22) give the stresses in the wire winding for all radii between R_2 and R_3.

(*b*) *Stresses in the tube*

The stresses in the tube due to wire winding may be found from the normal thick cylinder expressions when it is considered subject to an external pressure P_2 at radius R_2. The value of P_2 is that obtained from eqn. (10.21) with $r = R_2$.

If an additional internal pressure is applied to the wire-wound cylinder it may be treated as a single thick cylinder and the resulting stresses combined algebraically with those due to winding to obtain the resultant effect.

Examples

Example 10.1 (B)

A thick cylinder of 100 mm internal radius and 150 mm external radius is subjected to an internal pressure of 60 MN/m^2 and an external pressure of 30 MN/m^2. Determine the hoop and radial stresses at the inside and outside of the cylinder together with the longitudinal stress if the cylinder is assumed to have closed ends.

Solution (a): analytical

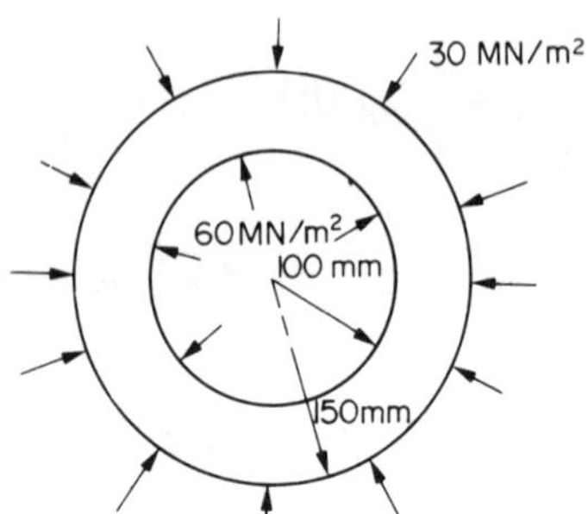

Fig. 10.19.

The internal and external pressures both have the effect of decreasing the thickness of the cylinder; the radial stresses at both the inside and outside radii are thus compressive, i.e. negative (Fig. 10.19).

$\therefore$ at $r = 0.1$ m, $\sigma_r = -60\ MN/m^2$

and at $r = 0.15$ m, $\sigma_r = -30\ MN/m^2$

Therefore, from eqn. (10.3), with stress units of MN/m^2,

$$-60 = A - 100B \quad (1)$$

and

$$-30 = A - 44.5B \quad (2)$$

Subtracting (2) from (1),

$$-30 = -55.5B$$

$$B = 0.54$$

Therefore, from (1),

$$A = -60 + 100 \times 0.54$$

$$A = -6$$

Therefore, at $r = 0.1$ m, from eqn. (10.4),

$$\sigma_H = A + \frac{B}{r^2} = -6 + 0.54 \times 100$$

$$= \mathbf{48\ MN/m^2}$$

and at $r = 0.15\,\text{m}$, $$\sigma_H = -6+0.54\times 44.5 = -6+24$$
$$= \mathbf{18\,MN/m^2}$$

From eqn. (10.7) the longitudinal stress is given by

$$\sigma_L = \frac{P_1R_1^2 - P_2R_2^2}{(R_2^2 - R_1^2)} = \frac{(60\times 0.1^2 - 30\times 0.15^2)}{(0.15^2 - 0.1^2)}$$

$$= \frac{10^2(60-30\times 2.25)}{1.25\times 10^2} = -\mathbf{6\,MN/m^2} \quad \text{i.e. compressive}$$

Solution (b): graphical

The graphical solution is shown in Fig. 10.20, where the boundaries of the cylinder are given by

$$\frac{1}{r^2} = 100 \textit{ for the inner radius where } r = 0.1\,\text{m}$$

$$\frac{1}{r^2} = 44.5 \text{ for the outer radius where } r = 0.15\,\text{m}$$

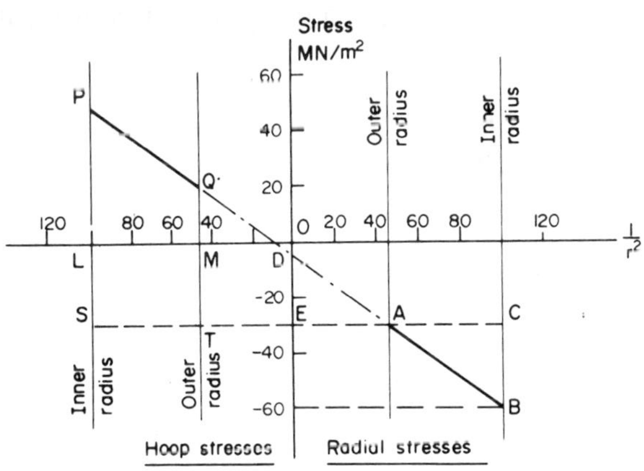

Fig. 10.20.

The two conditions which enable the Lamé line to be drawn are the same as those used above for the analytical solution,

i.e. $$\sigma_r = -60\,\text{MN/m}^2 \quad \text{at} \quad r = 0.1\,\text{m}$$
$$\sigma_r = -30\,\text{MN/m}^2 \quad \text{at} \quad r = 0.15\,\text{m}$$

The hoop stresses at these radii are then given by points P and Q on the graph. For complete accuracy these values should be calculated by proportions of the graph thus:

by similar triangles PAS and BAC

$$\frac{PS}{100+44.5} = \frac{CB}{100-44.5}$$

$$\therefore \qquad \frac{PL+LS}{144.5} = \frac{30}{55.5}$$

i.e. hoop stress at radius $r = 0.1$ m

$$= PL = \frac{30 \times 144.5}{55.5} - LS$$

$$= 78 - 30 = \mathbf{48\ MN/m^2}$$

Similarly, the hoop stress at radius $r = 0.15$ m is QM and given by the similar triangles QAT and BAC,

$$\text{i.e.} \qquad \frac{QM+MT}{44.5+44.5} = \frac{30}{55.5}$$

$$QM = \frac{30 \times 89}{55.5} - 30$$

$$= 48 - 30 = \mathbf{18\ MN/m^2}$$

The longitudinal stress σ_L = the intercept on the σ axis (which is negative)

$$= DO = OE - DE = 30 - DE$$

$$\text{Now} \qquad \frac{DE}{44.5} = \frac{30}{55.5}$$

$$\therefore \qquad DE = 24$$

$$\therefore \qquad \sigma_L = 30 - 24 = \mathbf{6\ MN/m^2} \quad \text{compressive}$$

Example 10.2 (B)

An external pressure of $10\ \text{MN/m}^2$ is applied to a thick cylinder of internal diameter 160 mm and external diameter 320 mm. If the maximum hoop stress permitted on the inside wall of the cylinder is limited to $30\ \text{MN/m}^2$, what maximum internal pressure can be applied assuming the cylinder has closed ends? What will be the change in outside diameter when this pressure is applied? $E = 207\ \text{GN/m}^2$, $\nu = 0.29$.

Solution (a): analytical

The conditions for the cylinder are:

When $\quad r = 0.08$ m, $\quad \sigma_r = -p \quad$ and $\quad \dfrac{1}{r^2} = 156$

when $\quad r = 0.16\ \text{m}, \quad \sigma_r = -10\ \text{MN/m}^2 \quad \text{and} \quad \frac{1}{r^2} = 39$

and when $\quad r = 0.08\ \text{m}, \quad \sigma_H = 30\ \text{MN/m}^2$

since the maximum hoop stress occurs at the inside surface of the cylinder.

Using the latter two conditions in eqns. (10.3) and (10.4) with units of MN/m^2,

$$-10 = A - 39B \tag{1}$$

$$30 = A + 156B \tag{2}$$

Subtracting (1) from (2),

$$40 = 195B \qquad \therefore\ B = 0.205$$

Substituting in (1),

$$A = -10 + (39 \times 0.205)$$

$$= -10 + 8 \qquad \therefore\ A = -2$$

Therefore, at $r = 0.08$, from eqn. (10.3),

$$\sigma_r = -p = A - 156B$$

$$= -2 - 156 \times 0.205$$

$$= -2 - 32 = \mathbf{-34\ MN/m^2}$$

i.e. the allowable internal pressure is 34 MN/m^2.

From eqn. (10.9) the change in diameter is given by

$$\Delta D = \frac{2r_0}{E}(\sigma_H - \nu\sigma_r - \nu\sigma_L)$$

Now at the outside surface

$$\sigma_r = -10\ \text{MN/m}^2 \quad \text{and} \quad \sigma_H = A + \frac{B}{r^2}$$

$$= -2 + (39 \times 0.205)$$

$$= -2 + 8 = 6\ \text{MN/m}^2$$

$$\sigma_L = \frac{P_1 R_1^2 - P_2 R_2^2}{(R_2^2 - R_1^2)} = \frac{(34 \times 0.08^2 - 10 \times 0.16^2)}{(0.16^2 - 0.08^2)}$$

$$= \frac{(34 \times 0.64 - 10 \times 2.56)}{(2.56 - 0.64)} = \frac{21.8 - 25.6}{1.92}$$

$$= -\frac{3.8}{1.92} = 1.98\ \text{MN/m}^2\ \text{compressive}$$

$$\therefore \qquad \Delta D = \frac{0.32}{207 \times 10^9}[6 - 0.29(-10) - 0.29(-1.98)]10^6$$

$$= \frac{0.32}{207 \times 10^3}(6 + 2.9 + 0.575)$$

$$= \mathbf{14.7\ \mu m}$$

Solution (b): graphical

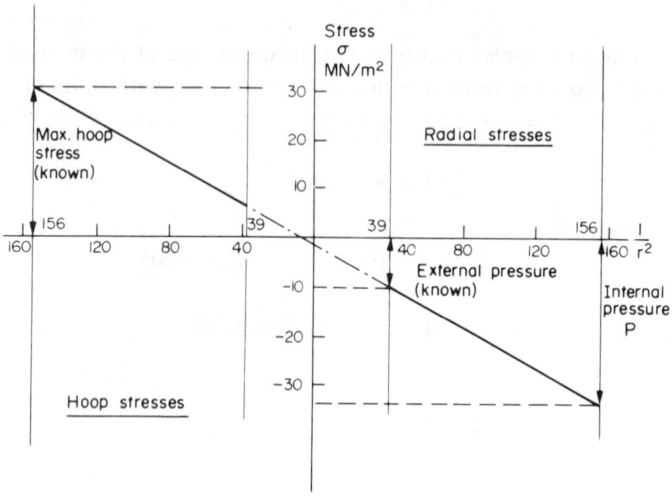

Fig. 10.21.

The graphical solution is shown in Fig. 10.21. The boundaries of the cylinder are as follows:

for $r = 0.08$ m, $\qquad \dfrac{1}{r^2} = 156$

and for $r = 0.16$ m, $\qquad \dfrac{1}{r^2} = 39$

The two fixed points on the graph which enable the line to be drawn are, therefore,

$$\sigma_r = -10 \text{ MN/m}^2 \text{ at } r = 0.16 \quad \text{and} \quad \sigma_H = 30 \text{ MN/m}^2 \text{ at } r = 0.08 \text{ m}$$

The allowable internal pressure is then given by the value of σ_r at $r = 0.08$ m $\left(\dfrac{1}{r^2} = 156\right)$, i.e. 34 MN/m^2.

Similarly, the hoop stress at the outside surface is given by the value of σ_H at $\dfrac{1}{r^2} = 39$, i.e. 6 MN/m^2, and the longitudinal stress by the intercept on the σ axis, i.e. 2 MN/m^2 compressive.

N.B.–In practice all these values should be calculated by proportions.

Example 10.3 (B)

(a) In an experiment on a thick cylinder of 100 mm external diameter and 50 mm internal diameter the hoop and longitudinal strains as measured by strain gauges applied to the outer surface of the cylinder were 240×10^{-6} and 60×10^{-6}, respectively, for an internal pressure of 90 MN/m^2, the external pressure being zero.

Determine the actual hoop and longitudinal stresses present in the cylinder if $E = 208\ GN/m^2$ and $\nu = 0.29$. Compare the hoop stress value so obtained with the theoretical value given by the Lame equations.

(b) Assuming that the above strain readings were obtained for a thick cylinder of 100 mm external diameter but unkonwn internal diameter calculate this internal diameter.

Solution

(a)

$$\varepsilon_H = \frac{1}{E}(\sigma_H - \nu\sigma_L) \quad \text{and} \quad \varepsilon_L = \frac{1}{E}(\sigma_L - \nu\sigma_H)$$

since $\sigma_r = 0$ at the outside surface of the cylinder for zero external pressure.

$$\therefore \quad 240 \times 10^{-6} \times 208 \times 10^9 = \sigma_H - 0.29\sigma_L = 50 \times 10^6 \quad (1)$$

$$60 \times 10^{-6} \times 208 \times 10^9 = \sigma_L - 0.29\sigma_H = 12.5 \times 10^6 \quad (2)$$

(1) × 0.29 $\quad 0.29\sigma_H - 0.084\sigma_L = 14.5 \times 10^6 \quad (3)$

(2) $\quad \sigma_L - 0.29\sigma_H = 12.5 \times 10^6$

(3) + (2) $\quad 0.916\sigma_L = 27 \times 10^6$

$\therefore \quad \boldsymbol{\sigma_L = 29.5\ MN/m^2}$

Substituting in (2) $\quad 0.29\sigma_H = 29.5 - 12.5 = 17 \times 10^6$

$$\boldsymbol{\sigma_H = 58.7\ MN/m^2}$$

The theoretical values of σ_H for an internal pressure of 90 MN/m^2 may be obtained from Fig. 10.22, the boundaries of the cylinder being given by $r = 0.05$ and $r = 0.025$,

i.e. $\quad \frac{1}{r^2} = 400$ and 1600 respectively

i.e. $\quad \boldsymbol{\sigma_H = 60\ MN/m^2}$ theoretically

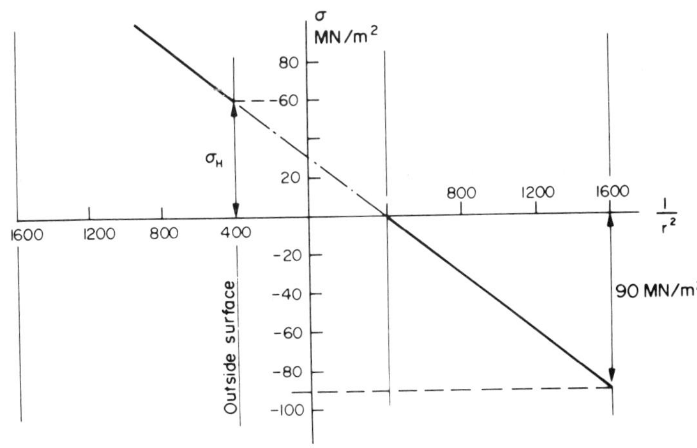

Fig. 10.22.

(b) From part (a) $\sigma_H = 58.7\ \text{MN/m}^2$ at $r = 0.05$

and $\sigma_r = 0$ at $r = 0.05$

$\therefore$
$$58.7 = A + 400B$$
$$0 = A - 400B$$

Adding: $58.7 = 2A \quad \therefore\ A = 29.35$

and since $A = 400B \quad \therefore\ B = 0.0734$

Therefore for the internal radius R_1 where $\sigma_r = 90\ \text{MN/m}^2$

$$-90 = 29.35 - \frac{0.0734}{R_1^2}$$
$$R_1^2 = \frac{0.0734}{119.35} = 0.000615$$
$$= 6.15 \times 10^{-4}$$

$\therefore$
$$R_1 = 2.48 \times 10^{-2}\ \text{m} = 24.8\ \text{mm}$$

$\therefore$ **Internal diameter = 49.6 mm**

For a graphical solution of part (b), see Fig. 10.23, where the known points which enable the Lame line to be drawn are, as above:

$$\sigma_H = 58.7 \quad \text{at}\ \frac{1}{r^2} = 400 \quad \text{and} \quad \sigma_r = 0 \quad \text{at}\ \frac{1}{r^2} = 400$$

It is thus possible to determine the value of $1/R_1^2$ which will produce $\sigma_r = -90\ \text{MN/m}^2$.

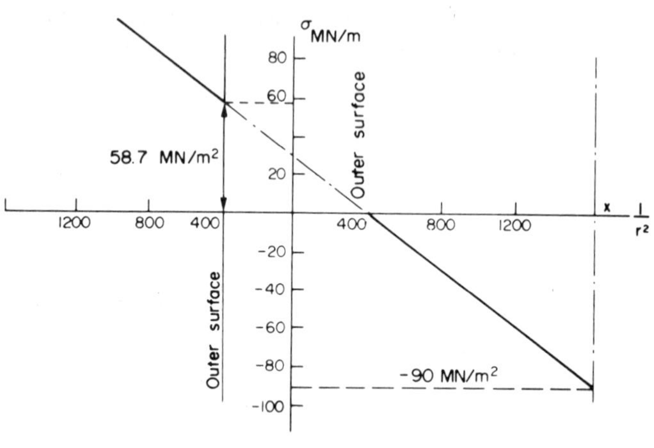

Fig. 10.23.

Let the required value of $\dfrac{1}{R_1^2} = x$

then by proportions $$\frac{90}{x - 400} = \frac{58.7}{800}$$

$$x - 400 = \frac{90 \times 800}{58.7} = 1225$$

$$\therefore \qquad x = 1625$$

$$\therefore \qquad R_1 = 24.8 \text{ mm}$$

i.e. required internal diameter = **49.6 mm**

Example 10.4 (B)

A compound cylinder is formed by shrinking a tube of 250 mm internal diameter and 25 mm wall thickness onto another tube of 250 mm external diameter and 25 mm wall thickness, both tubes being made of the same material. The stress set up at the junction owing to shrinkage is 10 MN/m². The compound tube is then subjected to an internal pressure of 80 MN/m². Compare the hoop stress distribution now obtained with that of a single cylinder of 300 mm external diameter and 50 mm thickness subjected to the same internal pressure.

Solution (a): analytical

A solution is obtained as described in §10.9, i.e. by considering the effects of shrinkage and internal pressure separately and combining the results algebraically.

Shrinkage only – outer tube

At $r = 0.15$, $\sigma_r = 0$ and at $r = 0.125$, $\sigma_r = \quad 10$ MN/m²

$$\therefore \qquad 0 = A - \frac{B}{0.15^2} = A - 44.5B \qquad (1)$$

$$-10 = A - \frac{B}{0.125^2} = A - 64B \qquad (2)$$

Subtracting (1) – (2), $10 = 19.5B \quad \therefore \; B = 0.514$

Substituting in (1), $A = 44.5B \quad \therefore \; A = 22.85$

$\therefore$ hoop stress at 0.15 m radius $= A + 44.5B = 45.7$ MN/m²

hoop stress at 0.125 m radius $= A + 64B = 55.75$ MN/m²

Shrinkage only – inner tube

At $r = 0.10$, $\sigma_r = 0$ and at $r = 0.125$, $\sigma_r = -10$ MN/m²

$$\therefore \qquad 0 = A - \frac{B}{0.1^2} = A - 100B \qquad (3)$$

$$-10 = A - \frac{B}{0.125^2} = A - 64B \qquad (4)$$

Subtracting (3) – (4), $10 = -36B \quad \therefore \; B = -0.278$

Substituting in (3), $A = 100B \quad \therefore \; A = -27.8$

$\therefore$ hoop stress at 0.125 m radius $= A + 64B = -45.6$ MN/m^2

hoop stress at 0.10 m radius $= A + 100B = -55.6$ MN/m^2

Considering internal pressure only (on complete cylinder)

At $r = 0.15$, $\sigma_r = 0$ and at $r = 0.10$, $\sigma_r = -80$

$$\therefore \quad 0 = A - 44.5B \tag{5}$$

$$-80 = A - 100B \tag{6}$$

Subtracting (5) – (6), $80 = 55.5B \quad \therefore \; B = 1.44$

From (5), $A = 44.5B \quad \therefore \; A = 64.2$

$\therefore$ At $r = 0.15$ m, $\sigma_H = A + 44.5B = 128.4$ MN/m^2

$r = 0.125$ m, $\sigma_H = A + 64B = 156.4$ MN/m^2

$r = 0.1$ m, $\sigma_H = A + 100B = 208.2$ MN/m^2

The resultant stresses for combined shrinkage and internal pressure are then:

outer tube: $r = 0.15$ $\sigma_H = 128.4 + 45.7 = \mathbf{174.1}$ **MN/m^2**

$r = 0.125$ $\sigma_H = 156.4 + 55.75 = \mathbf{212.15}$ **MN/m^2**

inner tube: $r = 0.125$ $\sigma_H = 156.4 - 45.6 = \mathbf{110.8}$ **MN/m^2**

$r = 0.1$ $\sigma_H = 208.2 - 55.6 = \mathbf{152.6}$ **MN/m^2**

Solution (b): graphical

The graphical solution is obtained in the same way by considering the separate effects of shrinkage and internal pressure as shown in Fig. 10.24.

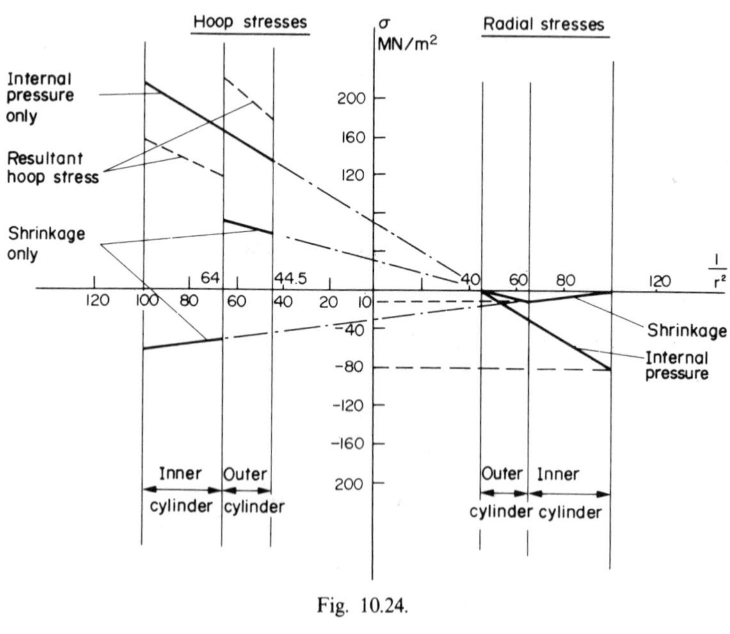

Fig. 10.24.

The final results are illustrated in Fig. 10.25 (values from the graph again being determined by proportion of the figure).

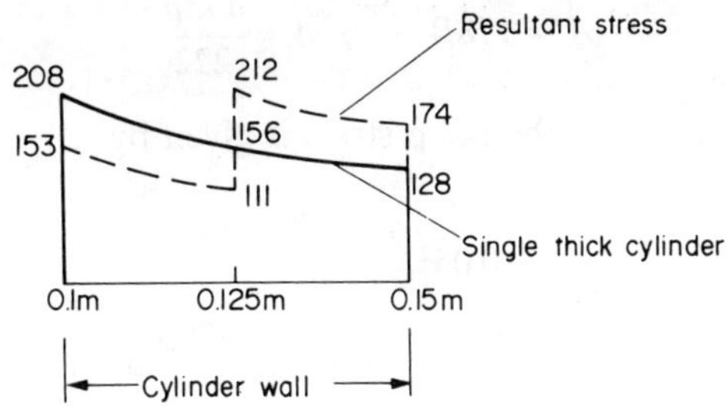

Fig. 10.25.

Example 10.5 (B)

A compound tube is made by shrinking one tube of 100 mm internal diameter and 25 mm wall thickness on to another tube of 100 mm external diameter and 25 mm wall thickness. The shrinkage allowance, *based on radius*, is 0.01 mm. If both tubes are of steel (with $E = 208\ \text{GN/m}^2$), calculate the radial pressure set up at the junction owing to shrinkage.

Solution

Let p be the required shrinkage pressure, then for the inner tube:

At $r = 0.025$, $\sigma_r = 0$ and at $r = 0.05$, $\sigma_r = -p$

$$0 = A - \frac{B}{0.025^2} = A - 1600B \qquad (1)$$

$$-p = A - \frac{B}{0.05^2} = A - 400B \qquad (2)$$

Subtracting (2) − (1),

$$-p = 1200B \qquad \therefore\ B = -p/1200$$

From (1),

$$A = 1600B \qquad \therefore\ A = -\frac{1600p}{1200} = -\frac{4p}{3}$$

Therefore at the common radius the hoop stress is given by eqn. (10.4),

$$\sigma_{H_i} = A + B/0.05^2$$

$$= -\frac{4p}{3} + 400\left(-\frac{p}{1200}\right) = -\frac{5p}{3} = -1.67p$$

For the outer tube:

at $r = 0.05$, $\sigma_r = -p$ and at $r = 0.075$, $\sigma_r = 0$

$$-p = A - B/0.05^2 = A - 400B \qquad (3)$$

$$0 = A - B/0.075^2 = A - 178B \qquad (4)$$

Subtracting (4) − (3), $\qquad p = 222B \quad \therefore \; B = p/222$

From (4) $\qquad A = 178B \quad \therefore \; A = \dfrac{178p}{222}$

Therefore at the common radius the hoop stress is given by

$$\sigma_{H_o} = A + \frac{B}{0.05^2}$$

$$= \frac{178p}{222} + \frac{p}{222} \times 400 = \frac{578p}{222} = 2.6p$$

Now from eqn. (10.14) the shrinkage allowance is

$$\frac{r}{E}[\sigma_{H_o} - \sigma_{H_i}]$$

$$\therefore \qquad 0.01 \times 10^{-3} = \frac{50 \times 10^{-3}}{208 \times 10^9}[2.6p - (-1.67p)]10^6$$

where p has units of MN/m^2

$$\therefore \qquad 4.27p = \frac{0.01 \times 208 \times 10^3}{50} = 41.6$$

$$\therefore \qquad \mathbf{p = 9.74\ MN/m^2}$$

Hoop and radial stresses in the compound cylinder owing to shrinkage and/or internal pressure can now be determined if desired using the procedure of the previous example.

Once again a graphical solution could have been employed to obtain the values of σ_{H_o} and σ_{H_i} in terms of the unknown pressure p which is set off to some convenient distance on the $r = 0.05$, i.e. $1/r^2 = 400$, line.

Example 10.6 (B)

Two steel rings of radial thickness 30 mm, common radius 70 mm and length 40 mm are shrunk together to form a compound ring. It is found that the axial force required to separate the rings, i.e. to push the inside ring out, is 150 kN. Determine the shrinkage pressure at the mating surfaces and the shrinkage allowance. $E = 208\ GN/m^2$. The coefficient of friction between the junction surfaces of the two rings is 0.15.

Solution

Let the pressure set up between the rings be p MN/m^2.
Then, normal force between rings $= p \times 2\pi r L = N$

$$= p \times 10^6 \times 2\pi \times 70 \times 10^{-3} \times 40 \times 10^{-3}$$

$$= 5600\pi p \text{ newtons.}$$

$\therefore$ friction force between rings $= \mu N = 0.15 \times 5600\pi p$

$$0.15 \times 5600\pi p = 150 \times 10^3$$

$$p = \frac{150 \times 10^3}{0.15 \times 5600\pi}$$

$$= \mathbf{57\ MN/m^2}$$

Now, *for the inner tube:*

$\sigma_r = -57$ at $r = 0.07$ and $\sigma_r = 0$ at $r = 0.04$

$$\therefore \quad -57 = A - B/0.07^2 = A - 204B \qquad (1)$$

$$0 = A - B/0.04^2 = A - 625B \qquad (2)$$

Subtracting (2) − (1), $57 = -421B \quad \therefore\ B = -0.135$

From (2), $A = 625B \quad \therefore\ A = -84.5$

Therefore at the common radius the hoop stress in the inner tube is given by

$$\sigma_{H_i} = A + \frac{B}{0.07^2} = A + 204B = -112.1\ \text{MN/m}^2$$

For the outer tube:

$\sigma_r = -57$ at $r = 0.07$ and $\sigma_r = 0$ at $r = 0.1$

$$\therefore \quad -57 = A - 204B \qquad (3)$$

$$0 = A - 100B \qquad (4)$$

Subtracting (4) − (3), $57 = 104B \quad \therefore\ B = 0.548$

From (4), $A = 100B \quad \therefore\ A = 54.8$

Therefore at the common radius the hoop stress in the outer tube is given by

$$\sigma_{H_o} = A + \frac{B}{0.07^2} = A + 204B = 166.8\ \text{MN/m}^2$$

$$\text{shrinkage allowance} = \frac{r}{E}(\sigma_{H_o} - \sigma_{H_i})$$

$$= \frac{70 \times 10^{-3}}{208 \times 10^9}[166.8 - (-112.1)]\,10^6$$

$$= \frac{70 \times 278.9}{208} \times 10^{-6}\ \text{m}$$

$$= 93.8 \times 10^{-6} = \mathbf{0.094\ mm}$$

Example 10.7 (B)

(a) A steel sleeve of 150 mm outside diameter is to be shrunk on to a solid steel shaft of 100 mm diameter. If the shrinkage pressure set up is 15 MN/m^2, find the initial difference between the inside diameter of the sleeve and the outside diameter of the shaft.

(b) What percentage error would be involved if the shaft were assumed to be incompressible? For steel, $E = 208\ \text{GN/m}^2$; $\nu = 0.3$.

Solution

(a) Treating the sleeve as a thick cylinder with internal pressure $15\ \text{MN/m}^2$,
at $r = 0.05$, $\sigma_r = -15\ \text{MN/m}^2$ and at $r = 0.075$, $\sigma_r = 0$

$$-15 = A - B/0.05^2 = A - 400B \quad (1)$$

$$0 = A - B/0.075^2 = A - 178B \quad (2)$$

Subtracting (2) − (1), $15 = 222B \quad \therefore\ B = 0.0676$

From (2), $A = 178B \quad \therefore\ A = 12.05$

Therefore the hoop stress in the sleeve at $r = 0.05$ m is given by

$$\sigma_{H_o} = A + B/0.05^2$$

$$= A + 400B = 39\ \text{MN/m}^2$$

The shaft will be subjected to a hoop stress which will be compressive and equal in value to the shrinkage pressure (see §10.12),

i.e. $$\sigma_{H_i} = -15\ \text{MN/m}^2$$

Thus the difference in radii or shrinkage allowance

$$= \frac{r}{E}(\sigma_{H_o} - \sigma_{H_i}) = \frac{50 \times 10^{-3}}{208 \times 10^9}[39 + 15]10^6$$

$$= \frac{50 \times 54}{208} \times 10^{-6} = 13 \times 10^{-6}\ \text{m}$$

$\therefore$ difference in diameters = **0.026 mm**

(b) If the shaft is assumed incompressible the difference in diameters will equal the necessary change in diameter of the sleeve to fit the shaft. This can be found from the diametral strain, i.e. from eqn. (10.9)

$$\Delta D = \frac{2r}{E}(\sigma_H - \nu\sigma_r) \quad \text{assuming } \sigma_L = 0$$

$$\therefore \quad \text{change of diameter} = \frac{100 \times 10^{-3}}{208 \times 10^9}[39 - 0.3(-15)]10^6$$

$$= \frac{43.5}{208} \times 10^{-4} = 20.9 \times 10^{-6}$$

$$= 0.0209\ \text{mm}$$

$$\therefore \quad \text{percentage error} = \frac{(0.026 - 0.0209)}{0.026} \times 100 = \mathbf{19.6\%}$$

Example 10.8 (C)

A thick cylinder of 100 mm external diameter and 50 mm internal diameter is wound with steel wire of 1 mm diameter, initially stressed to 20 MN/m^2 until the outside diameter is 120 mm. Determine the maximum hoop stress set up in the cylinder if an internal pressure of 30 MN/m^2 is now applied.

Solution

To find the stresses resulting from internal pressure only the cylinder and wire may be treated as a single thick cylinder of 50 mm internal diameter and 120 mm external diameter.

Now $\sigma_r = -30\ MN/m^2$ at $r = 0.025$ and $\sigma_r = 0$ at $r = 0.06$

$$\therefore \quad -30 = A - B/0.025^2 = A - 1600B \qquad (1)$$

$$0 = A - B/0.06^2 = A - 278B \qquad (2)$$

Subtracting (2) − (1), $30 = 1322B$ $\quad \therefore\ B = 0.0227$

From (2), $A = 278B$ $\quad \therefore\ A = 6.32$

$\therefore$ hoop stress at 25 mm radius $= A + 1600B = 42.7\ MN/m^2$

hoop stress at 50 mm radius $= A + 400B = 15.4\ MN/m^2$

The external pressure acting on the cylinder owing to wire winding is found from eqn. (10.21),

i.e.
$$\sigma_r = -\frac{(r^2 - R_1^2)}{2r^2} T \log_e \frac{(R_3^2 - R_1^2)}{(r^2 - R_1^2)} = p$$

$\therefore$ where $r = R_2 = 0.05$ m, $R_1 = 0.025$ and $R_3 = 0.06$

$$p = -\frac{(0.05^2 - 0.025^2)}{2 \times 0.05^2} T \log_e \frac{(0.06^2 - 0.025^2)}{(0.05^2 - 0.025^2)}$$

$$= -\frac{(25 - 6.25)}{50} 20 \log_e \frac{(36 - 6.25)}{(25 - 6.25)}\ MN/m^2$$

$$= -\frac{18.75 \times 20}{50} \log_e \frac{29.75}{18.75}$$

$$= -7.5 \log_e 1.585 = -7.5 \times 0.4606$$

$$= -3.45\ MN/m^2$$

Therefore for wire winding only the stresses in the tube are found from the conditions

$$\sigma_r = -3.45 \text{ at } r = 0.05 \quad \text{and} \quad \sigma_r = 0 \text{ at } r = 0.025$$

$$\therefore \quad -3.45 = A - 400B$$

$$0 = A - 1600B$$

Subtracting, $\quad -3.45 = 1200B \qquad -B = -2.88 \times 10^{-3}$

$$\therefore \; A = -4.6$$

$\therefore$ $\quad$ hoop stress at 25 mm radius $= A + 1600B = -9.2 \;\; \text{MN/m}^2$

hoop stress at 50 mm radius $= A + 400B = -5.75 \; \text{MN/m}^2$

The resultant stresses owing to winding and internal pressure are, therefore:

At $r = 25$ mm, $\quad \sigma_H = 42.7 - 9.2 \;= 33.5 \;\; \text{MN/m}^2$

At $r = 50$ mm, $\quad \sigma_H = 15.4 - 5.75 = \;\; 9.65 \; \text{MN/m}^2$

Thus the maximum hoop stress is $\mathbf{33.5 \; MN/m^2}$

Example 10.9 (C)

A thick cylinder of internal and external radii 300 mm and 500 mm respectively is subjected to a gradually increasing internal pressure P. Determine the value of P when:

(a) the material of the cylinder first commences to yield;
(b) yielding has progressed to mid-depth of the cylinder wall;
(c) the cylinder material suffers complete "collapse".

Take $\sigma_y = 600 \; \text{MN/m}^2$.

Solution—See Chapter 18

From eqn. (18.35) the initial yield pressure

$$= \frac{\sigma_y}{2R_2^2}[R_2^2 - R_1^2] = \frac{600}{2 \times 0.5^2}[0.5^2 - 0.3^2]$$

$$= \frac{600}{2 \times 25}[25 - 9] = \mathbf{192 \; MN/m^2}$$

The pressure required to cause yielding to a depth $R_p = 40$ mm is given by eqn. (18.36):

$$\sigma_r = \sigma_y\left[\log_e \frac{R_1}{R_p} - \frac{1}{2R_2^2}(R_2^2 - R_p^2)\right]$$

$$= 600\left[\log_e \frac{0.3}{0.4} - \frac{1}{2 \times 0.5^2}(0.5^2 - 0.4^2)\right]$$

$$= -600\left[\log_e 1.33 + \left(\frac{25 - 16}{50}\right)\right]$$

$$= -600(0.2852 + 0.18)$$

$$= -600 \times 0.4652 = -280 \; \text{MN/m}^2$$

i.e. the required internal pressure = **280 MN/m²**

For complete collapse from eqn. (18.34),

$$p = -\sigma_y \log_e \frac{R_1}{R_2} = \sigma_y \log_e \frac{R_2}{R_1}$$

$$= \sigma_y \log_e \frac{0.5}{0.3}$$

$$= 600 \times \log_e 1.67$$

$$= 600 \times 0.513 = \mathbf{308\ MN/m^2}$$

Problems

10.1 (B). A thick cylinder of 150 mm inside diameter and 200 mm outside diameter is subjected to an internal pressure of 15 MN/m². Determine the value of the maximum hoop stress set up in the cylinder walls. [53.4 MN/m².]

10.2 (B). A cylinder of 100 mm internal radius and 125 mm external radius is subjected to an external pressure of 14 bar (1.4 MN/m²). What will be the maximum stress set up in the cylinder? [−7.8 MN/m².]

10.3 (B). The cylinder of Problem 10.2 is now subjected to an additional internal pressure of 200 bar (20 MN/m²). What will be the value of the maximum stress? [84.7 MN/m².]

10.4 (B). A steel thick cylinder of external diameter 150 mm has two strain gauges fixed externally, one along the longitudinal axis and the other at right angles to read the hoop strain. The cylinder is subjected to an internal pressure of 75 MN/m² and this causes the following strains:

(a) hoop gauge: 455×10^{-6} tensile;

(b) longitudinal gauge: 124×10^{-6} tensile.

Find the internal diameter of the cylinder assuming that Young's modulus for steel is 208 GN/m² and Poisson's ratio is 0.283. [B.P.] [96.7 mm.]

10.5 (B) A compound tube of 300 mm external and 100 mm internal diameter is formed by shrinking one cylinder on to another, the common diameter being 200 mm. If the maximum hoop tensile stress induced in the outer cylinder is 90 MN/m² find the hoop stresses at the inner and outer diameters of both cylinders and show by means of a sketch how these stresses vary with the radius. [90, 55.35; −92.4, 57.8 MN/m².]

10.6 (B). A compound thick cylinder has a bore of 100 mm diameter, a common diameter of 200 mm and an outside diameter of 300 mm. The outer tube is shrunk on to the inner tube, and the radial stress at the common surface owing to shrinkage is 30 MN/m².

Find the maximum internal pressure the cylinder can receive if the maximum circumferential stress in the outer tube is limited to 110 MN/m². Determine also the resulting circumferential stress at the outer radius of the inner tube. [B.P.] [79, −18 MN/m².]

10.7 (B). Working from first principles find the interference fit per metre of diameter if the radial pressure owing to this at the common surface of a compound tube is 90 MN/m², the inner and outer diameters of the tube being 100 mm and 250 mm respectively and the common diameter being 200 mm. The two tubes are made of the same material, for which $E = 200$ GN/m². If the outside diameter of the inner tube is originally 200 mm, what will be the original inside diameter of the outer tube for the above conditions to apply when compound? [199.44 mm.]

10.8 (B). A compound cylinder is formed by shrinking a tube of 200 mm outside and 150 mm inside diameter on to one of 150 mm outside and 100 mm inside diameter. Owing to shrinkage the radial stress at the common surface is 20 MN/m². If this cylinder is now subjected to an internal pressure of 100 MN/m² (1000 bar), what is the magnitude and position of the maximum hoop stress? [164 MN/m² at inside of outer cylinder.]

10.9 (B). A thick cylinder has an internal diameter of 75 mm and an external diameter of 125 mm. The ends are closed and it carries an internal pressure of 60 MN/m². Neglecting end effects, calculate the hoop stress and radial stress at radii of 37.5 mm, 40 mm, 50 mm, 60 mm and 62.5 mm. Plot the values on a diagram to show the variation of these stresses through the cylinder wall. What is the value of the longitudinal stress in the cylinder?

[C.U.] [Hoop: 128, 116, 86.5, 70.2, 67.5 MN/m². Radial: −60, −48.7, −19, −2.9, 0 MN/m²; 33.8 MN/m².]

10.10 (B). A compound tube is formed by shrinking together two tubes with common radius 150 mm and thickness 25 mm. The shrinkage allowance is to be such that when an internal pressure of 30 MN/m^2 (300 bar) is applied the final maximum stress in each tube is to be the same. Determine the value of this stress. What must have been the difference in diameters of the tubes before shrinkage? $E = 210\,\text{GN/m}^2$. [83.1 MN/m^2; 0.025 mm.]

10.11 (B). A steel shaft of 75 mm diameter is pressed into a steel hub of 100 mm outside diameter and 200 mm long in such a manner that under an applied torque of 6 kN m relative slip is just avoided. Find the interference fit, assuming a 75 mm common diameter, and the maximum circumferential stress in the hub. $\mu = 0.3$. $E = 210\,\text{GN/m}^2$ [0.0183 mm; 40.4 MN/m^2.]

10.12 (B). A steel plug of 75 mm diameter is forced into a steel ring of 125 mm external diameter and 50 mm width. From a reading taken by fixing in a circumferential direction an electric resistance strain gauge on the external surface of the ring, the strain is found to be 1.49×10^{-4}. Assuming $\mu = 0.2$ for the mating surfaces, find the force required to push the plug out of the ring. Also estimate the greatest hoop stress in the ring. $E = 210\,\text{GN/m}^2$. [I.Mech.E.] [65.6 kN; 59 MN/m^2.]

10.13 (B). A steel cylindrical plug of 125 mm diameter is forced into a steel sleeve of 200 mm external diameter and 100 mm long. If the greatest circumferential stress in the sleeve is 90 MN/m^2, find the torque required to turn the sleeve, assuming $\mu = 0.2$ at the mating surfaces. [U.L.] [19.4 kN m.]

10.14 (B). A solid steel shaft of 0.2 m diameter has a bronze bush of 0.3 m outer diameter shrunk on to it. In order to remove the bush the whole assembly is raised in temperature uniformly. After a rise of 100°C the bush can just be moved along the shaft. Neglecting any effect of temperature in the axial direction, calculate the original interface pressure between the bush and the shaft.

For steel: $E = 208\,\text{GN/m}^2$, $\nu = 0.29$, $\alpha = 12 \times 10^{-6}$ per °C.
For bronze: $E = 112\,\text{GN/m}^2$, $\nu = 0.33$, $\alpha = 18 \times 10^{-6}$ per °C.

[C.E.I.] [20.2 MN/m^2.]

10.15 (B). (a) State the Lamé equations for the hoop and radial stresses in a thick cylinder subjected to an internal pressure and show how these may be expressed in graphical form. Hence, show that the hoop stress at the outside surface of such a cylinder subjected to an internal pressure P is given by

$$\sigma_H = \frac{2PR_1^2}{(R_2^2 - R_1^2)}$$

where R_1 and R_2 are the internal and external radii of the cylinder respectively.

(b) A steel tube is shrunk on to another steel tube to form a compound cylinder 60 mm internal diameter, 180 mm external diameter. The initial radial compressive stress at the 120 mm common diameter is 30 MN/m^2. Calculate the shrinkage allowance. $E = 200\,\text{GN/m}^2$.

(c) If the compound cylinder is now subjected to an internal pressure of 25 MN/m^2 calculate the resultant hoop stresses at the internal and external surfaces of the compound cylinder. [0.0768 mm; -48.75, $+54.25$ MN/m^2.]

10.16 (B). A bronze tube, 60 mm external diameter and 50 mm bore, fits closely inside a steel tube of external diameter 100 mm. When the assembly is at a uniform temperature of 15°C the bronze tube is a sliding fit inside the steel tube, that is, the two tubes are free from stress. The assembly is now heated uniformly to a temperature of 115°C.

(a) Calculate the radial pressure induced between the mating surfaces and the thermal circumferential stresses, in magnitude and nature, induced at the inside and outside surfaces of each tube. [10.9 MN/m^2; 23.3, 12.3, -71.2, -60.3 MN/m^2.]

(b) Sketch the radial and circumferential stress distribution across the combined wall thickness of the assembly when the temperature is 115° C and insert the numerical values. Use the tabulated data given below.

	Young's modulus (E)	Poisson's ratio (ν)	Coefficient of linear expansion (α)
Steel	200 GN/m^2	0.3	12×10^{-6}/°C
Bronze	100 GN/m^2	0.33	19×10^{-6}/°C

10.17 (B) A steel cylinder, 150 mm external diameter and 100 mm internal diameter, just fits over a brass cylinder of external diameter 100 mm and bore 50 mm. The compound cylinder is to withstand an internal pressure of such a magnitude that the pressure set up between the common junction surfaces is 30 MN/m^2 when the internal pressure is applied. The external pressure is zero. Determine:

(a) the value of the internal pressure;

(b) the hoop stress induced in the material of both tubes at the inside and outside surfaces.

Lamé's equations for thick cylinders may be assumed without proof, and neglect any longitudinal stress and strain.

For steel, $E = 207\ GN/m^2$ (2.07 Mbar) and $\nu = 0.28$.
For brass, $E = 100\ GN/m^2$ (1.00 Mbar) and $\nu = 0.33$.

Sketch the hoop and radial stress distribution diagrams across the combined wall thickness, inserting the peak values. [B.P.] [123 MN/m^2; 125.4, 32.2 MN/m^2; 78.2, 48.2 MN/m^2.]

10.18 (C). Assuming the Lamé equations for stresses in a thick cylinder, show that the radial and circumferential stresses in a solid shaft owing to the application of external pressure are equal at all radii.

A solid steel shaft having a diameter of 100 mm has a steel sleeve shrunk on to it. The maximum tensile stress in the sleeve is not to exceed twice the compressive stress in the shaft. Determine (a) the least thickness of the sleeve and (b) the maximum tensile stress in the sleeve after shrinkage if the shrinkage allowance, based on diameter, is 0.015 mm. $E = 210\ GN/m^2$. [I.Mech.E.][36.6 mm; 21 MN/m^2.]

10.19 (C). A steel tube of internal radius 25 mm and external radius 40 mm is wound with wire of 0.75 mm diameter until the external diameter of the tube and wire is 92 mm. Find the maximum hoop stress set up within the walls of the tube if the wire is wound with a tension of 15 MN/m^2 and an internal pressure of 30 MN/m^2 (300 bar) acts within the tube. [49 MN/m^2.]

10.20 (C). A thick cylinder of 100 mm internal diameter and 125 mm external diameter is wound with wire until the external diameter is increased by 30 %. If the initial tensile stress in the wire when being wound on the cylinder is 135 MN/m^2, calculate the maximum stress set up in the cylinder walls. [$-144.5\ MN/m^2$.]

CHAPTER 11

STRAIN ENERGY

Summary

The energy stored within a material when work has been done on it is termed the *strain energy* or *resilience*,

i.e. strain energy = work done

In general there are four types of loading which can be applied to a material:

1. *Direct load* (*tension or compression*)

$$\text{Strain energy } U = \int \frac{P^2\,ds}{2AE} \quad \text{or} \quad \frac{P^2 L}{2AE}$$

$$= \frac{\sigma^2 AL}{2E} = \frac{\sigma^2}{2E} \times \text{volume of bar}$$

2. *Shear load*

$$\text{Strain energy } U = \int \frac{Q^2\,ds}{2AG} \quad \text{or} \quad \frac{Q^2 L}{2AG}$$

$$= \frac{\tau^2}{2G} \times AL = \frac{\tau^2}{2G} \times \text{volume of bar}$$

3. *Bending*

$$\text{Strain energy } U = \int \frac{M^2\,ds}{2EI} \quad \text{or} \quad \frac{M^2 L}{2EI} \quad \text{if } M \text{ is constant}$$

4. *Torsion*

$$\text{Strain energy } U = \int \frac{T^2\,ds}{2GJ} \quad \text{or} \quad \frac{T^2 L}{2GJ} \quad \text{if } T \text{ is constant}$$

From 1 above, the strain energy or resilience when the tensile stress reaches the proof stress σ_p, i.e. the *proof resilience*, is

$$\frac{\sigma_p^2}{2E} \times \text{volume of bar}$$

and the *modulus of resilience* is

$$\frac{\sigma_p^2}{2E}$$

The *strain energy per unit volume* of a *three-dimensional principal stress system* is

$$U = \frac{1}{2E}\left[\sigma_1^2 + \sigma_2^2 + \sigma_3^2 - 2\nu(\sigma_1\sigma_2 + \sigma_2\sigma_3 + \sigma_3\sigma_1)\right]$$

The *volumetric or "dilatational" strain energy per unit volume* is then

$$\frac{(1-2\nu)}{6E}[(\sigma_1+\sigma_2+\sigma_3)^2]$$

and the *shear*, or *"distortional"*, *strain energy per unit volume* is

$$\frac{1}{12G}[(\sigma_1-\sigma_2)^2+(\sigma_2-\sigma_3)^2+(\sigma_3-\sigma_1)^2]$$

The *maximum instantaneous stress* in a uniform bar caused by a weight W falling through a distance h on to the bar is given by

$$\sigma = \frac{W}{A} \pm \sqrt{\left[\left(\frac{W}{A}\right)^2 + \frac{2WEh}{AL}\right]}$$

The *instantaneous extension* is then given by

$$\delta = \frac{\sigma L}{E}$$

If this is small compared to the height h, then

$$\sigma = \sqrt{\left(\frac{2WEh}{AL}\right)}$$

For *any shock-loaded system* the instantaneous deflection is given by

$$\delta = \delta_s\left[1 \pm \sqrt{\left(1+\frac{2h}{\delta_s}\right)}\right]$$

where δ_s is the deflection under an equal static load.

Castigliano's first theorem for deflection states that:

If the total strain energy expressed in terms of the external loads is partially differentiated with respect to one of the loads the result is the deflection of the point of application of that load and in the direction of that load (see Examples 11.5 and 11.6):

i.e. Deflection in direction of $W = \frac{\partial U}{\partial W} = \delta$

In applications where bending provides practically all of the strain energy,

$$\delta = \frac{\partial}{\partial W}\int\frac{M^2 ds}{2EI} = \int\frac{M}{EI}\frac{\partial M}{\partial W}ds$$

This is sometimes written in the form

$$\delta = \int\frac{Mm}{EI}ds$$

where $m = \frac{\partial M}{\partial W}$ = the bending moment resulting from a unit load only in the place of W. This method of solution is then termed the *unit load method.*

Castigliano's theorem also applies to ***angular movements***:

If the total strain energy expressed in terms of the external moments be partially differentiated with respect to one of the moments, the result is the angular deflection in radians of the point of application of that moment and in its direction

$$\theta = \int \frac{M}{EI} \frac{\partial M}{\partial M_i} ds$$

where M_i is the actual or imaginary moment at the point where θ is required.

Deflections due to shear

Beam loading	Shear deflection	
	Rectangular-section beam	I-section beam
Cantilever–concentrated end load W'	$\frac{6WL}{5AG}$	$\frac{WL}{AG}$
Cantilever–u.d.l.	$\frac{3wL^2}{5AG}$	$\frac{wL^2}{2AG} = \frac{WL}{2AG}$
Simply supported beam – central concentrated load W	$\frac{3WL}{10AG}$	$\frac{WL}{4AG}$
Simply supported beam – concentrated load dividing span into lengths a and b	$\frac{6Wab}{5AGL}$	
Simply supported beam–u.d.l.	$\frac{3wL^2}{20AG}$	$\frac{wL^2}{8AG} = \frac{WL}{8AG}$

Introduction

Energy is normally defined as the *capacity to do work* and it may exist in any of many forms, e.g. mechanical (potential or kinetic), thermal, nuclear, chemical, etc. The potential energy of a body is the form of energy which is stored by virtue of the work which has previously been done on that body, e.g. in lifting it to some height above a datum. Strain energy is a particular form of potential energy which is stored within materials which have been subjected to strain, i.e. to some change in dimension. The material is then capable of doing work, equivalent to the amount of strain energy stored, when it returns to its original unstrained dimension.

Strain energy is therefore defined as the energy which is stored within a material when work has been done on the material. Here it is assumed that the material remains elastic whilst work is done on it so that all the energy is recoverable and no permanent deformation occurs due to yielding of the material,

i.e. $\qquad$ strain energy U = work done

Thus for a gradually applied load the work done in straining the material will be given by the shaded area under the load–extension graph of Fig. 11.1.

$$U = \tfrac{1}{2}P\delta$$

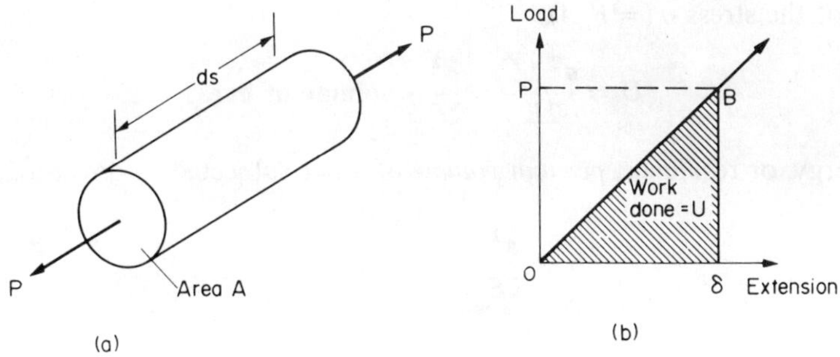

Fig. 11.1. Work done by a gradually applied load.

The strain energy per unit volume is often referred to as the *resilience*. The value of the resilience at the yield point or at the proof stress for non-ferrous materials is then termed the *proof resilience*.

The unshaded area above the line OB of Fig. 11.1 is called the *complementary energy*, a quantity which is utilised in some advanced energy methods of solution and is not considered within the terms of reference of this text.†

11.1. Strain energy – tension or compression

(*a*) *Neglecting the weight of the bar*

Consider a small element of a bar, length ds, shown in Fig. 11.1. If a graph is drawn of load against elastic extension the shaded area under the graph gives the work done and hence the strain energy,

i.e. $$\text{strain energy } U = \tfrac{1}{2}P\delta$$

Now $$\text{Young's modulus } E = \frac{\text{stress}}{\text{strain}} = \frac{P}{A} \times \frac{ds}{\delta}$$

$$\therefore \quad \delta = \frac{P\,ds}{AE}$$

$$\therefore \quad \text{for the bar element } U = \frac{P^2\,ds}{2AE}$$

$$\therefore \quad \text{total strain energy for a bar of length } L = \int_0^L \frac{P^2\,ds}{2AE}$$

Thus, assuming that the area of the bar remains constant along the length,

$$U = \frac{P^2 L}{2AE} \tag{11.1}$$

† See H. Ford and J. M. Alexander, *Advanced Mechanics of Materials* (Longmans, London, 1963).

or, in terms of the stress $\sigma\ (= P/A)$,

$$U = \frac{\sigma^2 AL}{2E} = \frac{\boldsymbol{\sigma^2}}{\mathbf{2E}} \times \textbf{volume of bar} \qquad (11.2)$$

i.e. strain energy, or resilience, *per unit volume* of a bar subjected to direct load, tensile or compressive

$$= \frac{\boldsymbol{\sigma^2}}{\mathbf{2E}} \qquad (11.3)$$

or, alternatively,

$$= \frac{1}{2}\sigma \times \frac{\sigma}{E} = \tfrac{1}{2}\sigma \times \varepsilon$$

i.e. **resilience = $\tfrac{1}{2}$ stress × strain**

(*b*) *Including the weight of the bar*

Consider now a bar of length L mounted vertically, as shown in Fig. 11.2. At any section AB the total load on the section will be the external load P together with the weight of the bar material below AB.

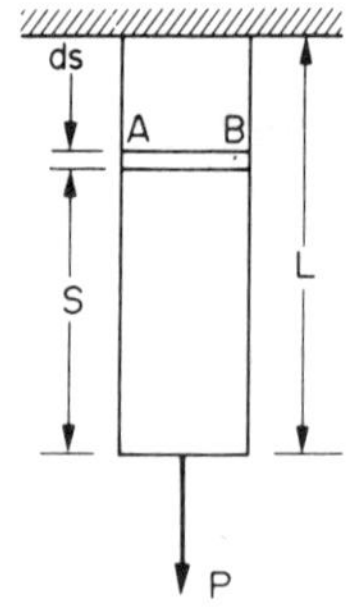

Fig. 11.2. Direct load – tension or compression.

Assuming a uniform cross-section of area A with density ρ,

$$\text{load on section } AB = P \pm \rho g As$$

the positive sign being used when P is tensile and the negative sign when P is compressive. Thus, for a tensile force P the extension of the element ds is given by the definition of Young's modulus E to be

$$\delta = \frac{\sigma ds}{E}$$

$$= \frac{(P + \delta g As)}{AE} ds$$

$\therefore$ work done $= \frac{1}{2} \times$ load $\times$ extension

$$= \tfrac{1}{2}(P+\rho g A s)\frac{(P+\rho g A s)}{AE}\,ds$$

$$= \frac{P^2}{2AE}\,ds + \frac{P\rho g}{E}\,s\,ds + \frac{(\rho g)^2 A}{2E}\,s^2\,ds$$

$\therefore$ total strain energy or work done

$$= \int_0^L \frac{P^2}{2AE}\,ds + \int_0^L \frac{P\rho g}{E}\,s\,ds + \int_0^L \frac{(\rho g)^2 A}{2E}\,s^2\,ds$$

$$= \frac{P^2 L}{2AE} + \frac{P\rho g L^2}{2E} + \frac{(\rho g)^2 A L^3}{6E} \qquad (11.4)$$

The last two terms are therefore the modifying terms to eqn. (11.1) to account for the body-weight effect of the bar.

11.2. Strain energy – shear

Consider the elemental bar now subjected to a shear load Q at one end causing deformation through the angle γ (the shear strain) and a shear deflection δ, as shown in Fig. 11.3.

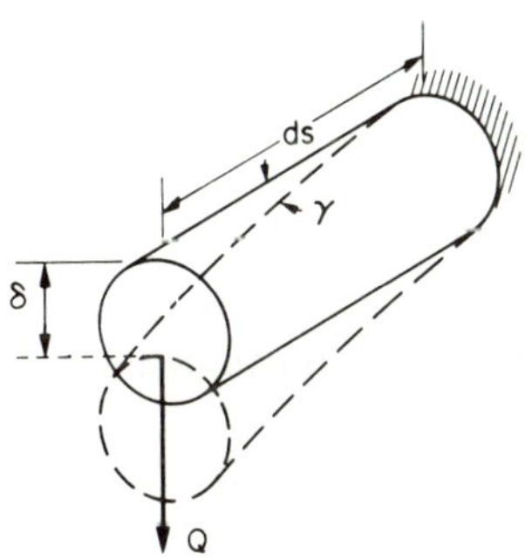

Fig. 11.3. Shear.

$$\text{Strain energy } U = \text{work done} = \tfrac{1}{2}Q\delta = \tfrac{1}{2}Q\gamma\,ds$$

Now $$G = \frac{\text{shear stress}}{\text{shear strain}} = \frac{\tau}{\gamma} = \frac{Q}{\gamma A}$$

$\therefore$ $$\gamma = \frac{Q}{AG}$$

$\therefore$ $$\text{shear strain energy} = \tfrac{1}{2}Q \times \frac{Q}{AG} \times ds = \frac{Q^2}{2AG}\,ds$$

∴ total strain energy resulting from shear

$$= \int_0^L \frac{Q^2 ds}{2AG} = \frac{Q^2 L}{2AG} \tag{11.5}$$

or, in terms of the shear stress $\tau = (Q/A)$,

$$U = \frac{\tau^2 AL}{2G} = \frac{\tau^2}{2G} \times \text{volume of bar} \tag{11.6}$$

11.3. Strain energy – bending

Let the element now be subjected to a constant bending moment M causing it to bend into an arc of radius R and subtending an angle $d\theta$ at the centre (Fig. 11.4). The beam will also have moved through an angle $d\theta$.

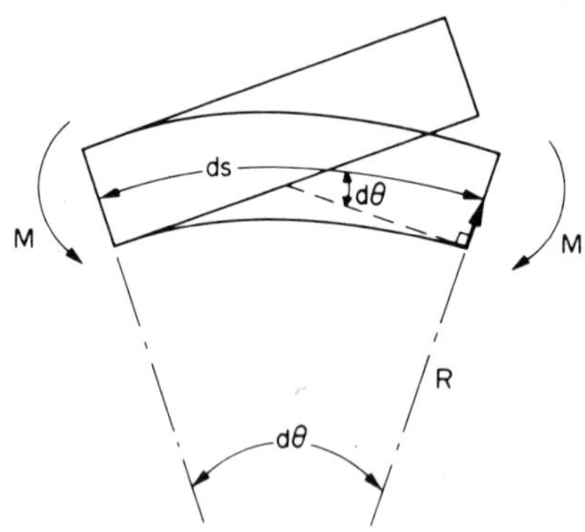

Fig. 11.4. Bending.

$$\text{Strain energy} = \text{work done} = \tfrac{1}{2} \times \text{moment} \times \text{angle turned through (in radians)}$$

$$= \tfrac{1}{2} M\, d\theta$$

But $\quad ds = R\, d\theta \quad \text{and} \quad \dfrac{M}{I} = \dfrac{E}{R}$

$$\therefore \quad d\theta = \frac{ds}{R} = \frac{M}{EI} ds$$

$$\therefore \quad \text{strain energy} = \tfrac{1}{2} M \times \frac{M}{EI} ds = \frac{M^2 ds}{2EI}$$

Total strain energy resulting from bending,

$$U = \int_0^L \frac{M^2 ds}{2EI} \tag{11.7}$$

If the bending moment is constant this reduces to

$$U = \frac{M^2 L}{2EI}$$

11.4. Strain energy – torsion

The element is now considered subjected to a torque T as shown in Fig. 11.5, producing an angle of twist $d\theta$ radians.

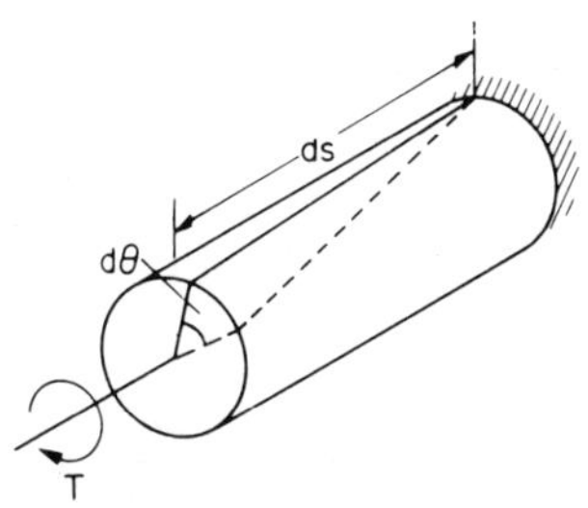

Fig. 11.5. Torsion.

$$\text{Strain energy} = \text{work done} = \tfrac{1}{2} T\, d\theta$$

But, from the simple torsion theory,

$$\frac{T}{J} = \frac{G\,d\theta}{ds} \quad \text{and} \quad d\theta = \frac{T\,ds}{GJ}$$

$\therefore$ total strain energy resulting from torsion,

$$U = \int_0^L \frac{T^2\,ds}{2GJ} = \frac{T^2 L}{2GJ} \qquad (11.8)$$

since in most practical applications T is constant.

For a hollow circular shaft eqn. (11.8) still applies

i.e. $$\text{Strain energy } U = \frac{T^2 L}{2GJ}$$

Now, from the simple bending theory

$$\frac{T}{J} = \frac{\tau}{r} = \frac{\tau_{max}}{R}$$

where R is the outer radius of the shaft and

$$J = \frac{\pi}{2}(R^4 - r^4)$$

$\therefore$ $$T = \frac{\pi}{2R}\tau_{max}(R^4 - r^4)$$

Substituting in the strain energy equation (11.8) we have:

$$U = \frac{\left[\dfrac{\pi\tau_{max}}{2R}(R^4 - r^4)\right]^2 L}{2G\dfrac{\pi}{2}(R^4 - r^4)}$$

$$= \frac{\tau_{max}^2}{4G}\frac{\pi(R^4 - r^4)L}{R^2}$$

$$= \frac{\tau_{max}^2}{4G}\frac{[R^2 + r^2]}{R^2} \times \text{volume of shaft}$$

or

$$\textbf{Strain energy/unit volume} = \frac{\tau_{max}^2}{4G}\frac{[R^2 + r^2]}{R^2} \qquad (11.8a)$$

It should be noted that in the four types of loading case considered above the strain energy expressions are all identical in form,

i.e.

$$\text{strain energy } U = \frac{(\text{applied "load"})^2 \times L}{2 \times \text{product of two related constants}}$$

the constants being related to the type of loading considered. In bending, for example, the relevant constants which appear in the bending theory are E and I, whilst for torsion G and J are more applicable. Thus the above standard equations for strain energy should easily be remembered.

11.5. Strain energy of a three-dimensional principal stress system

The reader is referred to §14.17 for the derivation of the following expression for the strain energy of a system of three principal stresses:

$$U = \frac{1}{2E}[\sigma_1^2 + \sigma_2^2 + \sigma_3^2 - 2\nu(\sigma_1\sigma_2 + \sigma_2\sigma_3 + \sigma_3\sigma_1)] \quad \textbf{per unit volume}$$

It is then shown in §14.17 that this total strain energy can be conveniently considered as made up of two parts:

(a) the *volumetric* or *dilatational* strain energy;
(b) the *shear* or *distortional* strain energy.

11.6. Volumetric or dilatational strain energy

This is the strain energy associated with a mean or hydrostatic stress of $\frac{1}{3}(\sigma_1 + \sigma_2 + \sigma_3) = \bar{\sigma}$ acting equally in all three mutually perpendicular directions giving rise to no distortion, merely a change in volume.

Then from eqn. (14.22),

$$\textbf{volumetric strain energy} = \frac{(1 - 2\nu)}{6E}[(\sigma_1 + \sigma_2 + \sigma_3)^2] \quad \textbf{per unit volume}$$

11.7. Shear or distortional strain energy

In order to consider the general principal stress case it has been shown necessary, in §14.6, to add to the mean stress $\bar{\sigma}$ in the three perpendicular directions, certain so-called deviatoric stress values to return the stress system to values of σ_1, σ_2 and σ_3. These *deviatoric stresses* are then associated directly with change of shape, i.e. distortion, without change in volume and the strain energy associated with this mechanism is shown to be given by

$$\textbf{shear strain energy} = \frac{1}{12G}[(\sigma_1 - \sigma_2)^2 + (\sigma_2 - \sigma_3)^2 + (\sigma_3 - \sigma_1)^2] \quad \text{per unit volume}$$

$$= \frac{1}{6G}[\sigma_1^2 + \sigma_2^2 + \sigma_3^2 - (\sigma_1\sigma_2 + \sigma_2\sigma_3 + \sigma_3\sigma_1)] \quad \textbf{per unit volume}$$

This equation is used as the basis of the Maxwell–von Mises theory of elastic failure which is discussed fully in Chapter 15.

11.8. Suddenly applied loads

If a load P is applied gradually to a bar to produce an extension δ the load–extension graph will be as shown in Fig. 11.1 and repeated in Fig. 11.6, the work done being given by $U = \frac{1}{2}P\delta$.

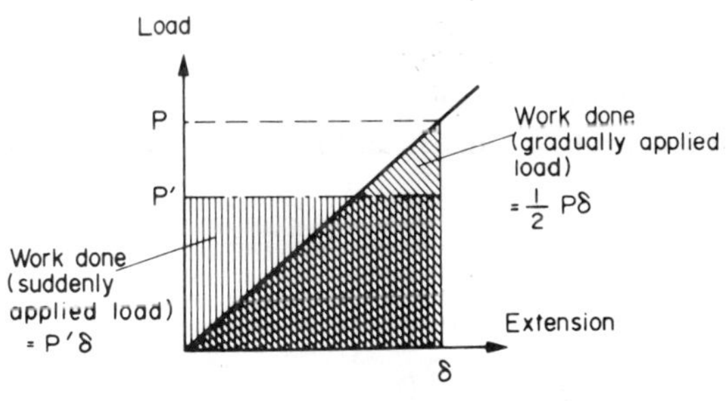

Fig. 11.6. Work done by a suddenly applied load.

If now a load P' is suddenly applied (i.e. applied with an instantaneous value, not gradually increasing from zero to P') to produce the same extension δ, the graph will now appear as a horizontal straight line with a work done or strain energy $= P'\delta$.

The bar will be strained by an equal amount δ in both cases and the energy stored must therefore be equal,

i.e. $$P'\delta = \tfrac{1}{2}P\delta$$

or $$P' = \tfrac{1}{2}P$$

Thus the suddenly applied load which is required to produce a certain value of instantaneous strain is half the equivalent value of static load required to perform the same function. It is then clear that vice versa *a load P which is suddenly applied will produce twice the effect of the same load statically applied.* Great care must be exercised, therefore, in the design of, for example, machine parts to exclude the possibility of sudden applications of load since associated stress levels are likely to be doubled.

11.9. Impact loads – axial load application

Consider now the bar shown vertically in Fig. 11.7 with a rigid collar firmly attached at the end. The load W is free to slide vertically and is suspended by some means at a distance h above the collar. When the load is dropped it will produce a maximum instantaneous extension δ of the bar, and will therefore have done work (neglecting the mass of the bar and collar)

$$= \text{force} \times \text{distance} = W(h+\delta)$$

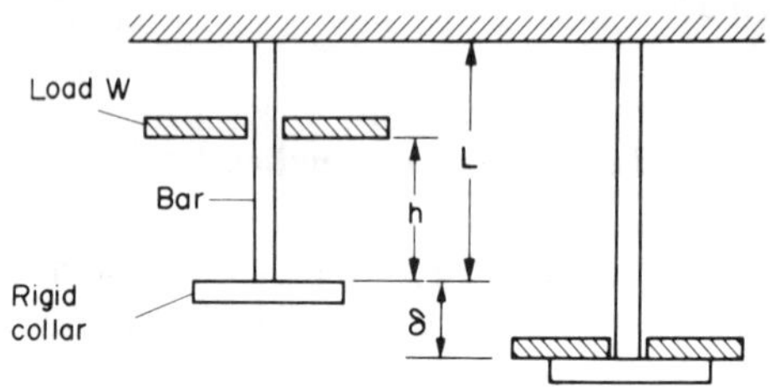

Fig. 11.7. Impact load – axial application.

This work will be stored as strain energy and is given by eqn. (11.2):

$$U = \frac{\sigma^2 AL}{2E}$$

where σ is the instantaneous stress set up.

$$\therefore \qquad \frac{\sigma^2 AL}{2E} = W(h+\delta) \qquad (11.9)$$

If the extension δ is small compared with h it may be ignored and then, approximately,

$$\sigma^2 = 2WEh/AL$$

i.e.

$$\sigma = \sqrt{\left(\frac{2WEh}{AL}\right)} \qquad (11.10)$$

If, however, δ is not small compared with h it must be expressed in terms of σ, thus

$$E = \frac{\text{stress}}{\text{strain}} = \frac{\sigma L}{\delta} \quad \text{and} \quad \delta = \frac{\sigma L}{E}$$

Therefore substituting in eqn. (11.9)

$$\frac{\sigma^2 AL}{2E} = Wh + \frac{W\sigma L}{E}$$

$$\therefore \qquad \frac{\sigma^2 AL}{2E} - \sigma\frac{WL}{E} - Wh = 0$$

$$\sigma^2 - \frac{2W}{A}\sigma - \frac{2WEh}{AL} = 0$$

Solving by "the quadratic formula" and ignoring the negative sign,

$$\sigma = \tfrac{1}{2}\left\{\frac{2W}{A} + \sqrt{\left[\left(\frac{2W}{A}\right)^2 + 4\left(\frac{2WEh}{AL}\right)\right]}\right.$$

i.e.

$$\sigma = \frac{W}{A} + \sqrt{\left[\left(\frac{W}{A}\right)^2 + \frac{2WEh}{AL}\right]} \qquad (11.11)$$

This is the *accurate* equation for the *maximum* stress set up, and should always be used if there is any doubt regarding the relative magnitudes of δ and h.

Instantaneous extensions can then be found from

$$\delta = \frac{\sigma L}{E}$$

If the load is not dropped but *suddenly applied* from effectively zero height, $h = 0$, and eqn. (11.11) reduces to

$$\sigma = \frac{W}{A} + \frac{W}{A} = \frac{2W}{A}$$

This verifies the work of §11.8 and confirms that stresses resulting from suddenly applied loads are twice those resulting from statically applied loads of the same magnitude. Inspection of eqn. (11.11) shows that stresses resulting from impact loads of similar magnitude will be even higher than this and any design work in applications where impact loading is at all possible should always include a safety factor well in excess of two.

11.10. Impact loads – bending applications

Consider the beam shown in Fig. 11.8 subjected to a shock load W falling through a height h and producing an instantaneous deflection δ.

$$\text{Work done by falling load} = W(h + \delta)$$

In these cases it is often convenient to introduce an *equivalent static load* W_E defined as that load which, when gradually applied, produces the same deflection as the shock load

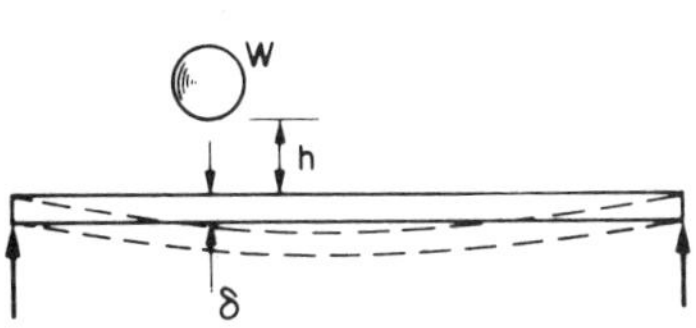

Fig. 11.8. Impact load – bending application.

which it replaces, then

$$\text{work done by equivalent static load} = \tfrac{1}{2}W_E\delta$$

$$W(h+\delta) = \tfrac{1}{2}W_E\delta \tag{11.12}$$

Thus if δ is obtained in terms of W_E using the standard deflection equations of Chapter 5 for the support conditions in question, the above equation becomes a quadratic equation in one unknown W_E. Hence W_E can be determined and the required stresses or deflections can be found on the equivalent beam system using the usual methods for static loading, i.e. the dynamic load case has been reduced to the equivalent static load condition.

Alternatively, if W produces a deflection δ_s when applied statically then, by proportion,

$$\frac{W_E}{\delta} = \frac{W}{\delta_s} \quad \text{or} \quad W_E = \frac{\delta}{\delta_s}W$$

Substituting in eqn. (11.12)

$$W(h+\delta) = \tfrac{1}{2}W \times \frac{\delta}{\delta_s} \times \delta$$

$$\therefore \quad \delta^2 - 2\delta_s\delta - 2\delta_s h = 0$$

$$\therefore \quad \delta = \delta_s \pm \sqrt{(\delta_s + 2\delta_s h)}$$

$$\boldsymbol{\delta = \delta_s\left[1 \pm \left(1 + \frac{2h}{\delta_s}\right)^{\frac{1}{2}}\right]} \tag{11.13}$$

The instantaneous deflection of any shock-loaded system is thus obtained from a knowledge of the static deflection produced by an equal load. Stresses are then calculated as before.

11.11. Castigliano's first theorem for deflection

Castigliano's first theorem states that:

> *If the total strain energy of a body or framework is expressed in terms of the external loads and is partially differentiated with respect to one of the loads the result is the deflection of the point of application of that load and in the direction of that load,*

i.e. if U is the total strain energy, the deflection in the direction of load $W = \partial U/\partial W$.

In order to prove the theorem, consider the beam or structure shown in Fig. 11.9 with forces P_A, P_B, P_C, etc., acting at points A, B, C, etc.

If a, b, c, etc., are the deflections in the direction of the loads then the total strain energy of the system is equal to the work done.

$$U = \tfrac{1}{2}P_A a + \tfrac{1}{2}P_B b + \tfrac{1}{2}P_C c + \ldots \tag{11.14}$$

N.B. *Limitations of theory*. The above simplified approach to impact loading suffers severe limitations. For example, the distribution of stress and strain under impact conditions will not strictly be the same as under static loading, and perfect elasticity of the bar will not be exhibited. These and other effects are discussed by Roark and Young in their advanced treatment of dynamic stresses: *Formulas for Stress & Strain*, 5th edition (McGraw Hill), Chapter 15.

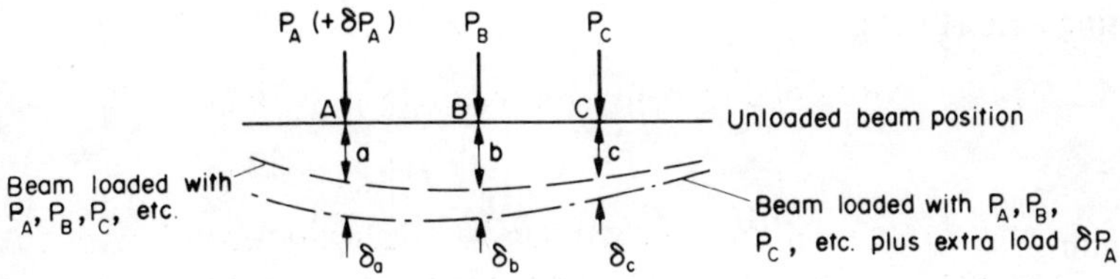

Fig. 11.9. Any beam or structure subjected to a system of applied concentrated loads P_A, P_B, P_C . . . P_N, etc.

If one of the loads, P_A, is now increased by an amount δP_A the changes in deflections will be δa, δb and δc, etc., as shown in Fig. 11.9.

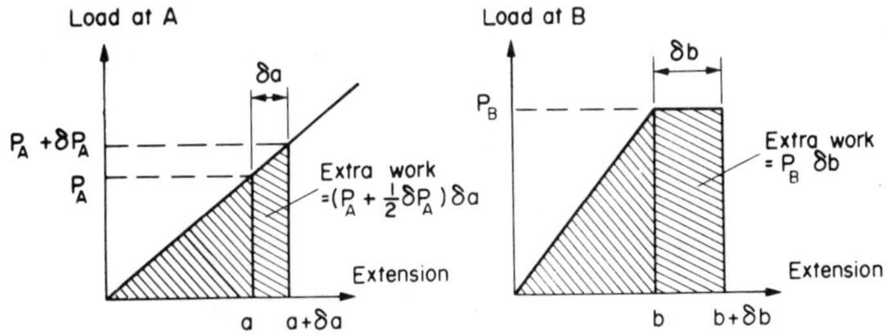

Fig. 11.10. Load–extension curves for positions *A* and *B*.

Extra work done at *A* (see Fig. 11.10)

$$= (P_A + \tfrac{1}{2}\delta P_A)\delta a$$

Extra work done at *B*, *C*, etc. (see Fig. 11.10)

$$= P_B\,\delta b,\ P_C\,\delta c, \text{ etc.}$$

Increase in strain energy

$$= \text{total extra work done}$$

$$\therefore \qquad \delta U = P_A\,\delta a + \tfrac{1}{2}\delta P_A\,\delta a + P_B\,\delta b + P_C\,\delta c + \ldots$$

and neglecting the product of small quantities

$$\delta U = P_A\,\delta a + P_B\,\delta b + P_C\,\delta c + \ldots \qquad (11.15)$$

But if the loads $P_A + \delta P_A$, P_B, P_C, etc., were applied gradually from zero the total strain energy would be

$$U + \delta U = \sum \tfrac{1}{2} \times \text{load} \times \text{extension}$$

$$U + \delta U = \tfrac{1}{2}(P_A + \delta P_A)(a + \delta a) + \tfrac{1}{2}P_B(b + \delta b) + \tfrac{1}{2}P_C(c + \delta c) + \ldots$$

$$= \tfrac{1}{2}P_A\,a + \tfrac{1}{2}P_A\,\delta a + \tfrac{1}{2}\,\delta P_A\,a + \tfrac{1}{2}\delta P_A\,\delta a + \tfrac{1}{2}P_B\,b + \tfrac{1}{2}P_B\,\delta b + \tfrac{1}{2}P_C\,c + \tfrac{1}{2}P_C\,\delta c + \ldots$$

Neglecting the square of small quantities ($\frac{1}{2}\delta P_A \delta a$) and subtracting eqn. (11.14),

$$\delta U = \tfrac{1}{2}\delta P_A\,a + \tfrac{1}{2}P_A\,\delta a + \tfrac{1}{2}P_B\,\delta b + \tfrac{1}{2}P_C\,\delta c + \ldots$$

or

$$2\delta U = \delta P_A\,a + P_A\,\delta a + P_B\,\delta b + P_C\,\delta c + \ldots$$

Subtracting eqn. (11.15),

$$\delta U = \delta P_A\, a \quad \therefore \quad \frac{\delta U}{\delta P_A} = a$$

or, in the limit,

$$\frac{\partial U}{\partial P_A} = a$$

i.e. the partial differential of the strain energy U with respect to P_A gives the deflection under and in the direction of P_A. Similarly,

$$\frac{\partial U}{\partial P_B} = b \quad \text{and} \quad \frac{\partial U}{\partial P_C} = c, \text{ etc.}$$

In most beam applications the strain energy, and hence the deflection, resulting from end loads and shear forces are taken to be negligible in comparison with the strain energy resulting from bending (torsion not normally being present),

$$\therefore \qquad U = \int \frac{M^2}{2EI}\, ds$$

$$\frac{\partial U}{\partial P} = \frac{\partial U}{\partial M} \times \frac{\partial M}{\partial P} = \int \frac{2M}{2EI}\, ds \times \frac{\partial M}{\partial P}$$

i.e.

$$\delta = \frac{\partial U}{\partial P} = \int \frac{M}{EI} \frac{\partial M}{\partial P}\, ds \qquad (11.16)$$

which is the usual form of Castigliano's first theorem. The integral is evaluated as it stands to give the deflection under an existing load P, the value of the bending moment M at some general section having been determined in terms of P. If no general expression for M in terms of P can be obtained to cover the whole beam then the beam, and hence the integral limits, can be divided into any number of convenient parts and the results added. In cases where the deflection is required at a point or in a direction in which there is no load applied, an imaginary load P is introduced in the required direction, the integral obtained in terms of P and then evaluated with P equal to zero.

The above procedures are illustrated in worked examples at the end of this chapter.

11.12. "Unit-load" method

It has been shown in §11.11 that in applications where bending provides practically all of the total strain energy of a system

$$\delta = \int \frac{M}{EI} \frac{\partial M}{\partial W}\, ds$$

Now W is an applied concentrated load and M will therefore include terms of the form Wx, where x is some distance from W to the point where the bending moment (B.M.) is required plus terms associated with the other loads. The latter will reduce to zero when partially differentiated with respect to W since they do not include W.

Now

$$\frac{\partial}{\partial W}(Wx) = x = 1 \times x$$

i.e. the partial differential of the B.M. term containing W is identical to the result achieved if W is replaced by unity in the B.M. expression. Using this information the Castigliano expression can be simplified to remove the partial differentiation procedure, thus

$$\delta = \int \frac{Mm}{EI}\, ds \tag{11.17}$$

where m is the B.M. resulting from a *unit load only* applied at the point of application of W and in the direction in which the deflection is required. The value of M remains the same as in the standard Castigliano procedure and is therefore the B.M. due to the *applied load system, **including W***.

This so-called "unit load" method is particularly powerful for cases where deflections are required at points where no external load is applied or in directions different from those of the applied loads. The method mentioned previously of introducing imaginary loads P and then subsequently assuming P is zero often gives rise to confusion. It is much easier to simply apply a unit load at the point, and in the direction, in which deflection is required regardless of whether external loads are applied there or not (see Example 11.6).

11.13. Application of Castigliano's theorem to angular movements

Castigliano's theorem can also be applied to angular rotations under the action of bending moments or torques. For the bending application the theorem becomes:

If the total strain energy, expressed in terms of the external moments, be partially differentiated with respect to one of the moments, the result is the angular deflection (*in radians*) *of the point of application of that moment and in its direction,*

i.e.
$$\theta = \int \frac{M}{EI} \frac{\partial M}{\partial M_i}\, ds \tag{11.18}$$

where M_i is the imaginary or applied moment at the point where θ is required.

Alternatively the "unit-load" procedure can again be used, this time replacing the applied or imaginary moment at the point where θ is required by a "unit moment". Castigliano's expression for slope or angular rotation then becomes

$$\theta = \int \frac{Mm}{EI} \cdot ds$$

where M is the bending moment at a general point due to the applied loads or moments and m is the bending moment at the same point due to the unit moment at the point where θ is required and in the required direction. See Example 11.8 for a simple application of this procedure.

11.14. Shear deflection

(*a*) *Cantilever carrying a concentrated end load*

In the majority of beam-loading applications the deflections due to bending are all that need be considered. For very short, deep beams, however, a secondary deflection, that due to shear, must also be considered. This may be determined using the strain energy formulae derived earlier in this chapter.

For bending,
$$U_B = \int_0^L \frac{M^2\,ds}{2EI}$$

For shear,
$$U_S = \int_0^L \frac{Q^2\,ds}{2AG} = \frac{\tau^2}{2G} \times \text{volume}$$

Consider, therefore, the cantilever, of solid rectangular section, shown in Fig. 11.11.

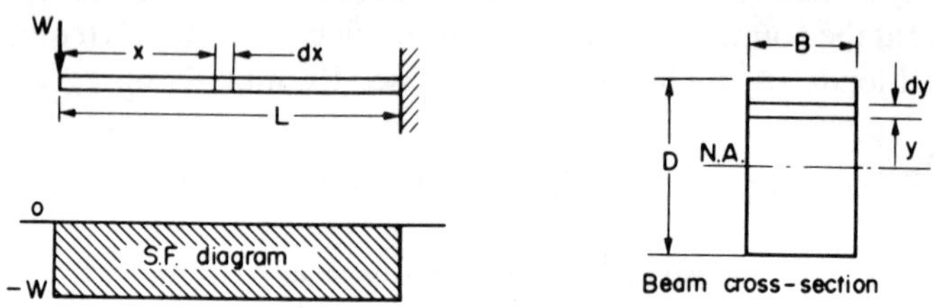

Fig. 11.11.

For the element of length dx

$$U_S = \int \frac{\tau^2}{2G} \times B\,dy\,dx$$

But
$$\tau = \frac{QA\bar{y}}{Ib} \quad \text{(see §7.1)}$$

$$= Q \times \frac{B\left(\dfrac{D}{2} - y\right)}{IB} \left[\frac{\left(\dfrac{D}{2} - y\right)}{2} + y\right]$$

$$= \frac{Q}{2I}\left(\frac{D^2}{4} - y^2\right)$$

∴
$$U_S = \frac{1}{2G}\int\left\{\frac{Q}{2I}\left(\frac{D^2}{4} - y^2\right)\right\}^2 B\,dx\,dy$$

$$= \frac{Bdx}{2G}\int_{-D/2}^{D/2}\left\{\frac{Q}{2I}\left(\frac{D^2}{4} - y^2\right)\right\}^2 dy$$

$$= \frac{Q^2B}{8GI^2}\,dx\left(\frac{D^5}{30}\right)$$

To obtain the total strain energy we must now integrate this along the length of the cantilever. In this case Q is constant and equal to W and the integration is simple.

$$U_s = \int_0^L \frac{W^2B}{8GI^2}\frac{D^5}{30}\,dx$$

$$= \frac{W^2B}{8GI^2}\frac{D^5}{30}L = \frac{W^2BLD^5}{240G}\left(\frac{12}{BD^3}\right)^2$$

$$= \frac{3W^2L}{5AG}$$

where $A = BD$.

Therefore deflection due to shear

$$\delta_s = \frac{\partial U_s}{\partial W} = \frac{\mathbf{6WL}}{\mathbf{5AG}} \qquad (11.19)$$

Similarly, since $M = -Wx$

$$U_B = \int_0^L \frac{(-Wx)^2}{2EI}\,ds = \frac{W^2L^3}{6EI}$$

Therefore deflection due to bending

$$\delta_B = \frac{\partial U}{\partial W} = \frac{\mathbf{WL^3}}{\mathbf{3EI}} \qquad (11.20)$$

Comparison of eqns. (11.19) and (11.20) then yields the relationship between the shear and bending deflections. For very short beams, where the length equals the depth, the shear deflection is almost twice that due to bending. For longer beams, however, the bending deflection is very much greater than that due to shear and the latter can usually be neglected, e.g. for $L = 10D$ the deflection due to shear is less than 1% of that due to bending.

(b) Cantilever carrying uniformly distributed load

Consider now the same cantilever but carrying a uniformly distributed load over its complete length as shown in Fig. 11.12.

The shear force at any distance x from the free end

$$Q = wx$$

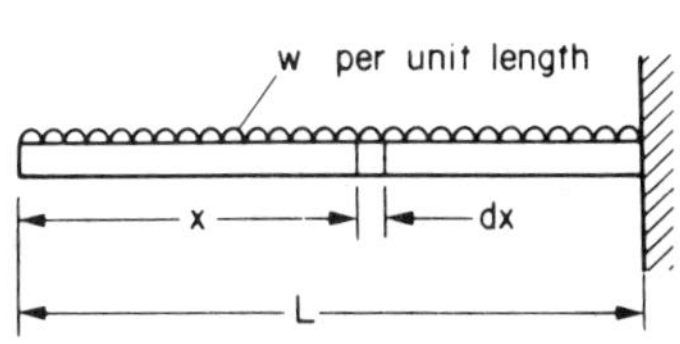

Fig. 11.12.

Therefore shear deflection over the length of the small element dx

$$= \frac{6}{5}\frac{(wx)}{AG}\,dx \qquad \text{from (11.19)}$$

Therefore total shear deflection

$$\delta_S = \int_0^L \frac{6}{5}\frac{wx\,dx}{AG} = \frac{3wL^2}{5AG} \tag{11.21}$$

(c) Simply supported beam carrying central concentrated load

In this case it is convenient to treat the beam as two cantilevers each of length equal to half the beam span and each carrying an end load half that of the central beam load (Fig. 11.13). The required central deflection due to shear will equal that of the end of each cantilever, i.e. from eqn. (11.19), with $W = W/2$ and $L = L/2$,

$$\delta_s = \frac{6}{5AG}\left(\frac{W}{2} \times \frac{L}{2}\right) = \frac{3WL}{10AG} \tag{11.22}$$

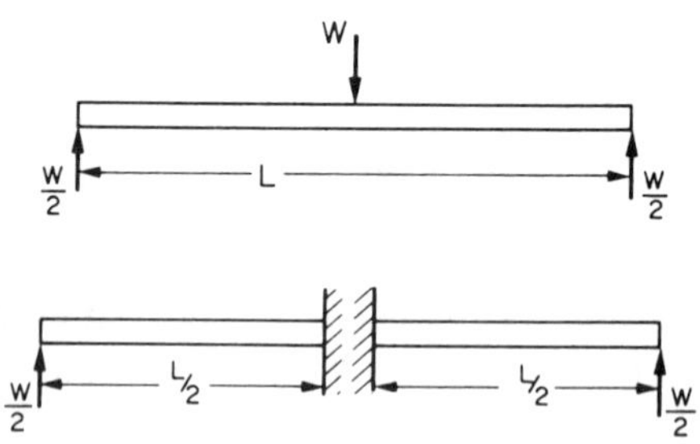

Fig. 11.13. Shear deflection of simply supported beam carrying central concentrated load–equivalent loading diagram.

(d) Simply supported beam carrying a concentrated load in any position

If the load divides the beam span into lengths a and b the reactions at each end will be Wa/L and Wb/L. The equivalent cantilever system is then shown in Fig. 11.14 and the shear

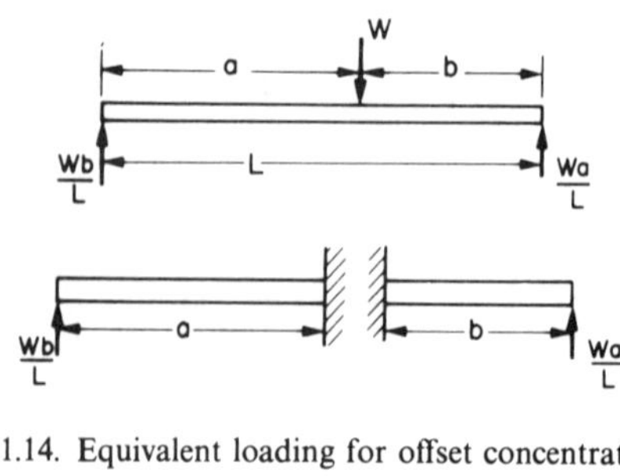

Fig. 11.14. Equivalent loading for offset concentrated load.

deflection under the load is equal to the end deflection of either cantilever and given by eqn. (11.19),

$$\delta_s = \frac{6}{5AG}\left(\frac{Wa}{L}\right)b \quad \text{or} \quad \delta_s = \frac{6}{5AG}\left(\frac{Wb}{L}\right)a$$

$$\therefore \qquad \boldsymbol{\delta_s = \frac{6Wab}{5AGL}} \qquad (11.23)$$

(e) Simply supported beam carrying uniformly distributed load

Using a similar treatment to that described above, the equivalent cantilever system is shown in Fig. 11.15, i.e. each cantilever now carries an end load of $wL/2$ in one direction and a uniformly distributed load w over its complete length $L/2$ in the opposite direction.

From eqns. (11.19) and (11.20)

$$\delta_s = \frac{6}{5AG}\left(\frac{wL}{2} \times \frac{L}{2}\right) - \frac{3}{5AG}\,w\left(\frac{L}{2}\right)^2$$

$$\boldsymbol{\delta_s = \frac{3wL^2}{20AG}} \qquad (11.24)$$

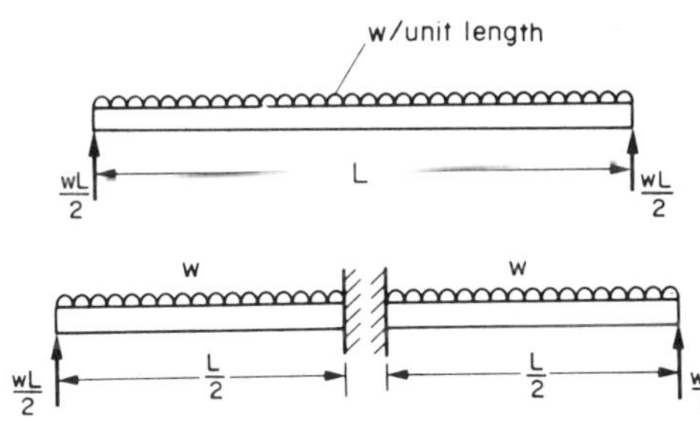

Fig. 11.15. Equivalent loading for uniformly loaded beam.

(f) I-section beams

If the shear force is assumed to be uniformly distributed over the web area A, a similar treatment to that described above yields the following approximate results:

cantilever with concentrated end load W	$\delta_s = \boldsymbol{\frac{WL}{AG}}$
cantilever with uniformly distributed load w	$\delta_s = \frac{wL^2}{2AG} = \boldsymbol{\frac{WL}{2AG}}$
simply supported beam with concentrated end load W	$\delta_s = \boldsymbol{\frac{WL}{4AG}}$
simply supported beam with uniformly distributed load w	$\delta_s = \frac{wL^2}{8AG} = \boldsymbol{\frac{WL}{8AG}}$

In the above expressions the effect of the flanges has been neglected and it therefore follows that the same formulae would apply for rectangular sections if it were assumed that the shear stress is evenly distributed across the section. The result of WL/AG for the cantilever carrying a concentrated end load is then directly comparable to that obtained in eqn. (11.19) taking full account of the variation of shear across the section, i.e. $6/5\ (WL/AG)$. Since the shear strain $\gamma = \delta/L$ it follows that both the deflection and associated shear strain is underestimated by 20% if the shear is assumed to be uniform.

(g) *Shear deflections at points other than loading points*

In the case of simply supported beams, deflections at points other than loading positions are found by simple proportion, deflections increasing linearly from zero at the supports (Fig. 11.16). For cantilevers, however, if the load is not at the free end, the above remains true between the load and the support but between the load and the free end the beam remains horizontal, i.e. there is no shear deflection. This, of course, must not be confused with deflections due to bending when there will always be some deflection of the end of a cantilever whatever the position of loading.

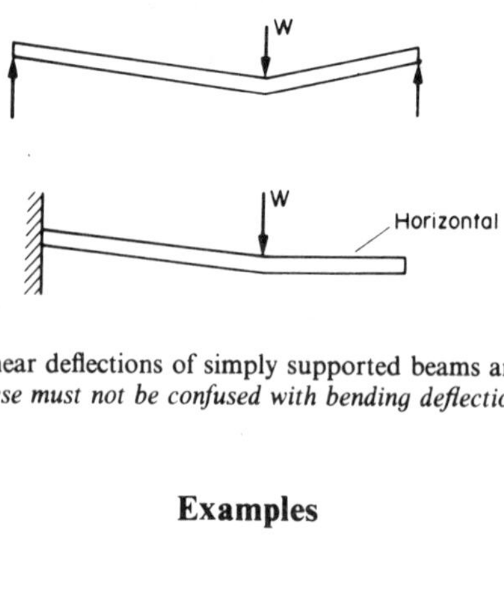

Fig. 11.16. Shear deflections of simply supported beams and cantilevers. *These must not be confused with bending deflections.*

Examples

Example 11.1

Determine the diameter of an aluminium shaft which is designed to store the same amount of strain energy per unit volume as a 50 mm diameter steel shaft of the same length. Both shafts are subjected to equal compressive axial loads.

What will be the ratio of the stresses set up in the two shafts?

$E_{\text{steel}} = 200\,\text{GN/m}^2;\quad E_{\text{aluminium}} = 67\,\text{GN/m}^2.$

Solution

$$\text{Strain energy per unit volume} = \frac{\sigma^2}{2E}$$

Since the strain energy/unit volume in the two shafts is equal,

then
$$\frac{\sigma_A^2}{2E_A} = \frac{\sigma_S^2}{2E_S}$$

$$\therefore \quad \frac{\sigma_A^2}{\sigma_S^2} = \frac{E_A}{E_S} = \frac{67}{200} = \tfrac{1}{3} \text{ (approximately)} \qquad (1)$$

$$\therefore \quad 3\sigma_A^2 = \sigma_S^2 \qquad (2)$$

Now
$$\sigma = \frac{P}{\text{area}} \quad \text{where } P \text{ is the applied load}$$

Therefore from (1)
$$\left[\frac{P}{\frac{\pi}{4}D_A^2}\right]^2 \times \left[\frac{\frac{\pi}{4}D_S^2}{P}\right]^2 = \frac{1}{3}$$

$$\therefore \quad \frac{D_S^4}{D_A^4} = \frac{1}{3}$$

$$\therefore \quad D_A^4 = 3 \times D_S^4 = 3 \times (50)^4$$
$$= 3 \times 625 \times 10^4$$

$$\therefore \quad D_A = \sqrt[4]{(1875 \times 10^4)} = \mathbf{65.8\ mm}$$

The required diameter of the aluminium shaft is 65.8 mm.

From (2)
$$3\sigma_A^2 = \sigma_S^2$$

$$\therefore \quad \frac{\sigma_S}{\sigma_A} = \sqrt{3}$$

Example 11.2

Two shafts are of the same material, length and weight. One is solid and 100 mm diameter, the other is hollow. If the hollow shaft is to store 25% more energy than the solid shaft when transmitting torque, what must be its internal and external diameters?

Assume the same maximum shear stress applies to both shafts.

Solution

Let A be the solid shaft and B the hollow shaft. If they are the same weight and the same material their volume must be equal.

$$\therefore \quad \frac{\pi}{4} D_A^2 \times L = \frac{\pi}{4}[D_B^2 - d_B^2]\, L$$

$$\therefore \quad D_A^2 = D_B^2 - d_B^2 = \frac{100^2}{10^6}\, m^2 = \mathbf{10 \times 10^{-3}\ m^2} \qquad (1)$$

Now for the same maximum shear stress

$$\tau = \frac{Tr}{J} = \frac{TD}{2J}$$

i.e.
$$\frac{T_A D_A}{J_A} = \frac{T_B D_B}{J_B}$$

$$\therefore \qquad \frac{T_A}{T_B} = \frac{D_B J_A}{D_A J_B} \qquad (2)$$

But the strain energy of $B = 1.25 \times$ strain energy of A.

$$\therefore \qquad \text{since } U = \frac{T^2 L}{2GJ}$$

then
$$\frac{T_B^2 L}{2GJ_B} = 1.25 \frac{T_A^2 L}{2GJ_A} \quad \text{or} \quad \frac{T_A^2}{T_B^2} = \frac{J_A}{1.25 J_B}$$

Therefore substituting from (2),

$$\frac{D_B^2}{D_A^2} = \frac{J_B}{1.25 J_A}$$

$$\therefore \qquad \frac{D_B^2}{D_A^2} = \frac{\frac{\pi}{32}[D_B^4 - d_B^4]}{1.25 \frac{\pi}{32} D_A^4} = \frac{D_B^4 - d_B^4}{1.25 D_A^4}$$

$$D_B^2 = \frac{D_B^4 - d_B^4}{1.25 D_A^2}$$

$$= \frac{D_B^4 - (D_B^2 - 10 \times 10^{-3})^2}{1.25 \times 10 \times 10^{-3}}$$

$$12.5 \times 10^{-3} D_B^2 = D_B^4 - D_B^4 + 20 \times 10^{-3} D_B^2 - 100 \times 10^{-6}$$

$$\therefore \qquad 7.5 \times 10^{-3} \times D_B^2 = 100 \times 10^{-6}$$

$$\therefore \qquad D_B^2 = \frac{100 \times 10^{-6}}{7.5 \times 10^{-3}} = 13.3 \times 10^{-3}$$

$$\mathbf{D_B = 115.47\,mm}$$

$$d_B^2 = D_B^2 - D_A^2 = \frac{13.3}{10^3} - \frac{10}{10^3} = \frac{3.3}{10^3}.$$

$$\therefore \qquad \mathbf{d_B = 57.74\,mm}$$

The internal and external diameters of the hollow tube are therefore 57.7 mm and 115.5 mm respectively.

Example 11.3

(a) What will be the instantaneous stress and elongation of a 25 mm diameter bar, 2.6 m long, suspended vertically, if a mass of 10 kg falls through a height of 300 mm on to a collar which is rigidly attached to the bottom end of the bar?

Take $g = 10\,\text{m/s}^2$.

(b) When used horizontally as a simply supported beam, a concentrated force of 1 kN applied at the centre of the support span produces a static deflection of 5 mm. The same load will produce a maximum bending stress of 158 MN/m^2.

Determine the magnitude of the instantaneous stress produced when a mass of 10 kg is allowed to fall through a height of 12 mm on to the beam at mid-span.

What will be the instantaneous deflection?

Solution

(a) From eqn. (11.9)

$$W\left(h+\frac{\sigma L}{E}\right) = \frac{\sigma^2}{2E} \times \text{volume} \quad \text{(Fig. 11.7)}$$

$$\text{volume of bar} = \tfrac{1}{4}\pi \times \frac{25^2}{10^6} \times 2.6 = 12.76 \times 10^{-4}$$

Then $$10 \times 10\left(0.3+\frac{2.6\sigma}{200 \times 10^9}\right) = \frac{\sigma^2 \times 12.76 \times 10^{-4}}{2 \times 200 \times 10^9}$$

$\therefore$ $$30+\frac{1.3\sigma}{10^9} = \frac{\sigma^2}{313 \times 10^{12}}$$

and $$30 \times 313 \times 10^{12} + \frac{1.3\sigma}{10^9} \times 313 \times 10^{12} = \sigma^2$$

Then $$\sigma^2 - 406.9 \times 10^3 \times \sigma - 9390 \times 10^{12} = 0$$

$$\sigma = \frac{406.9 \times 10^3 \pm \sqrt{(166 \times 10^9 + 37560 \times 10^{12})}}{2}$$

$$= \frac{406.9 \times 10^3 \pm 193.9 \times 10^6}{2}$$

$$= \mathbf{97.18\ MN/m^2}$$

If the instantaneous deflection is ignored (the term $\sigma L/E$ omitted) in the above calculation a very small difference in stress is noted in the answer,

i.e. $$W(h) = \frac{\sigma^2 \times \text{volume}}{2E}$$

$\therefore$ $$100 \times 0.3 = \frac{\sigma^2 \times 12.76 \times 10^{-4}}{2 \times 200 \times 10^9}$$

$\therefore$ $$\sigma^2 = \frac{30 \times 400 \times 10^9}{12.76 \times 10^{-4}} = 9404 \times 10^{12}$$

$\therefore$ $$\sigma = \mathbf{96.97\ MN/m^2}$$

This suggests that if the deflection δ is small in comparison to h (the distance through which the mass falls) it can, for all practical purposes, be ignored in the above calculation:

$$\text{deflection produced } (\delta) = \frac{\sigma L}{E} = \frac{97.18 \times 2.6 \times 10^6}{200 \times 10^9}$$

i.e. elongation of bar = **1.26 mm**

(b) Consider the loading system shown in Fig. 11.8. Let W_E be the equivalent force that produces the same deflection and stress when gradually applied as that produced by the falling mass.

Then
$$\frac{W_E}{\delta_{max}} = \frac{W_s}{\delta_s}$$

where W_s is a known load, gradually applied to the beam at mid-span, producing deflection δ_s and stress σ_s.

Then
$$\delta_{max} = \frac{W_E \delta_s}{W_s} = \frac{W_E \times 5 \times 10^{-3}}{1 \times 10^3}$$

∴
$$\delta_{max} = \frac{5}{10^6} W_E$$

Now
$$W(h + \delta_{max}) = \frac{W_E}{2} \delta_{max}$$

∴
$$100\left[\frac{12}{10^3} + \frac{5W_E}{10^6}\right] = \frac{W_E}{2} \times \frac{5W_E}{10^6}$$

$$1.2 + \frac{500\,W_E}{10^6} = \frac{2.5\,W_E^2}{10^6}$$

∴
$$W_E^2 - \frac{500\,W_E}{2.5} - \frac{1.2 \times 10^6}{2.5} = 0$$

and
$$W_E^2 - 200\,W_E - 0.48 \times 10^6 = 0$$

By factors,
$$W_E = 800\text{ N} \quad \text{or} \quad -600\text{ N}$$

∴
$$\boldsymbol{W_E = 800\text{ N}}$$

By proportion
$$\frac{\sigma_s}{W_s} = \frac{\sigma_{max}}{W_E}$$

and the maximum stress is given by

$$\sigma_{max} = \frac{\sigma_s}{W_s} \times W_E = \frac{158 \times 10^6 \times 800}{1 \times 10^3} = \mathbf{126.4\ MN/m^2}$$

And since
$$\frac{W_E}{\delta} = \frac{W_s}{\delta_s}$$

the deflection is given by

$$\delta = \frac{W_E}{W_s} \times \delta_s$$

$$= \frac{800 \times 5 \times 10^{-3}}{1 \times 10^3} = 4 \times 10^{-3}$$

$$= \mathbf{4\,mm}$$

Example 11.4

A horizontal steel beam of I-section rests on a rigid support at one end, the other end being supported by a vertical steel rod of 20 mm diameter whose upper end is rigidly held in a support 2.3 m above the end of the beam (Fig. 11.17). The beam is a 200 × 100 mm B.S.B. for which the relevant I-value is $23 \times 10^{-6}\,\text{m}^4$ and the distance between its two points of support is 3 m. A load of 2.25 kN falls on the beam at mid-span from a height of 20 mm above the beam.

Determine the maximum stresses set up in the beam and rod, and find the deflection of the beam at mid-span measured from the unloaded position. Assume $E = 200\,\text{GN/m}^2$ for both beam and rod.

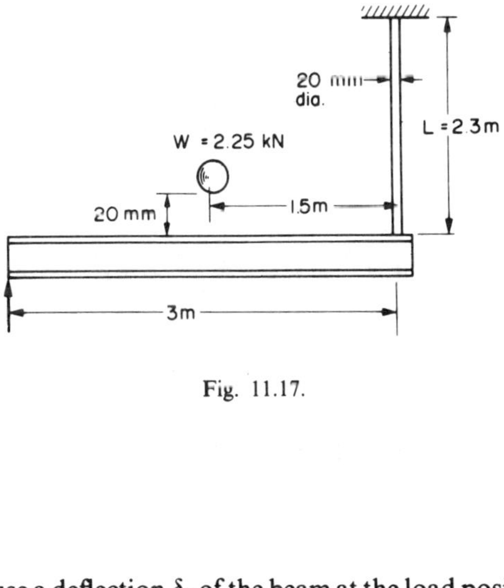

Fig. 11.17.

Solution

Let the shock load cause a deflection δ_B of the beam at the load position and an extension δ_R of the rod. Then if W_E is the equivalent static load which produces the deflection δ_B and P is the maximum tension in the rod,

$$\text{total strain energy} = \frac{P^2 L_R}{2AE} + \frac{1}{2} W_E \delta_B$$

$$= \text{work done by falling mass}$$

Now the mass falls through a distance

$$h+\delta_B+\frac{\delta_R}{2}$$

where $\delta_R/2$ is the effect of the rod extension on the mid-point of the beam. (This assumes that the beam remains straight and rotates about the fixed support position.)

$\therefore$ work done by falling mass $= W\left(h+\delta_B+\dfrac{\delta_R}{2}\right)$

If P = reaction at one end of beam

then $$P = \frac{W_E}{2}$$

$$\therefore \quad W\left(h+\delta_B+\frac{\delta_R}{2}\right) = \frac{W_E^2 L_B}{8AE}+\frac{W_E\delta_B}{2} \tag{1}$$

For a centrally loaded beam $$\delta = \frac{WL^3}{48EI}$$

$$\therefore \quad \delta_B = \frac{W_E \times 3^3}{48\times 200\times 10^9 \times 23 \times 10^{-6}} = \frac{W_E}{8.18\times 10^6} \tag{2}$$

For an axially loaded rod $$\delta_R = \frac{WL}{AE}$$

$$\therefore \quad \delta_R = \frac{W_E \times 2.3}{\frac{\pi}{4}\times 20^2 \times 10^{-6}\times 200\times 10^9} = \frac{W_E}{27.3\times 10^6} \tag{3}$$

Substituting (2) and (3) in (1),

$$2.25\times 10^3\left[\frac{20}{10^3}+\frac{W_E}{8.18\times 10^6}+\frac{W_E}{54.6\times 10^6}\right] = \frac{W_E^2\times 2.3}{8\,(\frac{\pi}{4}\times 20^2\times 10^{-6})\times 200\times 10^9} + \frac{W_E^2}{2\times 8.18\times 10^6}$$

$$45+\frac{2.25\times 10^3 W_E}{8.18\times 10^6}+\frac{2.25\times 10^3 W_E}{54.6\times 10^6} = \frac{W_E^2\times 2.3}{8\times 314\times 10^{-6}\times 200\times 10^9} + \frac{W_E^2}{16.36\times 10^6}$$

$$45+275\times 10^{-6}\,W_E+41.2\times 10^{-6}\,W_E = 4.58\times 10^{-9}\,W_E^2+61.1\times 10^{-9}\,W_E^2$$

$$45+316.2\,W_E\times 10^{-6} = 65.68\times 10^{-9}\,W_E^2$$

Then $$W_E^2-\frac{316.2\times 10^{-6}}{65.68\times 10^{-9}}\,W_E-\frac{45}{65.68\times 10^{-9}} = 0$$

$$\therefore \quad W_E^2-4.8\times 10^3\,W_E-685\times 10^6 = 0$$

and $$W_E = \frac{4.8 \times 10^3 \pm \sqrt{(23 \times 10^6 + 2740 \times 10^6)}}{2}$$

$$= \frac{4.8 \times 10^3 \pm \sqrt{(2763 \times 10^6)}}{2}$$

$$= \frac{4.8 \times 10^3 \pm 52.59 \times 10^3}{2}$$

$$= \frac{57.3 \times 10^3}{2}$$

$$= 28.65 \times 10^3 \text{ N}$$

$$\text{Maximum bending moment} = \frac{W_E L}{4}$$

$$= \frac{28.65 \times 10^3 \times 3}{4}$$

$$= 21.5 \times 10^3 \text{ N}$$

$$\text{Then maximum bending stress} = \frac{My}{I}$$

$$= \frac{21.5 \times 10^3 \times 100 \times 10^{-3}}{23 \times 10^{-6}}$$

$$= 93.9 \times 10^6 \text{ N/m}^2$$

$$\text{Maximum stress in rod} = \frac{\frac{1}{2} W_E}{\text{area}}$$

$$= \frac{28.65 \times 10^3}{2 \times \frac{\pi}{4} \times 20^2 \times 10^{-6}}$$

$$= \mathbf{45.9 \times 10^6 \text{ N/m}^2}$$

$$\text{Deflection of beam } \delta_B = \frac{W_E}{8.18 \times 10^6}$$

$$= \frac{28.65 \times 10^3}{8.18 \times 10^6}$$

$$= \mathbf{3.52 \times 10^{-3} \text{ m}}$$

This is the extension at mid-span and neglects the extension of the rod.

$$\text{Extension of rod} = \frac{\sigma L}{E} = \frac{PL}{AE} = \frac{W_E L}{2AE}$$

$$= \frac{28.8 \times 10^3 \times 2.3}{2 \times 314 \times 10^{-6} \times 200 \times 10^9}$$

$$= \mathbf{0.527 \times 10^{-3} \text{ m}}$$

Assuming, as stated earlier, that the beam remains straight and that the beam rotates about the fixed end, then the effect of the rod extension at the mid-span

$$= \frac{\delta_R}{2} = \frac{0.527 \times 10^{-3}}{2} = 0.264 \times 10^{-3}\ \text{m}$$

Then, total deflection at mid-span $= \delta_B + \delta_R/2$

$$= 3.52 \times 10^{-3} + 0.264 \times 10^{-3}$$

$$= \mathbf{3.784 \times 10^{-3}\ m}$$

Example 11.5

Using Castigliano's first theorem, obtain the expressions for (a) the deflection under a single concentrated load applied to a simply supported beam as shown in Fig. 11.18, (b) the deflection at the centre of a simply supported beam carrying a uniformly distributed load.

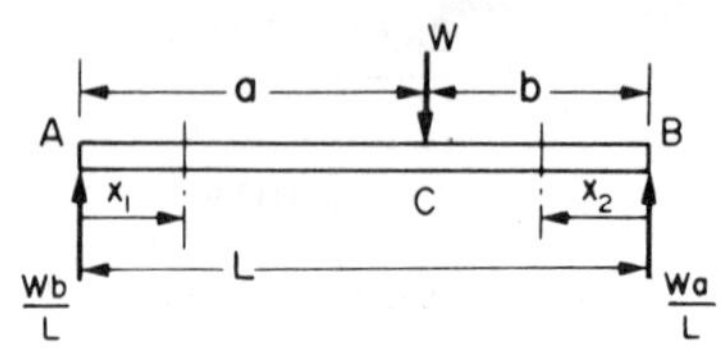

Fig. 11.18.

Solution

(a) For the beam shown in Fig. 11.18

$$\delta = \int_B^A \frac{M}{EI}\frac{\partial M}{\partial W}\,ds$$

$$= \int_A^C \frac{M}{EI}\frac{\partial M}{\partial W}\,ds + \int_C^B \frac{M}{EI}\frac{\partial M}{\partial W}\,ds$$

$$= \frac{1}{EI}\int_0^a \frac{Wbx_1}{L} \times \frac{bx_1}{L} \times dx_1 + \frac{1}{EI}\int_0^b \frac{Wax_2}{L} \times \frac{ax_2}{L} \times dx_2$$

$$= \frac{Wb^2}{L^2EI}\int_0^a x_1^2\,dx_1 + \frac{Wa^2}{L^2EI}\int_0^b x_2^2\,dx_2$$

$$= \frac{Wb^2a^3}{3L^2EI} + \frac{Wa^2b^3}{3L^2EI} = \frac{Wa^2b^2}{3L^2EI}(a+b) = \mathbf{\frac{Wa^2b^2}{3LEI}}$$

(b) For the u.d.l. beam shown in Fig. 11.19a an imaginary load P must be introduced at mid-span; then the mid-span deflection will be

$$\delta = \int_0^L \frac{M}{EI}\frac{\partial M}{\partial W}\,ds = 2\int_0^{L/2} \frac{M}{EI}\frac{\partial M}{\partial W}\,ds$$

but $$M_{xx} = \frac{(wL+P)}{2}x - \frac{wx^2}{2} \quad \text{and} \quad \frac{\partial M}{\partial W} = \frac{x}{2}$$

Then $$\delta = \frac{2}{EI}\int_0^{L/2}\left[\frac{(wL+P)}{2}x - \frac{wx^2}{2}\right]\frac{x}{2}\,dx$$

$$= \frac{1}{2EI}\int_0^{L/2}(wLx^2 - wx^3)\,dx \quad \text{since } P = 0$$

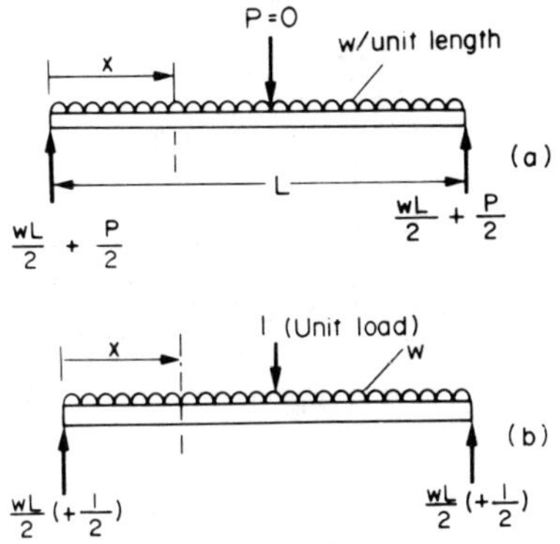

Fig. 11.19.

Alternatively, using a unit load applied vertically at mid-span (Fig. 11.19b),

$$\delta = \int_0^L \frac{Mm}{EI}\,ds = 2\int_0^{L/2}\frac{Mm}{EI}\,ds$$

where $$M = \frac{wL}{2} - \frac{wx^2}{2} \quad \text{and} \quad m = \frac{x}{2}$$

Then $$\delta = \frac{2}{EI}\int_0^{L/2}\left(\frac{wLx}{2} - \frac{wx^2}{2}\right)\frac{x}{2}\,dx$$

$$= \frac{1}{2EI}\int_0^{L/2}(wLx^2 - wx^3)\,dx$$

as before. Thus, in each case,

$$\delta = \frac{w}{2EI}\left[\frac{Lx^3}{3} - \frac{x^4}{4}\right]_0^{L/2}$$

$$= \frac{wL^4}{2EI}\left[\frac{1}{24} - \frac{1}{64}\right]$$

$$= \frac{wL^4}{2EI}\left[\frac{8-3}{192}\right] = \mathbf{\frac{5WL^4}{384EI}}$$

Example 11.6

Determine by the methods of unit load and Castigliano's first theorem, (a) the vertical deflection of point A of the bent cantilever shown in Fig. 11.20 when loaded at A with a vertical load of 600 N. (b) What will then be the horizontal movement of A?

The cantilever is constructed from 50 mm diameter bar throughout, with $E = 200\ \text{GN/m}^2$.

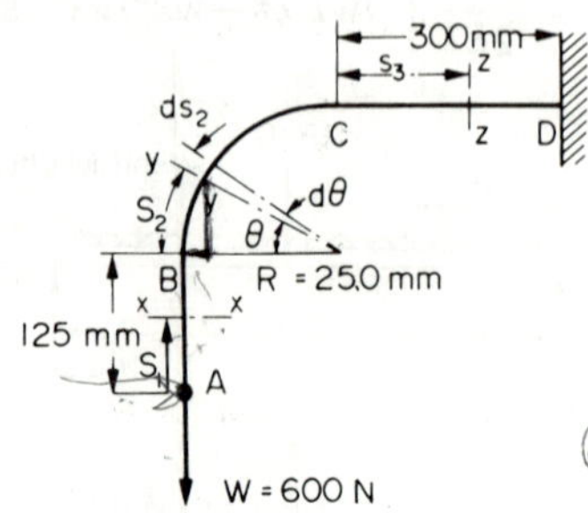

Fig. 11.20.

Solution

The total deflection of A can be considered in three parts, resulting from AB, BC, and CD. Since the question requires solution by two similar methods, they will be worked in parallel.

(a) *For vertical deflection*

Castigliano	Unit load
$\delta = \int \frac{M}{EI}\frac{\partial M}{\partial W}\,ds$	$\delta = \int \frac{Mm}{EI}\,ds$ where m = bending moment resulting from a unit load at A.
For AB $M_{xx} = 0$. Hence vertical deflection resulting from $AB = 0$ by both methods.	
For CD $M_{zz} = W(0.25 + s_3)$ $\frac{\partial M}{\partial W} = 0.25 + s_3$ $\delta_{CD} = \int_0^{0.3} \frac{W(0.25+s_3)(0.25+s_3)\,ds_3}{EI}$	$M_{zz} = W(0.25 + s_3)$ $m = 1(0.25) + s_3)$ $\therefore\ \delta_{CD} = \int_0^{0.3} \frac{W(0.25+s_3)(0.25+s_3)\,ds_3}{EI}$

Thus the same equation is achieved by both methods.

Castigliano	Unit load

$$\therefore \quad \delta_{CD} = \frac{W}{EI}\int_0^{0.3} (0.0625 + 0.5\,s_3 + s_3^2)\,ds_3$$

$$= \frac{W}{EI}\left[0.0625\,s_3 + \frac{0.5\,s_3^2}{2} + \frac{s_3^3}{3}\right]_0^{0.3}$$

$$= \frac{W}{EI}[0.01875 + 0.0225 + 0.009]$$

$$= \frac{600}{EI} \times 0.05025 = \frac{\mathbf{30.15}}{\mathbf{EI}}$$

Castigliano	Unit load
For *BC*	
$M_{yy} = W(0.25 - 0.25\cos\theta)$	$M_{yy} = W(0.25 - 0.25\cos\theta)$
$\frac{\partial M}{\partial W} = 0.25 - 0.25\cos\theta$	$m = 1(0.25 - 0.25\cos\theta)$
$ds_2 = 0.25\,d\theta$	$ds_2 = 0.25\,d\theta$

Once again the same equation for deflection is obtained

i.e.
$$\delta_{BC} = \int_0^{\pi/2} \frac{W(0.25 - 0.25\cos\theta)}{EI}(0.25 - 0.25\cos\theta)\,0.25\,d\theta$$

$$= \frac{(0.25)^3\,W}{EI}\int_0^{\pi/2}(1 - 2\cos\theta + \cos^2\theta)\,d\theta$$

but
$$\cos^2\theta = \frac{1 + \cos 2\theta}{2}$$

$$\therefore \quad \delta_{BC} = \frac{(0.25)^3\,W}{EI}\int_0^{\pi/2}\left(1 - 2\cos\theta + \frac{1+\cos 2\theta}{2}\right)d\theta$$

$$= \frac{(0.25)^3\,W}{EI}\left[\theta - 2\sin\theta + \frac{\theta}{2} + \frac{\sin 2\theta}{4}\right]_0^{\pi/2}$$

$$= \frac{(0.25)^3\,W}{EI}\left[\frac{\pi}{2} - 2 + \frac{\pi}{4}\right]$$

$$= \frac{(0.25)^3 \times 600}{EI}\left[\tfrac{3}{4}\pi - 2\right]$$

$$= \frac{\mathbf{3.34}}{\mathbf{EI}}$$

Total vertical deflection at *A*

$$= \frac{30.15 + 3.34}{EI} = \frac{33.49 \times 64 \times 10^{12}}{200 \times 10^9 \times \pi \times 50^4} = \mathbf{0.546\ mm}$$

Castigliano	Unit load

Again, working in parallel with Castigliano and unit load methods:–

(b) For the horizontal deflection using Castigliano's method an imaginary load P must be applied horizontally since there is no external load in this direction at A (Fig. 11.21).

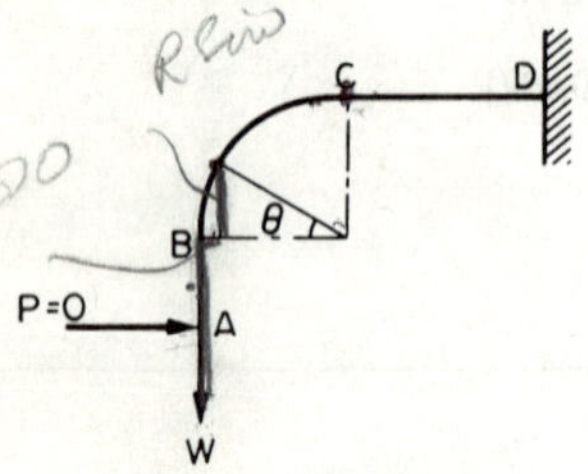

Fig. 11.21.

Then $\delta_H = \int \frac{M}{EI}\frac{\partial M}{\partial P}\, ds$, with $P = 0$

For AB

$$M_{xx} = P \times s_1 + W \times 0 = Ps_1$$

$$\therefore \quad \frac{\partial M}{\partial P} = s_1$$

$$\therefore \quad \delta_{AB} = \int \frac{Ps_1}{EI} \times s_1\, ds_1$$

but $P = 0$

$$\therefore \quad \delta_{AB} = 0$$

For the unit load method a unit load must be applied at A in the direction in which the deflection is required as shown in Fig. 11.22.

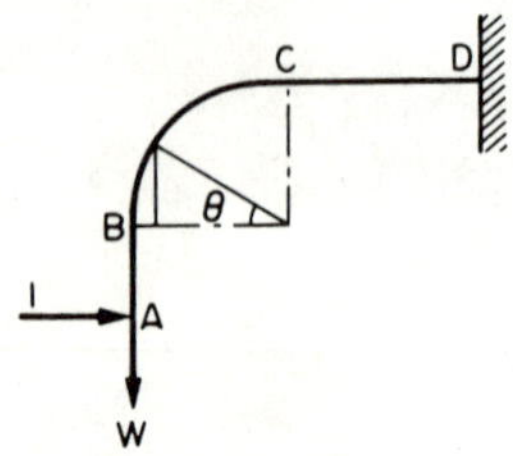

Fig. 11.22.

Then $\delta_H = \int \frac{Mm}{EI}\, ds$

$$M_{xx} = W \times 0 = 0$$

$$m = 1 \times s_1$$

$$\therefore \quad \delta_{AB} = 0$$

For BC

$$M_{yy} = W(0.25 - 0.25\cos\theta)$$
$$+ P(0.125 + 0.25\sin\theta)$$

$$\frac{\partial M}{\partial P} = 0.125 + 0.25\sin\theta$$

$$ds_2 = 0.25\, d\theta$$

$$\therefore \quad \delta_{BC} = \int_0^{\pi/2} \frac{W}{EI}(0.25 - 0.25\cos\theta) \times (0.125 + 0.25\sin\theta)0.25\, d\theta$$

since $P = 0$

$$M_{yy} = W(0.25 - 0.25\cos\theta)$$

$$m = 1(0.125 + 0.25\sin\theta)$$

$$ds_2 = 0.25\, d\theta$$

$$\therefore \quad \delta_{BC} = \int_0^{\pi/2} \frac{W}{EI}(0.25 - 0.25\cos\theta) \times (0.125 - 0.25\sin\theta)0.25\, d\theta$$

Thus, once again, the same equation is obtained. This is always the case and there is little difference in the amount of work involved in the two methods.

$$\therefore \quad \delta_{BC} = \frac{W \times 0.25^3}{EI}\int_0^{\pi/2} (1 - \cos\theta)\,(0.5 + \sin\theta)\, d\theta$$

$$= \frac{0.25^3\, W}{EI}\int_0^{\pi/2} \left(0.5 - \frac{\cos\theta}{2} + \sin\theta - \sin\theta\cos\theta\right) d\theta$$

Castigliano	Unit load

but $\sin\theta\cos\theta = \frac{1}{2}\sin^2\theta$

$$\therefore \quad \delta_{BC} = \frac{0.25^3\,W}{EI}\int_0^{\pi/2}\left(\frac{1}{2}-\frac{\cos\theta}{2}+\sin\theta-\frac{\sin 2\theta}{2}\right)d\theta$$

$$= \frac{0.25^3\,W}{EI}\left[\frac{\theta}{2}-\frac{\sin\theta}{2}-\cos\theta+\frac{\cos 2\theta}{4}\right]_0^{\pi/2}$$

$$= \frac{0.25^3\,W}{EI}\left[\left(\tfrac{\pi}{4}-\tfrac{1}{2}-\tfrac{1}{4}\right)-\left(-1+\tfrac{1}{4}\right)\right]$$

$$= \frac{0.25^3\times 600}{EI}\left(\frac{\pi}{4}\right)=\frac{7.36}{EI}$$

For CD, *using unit load method,*

$$M_{zz} = W(0.25+s_3) \qquad m = 1(0.125+0.25) = 0.375$$

$$\delta_{CD} = \frac{1}{EI}\int_0^{0.3} W(0.25+s_3)(0.375)\,ds_3$$

$$= \frac{0.375\,W}{EI}\int_0^{0.3}(0.25+s_3)\,ds_3$$

$$= \frac{0.375\,W}{EI}\left[0.25\,s_3+\frac{s_3^2}{2}\right]_0^{0.3}$$

$$= \frac{0.375\,W}{EI}[0.075+0.045]$$

$$= \frac{0.375\times 600}{EI}\times(0.12) = \frac{27}{EI}$$

Therefore total horizontal deflection

$$= \frac{7.36+27}{EI} = \frac{34.36\times 64\times 10^{12}}{200\times 10^9\times\pi\times 50^4}$$

$$= \mathbf{0.56\ mm}$$

Example 11.7

The frame shown in Fig. 11.23 is constructed from rectangular bar 25 mm wide by 12 mm thick. The end A is constrained by guides to move in a vertical direction and carries a vertical load of 400 N. For the frame material $E = 200\ \mathrm{GN/m^2}$.

Determine (a) the horizontal reaction at the guides, (b) the vertical deflection of A.

Solution

(a) Consider the frame of Fig. 11.23. If A were not constrained in guides it would move in some direction (shown dotted) which would have both horizontal and vertical components. If

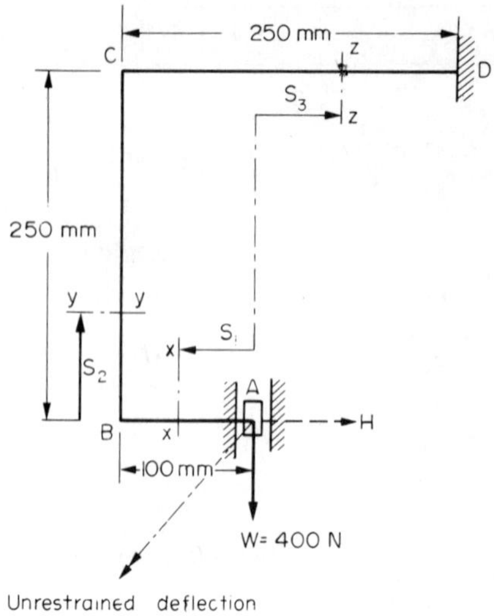

Fig. 11.23.

the horizontal movement is restricted by guides a horizontal reaction H must be set up as shown. Its value is determined by equating the horizontal deflection of A to zero,

i.e.
$$\int \frac{M}{EI}\frac{\partial M}{\partial H}ds = 0$$

For AB

$$M_{xx} = Ws_1 \quad \text{and} \quad \frac{\partial M}{\partial H} = 0$$

$\therefore$
$$\delta_{AB} = 0$$

For BC

$$M_{yy} = 0.1W - Hs_2 \quad \text{and} \quad \frac{\partial M}{\partial H} = -s_2$$

$\therefore$
$$\delta_{BC} = \int_0^{0.25} \frac{(0.1W - Hs_2)}{EI}(-s_2)ds_2$$

$$= \frac{1}{EI}\int_0^{0.25}(-0.1Ws_2 + Hs_2^2)ds_2$$

$$= \frac{1}{EI}\left[-\frac{0.1Ws_2^2}{2} + \frac{Hs_2^3}{3}\right]_0^{0.25}$$

$$= \frac{1}{EI}\left[-\frac{0.00625W}{2} + \frac{0.015625H}{3}\right]$$

$$= \frac{1}{EI\times 10^3}(-3.125W + 5.208H)$$

For *CD*

$$M_{zz} = Ws_3 + 0.25H \quad \text{and} \quad \frac{\partial M}{\partial H} = 0.25$$

$$\therefore \qquad \delta_{CD} = \int_{-0.10}^{0.15} \frac{(Ws_3 + 0.25H)}{EI} 0.25 ds_3$$

$$= \frac{1}{EI} \int_{-0.10}^{0.15} (0.25\,Ws_3 + 0.0625H) ds_3$$

$$= \frac{1}{EI} \left[\frac{0.25\,Ws_3^2}{2} + 0.0625Hs_3 \right]_{-0.10}^{0.15}$$

$$= \frac{1}{EI} \left\{ \left[\frac{0.25\,W}{2} \times 0.0225 + 0.0625H \times 0.15 \right] - \left[\frac{0.25\,W}{2} \times 0.01 + 0.0625H(-0.1) \right] \right\}$$

$$= \frac{1}{EI \times 10^3} \{(1.25 \times 2.25\,W + 6.25 \times 1.5H) - (1.25\,W - 6.25H)\}$$

$$= \frac{1}{EI \times 10^3} \{(2.81\,W + 9.375H) - (1.25\,W - 6.25H)\}$$

$$= \frac{1}{EI \times 10^3} (1.56\,W + 15.625H)$$

Now the total horizontal deflection of $A = 0$

$$\therefore \qquad -3.125\,W + 5.208H + 1.56\,W + 15.625H = 0$$

$$-1.565\,W + 20.833H = 0$$

$$\therefore \qquad H = \frac{1.565 \times 400}{20.833} = \mathbf{30\,N}$$

Since a positive sign has been obtained, H must be in the direction assumed.

(b) For vertical deflection

$$\delta = \int \frac{M}{EI} \frac{\partial M}{\partial W} ds$$

For *AB*

$$M_{xx} = Ws_1 \quad \text{and} \quad \frac{\partial M}{\partial W} = s_1$$

$$\therefore \qquad \delta_{AB} = \int_0^{0.1} \frac{Ws_1 \times s_1}{EI} ds_1$$

$$= \frac{400}{EI}\left[\frac{s_1^3}{3}\right]_0^{0.1}$$

$$= \frac{0.4}{3EI} = \frac{\mathbf{0.133}}{\mathbf{EI}}$$

For BC $\quad M_{yy} = W \times 0.1 - 30s_2 \quad \text{and} \quad \frac{\partial M}{\partial W} = 0.1$

$$\therefore \quad \delta_{BC} = \int_0^{0.25} \frac{(0.1\,W - 30s_2)}{EI} \times 0.1 ds_2$$

$$= \frac{1}{EI}\int_0^{0.25} (0.01 \times 400 - 3s_2) ds_2$$

$$= \frac{1}{EI}\left[4s_2 - \frac{3s_2^2}{2}\right]_0^{0.25}$$

$$= \frac{1}{EI}\left[1 - \frac{3 \times 0.0625}{2}\right]$$

$$= \frac{\mathbf{0.906}}{\mathbf{EI}}$$

For CD

$$M_{zz} = Ws_3 + 0.25H \quad \text{and} \quad \frac{\partial M}{\partial W} = s_3$$

$$\therefore \quad \delta_{CD} = \int_{-0.10}^{+0.15} \frac{(Ws_3 + 0.25H)}{EI} s_3 ds_3$$

$$= \frac{1}{EI}\int_{-0.1}^{+0.15} (Ws_3^2 + 0.25Hs_3) ds_3$$

$$= \frac{1}{EI}\left[\frac{400 \times s_3^3}{3} + \frac{0.25Hs_3^2}{2}\right]_{-0.1}^{0.15}$$

$$= \frac{1}{EI}\left[\frac{400}{3}(3.375 \times 10^{-3} + 1 \times 10^{-3}) + \frac{0.25 \times 30}{2}(22.5 \times 10^{-3} - 10 \times 10^{-3})\right]$$

$$= \frac{1}{EI}\left[\frac{400}{3} \times 4.375 \times 10^{-3} + \frac{0.25 \times 30}{2} \times 12.5 \times 10^{-3}\right]$$

$$= \frac{1}{EI}[0.583 + 0.047]$$

$$= \frac{\mathbf{0.63}}{\mathbf{EI}}$$

Total vertical deflection of A

$$= \frac{1}{EI}(0.133 + 0.906 + 0.63)$$

$$= \frac{1.669}{EI}$$

$$= \frac{1.669 \times 12 \times 10^{12}}{200 \times 10^9 \times 25 \times 12^3} = \mathbf{2.32\ mm}$$

Example 11.8 (B)

Derive the equation for the slope at the free end of a cantilever carrying a uniformly distributed load over its full length.

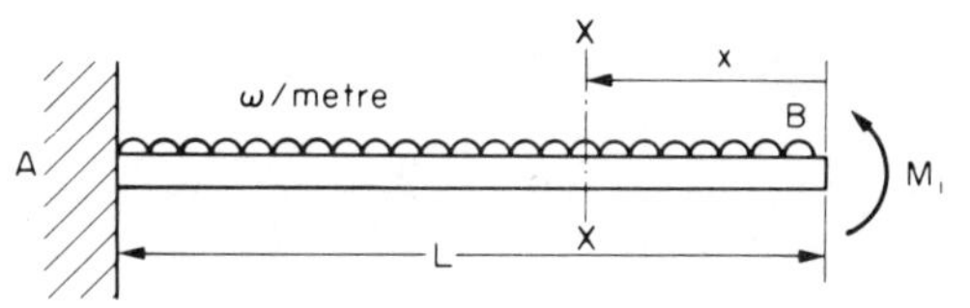

Fig. 11.24.

Solution (*a*)

Using Castigliano's procedure, apply an imaginary moment M_i in a positive direction at point B where the slope, i.e. rotation, is required.
BM at XX due to applied loading and imaginary couple

$$M = M_i - \frac{wx^2}{2}$$

$$\frac{\partial M}{\partial M_i} = 1$$

from Castigliano's theorem

$$\theta = \int_0^L \frac{M}{EI} \cdot \frac{\partial M}{\partial M_i} \cdot dx$$

$$= \frac{1}{EI}\int_0^L \left(M_i - \frac{wx^2}{2}\right)(1)\,dx$$

which, with $M_i = 0$ in the absence of any applied moment at B, becomes

$$\theta = \frac{-w}{2EI}\int_0^L x^2 . dx = \frac{\mathbf{w L^3}}{\mathbf{6EI}} \textbf{ radian}$$

The negative sign indicates that rotation of the free end is in the opposite direction to that taken for the imaginary moment, i.e. the beam will slope downwards at B as should have been expected.

Alternative solution (b)

Using the "unit-moment" procedure, apply a unit moment at the point B where rotation is required and since we know that the beam will slope downwards the unit moment can be applied in the appropriate direction as shown.

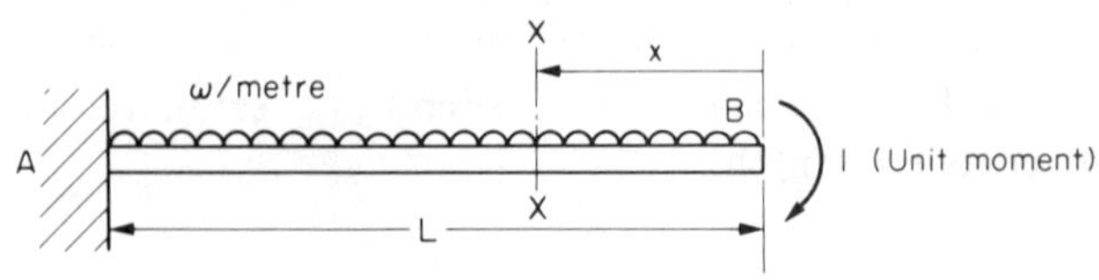

Fig. 11.25.

$$\text{B.M. at } XX \text{ due to applied loading} = M = -\frac{wx^2}{2}$$

$$\text{B.M. at } XX \text{ due to unit moment} = m = -1$$

The required rotation, or slope, is now given by

$$\theta = \int_0^L \frac{Mm}{EI} \cdot dx$$

$$= \frac{1}{EI} \int_0^L \left(-\frac{wx^2}{2}\right)(-1)\, dx.$$

$$= \frac{w}{2EI} \int_0^L x^2\, dx = \frac{wL^3}{6EI} \textbf{ radian.}$$

The answer is thus the same as before and a positive value has been obtained indicating that rotation will occur in the direction of the applied unit moment (i.e. opposite to M_i in the previous solution).

Problems

11.1 (A). Define what is meant by "resilience" or "strain energy". Derive an equation for the strain energy of a uniform bar subjected to a tensile load of P newtons. Hence calculate the strain energy in a 50 mm diameter bar, 4 m long, when carrying an axial tensile pull of 150 kN. $E = 208$ GN/m^2. [110.2 N m.]

11.2 (A). (a) Derive the formula for strain energy resulting from bending of a beam (neglecting shear).

(b) A beam, simply supported at its ends, is of 4 m span and carries, at 3 m from the left-hand support, a load of 20 kN. If I is 120×10^{-6} m^4 and $E = 200$ GN/m^2, find the deflection under the load using the formula derived in part (a). [0.625 mm.]

11.3 (A) Calculate the strain energy stored in a bar of circular cross-section, diameter 0.2 m, length 2 m:
(a) when subjected to a tensile load of 25 kN,
(b) when subjected to a torque of 25 kNm,
(c) when subjected to a uniform bending moment of 25 kNm.
For the bar material $E = 208\ \text{GN/m}^2$, $G = 80\ \text{GN/m}^2$. [0.096, 49.7, 38.2 N m.]

11.4 (A/B). Compare the strain energies of two bars of the same material and length and carrying the same gradually applied compressive load if one is 25 mm diameter throughout and the other is turned down to 20 mm diameter over half its length, the remainder being 25 mm diameter.

If both bars are subjected to pure torsion only, compare the torsional strain energies stored if the shear stress in both bars is limited to $75\ \text{MN/m}^2$. [0.78, 2.22.]

11.5 (A/B). Two shafts, one of steel and the other of phosphor bronze, are of the same length and are subjected to equal torques. If the steel shaft is 25 mm diameter, find the diameter of the phosphor-bronze shaft so that it will store the same amount of energy per unit volume as the steel shaft. Also determine the ratio of the maximum shear stresses induced in the two shafts. Take the modulus of rigidity for phosphor bronze as $50\ \text{GN/m}^2$ and for steel as $80\ \text{GN/m}^2$. [27.04 mm, 1.26.]

11.6 (A/B). Show that the torsional strain energy of a solid circular shaft transmitting power at a constant speed is given by the equation:

$$U = \frac{\tau^2}{4G} \times \text{volume}.$$

Such a shaft is 0.06 m in diameter and has a flywheel of mass 30 kg and radius of gyration 0.25 m situated at a distance of 1.2 m from a bearing. The flywheel is rotating at 200 rev/min when the bearing suddenly seizes. Calculate the maximum shear stress produced in the shaft material and the instantaneous angle of twist under these conditions. Neglect the shaft inertia. For the shaft material $G = 80\ \text{GN/m}^2$. [B.P.] [$196.8\ \text{MN/m}^2$, 5.64°.]

11.7 (A/B). A solid shaft carrying a flywheel of mass 100 kg and radius of gyration 0.4 m rotates at a uniform speed of 75 rev/min. During service, a bearing 3 m from the flywheel suddenly seizes producing a fixation of the shaft at this point. Neglecting the inertia of the shaft itself determine the necessary shaft diameter if the instantaneous shear stress produced in the shaft does not exceed $180\ \text{MN/m}^2$. For the shaft material $G = 80\ \text{GN/m}^2$. Assume all kinetic energy of the shaft is taken up as strain energy without any losses. [22.7 mm.]

11.8 (A/B). A multi-bladed turbine disc can be assumed to have a combined mass of 150 kg with an effective radius of gyration of 0.59 m. The disc is rigidly attached to a steel shaft 2.4 m long and, under service conditions, rotates at a speed of 250 rev/min. Determine the diameter of shaft required in order that the maximum shear stress set up in the event of sudden seizure of the shaft shall not exceed $200\ \text{MN/m}^2$. Neglect the inertia of the shaft itself and take the modulus of rigidity G of the shaft material to be $85\ \text{GN/m}^2$. [284 mm.]

11.9 (A/B). Develop from first principles an expression for the instantaneous stress set up in a vertical bar by a weight W falling from a height h on to a stop at the end of the bar. The instantaneous extension x may not be neglected.

A weight of 500 N can slide freely on a vertical steel rod 2.5 m long and 20 mm diameter. The rod is rigidly fixed at its upper end and has a collar at the lower end to prevent the weight from dropping off. The weight is lifted to a distance of 50 mm above the collar and then released. Find the maximum instantaneous stress produced in the rod. $E = 200\ \text{GN/m}^3$. [$114\ \text{MN/m}^2$.]

11.10 (A/B). A load of 2 kN falls through 25 mm on to a stop at the end of a vertical bar 4 m long, $600\ \text{mm}^2$ cross-sectional area and rigidly fixed at its other end. Determine the instantaneous stress and elongation of the bar. $E = 200\ \text{GN/m}^2$. [$94.7\ \text{MN/m}^2$, 1.9 mm.]

11.11 (A/B). A load of 2.5 kN slides freely on a vertical bar of 12 mm diameter. The bar is fixed at its upper end and provided with a stop at the other end to prevent the load from falling off. When the load is allowed to rest on the stop the bar extends by 0.1 mm. Determine the instantaneous stress set up in the bar if the load is lifted and allowed to drop through 12 mm on to the stop. What will then be the extension of the bar? [$365\ \text{MN/m}^2$, 1.65 mm.]

11.12 (A/B). A bar of a certain material, 40 mm diameter and 1.2 m long, has a collar securely fitted to one end. It is suspended vertically with the collar at the lower end and a mass of 2000 kg is gradually lowered on to the collar producing an extension in the bar of 0.25 mm. Find the height from which the load could be dropped on to the collar if the maximum tensile stress in the bar is to be $100\ \text{MN/m}^2$. Take $g = 9.81\ \text{m/s}^2$. The instantaneous extension cannot be neglected. [U.L.] [3.58 mm]

11.13 (A/B). A stepped bar is 2 m long. It is 40 mm diameter for 1.25 m of its length and 25 mm diameter for the remainder. If this bar hangs vertically from a rigid structure and a ring weight of 200 N falls freely from a height of 75 mm on to a stop formed at the lower end of the bar, neglecting all external losses, what would be the maximum instantaneous stress induced in the bar, and the maximum extension? $E = 200\ \text{GN/m}^2$.

[$99.3\ \text{MN/m}^2$, 0.615 mm.]

11.14 (B). A beam of uniform cross-section, with centroid at mid-depth and length 7 m, is simply supported at its ends and carries a point load of 5 kN at 3 m from one end. If the maximum bending stress is not to exceed 90 MN/m^2 and the beam is 150 mm deep, (i) working from first principles find the deflection under the load, (ii) what load dropped from a height of 75 mm on to the beam at 3 m from one end would produce a stress of 150 MN/m^2 at the point of application of the load? $E = 200\ GN/m^2$. [24 mm; 1.45 kN.]

11.15 (B). A steel beam of length 7 m is built in at both ends. It has a depth of 500 mm and the second moment of area is $300 \times 10^{-6}\ m^4$. Calculate the load which, falling through a height of 75 mm on to the centre of the span, will produce a maximum stress of 150 MN/m^2. What would be the maximum deflection if the load were gradually applied? $E = 200\ GN/m^2$. [B.P.] [7.77 kN, 0.23 mm.]

11.16 (B). When a load of 20 kN is gradually applied at a certain point on a beam it produces a deflection of 13 mm and a maximum bending stress of 75 MN/m^2. From what height can a load of 5 kN fall on to the beam at this point if the maximum bending stress is to be 150 MN/m^2? [U.L.] [78 mm.]

11.17 (B). Show that the vertical and horizontal deflections of the end B of the quadrant shown in Fig. 11.26 are, respectively,

$$\frac{WR^3}{EI}\left[\frac{3\pi}{4} - 2\right] \text{ and } \frac{WR^3}{2EI}.$$

What would the values become if W were applied horizontally instead of vertically? $\left[\frac{WR^3}{EI}\left(\frac{\pi}{4}\right);\ \frac{WR^3}{2EI}.\right]$

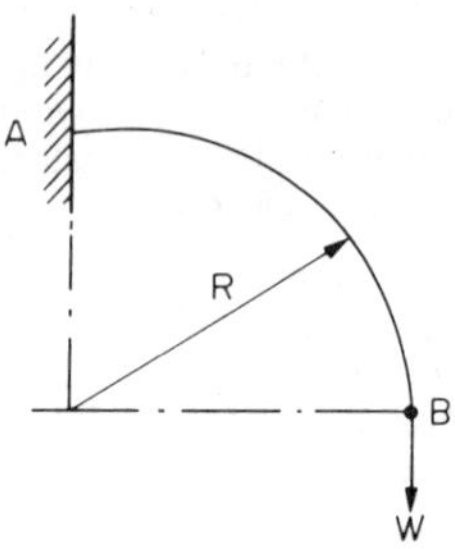

Fig. 11.26.

11.18 (B). A semi-circular frame of flexural rigidity EI is built in at A and carries a vertical load W at B as shown in Fig. 11.27. Calculate the magnitudes of the vertical and horizontal deflections at B and hence the magnitude and direction of the resultant deflection.

$$\left[\frac{3\pi}{2}\frac{WR^3}{EI};\ 2\frac{WR^3}{EI};\ 5.12\frac{WR^3}{EI} \text{ at } 23^\circ \text{ to vertical.}\right]$$

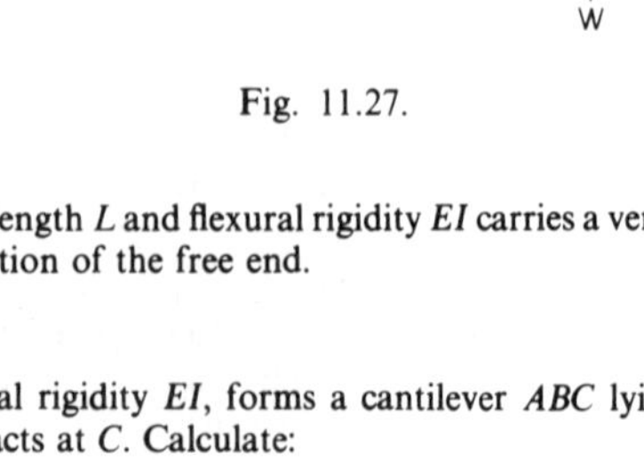

Fig. 11.27.

11.19 (B). A uniform cantilever, length L and flexural rigidity EI carries a vertical load W at mid-span. Calculate the magnitude of the vertical deflection of the free end. $\left[5\frac{WL^3}{48EI}\right]$

11.20 (B). A steel rod, of flexural rigidity EI, forms a cantilever ABC lying in a vertical plane as shown in Fig. 11.28. A horizontal load of P acts at C. Calculate:

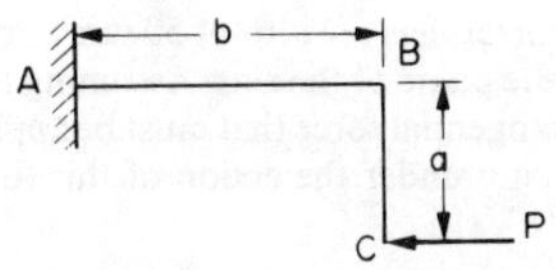

Fig. 11.28.

(a) the horizontal deflection of C;
(b) the vertical deflection of C;
(c) the slope at B.
Consider the strain energy resulting from bending only. [U.E.I.] $\left[\frac{Pa^2}{3EI}[a+3b]; \frac{Pab^2}{2EI}; \frac{Pab}{EI}.\right]$

11.21 (B). Derive the formulae for the slope and deflection at the free end of a cantilever when loaded at the end with a concentrated load W. Use a strain energy method for your solution.

A cantilever is constructed from metal strip 25 mm deep throughout its length of 750 mm. Its width, however, varies uniformly from zero at the free end to 50 mm at the support. Determine the deflection of the free end of the cantilever if it carries uniformly distributed load of 300 N/m across its length. $E = 200$ GN/m^2. [1.2 mm.]

11.22 (B). Determine the vertical deflection of point A on the bent cantilever shown in Fig. 11.29 when loaded at A with a vertical load of 25 N. The cantilever is built in at B, and EI may be taken as constant throughout and equal to 450 N m^2. [B.P.] [0.98 mm.]

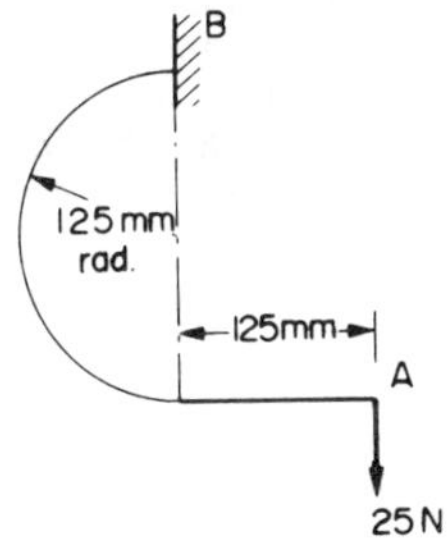

Fig. 11.29.

11.23 (B). What will be the horizontal deflection of A in the bent cantilever of Problem 11.22 when carrying the vertical load of 25 N? [0.56 mm.]

11.24 (B). A steel ring of mean diameter 250 mm has a square section 2.5 mm by 2.5 mm. It is split by a narrow radial saw cut. The saw cut is opened up farther by a tangential separating force of 0.2 N. Calculate the extra separation at the saw cut. $E = 200$ GN/m^2. [U.E.I.] [5.65 mm.]

11.25 (B). Calculate the strain energy of the gantry shown in Fig. 11.30 and hence obtain the vertical deflection of the point C. Use the formula for strain energy in bending $U = \int \frac{M^2}{2EI} dx$, where M is the bending moment, E is Young's modulus, I is second moment of area of the beam section about axis XX. The beam section is as shown in Fig. 11.30. Bending takes place along AB and BC about the axis XX. $E = 210$ GN/m^2. [U.L.C.I.] [53.9 mm.]

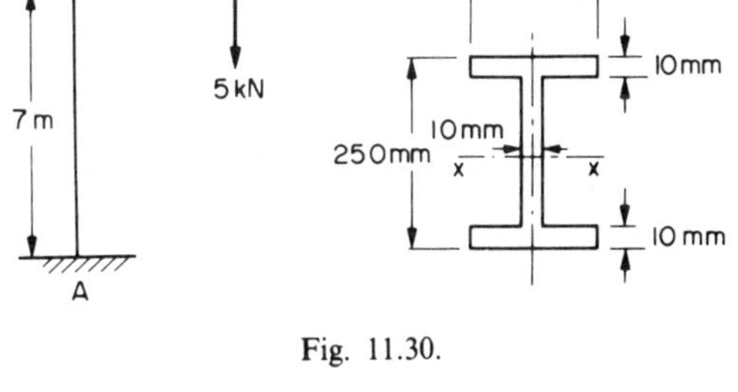

Fig. 11.30.

11.26 (B). A steel ring, of 250 mm diameter, has a width of 50 mm and a radial thickness of 5 mm. It is split to leave a narrow gap 5 mm wide normal to the plane of the ring. Assuming the radial thickness to be small compared with the radius of ring curvature, find the tangential force that must be applied to the edges of the gap to just close it. What will be the maximum stress in the ring under the action of this force? $E = 200\ \text{GN/m}^2$.

[I.Mech.E.] [28.3 N; 34 MN/m^2.]

11.27 (B). Determine, for the cranked member shown in Fig. 11.31:
(a) the magnitude of the force P necessary to produce a vertical movement of P of 25 mm;
(b) the angle, in degrees, by which the tip of the member diverges when the force P is applied.
The member has a uniform width of 50 mm throughout. $E = 200\ \text{GN/m}^2$. [B.P.] [6.58 kN; 4.1°.]

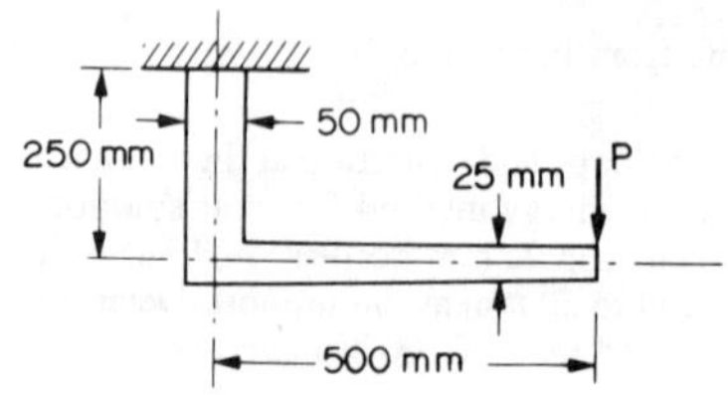

Fig. 11.31.

11.28 (C). A 12 mm diameter steel rod is bent to form a square with sides $2a = 500$ mm long. The ends meet at the mid-point of one side and are separated by equal opposite forces of 75 N applied in a direction perpendicular to the plane of the square as shown in perspective in Fig. 11.32. Calculate the amount by which they will be out of alignment. Consider only strain energy due to bending. $E = 200\ \text{GN/m}^2$. [38.3 mm.]

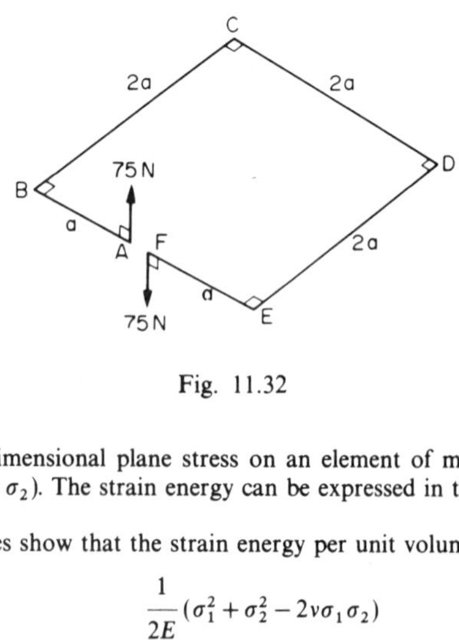

Fig. 11.32

11.29 (B/C). A state of two-dimensional plane stress on an element of material can be represented by the principal stresses σ_1 and σ_2 ($\sigma_1 > \sigma_2$). The strain energy can be expressed in terms of the strain energy per unit volume. Then:

(a) working from first principles show that the strain energy per unit volume is given by the expression

$$\frac{1}{2E}(\sigma_1^2 + \sigma_2^2 - 2\nu\sigma_1\sigma_2)$$

for a material which follows Hooke's law where E denotes Young's modulus and ν denotes Poisson's ratio, and

(b) by considering the relations between each of $\sigma_x, \sigma_y, \tau_{xy}$ respectively and the principal stresses, where x and y are two other mutually perpendicular axes in the same plane, show that the expression

$$\frac{1}{2E}[\sigma_x^2 + \sigma_y^2 - 2\nu\sigma_x\sigma_y + 2(1+\nu)\tau_{xy}^2]$$

is identical with the expression given above. [City U.]

CHAPTER 12

SPRINGS

Summary

Close-coiled springs

(a) Under axial load W

Maximum shear stress set up in the material of the spring

$$= \tau_{max} = \frac{2WR}{\pi r^3} = \frac{8WD}{\pi d^3}$$

Total deflection of the spring for n turns

$$= \delta = \frac{4WR^3n}{Gr^4} = \frac{8WD^3n}{Gd^4}$$

where r is the radius of the wire and R the mean radius of the spring coils.

i.e. $$\text{Spring rate} = \frac{W}{\delta} = \frac{Gd^4}{8nD^3}$$

(b) Under axial torque T

$$\text{Maximum bending stress set up} = \sigma_{max} = \frac{4T}{\pi r^3} = \frac{32T}{\pi d^3}$$

$$\text{Wind-up angle} = \theta = \frac{8TRn}{Er^4} = \frac{64TDn}{Ed^4}$$

$$\therefore \text{Torque per turn} = \frac{T}{\theta/2\pi} = \frac{\pi Ed^4}{32Dn}$$

The stress formulae given in (a) and (b) may be modified in practice by the addition of 'Wahl' correction factors.

Open-coiled springs

(a) Under axial load W

$$\text{Deflection } \delta = 2\pi nWR^3 \sec\alpha\left[\frac{\cos^2\alpha}{GJ} + \frac{\sin^2\alpha}{EI}\right]$$

$$\text{Angular rotation } \theta = 2\pi nWR^2 \sin\alpha\left[\frac{1}{GJ} - \frac{1}{EI}\right]$$

(*b*) *Under axial torque T*

$$\text{Wind-up angle } \theta = 2\pi nRT \sec\alpha\left[\frac{\sin^2\alpha}{GJ} + \frac{\cos^2\alpha}{EI}\right]$$

where α is the helix angle of the spring.

$$\text{Axial deflection } \delta = 2\pi nTR^2 \sin\alpha\left[\frac{1}{GJ} - \frac{1}{EI}\right]$$

Springs in series

$$\text{Stiffness } S = \frac{S_1 S_2}{(S_1 + S_2)}$$

Springs in parallel

$$\text{Stiffness } S = S_1 + S_2$$

Leaf or carriage springs

(*a*) *Semi-elliptic*

Under a central load W:

$$\text{maximum bending stress} = \frac{3WL}{2nbt^2}$$

$$\text{deflection } \delta = \frac{3WL^3}{8Enbt^3}$$

where L is the length of spring, b is the breadth of each plate, t is the thickness of each plate, and n is the number of plates.

$$\text{Proof load } W_p = \frac{8Enbt^3}{3L^3}\delta_p$$

where δ_p is the initial central "deflection".

$$\text{Proof or limiting stress } \sigma_p = \frac{4tE}{L^2}\delta_p$$

(*b*) *Quarter-elliptic*

$$\text{Maximum bending stress} = \frac{6WL}{nbt^2}$$

$$\text{Deflection } \delta = \frac{6WL^3}{Enbt^3}$$

Plane spiral springs

$$\text{Maximum bending stress} = \frac{6Ma}{RBt^2}$$

or, assuming $a = 2R$,

$$\text{maximum bending stress} = \frac{12M}{Bt^2}$$

$$\text{wind-up angle } \theta = \frac{ML}{EI}$$

where M is the applied moment to the spring spindle, R is the radius of spring from spindle to pin, a is the maximum dimension of the spring from the pin, B is the breadth of the material of the spring, t is the thickness of the material of the spring, L is equal to $\frac{1}{2}(\pi n)(a+b)$, and b is the diameter of the spindle.

Introduction

Springs are energy-absorbing units whose function it is to store energy and to release it slowly or rapidly depending on the particular application. In motor vehicle applications the springs act as buffers between the vehicle itself and the external forces applied through the wheels by uneven road conditions. In such cases the shock loads are converted into strain energy of the spring and the resulting effect on the vehicle body is much reduced. In some cases springs are merely used as positioning devices whose function it is to return mechanisms to their original positions after some external force has been removed.

From a design point of view "good" springs store and release energy but do not significantly absorb it. Should they do so then they will be prone to failure.

Throughout this chapter reference will be made to strain energy formulae derived in Chapter 11 and it is suggested that the reader should become familiar with the equations involved.

12.1. Close-coiled helical spring subjected to axial load W

(a) *Maximum stress*

A close-coiled helical spring is, as the name suggests, constructed from wire in the form of a helix, each turn being so close to the adjacent turn that, for the purposes of derivation of formulae, the helix angle is considered to be so small that it may be neglected, i.e. each turn may be considered to lie in a horizontal plane if the central axis of the spring is vertical. Discussion throughout the subsequent section on both close-coiled and open-coiled springs will be limited to those constructed from wire of circular cross-section and of constant coil diameter.

Consider, therefore, one half-turn of a close-coiled helical spring shown in Fig. 12.1. Every cross-section will be subjected to a torque WR tending to twist the section, a bending moment tending to alter the curvature of the coils and a shear force W. Stresses set up owing to the shear force are usually insignificant and with close-coiled springs the bending stresses

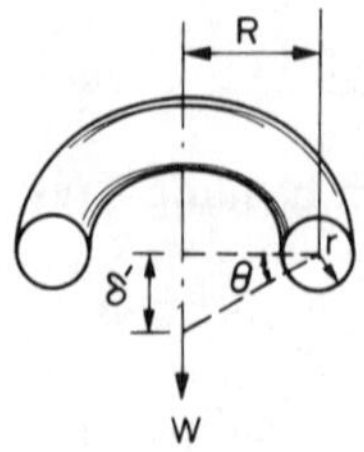

Fig. 12.1. Close-coiled helical spring subjected to axial load W.

are found to be negligible compared with the torsional stresses. Thus the maximum stress in the spring material may be determined to a good approximation using the torsion theory.

$$\tau_{max} = \frac{Tr}{J} = \frac{WRr}{\pi r^4/2}$$

i.e.

$$\textbf{maximum stress} = \frac{2WR}{\pi r^3} = \frac{8WD}{\pi d^3} \qquad (12.1)$$

(*b*) *Deflection*

Again, for one half-turn, if one cross-section twists through an angle θ relative to the other, then from the torsion theory

$$\theta = \frac{TL}{GJ} = \frac{WR(\pi R)}{G} \times \frac{2}{\pi r^4} = \frac{2WR^2}{Gr^4}$$

But

$$\delta' = R\theta = \frac{2WR^3}{Gr^4}$$

$\therefore$

$$\textbf{total deflection } \delta = 2n\delta' = \frac{4WR^3 n}{Gr^4} = \frac{8WD^3 n}{Gd^4} \qquad (12.2)$$

$$\textbf{Spring rate} = \frac{W}{\delta'} = \frac{Gd^4}{8nD^3}$$

12.2. Close-coiled helical spring subjected to axial torque T

(*a*) ***Maximum stress***

In this case the material of the spring is subjected to pure bending which tends to reduce the radius R of the coils (Fig. 12.2). The bending moment is constant throughout the spring and equal to the applied axial torque T. The maximum stress may thus be determined from the bending theory

$$\sigma_{max} = \frac{My}{I} = \frac{Tr}{\pi r^4/4}$$

i.e. $$\textbf{maximum bending stress} = \frac{4T}{\pi r^3} = \frac{32T}{\pi d^3} \qquad (12.3)$$

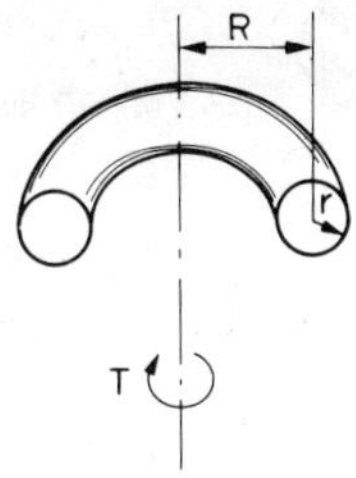

Fig. 12.2. Close-coiled helical spring subjected to axial torque T.

(b) Deflection (wind-up angle)

Under the action of an axial torque the deflection of the spring becomes the "wind-up angle" of the spring, i.e. the angle through which one end turns relative to the other. This will be equal to the total change of slope along the wire, which, according to Mohr's area–moment theorem (see § 5.7), is the area of the M/EI diagram between the ends.

$$\therefore \qquad \theta = \int_0^L \frac{M\,dL}{EI} = \frac{TL}{EI}$$

where L = total length of the wire = $2\pi Rn$.

$$\therefore \qquad \theta = \frac{T2\pi Rn}{E} \times \frac{4}{\pi r^4}$$

i.e. $$\textbf{wind-up angle } \theta = \frac{8TRn}{Er^4} \qquad (12.4)$$

N.B. The stress formulae derived above are slightly inaccurate in practice, particularly for small D/d ratios, since they ignore the higher stress produced on the inside of the coil due to the high curvature of the wire. "Wahl" correction factors are therefore introduced – see page 307.

12.3. Open-coiled helical spring subjected to axial load W

(a) Deflection

In an open-coiled spring the coils are no longer so close together that the effect of the helix angle α can be neglected and the spring is subjected to comparable bending and twisting effects. The axial load W can now be considered as a direct load W acting on the spring at the mean radius R, together with a couple WR about AB (Fig. 12.3). This couple has a component about AX of $WR \cos \alpha$ tending to twist the section, and a component about AY

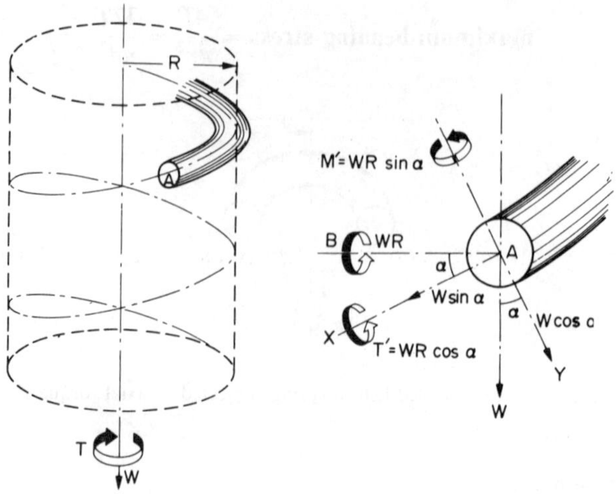

Fig. 12.3. Open-coiled helical spring.

of $WR \sin \alpha$ tending to reduce the curvature of the coils, i.e. a bending effect. Once again the shearing effect of W across the spring section is neglected as being very small in comparison with the other effects.

Thus $$T' = WR\cos\alpha \quad \text{and} \quad M' = WR\sin\alpha$$

Now, the total strain energy, neglecting shear,

$$U = \frac{T^2 L}{2GJ} + \frac{M^2 L}{2EI} \quad \text{(see §§ 11.3 and 11.4)}$$

$$= \frac{L(WR\cos\alpha)^2}{2GJ} + \frac{L(WR\sin\alpha)^2}{2EI}$$

$$= \frac{LW^2R^2}{2}\left[\frac{\cos^2\alpha}{GJ} + \frac{\sin^2\alpha}{EI}\right] \tag{12.5}$$

and this must equal the total work done $\frac{1}{2}W\delta$.

$$\therefore \quad \tfrac{1}{2}W\delta = \frac{LW^2R^2}{2}\left[\frac{\cos^2\alpha}{GJ} + \frac{\sin^2\alpha}{EI}\right]$$

From the helix form of Fig. 12.4

$$2\pi Rn = L\cos\alpha$$

$$\therefore \quad L = 2\pi Rn\sec\alpha$$

$$\therefore \quad \textbf{deflection } \boldsymbol{\delta = 2\pi n\, WR^3 \sec\alpha\left[\frac{\cos^2\alpha}{GJ} + \frac{\sin^2\alpha}{EI}\right]} \tag{12.6}$$

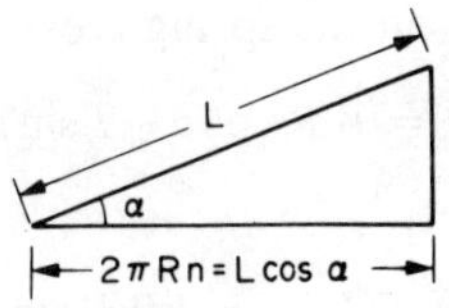

Fig. 12.4.

Since the stiffness of a spring S is normally defined as the value of W required to produce unit deflection,

$$\text{stiffness } S = \frac{W}{\delta}$$

$$\therefore \qquad \frac{1}{S} = \frac{\delta}{W} = 2\pi n R^3 \sec\alpha \left[\frac{\cos^2\alpha}{GJ} + \frac{\sin^2\alpha}{EI}\right] \qquad (12.7)$$

Alternatively, the deflection in the direction of W is given by Castigliano's theorem (see § 11.11) as

$$\delta = \frac{\partial U}{\partial W} = \frac{\partial}{\partial W}\left[\frac{LW^2R^2}{2}\left(\frac{\cos^2\alpha}{GJ} + \frac{\sin^2\alpha}{EI}\right)\right]$$

$$= LWR^2\left[\frac{\cos^2\alpha}{GJ} + \frac{\sin^2\alpha}{EI}\right]$$

and with $L = 2\pi Rn \sec\alpha$

$$\delta = 2\pi n WR^3 \sec\alpha \left[\frac{\cos^2\alpha}{GJ} + \frac{\sin^2\alpha}{EI}\right] \qquad (12.8)$$

This is the same equation as obtained previously and illustrates the flexibility and ease of application of Castigliano's energy theorem.

(*b*) *Maximum stress*

The material of the spring is subjected to combined bending and torsion, the maximum stresses in each mode of loading being determined from the appropriate theory.

From the bending theory

$$\sigma = \frac{My}{I} \quad \text{with} \quad M = WR\sin\alpha$$

and from the torsion theory

$$\tau = \frac{Tr}{J} \quad \text{with} \quad T = WR\cos\alpha$$

The principal stresses at any point can then be obtained analytically or graphically using the procedures described in § 13.4.

(*c*) *Angular rotation*

Consider an imaginary axial torque T applied to the spring, together with W producing an angular rotation θ of one end of the spring relative to the other.

The combined twisting moment on the spring cross-section is then

$$\bar{T} = WR\cos\alpha + T\sin\alpha$$

and the combined bending moment

$$\bar{M} = T\cos\alpha - WR\sin\alpha$$

The total strain energy of the system is then

$$U = \frac{\bar{T}^2 L}{2GJ} + \frac{\bar{M}^2 L}{2EI}$$

$$= \frac{(WR\cos\alpha + T\sin\alpha)^2 L}{2GJ} + \frac{(T\cos\alpha - WR\sin\alpha)^2 L}{2EI}$$

Now from Castigliano's theorem the angle of twist in the direction of the axial torque T is given by $\theta = \dfrac{\partial U}{\partial T}$ and since $T = 0$ all terms including T may be ignored.

$$\therefore \qquad \theta = \frac{2WR\cos\alpha\sin\alpha\, L}{2GJ} + \frac{(-2WR\sin\alpha\cos\alpha)\, L}{2EI}$$

$$= WRL\cos\alpha\sin\alpha\left[\frac{1}{GJ} - \frac{1}{EI}\right]$$

i.e.

$$\boldsymbol{\theta = 2\pi n WR^2 \sin\alpha\left[\frac{1}{GJ} - \frac{1}{EI}\right]} \qquad (12.9)$$

12.4. Open-coiled helical spring subjected to axial torque T

(a) *Wind-up angle*

When an axial torque T is applied to an open-coiled helical spring it has components as shown in Fig. 12.5, i.e. a torsional component $T\sin\alpha$ about AX and a flexural (bending) component $T\cos\alpha$ about AY, the latter tending to increase the curvature of the coils.

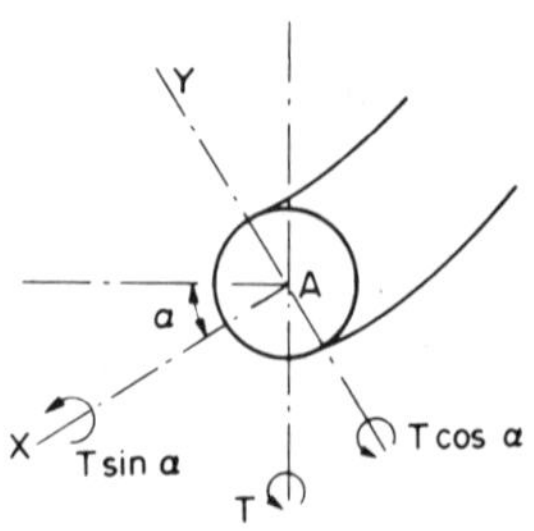

Fig. 12.5. Open-coiled helical spring subjected to axial torque T.

As for the close-coiled spring the total strain energy is given by

$$\text{strain energy } U = \frac{T^2L}{2GJ} + \frac{M^2L}{2EI}$$

$$= \frac{L}{2}\left[\frac{(T\sin\alpha)^2}{GJ} + \frac{(T\cos\alpha)^2}{EI}\right]$$

$$= \frac{T^2L}{2}\left[\frac{\sin^2\alpha}{GJ} + \frac{\cos^2\alpha}{EI}\right] \qquad (12.10)$$

and this is equal to the work done by T, namely, $\frac{1}{2}T\theta$, where θ is the angle turned through by one end relative to the other, i.e. the wind-up angle of the spring.

$$\therefore \qquad \tfrac{1}{2}T\theta = \tfrac{1}{2}T^2L\left[\frac{\sin^2\alpha}{GJ} + \frac{\cos^2\alpha}{EI}\right]$$

and, with $L = 2\pi Rn \sec\alpha$ as before,

$$\textbf{wind-up angle } \boldsymbol{\theta = 2\pi nRT\sec\alpha\left[\frac{\sin^2\alpha}{GJ} + \frac{\cos^2\alpha}{EI}\right]} \qquad (12.11)$$

(b) *Maximum stress*

The maximum stress in the spring material will be found by the procedure outlined in § 12.3(b) with a bending moment of $T\cos\alpha$ and a torque of $T\sin\alpha$ applied to the section.

(c) *Axial deflection*

Assuming an imaginary axial load W applied to the spring the total strain energy is given by eqn. (11.5) as

$$U = \frac{(WR\cos\alpha + T\sin\alpha)^2L}{2GJ} + \frac{(T\cos\alpha - WR\sin\alpha)^2L}{2EI}$$

Now from Castigliano's theorem the deflection in the direction of W is given by

$$\delta = \frac{\partial U}{\partial W}$$

$$= TRL\cos\alpha\sin\alpha\left[\frac{1}{GJ} - \frac{1}{EI}\right] \quad \text{when } W = 0$$

$$\therefore \qquad \textbf{deflection } \boldsymbol{\delta = 2\pi nTR^2\sin\alpha\left[\frac{1}{GJ} - \frac{1}{EI}\right]} \qquad (12.12)$$

12.5. Springs in series

If two springs of different stiffness are joined end-on and carry a common load W, they are said to be *connected in series* and the combined stiffness and deflection are given by the following equations.

$$\text{Deflection} = \frac{W}{S} = \delta_1 + \delta_2 = \frac{W}{S_1} + \frac{W}{S_2}$$

$$= W\left[\frac{1}{S_1} + \frac{1}{S_2}\right] \tag{12.13}$$

$$\therefore \quad \frac{1}{S} = \frac{1}{S_1} + \frac{1}{S_2}$$

and

$$\textbf{stiffness } S = \frac{S_1 S_2}{S_1 + S_2} \tag{12.14}$$

12.6. Springs in parallel

If two springs are joined in such a way that they have a common deflection δ they are said to be *connected in parallel*. In this case the load carried is shared between the two springs and

$$\text{total load } W = W_1 + W_2 \tag{1}$$

Now

$$\delta = \frac{W}{S} = \frac{W_1}{S_1} = \frac{W_2}{S_2} \tag{12.15}$$

so that

$$W_1 = \frac{S_1 W}{S} \quad \text{and} \quad W_2 = \frac{S_2 W}{S}$$

Substituting in eqn. (1)

$$W = \frac{S_1 W}{S} + \frac{S_2 W}{S}$$

$$= \frac{W}{S}\left[S_1 + S_2\right]$$

i.e.

$$\textbf{combined stiffness } S = S_1 + S_2 \tag{12.16}$$

12.7. Limitations of the simple theory

Whilst the simple torsion theory can be applied successfully to bars with small curvature without significant error the theory becomes progressively more inappropriate as the curvatures increase and become high as in most helical springs. The stress and deflection equations derived in the preceding sections, are, therefore, slightly inaccurate in practice, particularly for small D/d ratios. For accurate assessment of stresses and deflections account should be taken of the influence of curvature and slope by applying factors due to Wahl† and Ancker and Goodier‡. These are discussed in Roark and Young§ where the more accurate

† A. M. Wahl, *Mechanical Springs*, 2nd edn. (McGraw-Hill, New York 1963).
‡ C. J. Ancker (Jr) and J. N. Goodier, "Pitch and curvature correction for helical springs", *ASME J. Appl. Mech.*, **25**(4), Dec. 1958.
§ R. J. Roark and W. C. Young, *Formulas for Stress and Strain*, 5th edn. (McGraw-Hill, Kogakusha, 1965).

expressions for circular, square and rectangular section springs are introduced. For the purposes of this text it is considered sufficient to indicate the use of these factors on circular section wire.

For example, Ancker and Goodier write the stress and deflection equations for circular section springs subjected to an axial load W in the following form (which can be related directly to eqns. (12.1) and (12.2)).

Maximum stress $$\tau_{max} = K_1\left(\frac{2WR}{\pi r^3}\right) = K_1\left(\frac{8WD}{\pi d^3}\right)$$

and deflection $$\delta = K_2\left(\frac{4WR^3n}{Gr^4}\right) = K_2\left(\frac{8WD^3n}{Gd^4}\right)$$

where $$K_1 = \left[1 + \frac{5}{8}\left(\frac{d}{R}\right) + \frac{7}{32}\left(\frac{d}{R}\right)^2\right]$$

and $$K_2 = \left[1 - \frac{3}{64}\left(\frac{d}{R}\right)^2 + \frac{(3+\nu)}{2(1+\nu)}(\tan\alpha)^2\right]$$

where α is the pitch angle of the spring.

In an exactly similar way Wahl also proposes the introduction of correction factors which are related to the so-called spring index $C = D/d$.

Thus, for central load W:

maximum stress $$\tau_{max} = K\left[\frac{8WD}{\pi d^3}\right]$$

with $$K = \frac{(4C-1)}{(4C-4)} + \frac{0.615}{C}$$

The British Standard for spring design, BS1726, quotes a simpler equation for K, namely:

$$K = \left[\frac{C+0.2}{C-1}\right]$$

The Standard also makes the point that the influence of the correction factors is often small in comparison with the uncertainty regarding what should be selected as the true number of working coils (depending on thc method of support, etc).

Values of K for different ratios of spring index are given in Fig. 12.6 on page 308.

12.8. Extension springs – initial tension.

The preceding laws and formulae derived for compression springs apply equally to extension springs except that the latter are affected by initial tension. When springs are closely wound a force is required to hold the coils together and this can seldom be controlled to a greater accuracy than $\pm$ 10%. This does not increase the ultimate load capacity but must be included in the stress calculation. As an approximate guide, the initial tension obtained in hand-coiled commercial-quality springs is taken to be equivalent to the rate of the spring, although this can be far exceeded if special coiling methods are used.

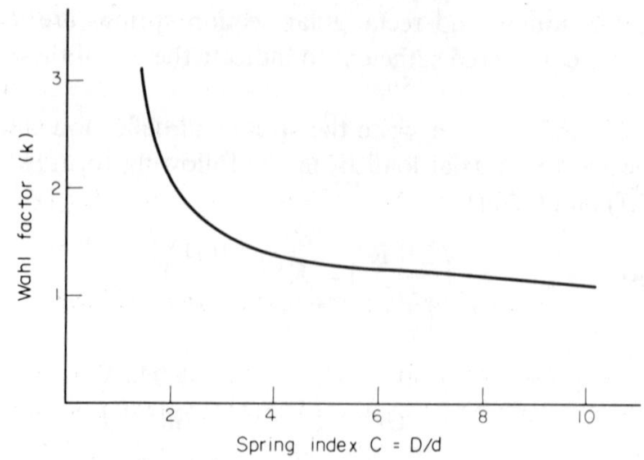

Fig. 12.6. Wahl correction factors for maximum shear stress.

12.9. Allowable stresses

As a rough approximation, the torsional elastic limit of commercial wire materials is taken to be 40% of the tensile strength. This is applied equally to ferrous and non-ferrous materials such as phosphor bronze and brass.

Typical values of allowable stress for hard-drawn spring steel piano wire based on the above assumption are given in Table 12.1.† These represent the corrected stress and generally should not be exceeded unless exceptionally high grade materials are used.

TABLE 12.1. Allowable stresses for hard-drawn steel spring wire

Wire size *SWG*	*Allowable stress* (MN/m^2)	
	Compression/Extension	*Torsion*
44–39	1134	1409
48–35	1079	1340
34–31	1031	1272
30–28	983	1203
27–24	928	1169
23–18	859	1066
17–13	770	963
12–10	688	859
9–7	619	756
6–5	550	688
4–3	516	619

Care must be exercised in the application of the quoted values bearing in mind the presence of any irregularities in the form or clamping method and the duty the spring is to perform. For example the quoted values may be far too high for springs to operate at high frequency, particularly in the presence of stress raisers, when fatigue failure would soon result. Under

† Spring Design, *Engineering Materials And Design*, Feb. 1980.

such conditions a high-grade annealed spring steel suitably heat-treated should be considered.

A useful comparison of the above theories together with further ones due to Rover, Honegger, Göhner and Bergsträsser is given in the monograph† *Helical Springs*, which then goes on to consider the effect of pitch angle, failure considerations, vibration frequency and spring surge (speed of propagation of wave along the axis of a spring).

12.10. Leaf or carriage spring: semi-elliptic

The principle of using a beam in bending as a spring has been known for many years and widely used in motor-vehicle applications. If the beam is arranged as a simple cantilever, as in Fig. 12.7a, it is called a *quarter-elliptic* spring, and if as a simply supported beam with central load, as in Fig. 12.7b, it is termed a *half* or *semi-elliptic* spring. The latter will be discussed first.

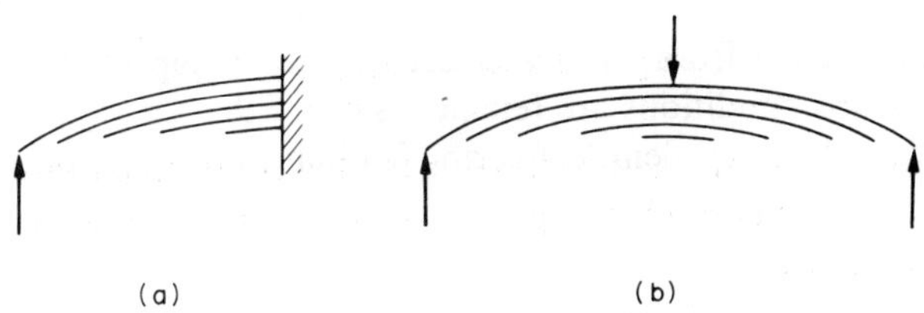

Fig. 12.7. (a) Quarter-elliptic, (b) semi-elliptic, carriage springs.

(*a*) *Maximum stress*

Consider the semi-elliptic leaf spring shown in Fig. 12.8. With a constant thickness t this design of spring gives a uniform stress throughout and is therefore economical in both material and weight.

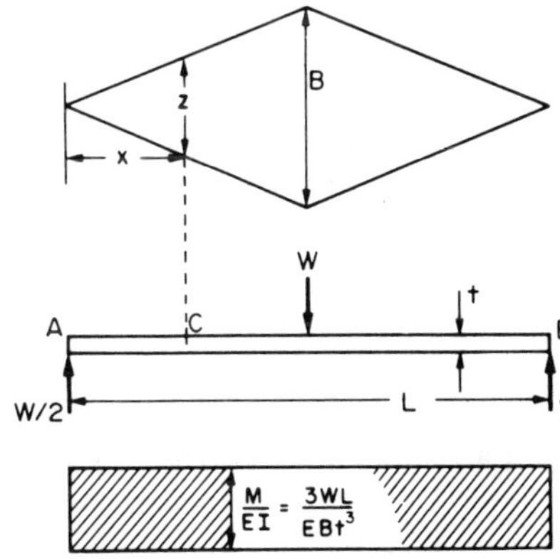

Fig. 12.8. Semi-elliptic leaf spring.

† J. R. Finniecome, *Helical Springs*. Mechanical World Monograph 56 (Emmott & Co., Manchester 1949).

By proportions $\frac{z}{x} = \frac{B}{L/2} \quad \therefore \; z = \frac{2Bx}{L}$

$$\text{Bending moment at } C = \frac{Wx}{2} \quad \text{and} \quad I = \frac{zt^3}{12} = \frac{2Bxt^3}{12L}$$

Therefore from the bending theory the stress set up at any section is given by

$$\sigma = \frac{My}{I} = \frac{Wx}{2} \times \frac{t}{2} \times \frac{12L}{2Bxt^3}$$

$$= \frac{3WL}{2Bt^2}$$

i.e. the bending stress in a semi-elliptic leaf spring is independent of x and equal to

$$\frac{\mathbf{3WL}}{\mathbf{2Bt^2}} \qquad (12.17)$$

If the spring is constructed from strips and placed one on top of the other as shown in Fig. 12.9, uniform stress conditions are retained, since if the strips are cut along XX and replaced side by side, the equivalent leaf spring is obtained as shown.

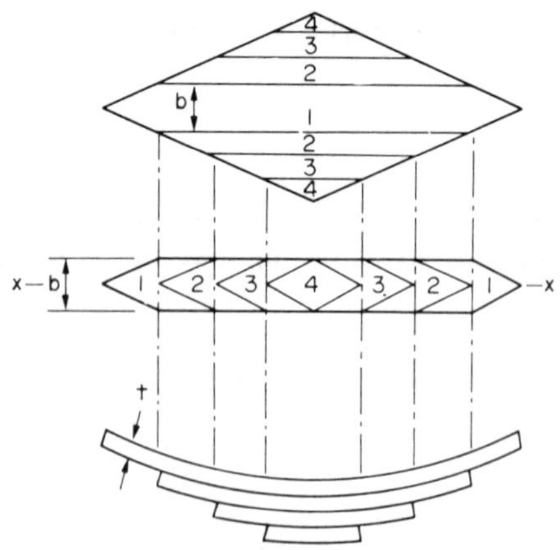

Fig. 12.9. Semi-elliptic carriage spring showing initial pre-forming.

Such a spring is then termed a *carriage spring* with n strips of width b, i.e. $B = nb$. Therefore the bending stress in a semi-elliptic carriage spring is

$$\frac{\mathbf{3WL}}{\mathbf{2nbt^2}} \qquad (12.18)$$

The diamond shape of the leaf spring could also be obtained by varying the thickness, but this type of spring is difficult to manufacture and has been found unsatisfactory in practice.

(*b*) *Deflection*

From the simple bending theory

$$\frac{M}{I} = \frac{E}{R} \qquad \therefore R = \frac{EI}{M}$$

$$R = E \times \frac{2Bxt^3}{12L} \times \frac{2}{Wx} = \frac{EBt^3}{3WL} \tag{12.19}$$

i.e. for a given spring and given load, R is constant and the spring bends into the arc of a circle.

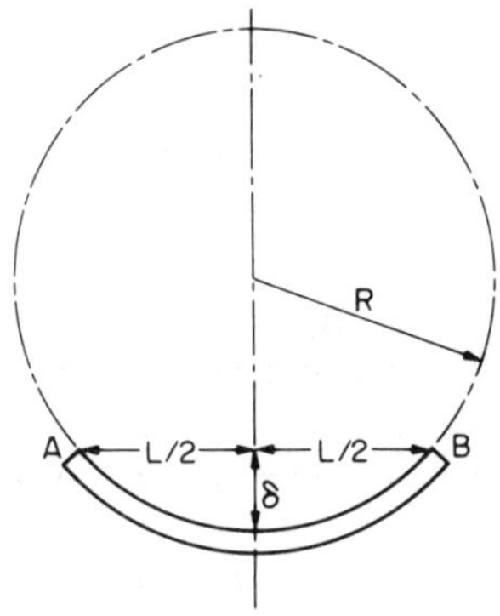

Fig. 12.10.

From the properties of intersecting chords (Fig. 12.10)

$$\delta(2R - \delta) = \frac{L}{2} \times \frac{L}{2}$$

Neglecting δ^2 as the product of small quantities

$$\delta = \frac{L^2}{8R}$$

$$= \frac{L^2}{8} \times \frac{3WL}{EBt^3}$$

i.e. deflection of a semi-elliptic *leaf* spring

$$\boldsymbol{\delta = \frac{3WL^3}{8EBt^3}} \tag{12.20}$$

But $B = nb$, so that the deflection of a semi-elliptic *carriage* spring is given by

$$\boldsymbol{\delta = \frac{3WL^3}{8Enbt^3}} \tag{12.21}$$

(*c*) *Proof load*

The proof load of a leaf or carriage spring is the load which is required to straighten the plates from their initial preformed position. From eqn. 12.18 the maximum bending stress for any given load W is

$$\sigma = \frac{3WL}{2nbt^2}$$

Thus if σ_p denotes the stress corresponding to the application of the proof load W_p

$$W_p = \frac{2nbt^2}{3L}\sigma_p \tag{12.22}$$

Now from eqn. (12.19) and inserting $B = nb$, the load W which would produce bending of a flat carriage spring to some radius R is given by

$$W = \frac{Enbt^3}{3RL}$$

Conversely, therefore, the load which is required to straighten a spring from radius R will be of the same value,

i.e. $$W_p = \frac{Enbt^3}{3RL}$$

Substituting for $$R = \frac{L^2}{8\delta}$$

$\therefore$ $$\textbf{proof load } \boldsymbol{W_p = \frac{8Enbt^3}{3L^3}\delta_p} \tag{12.23}$$

where δ_p is the initial central "deflection" of the spring.

Equating eqns. (12.22) and (12.23),

$$\frac{2nbt^2}{3L}\sigma_p = \frac{8Enbt^3}{3L^3}\delta_p$$

i.e. $$\textbf{proof stress } \boldsymbol{\sigma_p = \frac{4tE}{L^2}\delta_p} \tag{12.24}$$

For a given spring material the limiting value of σ_p will be known as will the value of E. The above equation therefore yields the correct relationship between the thickness and initial curvature of the spring plates.

12.11. Leaf or carriage spring: quarter-elliptic

(a) Maximum stress

Consider the *quarter-elliptic* leaf and carriage springs shown in Fig. 12.11. In this case the equations for the semi-elliptic spring of the previous section are modified to

$$z = \frac{Bx}{L} \quad \text{and} \quad \text{B.M. at } C = Wx$$

$\therefore$ $$I = \frac{zt^3}{12} = \frac{Bxt^3}{12L}$$

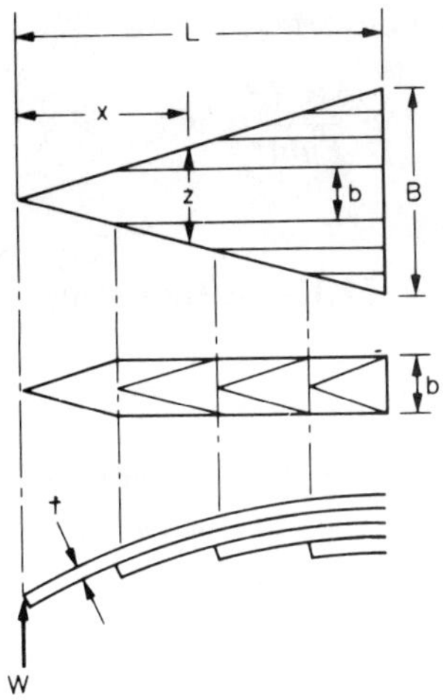

Fig. 12.11. Quarter-elliptic leaf and carriage springs.

Now $$\sigma = \frac{My}{I} = \frac{Wxt}{2} \times \frac{12L}{Bxt^3} = \frac{6WL}{Bt^2}$$

Therefore the maximum bending stress for a quarter-elliptic *leaf* spring

$$= \frac{6WL}{Bt^2} \qquad (12.25)$$

and the maximum bending stress for a quarter-elliptic *carriage* spring

$$= \frac{6WL}{nbt^2} \qquad (12.26)$$

(*b*) *Deflection*

With B.M. at $C = Wx$ and replacing $L/2$ by L in the proof of §12.7(b),

$$\delta = \frac{L^2}{2R}$$

and $$R = \frac{EI}{M} = \frac{E}{Wx} \times \frac{Bxt^3}{12L} = \frac{Ebt^3}{12WL}$$

$\therefore$ $$\delta = \frac{L^2}{2} \times \frac{12WL}{EBt^3} = \frac{6WL^3}{EBt^3}$$

Therefore deflection of a quarter-elliptic *leaf* spring

$$= \frac{6WL^3}{EBt^3} \qquad (12.27)$$

and deflection of a quarter-elliptic *carriage* spring

$$= \frac{6WL^3}{Enbt^3} \tag{12.28}$$

12.12. Spiral spring

(a) Wind-up angle

Spiral springs are normally constructed from thin rectangular-section strips wound into a spiral in one plane. They are often used in clockwork mechanisms, the winding torque or moment being applied to the central spindle and the other end firmly anchored to a pin at the outside of the spiral. Under the action of this central moment all sections of the spring will be subjected to uniform bending which tends to reduce the radius of curvature at all points.

Consider now the spiral spring shown in Fig. 12.12.

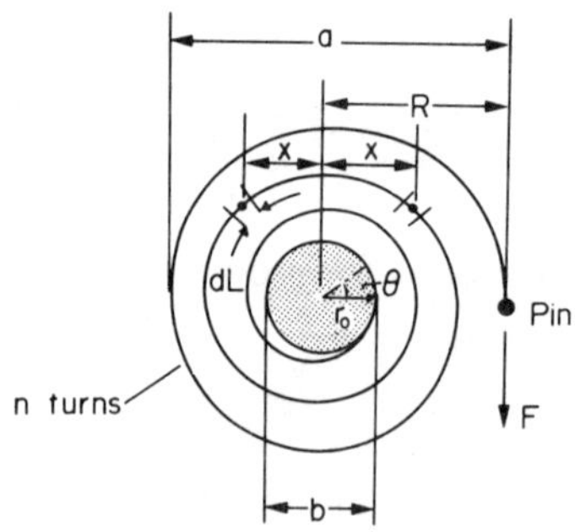

Fig. 12.12. Spiral spring.

Let M = winding moment applied to the spring spindle, R = radius of spring from spindle to pin, a = maximum dimension of the spring from the pin, B = breadth of the material of the spring, t = thickness of the material of the spring, and b = diameter of the spindle.

Assuming the polar equation of the spiral to be that of an Archimedean spiral,

$$r = r_0 + \left(\frac{A}{2\pi}\right)\theta \quad \text{where } A \text{ is some constant}$$

When $\qquad \theta = 0, \quad r = r_0 = \dfrac{b}{2}$

and for the nth turn, $\theta = 2n\pi$ and

$$r = \frac{a}{2} = \frac{b}{2} + \left(\frac{A}{2\pi}\right)2n\pi$$

$\therefore$

$$A = \frac{(a-b)}{2n}$$

i.e. the equation to the spiral is

$$r = \frac{b}{2} + \frac{(a-b)}{4\pi n}\theta \tag{12.29}$$

When a torque or winding couple M is applied to the spindle a resistive force F will be set up at the pin such that

$$\text{winding couple } M = F \times R$$

Consider now two small elements of material of length dl at distance x to each side of the centre line (Fig. 12.12).

For small deflections, from Mohr's area–moment method the change in slope between two points is

$$\left(\frac{M}{EI}\right)dL \qquad \text{(see §5.7)}$$

For the portion on the left,

$$\text{change in slope} = d\theta_1 = \frac{F(R+x)\,dL}{EI}$$

and similarly for the right-hand portion,

$$\text{change in slope} = d\theta_2 = \frac{F(R-x)\,dL}{EI}$$

The sum of these changes in slope is thus

$$d\theta_1 + d\theta_2 = \frac{F(R+x)\,dL}{EI} + \frac{F(R-x)\,dL}{EI}$$

$$= \frac{2FR\,dL}{EI}$$

If this is integrated along the length of the spring the result obtained will be twice the total change in slope along the spring, i.e. twice the angle of twist.

$$\therefore \qquad \text{angle of twist} = \tfrac{1}{2}\int_0^L \frac{2FR\,dL}{EI} = \frac{FRL}{EI} = \frac{ML}{EI}$$

where M is the applied winding moment and L the total length of the spring.

Now
$$L = \int_0^L dL = \int_0^{2n\pi} r\,d\theta = \int_0^{2n\pi} \frac{b}{2} + \frac{(a-b)}{4\pi n}\theta\,d\theta$$

$$= \left[\frac{b\theta}{2} + \frac{(a-b)}{4\pi n}\frac{\theta^2}{2}\right]_0^{2n\pi} = \left[\frac{2nb\pi}{2} + \frac{(a-b)}{4\pi n}\frac{(2n\pi)^2}{2}\right]$$

$$= \pi n\left[b + \frac{(a-b)}{2}\right]$$

$$= \frac{\pi n}{2}\left[a+b\right] \tag{12.30}$$

Therefore the wind-up angle of a spiral spring is

$$\theta = \frac{M}{EI}\left[\frac{\pi n}{2}(a+b)\right] \tag{12.31}$$

(*b*) *Maximum stress*

The maximum bending stress set up in the spring will be at the point of greatest bending moment, since the material of the spring is subjected to pure bending.

$$\text{Maximum bending moment} = F \times a$$

$$\therefore \qquad \text{maximum bending stress} = \frac{My}{I} = \frac{Fa(t/2)}{I}$$

But, for rectangular-section spring material of breadth B and thickness t,

$$I = \frac{Bt^3}{12}$$

$$\therefore \qquad \sigma_{max} = \frac{Fat}{2} \times \frac{12}{Bt^3} = \frac{6Fa}{Bt^2}$$

Now the applied moment $\qquad M = F \times R$

$$\therefore \qquad \textbf{maximum bending stress } \boldsymbol{\sigma_{max}} = \boldsymbol{\frac{6Ma}{RBt^2}} \tag{12.32}$$

or, assuming $a = 2R$,

$$\sigma_{max} = \frac{12M}{Bt^2} \tag{12.33}$$

Examples

Example 12.1

A close-coiled helical spring is required to absorb 2.25×10^3 joules of energy. Determine the diameter of the wire, the mean diameter of the spring and the number of coils necessary if:

(a) the maximum stress is not to exceed 400 MN/m^2;
(b) the maximum compression of the spring is limited to 250 mm;
(c) the mean diameter of the spring can be assumed to be eight times that of the wire.

How would the answers change if appropriate Wahl factors are introduced?
For the spring material $G = 70$ GN/m^2.

Solution

The spring is required to absorb 2.25×10^3 joules or 2.25 kN m of energy.

$$\therefore \qquad \text{work done} = \tfrac{1}{2} W\delta = 2.25 \times 10^3$$

But δ is limited to 250 mm.

$$\therefore \quad \tfrac{1}{2}W \times 250 \times 10^{-3} = 2.25 \times 10^{3}$$

$$W = \frac{2.25 \times 10^{3} \times 2}{250 \times 10^{-3}} = 18\,\text{kN}$$

Thus the maximum load which can be carried by the spring is 18 kN.

Now the maximum stress is not to exceed 400 MN/m^2; therefore from eqn. (12.1),

$$\frac{2WR}{\pi r^{3}} = 400 \times 10^{6}$$

But $R = 8r$

$$\therefore \quad \frac{2 \times 18 \times 10^{3} \times 8r}{\pi r^{3}} = 400 \times 10^{6}$$

$$r^{2} = \frac{2 \times 18 \times 10^{3} \times 8}{\pi \times 400 \times 10^{6}} = 229 \times 10^{-6}$$

$$r = 15.1 \times 10^{-3} = 15.1\,\text{mm}$$

The required diameter of the wire, for practical convenience, is, therefore,

$$2 \times 15 = \mathbf{30\,mm}$$

and, since $R = 8r$, the required mean diameter of the coils is

$$8 \times 30 = \mathbf{240\,mm}$$

Now total deflection

$$\delta = \frac{4WR^{3}n}{Gr^{4}} = 250\,\text{mm}$$

$$n = \frac{250 \times 10^{-3} \times 70 \times 10^{9} \times (15 \times 10^{-3})^{4}}{4 \times 18 \times 10^{3} \times (120 \times 10^{-3})^{3}}$$

$$= 7.12$$

Again from practical considerations, the number of complete coils necessary = 7. (If 8 coils were chosen the maximum deflection would exceed 250 mm.)

The effect of introducing Wahl correction factors is determined as follows:
From the given data $C = D/d = 8$ $\therefore$ From Fig. 12.6 $K = 1.184$.

Now $$\tau_{max} = K\left[\frac{8WD}{\pi d^{3}}\right] = K\left[\frac{2WR}{\pi r^{3}}\right] = 400 \times 10^{6}$$

$$\therefore \quad 400 \times 10^{6} = \frac{1.184 \times 2 \times 18 \times 10^{3} \times 8r}{\pi r^{3}}$$

$$\therefore \quad r^{2} = \frac{1.184 \times 2 \times 18 \times 10^{3} \times 8}{\pi \times 400 \times 10^{6}} = 271.35 \times 10^{-6}$$

$$\therefore \quad r = 16.47 \times 10^{-3} = 16.47\,\text{mm}$$

i.e. for practical convenience $d = 2 \times 16.5 =$ **33 mm,**
and since $D = 8d$, $D = 8 \times 33 =$ **264 mm.**

Total deflection
$$\delta = \frac{4WR^3n}{Gr^4} = 250\,\text{mm}.$$

$$\therefore \qquad n = \frac{250 \times 10^{-3} \times 70 \times 10^9 \times (16.5 \times 10^{-3})^4}{4 \times 18 \times 10^3 \times (132 \times 10^{-3})^3}$$

$$= 7.83.$$

Although this is considerably greater than the value obtained before, the number of complete coils required remains at **7** if maximum deflection is strictly limited to 250 mm.

Example 12.2

A compression spring is required to carry a load of 1.5 kN with a limiting shear stress of 250 MN/m^2. If the spring is to be housed in a cylinder of 70 mm diameter estimate the size of spring wire required. Use appropriate Wahl factors in your solution.

Solution

Maximum shear stress
$$\tau_{max} = K\left[\frac{8WD}{\pi d^3}\right]$$

i.e.
$$d^3 = \frac{8WDK}{\pi\tau_{max}}$$

$$\therefore \qquad d = \sqrt[3]{\frac{8 \times 1.5 \times 10^3\,DK}{\pi \times 250 \times 10^6}}$$

$$= 2.481 \times 10^{-2}\sqrt[3]{DK} \qquad (1)$$

Unfortunately, this cannot readily be solved for d since K is dependent on d, and D the mean diameter is not known except so far as its maximum value is limited to $(70 - d)$ mm.

If, therefore, as a first approximation, D is taken to be 70 mm and K is assumed to be 1, a rough order of magnitude is obtained for d from the above equation (1).

i.e.
$$d = 2.481 \times 10^{-2}\,(70 \times 10^{-3} \times 1)^{\frac{1}{3}}$$

$$= 10.22\ \text{mm}.$$

It is now appropriate to apply a graphical solution to the determination of the precise value of d using assumed values of d close to the above rough value, reading the appropriate value of K from Fig. 12.6 and calculating the corresponding d value from eqn. (1).

Assumed d	D $(= 70 - d)$	C $(= D/d)$	K	*Calculated* d (*from eqn.* (1))
10	60	6.0	1.25	10.46
10.5	59.5	5.67	1.27	10.49
11.0	59	5.36	1.29	10.51
11.5	58.5	5.09	1.304	10.522

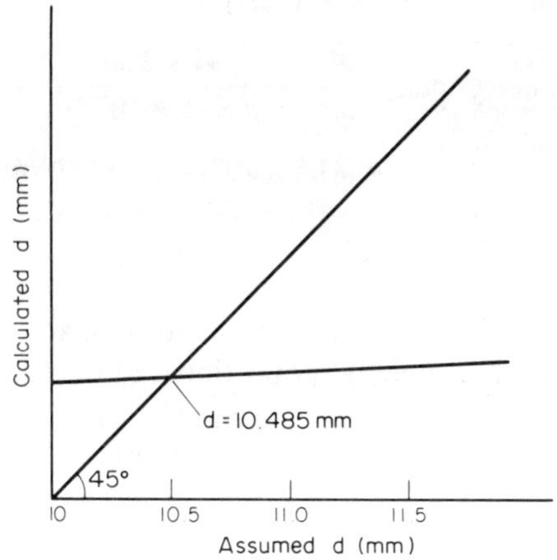

Fig. 12.13.

Plotting the assumed and calculated values gives the nearly horizontal line of Fig. 12.13. The other line is that of the required solution, i.e. it represents all points along which the assumed and calculated d values are the same (i.e. at 45° to the axes). Thus, where this line crosses the previously plotted line is the required value of d, namely 10.485 mm.

The spring wire must therefore have a minimum diameter of **10.485 mm** and a mean diameter of $70 - 10.485 =$ **59.51 mm.**

Example 12.3

A close-coiled helical spring, constructed from wire of 10 mm diameter and with a mean coil diameter of 50 mm, is used to join two shafts which transmit 1 kilowatt of power at 4000 rev/min. If the number of turns of the spring is 10 and the modulus of elasticity of the spring material is 210 GN/m^2 determine:

(a) the relative angle of twist between the two ends of the spring;
(b) the maximum stress set up in the spring material.

Solution

$$\text{Power} = T\omega = 1000\text{ W}$$

$$T = \frac{1000 \times 60}{4000 \times 2\pi} = 2.39\text{ N m}$$

Now the wind-up angle of the spring, from eqn. (12.4),

$$= \frac{8TRn}{Er^4}$$

$$\therefore \qquad \theta = \frac{8 \times 2.39 \times 25 \times 10^{-3} \times 10}{210 \times 10^9 \times (5 \times 10^{-3})^4}$$

$$= 0.036\text{ radian} = \mathbf{2.1^\circ}$$

The maximum stress is then given by eqn. (12.3),

$$\sigma_{max} = \frac{4T}{\pi r^3} = \frac{4 \times 2.39}{\pi \times (5 \times 10^{-3})^3}$$

$$= 24.3 \times 10^6 = \mathbf{24.3\ MN/m^2}$$

Example 12.4

Show that the ratio of extension per unit axial load to angular rotation per unit axial torque of a close-coiled helical spring is directly proportional to the square of the mean diameter, and hence that the constant of proportionality is $\frac{1}{4}(1+\nu)$.

If Poisson's ratio $\nu = 0.3$, determine the angular rotation of a close-coiled helical spring of mean diameter 80 mm when subjected to a torque of 3 N m, given that the spring extends 150 mm under an axial load of 250 N.

Solution

From eqns. (12.2) and (12.4)

$$\delta = \frac{4WR^3n}{Gr^4} \quad \text{and} \quad \theta = \frac{8TRn}{Er^4}$$

$$\therefore \quad \frac{\delta}{W} = \frac{4R^3n}{Gr^4} \quad \text{and} \quad \frac{\theta}{T} = \frac{8Rn}{Er^4}$$

$$\therefore \quad \frac{\delta/W}{\theta/T} = \frac{4R^3n}{Gr^4} \times \frac{Er^4}{8Rn} = \frac{R^2E}{2G} = \frac{D^2E}{8G}$$

But $E = 2G(1+\nu)$

$$\therefore \quad \frac{\delta/W}{\theta/T} = \frac{D^2}{8} \times \frac{2G(1+\nu)}{G} = \tfrac{1}{4}(1+\nu)D^2 \qquad (1)$$

Thus the ratio is directly proportional to D^2 and the constant of proportionality is $\frac{1}{4}(1+\nu)$.

From eqn. (1)
$$\frac{T\delta}{W\theta} = \tfrac{1}{4}(1+\nu)D^2$$

$$\therefore \quad \frac{3 \times 150 \times 10^{-3}}{250 \times \theta} = \tfrac{1}{4}(1+0.3)(80 \times 10^{-3})^2$$

$$\theta = \frac{3 \times 150 \times 10^{-3} \times 4}{250 \times 1.3 \times 6400 \times 10^{-6}}$$

$$= 0.865 \text{ radian} = \mathbf{49.6^\circ}$$

The required angle of rotation is 49.6°.

Example 12.5

(a) Determine the load required to produce an extension of 8 mm on an open coiled helical spring of 10 coils of mean diameter 76 mm, with a helix angle of 20° and manufactured from wire of 6 mm diameter. What will then be the bending and shear stresses in the surface of the wire? For the material of the spring, $E = 210\,\text{GN/m}^2$ and $G = 70\,\text{GN/m}^2$.

(b) What would be the angular twist at the free end of the above spring when subjected to an axial torque of 1.5 N m?

Solution

(a) From eqn. (12.6) the extension of an open-coiled helical spring is given by

$$\delta = 2\pi n W R^3 \sec\alpha \left[\frac{\cos^2\alpha}{GJ} + \frac{\sin^2\alpha}{EI}\right]$$

Now $$I = \frac{\pi d^4}{64} = \frac{\pi \times (6\times 10^{-3})^4}{64} = 63.63 \times 10^{-12}\,\text{m}^4$$

and $$J = \frac{\pi d^4}{32} = 127.26 \times 10^{-12}\,\text{m}^4$$

$$\therefore \quad 8\times 10^{-3} = 2\pi \times 10 \times W \times (38\times 10^{-3})^3 \sec 20^\circ \left[\frac{\cos^2 20^\circ}{70\times 10^9 \times 127.26\times 10^{-12}} + \frac{\sin^2 20^\circ}{210 \times 10^9 \times 63.63 \times 10^{-12}}\right]$$

$$= \frac{20\pi W \times 38^3 \times 10^{-9}}{0.9397}\left[\frac{(0.9397)^2}{8.91} + \frac{(0.342)^2}{13.36}\right]$$

$$= \frac{20\pi W \times 38^3 \times 10^{-9}}{0.9397}[0.1079]$$

$$\therefore \quad W = \frac{8\times 10^{-3}\times 0.9397}{20\pi \times 38^3 \times 10^{-9}\times 0.1079}$$

$$= \mathbf{20\,N}$$

The bending moment acting on the spring is

$$WR\sin\alpha = 20 \times 38 \times 10^{-3} \times 0.342$$

$$= 0.26\,\text{N m}$$

$$\therefore \quad \text{bending stress} = \frac{My}{I} = \frac{0.26 \times 3 \times 10^{-3}}{63.63\times 10^{-12}} = \mathbf{12.3\,MN/m^2}$$

Similarly, the torque on the spring material is

$$WR\cos\alpha = 20 \times 38\times 10^{-3} \times 0.9397$$

$$= 0.714\,\text{N m}$$

$$\therefore \quad \text{shear stress} = \frac{Tr}{J} = \frac{0.714 \times 3 \times 10^{-3}}{127.26 \times 10^{-12}}$$

$$= \mathbf{16.8\,MN/m^2}$$

(b) The wind-up angle of the spring under the action of an axial torque is given by eqn. (12.11):

$$\theta = 2\pi nRT\sec\alpha\left[\frac{\sin^2\alpha}{GJ}+\frac{\cos^2\alpha}{EI}\right]$$

$$= \frac{2\pi \times 10 \times 38 \times 10^{-3} \times 1.5}{0.9397}\left[\frac{(0.342)^2}{8.91}+\frac{(0.9397)^2}{13.36}\right]$$

$$= \frac{2\pi \times 10 \times 38 \times 10^{-3} \times 1.5}{0.9397}[0.0792]$$

$$= 0.302 \text{ radian} = \mathbf{17.3^\circ}$$

Example 12.6

Calculate the thickness and number of leaves of a semi-elliptic carriage spring which is required to support a central load of 2 kN on a span of 1 m if the maximum stress is limited to 225 MN/m^2 and the central deflection to 75 mm. The breadth of each leaf can be assumed to be 100 mm.

For the spring material $E = 210\,\text{GN/m}^2$.

Solution

From eqn. (12.18),

$$\text{maximum stress} = \frac{3WL}{2nbt^2} = 225 \times 10^6$$

$$\therefore \quad \frac{3 \times 2000 \times 1}{2 \times n \times 100 \times 10^{-3} t^2} = 225 \times 10^6$$

$$nt^2 = \frac{3 \times 2000}{2 \times 100 \times 10^{-3} \times 225 \times 10^6} = 0.133 \times 10^{-3}$$

And from eqn. (12.21),

$$\text{Deflection } \delta = \frac{3WL^3}{8Enbt^3}$$

$$\therefore \quad 75 \times 10^{-3} = \frac{3 \times 2000 \times 1}{8 \times 210 \times 10^9 \times n \times 100 \times 10^{-3} \times t^3}$$

$$\therefore \quad nt^3 = \frac{3 \times 2000}{75 \times 10^{-3} \times 8 \times 210 \times 10^8}$$

$$= 0.476 \times 10^{-6}$$

$$\therefore \quad \frac{nt^3}{nt^2} = t = \frac{0.476 \times 10^{-6}}{0.133 \times 10^{-3}}$$

$$t = 3.58 \times 10^{-3} = \mathbf{3.58\ mm}$$

and, since $nt^2 = 0.133 \times 10^{-3}$,

$$n = \frac{0.133 \times 10^{-3}}{(3.58 \times 10^{-3})^2}$$

$$= \mathbf{10.38}$$

The nearest whole number of leaves is therefore 10. However, with $n = 10$, the stress limit would be exceeded and this should be compensated for by increasing the thickness t in the ratio $\sqrt{\left(\frac{10.38}{10}\right)} = 1.02$,

i.e. $$t = \mathbf{3.65\,mm}$$

Example 12.7

A flat spiral spring is pinned at the outer end and a winding couple is applied to a spindle attached at the inner end as shown in Fig. 12.11, with $a = 150\,\text{mm}$, $b = 40\,\text{mm}$ and $R = 75\,\text{mm}$. The material of the spring is rectangular in cross-section, 12 mm wide and 2.5 mm thick, and there are 5 turns. Determine:

(a) the angle through which the spindle turns;
(b) the maximum bending stress produced in the spring material when a torque of 1.5 N m is applied to the winding spindle.

For the spring material, $E = 210\,\text{GN/m}^2$.

Solution

(a)

$$\text{The angle of twist} = \frac{ML}{EI}$$

where $$L = \frac{\pi n}{2}(a+b)$$

$$= \frac{\pi \times 5}{2}(150+40)10^{-3}$$

$$= 1492.3 \times 10^{-3}\,\text{m} = 1.492\,\text{m}$$

$\therefore$ $$\text{angle of twist} = \frac{1.5 \times 1.492 \times 12}{210 \times 10^9 \times 12 \times 2.5^3 \times 10^{-12}}$$

$$= 0.682\,\text{radian}$$

$$= \mathbf{39.1^\circ}$$

$$\text{Maximum bending moment} = F \times a$$

where $$\text{applied moment} = F \times R = 1.5\,\text{N m}$$

i.e. $$F = \frac{1.5}{75 \times 10^{-3}} = 20\,\text{N}$$

$\therefore$ $$\text{maximum bending moment} = 20 \times 150 \times 10^{-3} = 3\,\text{N m}$$

$\therefore$ $$\text{maximum bending stress} = \frac{My}{I} = \frac{3 \times (t/2)}{I}$$

$$= \frac{3 \times 1.25 \times 10^{-3} \times 12}{12 \times 2.5^3 \times 10^{-12}}$$

$$= 240 \times 10^6 = \mathbf{240\,MN/m^2}$$

Problems

(Take $E = 210\,\text{GN/m}^2$ and $G = 70\,\text{GN/m}^2$ throughout)

12.1 (A/B). A close-coiled helical spring is to have a stiffness of 90 kN/m and to exert a force of 3 kN; the mean diameter of the coils is to be 75 mm and the maximum stress is not to exceed 240 MN/m^2. Calculate the required number of coils and the diameter of the steel rod from which the spring should be made. [E.I.E.] [8, 13.5 mm.]

12.2 (A/B). A close-coiled helical spring is fixed at one end and subjected to axial twist at the other. When the spring is in use the axial torque varies from 0.75 N m to 3 N m, the working angular deflection between these torques being 35°. The spring is to be made from rod of circular section, the maximum permissible stress being 150 MN/m^2. The mean diameter of the coils is eight times the rod diameter. Calculate the mean coil diameter, the number of turns and the wire diameter. [B.P.] [48, 6 mm; 24.]

12.3 (A/B). A close-coiled helical compression spring made from round wire fits over the spindle of a plunger and has to work inside a tube. The spindle diameter is 12 mm and the tube is of 25 mm outside diameter and 0.15 mm thickness. The maximum working length of the spring has to be 120 mm and the minimum length 90 mm. The maximum force exerted by the spring has to be 350 N and the minimum force 240 N. If the shearing stress in the spring is not to exceed 600 MN/m^2 find:
(a) the free length of the spring (i.e. before assembly);
(b) the mean coil diameter;
(c) the wire diameter;
(d) the number of free coils. [185.4, 18.3, 3 mm; 32.]

12.4 (A/B). A close-coiled helical spring of circular wire and mean diameter 100 mm was found to extend 45 mm under an axial load of 50 N. The same spring when firmly fixed at one end was found to rotate through 90° under a torque of 5.7 N m. Calculate the value of Poisson's ratio for the material. [C.U.] [0.3.]

12.5 (B). Show that the total strain energy stored in an open-coiled helical spring by an axial load W applied together with an axial couple T is

$$U = (T\cos\theta - WR\sin\theta)^2\frac{L}{2EI} + (WR\cos\theta + T\sin\theta)^2\frac{L}{2GJ}$$

where θ is the helix angle and L the total length of wire in the spring, and the sense of the couple is in a direction tending to wind up the spring. Hence, or otherwise, determine the rotation of one end of a spring of helix angle 20° having 10 turns of mean radius 50 mm when an axial load of 25 N is applied, the other end of the spring being securely fixed. The diameter of the wire is 6 mm. [B.P.] [26°.]

12.6 (B). Deduce an expression for the extension of an open-coiled helical spring carrying an axial load W. Take α as the inclination of the coils, d as the diameter of the wire and R as the mean radius of the coils. Find by what percentage the axial extension is underestimated if the inclination of the coils is neglected for a spring in which $\alpha = 25°$. Assume n and R remain constant. [U.L.] [3.6%.]

12.7 (B). An open-coiled spring carries an axial vertical load W. Derive expressions for the vertical displacement and angular twist of the free end. Find the mean radius of an open-coiled spring (angle of helix 30°) to give a vertical displacement of 23 mm and an angular rotation of the loaded end of 0.02 radian under an axial load of 40 N. The material available is steel rod of 6 mm diameter. [U.L.] [182 mm.]

12.8 (B). A compound spring comprises two close-coiled helical springs having exactly the same initial length when unloaded. The outer spring has 16 coils of 12 mm diameter bar coiled to a mean diameter of 125 mm and the inner spring has 24 coils with a mean diameter of 75 mm. The working stress in each spring is to be the same. Find (a) the diameter of the steel bar for the inner spring and (b) the stiffness of the compound spring.
[I.Mech.E.] [6.48 mm; 7.33 kN/m.]

12.9 (B). A composite spring has two close-coiled helical springs connected in series; each spring has 12 coils at a mean diameter of 25 mm. Find the diameter of the wire in one of the springs if the diameter of wire in the other spring is 2.5 mm and the stiffness of the composite spring is 700 N/m. Estimate the greatest load that can be carried by the composite spring and the corresponding extension for a maximum shearing stress of 180 MN/m^2.
[U.L.] [44.2 N; 63.2 mm.]

12.10 (B). (a) Derive formulae in terms of load, leaf width and thickness, and number of leaves for the maximum deflection and maximum stress induced in a cantilever leaf spring. (b) A cantilever leaf spring is 750 mm long and the leaf width is to be 8 times the leaf thickness. If the bending stress is not to exceed 210 MN/m^2 and the spring is not to deflect more than 50 mm under a load of 5 kN, find the leaf thickness, the least number of leaves required, the deflection and the stress induced in the leaves of the spring.
[11.25 mm, say 12 mm; 9.4, say 10; 47 mm, 197.5 MN/m^2.]

12.11 (B). Make a sketch of a leaf spring showing the shape to which the ends of the plate should be made and give the reasons for doing this. A leaf spring which carries a central load of 9 kN consists of plates each 75 mm wide and 7 mm thick. If the length of the spring is 1 m, determine the least number of plates required if the maximum stress owing to bending is limited to 210 MN/m^2 and the maximum deflection must not exceed 30 mm. Find, for the number of plates obtained, the actual values of the maximum stress and maximum deflection and also the radius to which the plates should be formed if they are to straighten under the given load.
[U.L.] [14; 200 MN/m^2, 29.98 mm; 4.2 m.]

12.12 (B). A semi-elliptic laminated carriage spring is 1 m long and 75 mm wide with leaves 10 mm thick. It has to carry a central load of 6 kN with a deflection of 25 mm. Working from first principles find (a) the number of leaves, (b) the maximum induced stress. [6; 200 MN/m^2.]

12.13 (B). A semi-elliptic leaf spring has a span of 720 mm and is built up of leaves 10 mm thick and 45 mm wide. Find the number of leaves required to carry a load of 5 kN at mid-span if the stress is not to exceed 225 MN/m^2, nor the deflection 12 mm. Calculate also the radius of curvature to which the spring must be initially bent if it must just flatten under the application of the above load. [7; 6.17 m.]

12.14 (B) An open-coiled helical spring has 10 coils of 12 mm diameter steel bar wound with a mean diameter of 150 mm. The helix angle of the coils is 32°. Find the axial extension produced by a load of 250 N. Any formulae used must be established by the application of fundamental principles relating to this type of spring.
[U.L.] [49.7 mm.]

12.15 (B). An open-coiled spring carries an axial load W. Show that the deflection is related to W by

$$\delta = \frac{8WnD^3}{Gd^4} \times K$$

where K is a correction factor which allows for the inclination of the coils, n = number of effective coils, D = mean coil diameter, and d = wire diameter.

A close-coiled helical spring is wound from 6 mm diameter steel wire into a coil having a mean diameter of 50 mm. If the spring has 20 effective turns and the maximum shearing stress is limited to 225 MN/m^2, what is the greatest safe deflection obtainable? [U.Birm.] [84.2 mm.]

12.16 (B/C). A flat spiral spring, as shown in Fig. 12.11, has the following dimensions: a = 150 mm, b = 25 mm, R = 80 mm. Determine the maximum value of the moment which can be applied to the spindle if the bending stress in the spring is not to exceed 150 MN/m^2. Through what angle does the spindle turn in producing this stress? The spring is constructed from steel strip 25 mm wide × 1.5 mm thick and has six turns. [0.75 N m, 48°.]

12.17 (B/C). A strip of steel of length 6 m, width 12 mm and thickness 2.5 mm is formed into a flat spiral around a spindle, the other end being attached to a fixed pin. Determine the couple which can be applied to the spindle if the maximum stress in the steel is limited to 300 MN/m^2. What will then be the energy stored in the spring?
[1.875 N m, 3.2 J.]

12.18 (B/C). A flat spiral spring is 12 mm wide, 0.3 mm thick and 2.5 m long. Assuming the maximum stress of 900 MN/m^2 to occur at the point of greatest bending moment, calculate the torque, the work stored and the number of turns to wind up the spring. [U.L.] [0.081, 1.45 J; 5.68.]

CHAPTER 13

COMPLEX STRESSES

Summary

The normal stress σ and shear stress τ on oblique planes resulting from direct loading are

$$\sigma = \sigma_y \sin^2\theta \quad \text{and} \quad \tau = \tfrac{1}{2}\sigma_y \sin 2\theta$$

The stresses on oblique planes owing to a complex stress system are:

$$\text{normal stress} = \tfrac{1}{2}(\sigma_x+\sigma_y)+\tfrac{1}{2}(\sigma_x-\sigma_y)\cos 2\theta+\tau_{xy}\sin 2\theta$$

$$\text{shear stress} = \tfrac{1}{2}(\sigma_x-\sigma_y)\sin 2\theta-\tau_{xy}\cos 2\theta$$

The *principal stresses* (i.e. the maximum and minimum direct stresses) are then

$$\sigma_1 = \tfrac{1}{2}(\sigma_x+\sigma_y)+\tfrac{1}{2}\sqrt{[(\sigma_x-\sigma_y)^2+4\tau_{xy}^2]}$$

$$\sigma_2 = \tfrac{1}{2}(\sigma_x+\sigma_y)-\tfrac{1}{2}\sqrt{[(\sigma_x-\sigma_y)^2+4\tau_{xy}^2]}$$

and these occur on planes at an angle θ to the plane on which σ_x acts, given by either

$$\tan 2\theta = \frac{2\tau_{xy}}{(\sigma_x-\sigma_y)} \quad \text{or} \quad \tan\theta = \frac{\sigma_p-\sigma_x}{\tau_{xy}}$$

where $\sigma_p = \sigma_1$, or σ_2, the planes being termed *principal planes.* The principal planes are always at 90° to each other, and the *planes of maximum shear* are then located at 45° to them.

The *maximum shear stress* is

$$\tau_{max} = \tfrac{1}{2}\sqrt{[(\sigma_x-\sigma_y)^2+4\tau_{xy}^2]} = \tfrac{1}{2}(\sigma_1-\sigma_2)$$

In problems where the principal stress in the third dimension σ_3 either is known or can be assumed to be zero, the true maximum shear stress is then

$$\tfrac{1}{2}(\text{greatest principal stress} - \text{least principal stress})$$

$$\text{Normal stress on plane of maximum shear} = \tfrac{1}{2}(\sigma_x+\sigma_y)$$

$$\text{Shear stress on plane of maximum direct stress (principal plane)} = 0$$

Most problems can be solved graphically by *Mohr's stress circle.* All questions which are capable of solution by this method have been solved both analytically and graphically.

13.1. Stresses on oblique planes

Consider the general case, shown in Fig. 13.1, of a bar under direct load F giving rise to stress σ_y vertically.

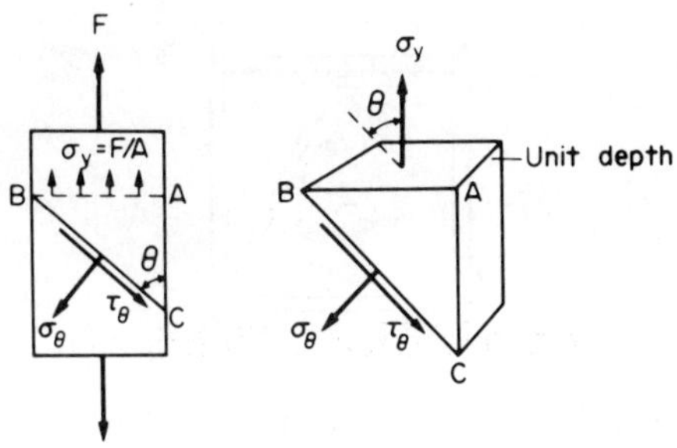

Fig. 13.1. Bar subjected to direct stress, showing stresses acting on any inclined plane.

Let the block be of *unit depth*; then considering the equilibrium of *forces* on the triangular portion ABC:

resolving forces perpendicular to BC,

$$\sigma_\theta \times BC \times 1 = \sigma_y \times AB \times 1 \times \sin\theta$$

But $AB = BC \sin\theta$,

$$\therefore \qquad \boldsymbol{\sigma_\theta = \sigma_y \sin^2\theta} \qquad (13.1)$$

Now resolving forces parallel to BC,

$$\tau_\theta \times BC \times 1 = \sigma_y \times AB \times 1 \times \cos\theta$$

Again $AB = BC \sin\theta$,

$$\therefore \qquad \tau_\theta = \sigma_y \sin\theta \cos\theta$$

$$= \tfrac{1}{2}\boldsymbol{\sigma_y \sin 2\theta} \qquad (13.2)$$

The stresses on the inclined plane, therefore, are not simply the resolutions of σ_y perpendicular and tangential to that plane. The direct stress σ_θ has a maximum value of σ_y when $\theta = 90°$ whilst the shear stress τ_θ has a maximum value of $\frac{1}{2}\sigma_y$ when $\theta = 45°$.

Thus any material whose yield stress in shear is less than half that in tension or compression will yield initially in shear under the action of direct tensile or compressive forces.

This is evidenced by the typical "cup and cone" type failure in tension tests of ductile specimens such as low carbon steel where failure occurs initially on planes at 45° to the specimen axis. Similar effects occur in compression tests on, for example, timber where failure is again due to the development of critical shear stresses on 45° planes.

13.2. Material subjected to pure shear

Consider the element shown in Fig. 13.2 to which shear stresses have been applied to the sides AB and DC. *Complementary shear stresses* of equal value but of opposite effect are then set up on sides AD and BC in order to prevent rotation of the element. Since the applied and complementary shears are of equal value on the x and y planes, they are both given the symbol τ_{xy}.

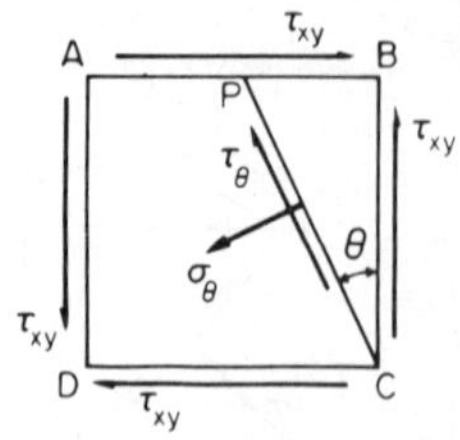

Fig. 13.2. Stresses on an element subjected to pure shear.

Consider now the equilibrium of portion PBC.
Resolving normal to PC *assuming unit depth,*

$$\sigma_\theta \times PC = \tau_{xy} \times BC \sin\theta + \tau_{xy} \times PB \cos\theta$$

$$= \tau_{xy} \times PC \cos\theta \sin\theta + \tau_{xy} \times PC \sin\theta \cos\theta$$

$$\therefore \qquad \boldsymbol{\sigma_\theta = \tau_{xy} \sin 2\theta} \qquad (13.3)$$

The maximum value of σ_θ is τ_{xy} when $\theta = 45°$.
Similarly, resolving forces parallel to PC,

$$\tau_\theta \times PC = \tau_{xy} \times PB \sin\theta - \tau_{xy} BC \cos\theta$$

$$= \tau_{xy} \times PC \sin^2\theta - \tau_{xy} \times PC \cos^2\theta$$

$$\therefore \qquad \boldsymbol{\tau_\theta = -\tau_{xy} \cos 2\theta} \qquad (13.4)$$

The negative sign means that the sense of τ_θ *is opposite to that assumed in Fig.* 13.2.

The maximum value of τ_θ is τ_{xy} when $\theta = 0°$ or $90°$ and it has a value of zero when $\theta = 45°$, i.e. on the planes of maximum direct stress.

Further consideration of eqns. (13.3) and (13.4) shows that the system of pure shear stresses produces an equivalent direct stress system as shown in Fig. 13.3, one set compressive and one tensile, each at 45° to the original shear directions, and equal in magnitude to the applied shear.

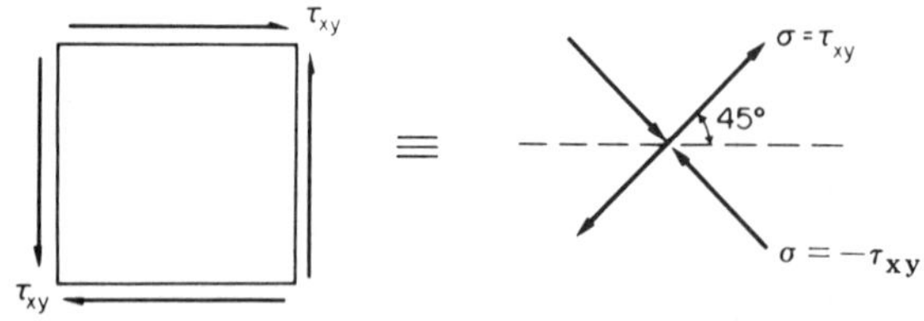

Fig. 13.3. Direct stresses due to shear.

This has great significance in the measurement of shear stresses or torques on shafts using strain gauges where the gauges are arranged to record the direct strains at 45° *to the shaft axis (see page* 609).

Practical evidence of the theory is also provided by the failure of brittle materials in shear. A shaft of a brittle material subjected to torsion will fail under direct stress on planes at 45° to the shaft axis. (This can be demonstrated easily by twisting a piece of blackboard chalk in one's hands; see Fig. 8.8a on page 185.) Tearing of a wet cloth when it is being wrung out is also attributed to the direct stresses introduced by the applied torsion.

13.3. Material subjected to two mutually perpendicular direct stresses

Consider the rectangular element of *unit depth* shown in Fig. 13.4 subjected to a system of two direct stresses, both tensile, at right angles, σ_x and σ_y.

For equilibrium of the portion ABC, resolving perpendicular to AC,

$$\sigma_\theta \times AC \times 1 = \sigma_x \times BC \times 1 \times \cos\theta + \sigma_y \times AB \times 1 \times \sin\theta$$

$$= \sigma_x \times AC \cos^2\theta + \sigma_y \times AC \sin^2\theta$$

$$\therefore \quad \sigma_\theta = \tfrac{1}{2}\sigma_x(1+\cos 2\theta) + \tfrac{1}{2}\sigma_y(1-\cos 2\theta)$$

i.e.

$$\boldsymbol{\sigma_\theta = \tfrac{1}{2}(\sigma_x+\sigma_y) + \tfrac{1}{2}(\sigma_x-\sigma_y)\cos 2\theta} \tag{13.5}$$

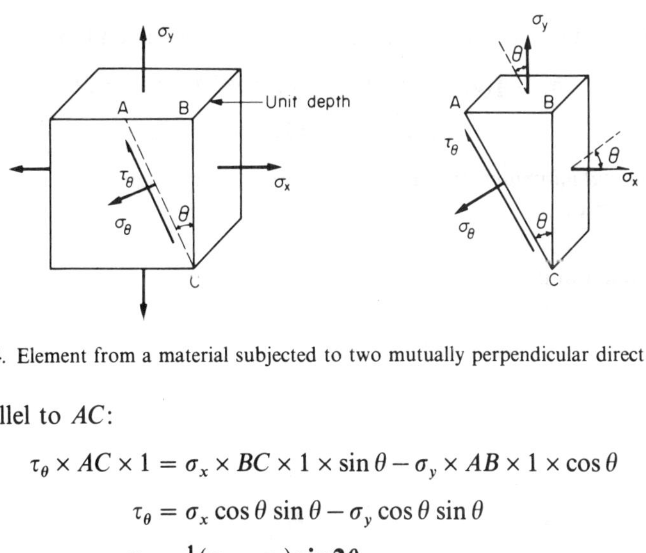

Fig. 13.4. Element from a material subjected to two mutually perpendicular direct stresses.

Resolving parallel to AC:

$$\tau_\theta \times AC \times 1 = \sigma_x \times BC \times 1 \times \sin\theta - \sigma_y \times AB \times 1 \times \cos\theta$$

$$\tau_\theta = \sigma_x \cos\theta \sin\theta - \sigma_y \cos\theta \sin\theta$$

$$\therefore \quad \boldsymbol{\tau_\theta = \tfrac{1}{2}(\sigma_x - \sigma_y)\sin 2\theta} \tag{13.6}$$

The maximum direct stress will equal σ_x or σ_y, whichever is the greater, when $\theta = 0$ or $90°$.

The maximum shear stress *in the plane of the applied stresses* (see §13.8) occurs when $\theta = 45°$,

i.e.

$$\boldsymbol{\tau_{max} = \tfrac{1}{2}(\sigma_x - \sigma_y)} \tag{13.7}$$

13.4. Material subjected to combined direct and shear stresses

Consider the complex stress system shown in Fig. 13.5 acting on an element of material.

The stresses σ_x and σ_y may be compressive or tensile and may be the result of direct forces or bending. The shear stresses may be as shown or completely reversed and occur as a result of either shear forces or torsion.

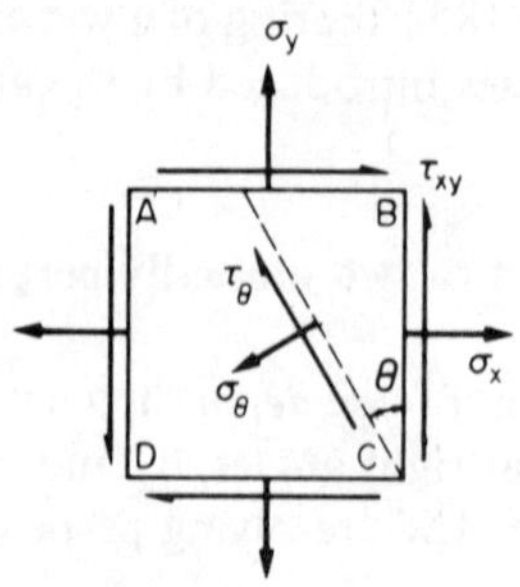

Fig. 13.5. Two-dimensional complex stress system.

The diagram thus represents a complete stress system for any condition of applied load in two dimensions and represents an addition of the stress systems previously considered in §§13.2 and 13.3.

The formulae obtained in these sections may therefore be combined to give

$$\boldsymbol{\sigma_\theta = \tfrac{1}{2}(\sigma_x + \sigma_y) + \tfrac{1}{2}(\sigma_x - \sigma_y)\cos 2\theta + \tau_{xy}\sin 2\theta} \qquad (13.8)$$

and

$$\boldsymbol{\tau_\theta = \tfrac{1}{2}(\sigma_x - \sigma_y)\sin 2\theta - \tau_{xy}\cos 2\theta} \qquad (13.9)$$

The *maximum and minimum stresses* which occur on any plane in the material can now be determined as follows:

For σ_θ to be a maximum or minimum $\dfrac{d\sigma_\theta}{d\theta} = 0$

Now

$$\sigma_\theta = \tfrac{1}{2}(\sigma_x + \sigma_y) + \tfrac{1}{2}(\sigma_x - \sigma_y)\cos 2\theta + \tau_{xy}\sin 2\theta$$

$\therefore$

$$\frac{d\sigma_\theta}{d\theta} = -(\sigma_x - \sigma_y)\sin 2\theta + 2\tau_{xy}\cos 2\theta = 0$$

or

$$\boldsymbol{\tan 2\theta = \frac{2\tau_{xy}}{(\sigma_x - \sigma_y)}} \qquad (13.10)$$

$\therefore$ from Fig. 13.6

$$\sin 2\theta = \frac{2\tau_{xy}}{\sqrt{[(\sigma_x - \sigma_y)^2 + 4\tau_{xy}^2]}}$$

$$\cos 2\theta = \frac{(\sigma_x - \sigma_y)}{\sqrt{[(\sigma_x - \sigma_y)^2 + 4\tau_{xy}^2]}}$$

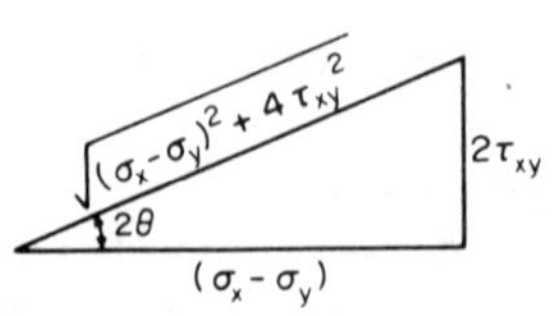

Fig. 13.6.

Therefore substituting in eqn. (13.8), the maximum and minimum direct stresses are given by

$$\sigma_1 \text{ or } \sigma_2 = \tfrac{1}{2}(\sigma_x+\sigma_y)+\tfrac{1}{2}\frac{(\sigma_x-\sigma_y)(\sigma_x-\sigma_y)}{\sqrt{[(\sigma_x-\sigma_y)^2+4\tau_{xy}^2]}}+\frac{\tau_{xy}\times 2\tau_{xy}}{\sqrt{[(\sigma_x-\sigma_y)^2+4\tau_{xy}^2]}}$$

$$= \tfrac{1}{2}(\sigma_x+\sigma_y)\pm\tfrac{1}{2}\sqrt{[(\sigma_x-\sigma_y)^2+4\tau_{xy}^2]} \quad (13.11)$$

These are then termed the *principal stresses* of the system.

The solution of eqn. (13.10) yields two values of 2θ separated by 180°, i.e. two values of θ separated by 90°. Thus the two principal stresses occur on mutually perpendicular planes termed *principal planes*, and substitution for θ from eqn. (13.10) into the shear stress expression eqn. (13.9) will show that $\tau_\theta = 0$ on the principal planes.

The complex stress system of Fig. 13.5 can now be reduced to the equivalent system of principal stresses shown in Fig. 13.7.

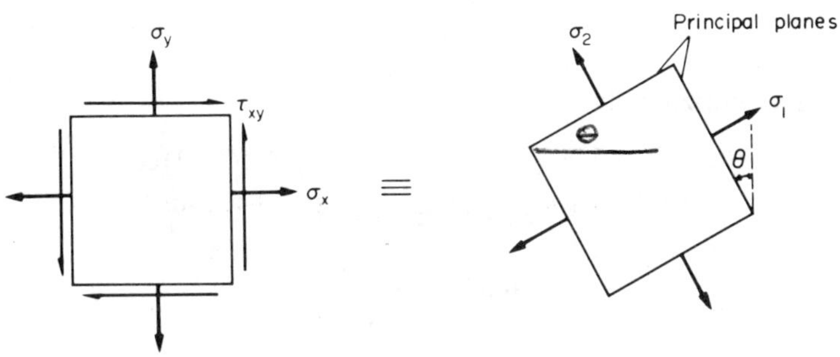

Fig. 13.7. Principal planes and stresses.

From eqn. (13.7) the maximum shear stress present in the system is given by

$$\tau_{max} = \tfrac{1}{2}(\sigma_1-\sigma_2) \quad (13.12)$$

$$= \tfrac{1}{2}\sqrt{[(\sigma_x-\sigma_y)^2+4\tau_{xy}^2]} \quad (13.13)$$

and this occurs **on planes at 45° to the principal planes.**

This result could have been obtained using a similar procedure to that used for determining the principal stresses, i.e. by differentiating expression (13.9), equating to zero and substituting the resulting expression for θ.

13.5. Principal plane inclination in terms of the associated principal stress

It has been stated in the previous section that expression (13.10), namely

$$\tan 2\theta = \frac{2\tau_{xy}}{(\sigma_x-\sigma_y)}$$

yields two values of θ, i.e. the inclination of the two principal planes on which the principal stresses σ_1 and σ_2 act. It is uncertain, however, which stress acts on which plane unless eqn. (13.8) is used, substituting *one* value of θ obtained from eqn. (13.10) and observing which one of the two principal stresses is obtained. The following alternative solution is therefore to be preferred.

Consider once again the equilibrium of a triangular block of material of unit depth (Fig. 13.8); this time *AC* is a principal plane on which a principal stress σ_p acts, and the shear stress is zero (from the property of principal planes).

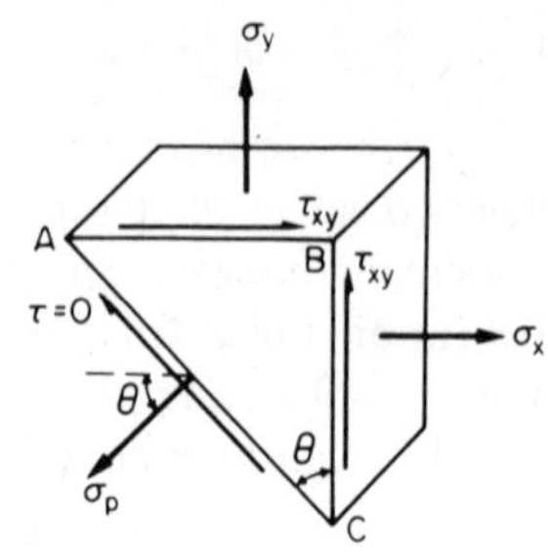

Fig. 13.8.

Resolving forces horizontally,

$$(\sigma_x \times BC \times 1) + (\tau_{xy} \times AB \times 1) = (\sigma_p \times AC \times 1)\cos\theta$$

$$\sigma_x + \tau_{xy}\tan\theta = \sigma_p$$

$$\therefore \qquad \tan\theta = \frac{\sigma_p - \sigma_x}{\tau_{xy}} \qquad (13.14)$$

Thus we have an equation for the inclination of the principal planes *in terms of the principal stress*. If, therefore, the principal stresses are determined and substituted in the above equation, each will give the corresponding angle of the plane on which it acts and there can then be no confusion.

The above formula has been derived with two tensile direct stresses and a shear stress system, as shown in the figure; should any of these be reversed in action, then the appropriate minus sign must be inserted in the equation.

13.6. Graphical solution – Mohr's stress circle

Consider the complex stress system of Fig. 13.5 (p. 330). As stated previously this represents a complete stress system for any condition of applied load in two dimensions.

In order to find graphically the direct stress σ_θ and shear stress τ_θ on any plane inclined at θ to the plane on which σ_x acts, proceed as follows:

(1) Label the block *ABCD*.

(2) Set up axes for direct stress (as abscissa) and shear stress (as ordinate) (Fig. 13.9).

(3) Plot the stresses acting on two *adjacent* faces, e.g. *AB* and *BC*, using the following sign conventions:

direct stresses: tensile, positive; compressive, negative;

shear stresses: tending to turn block clockwise, positive; tending to turn block counterclockwise, negative.

This gives two points on the graph which may then be labelled $\overline{AB}$ and $\overline{BC}$ respectively to denote stresses on these planes.

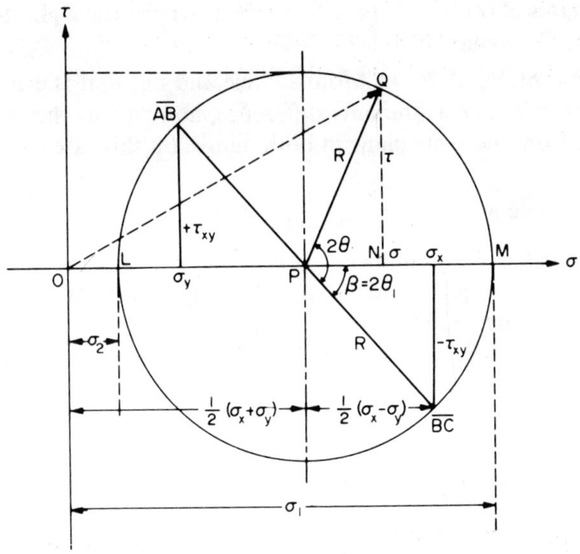

Fig. 13.9. Mohr's stress circle.

(4) Join $\overline{AB}$ and $\overline{BC}$.

(5) The point P where this line cuts the σ axis is then the centre of Mohr's circle, and the line is the diameter; therefore the circle can now be drawn.

Every point on the circumference of the circle then represents a state of stress on some plane through C.

Proof

Consider any point Q on the circumference of the circle, such that PQ makes an angle 2θ with $\overline{BC}$, and drop a perpendicular from Q to meet the σ axis at N.

Coordinates of Q:

$$ON = OP + PN = \tfrac{1}{2}(\sigma_x + \sigma_y) + R\cos(2\theta - \beta)$$

$$= \tfrac{1}{2}(\sigma_x + \sigma_y) + R\cos 2\theta\cos\beta + R\sin 2\theta\sin\beta$$

But $$R\cos\beta = \tfrac{1}{2}(\sigma_x - \sigma_y) \quad \text{and} \quad R\sin\beta = \tau_{xy}$$

$\therefore$ $$ON = \tfrac{1}{2}(\sigma_x + \sigma_y) + \tfrac{1}{2}(\sigma_x - \sigma_y)\cos 2\theta + \tau_{xy}\sin 2\theta$$

On inspection this is seen to be eqn. (13.8) for the direct stress σ_θ on the plane inclined at θ to BC in Fig. 13.5.

Similarly,

$$QN = R\sin(2\theta - \beta)$$

$$= R\sin 2\theta\cos\beta - R\cos 2\theta\sin\beta$$

$$= \tfrac{1}{2}(\sigma_x - \sigma_y)\sin 2\theta - \tau_{xy}\cos 2\theta$$

Again, on inspection this is seen to be eqn. (13.9) for the shear stress τ_θ on the plane inclined at θ to BC.

Thus the coordinates of Q are the normal and shear stresses on a plane inclined at θ to BC in the original stress system.

N.B.–Single angle $\overline{BCPQ}$ is 2θ on Mohr's circle and not θ, it is evident that *angles are doubled on Mohr's circle.* This is the only difference, however, as they are measured in the same direction and from the same plane in both figures (in this case counterclockwise from $\overline{BC}$).

Further points to note are:

(1) The direct stress is a maximum when Q is at M, i.e. OM is the length representing the maximum principal stress σ_1 and $2\theta_1$ gives the angle of the plane θ_1 from BC. Similarly, OL is the other principal stress.

(2) The maximum shear stress is given by the highest point on the circle and is represented by the radius of the circle. This follows since shear stresses and complementary shear stresses have the same value; *therefore the centre of the circle will always lie on the σ axis midway between σ_x and σ_y.*

(3) From the above point the direct stress on the plane of maximum shear must be midway between σ_x and σ_y, i.e. $\frac{1}{2}(\sigma_x + \sigma_y)$.

(4) The shear stress on the principal planes is zero.

(5) Since the resultant of two stresses at 90° can be found from the parallelogram of vectors as the diagonal, as shown in Fig. 13.10, the resultant stress on the plane at θ to BC is given by OQ on Mohr's circle.

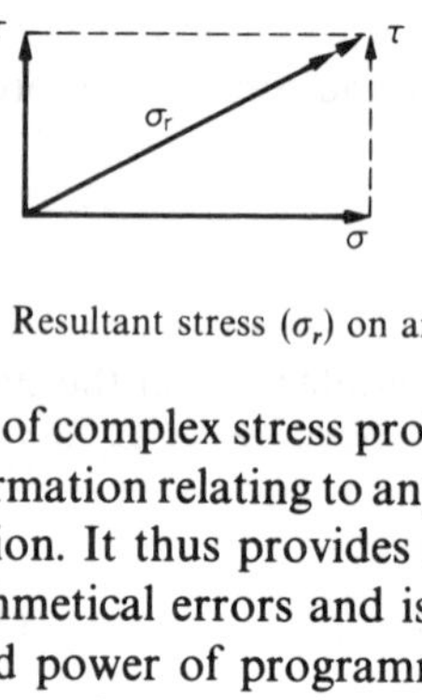

Fig. 13.10. Resultant stress (σ_r) on any plane.

The graphical method of solution of complex stress problems using Mohr's circle is a very powerful technique since all the information relating to any plane within the stressed element is contained in the single construction. It thus provides a convenient and rapid means of solution which is less prone to arithmetical errors and is highly recommended.

With the growing availability and power of programmable calculators and microcomputers it may be that the practical use of Mohr's circle for the analytical determination of stress (and strain–see Chapter 14) values will become limited. It will remain, however, a highly effective medium for the teaching and understanding of complex stress systems.

A free-hand sketch of the Mohr circle construction, for example, provides a convenient mechanism for the derivation (by simple geometric relationships) of the principal stress equations (13.11) or of the equations for the shear and normal stresses on any inclined plane in terms of the principal stresses as shown in Fig. 13.11.

13.7. Alternative representations of stress distributions at a point

The way in which the stress at a point varies with the angle at which a plane is taken through the point may be better understood with the aid of the following alternative graphical representations.

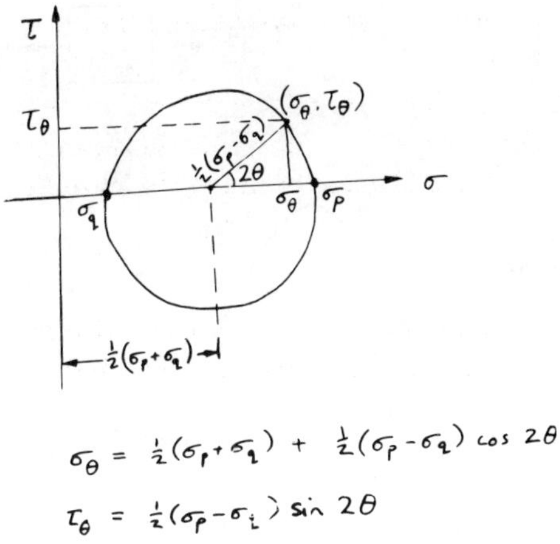

Fig. 13.11. Free-hand sketch of Mohr's stress circle.

Equations (13.8) and (13.9) give the values of the direct stress σ_θ and shear stress τ_θ on any plane inclined at an angle θ to the plane on which the direct stress σ_x acts within a two-dimensional complex stress system, viz:

$$\sigma_\theta = \tfrac{1}{2}(\sigma_x + \sigma_y) + \tfrac{1}{2}(\sigma_x - \sigma_y)\cos 2\theta + \tau_{xy}\sin 2\theta$$

$$\tau_\theta = \tfrac{1}{2}(\sigma_x - \sigma_y)\sin 2\theta - \tau_{xy}\cos 2\theta$$

(a) *Uniaxial stresses*

For the special case of a single uniaxial stress σ_x as in simple tension or on the surface of a beam in bending, $\sigma_y = \tau_{xy} = 0$ and the equations (13.8) and (13.9) reduce to

$$\sigma_\theta = \tfrac{1}{2}\sigma_x(1 + \cos 2\theta) = \sigma_x \cos^2\theta.$$

N.B. If the single stress were selected as σ_y then the relationship would have reduced to that of eqn. (13.1), i.e.

$$\sigma_\theta = \sigma_y \sin^2\theta.$$

Similarly:

$$\tau_\theta = \tfrac{1}{2}\sigma_x \sin 2\theta.$$

Plotting these equations on simple Cartesian axes produces the stress distribution diagrams of Fig. 13.12, both sinusoidal in shape with shear stress "shifted" by 45° from the normal stress.

Principal stresses σ_p and σ_q occur, as expected, at 90° intervals and the amplitude of the normal stress curve is given by the difference between the principal stress values. It should also be noted that shear stress is proportional to the derivative of the normal stress with respect to θ, i.e. τ_θ is a maximum where $d\sigma_\theta/d\theta$ is a maximum and τ_θ is zero where $d\sigma_\theta/d\theta$ is zero, etc.

Alternatively, plotting the same equations on polar graph paper, as in Fig. 13.13, gives an even more readily understood pictorial representation of the stress distributions showing a peak value of direct stress in the direction of application of the applied stress σ_x falling to zero

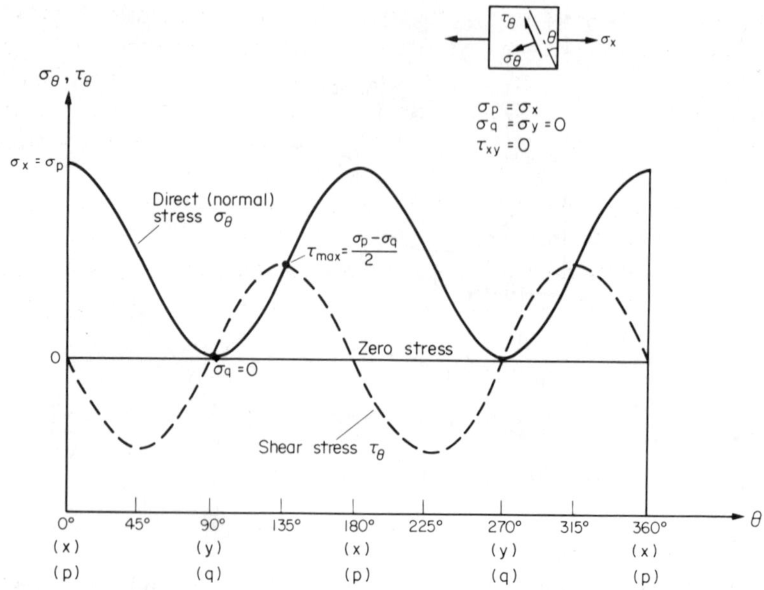

Fig. 13.12. Cartesian plot of stress distribution at a point under uniaxial applied stress.

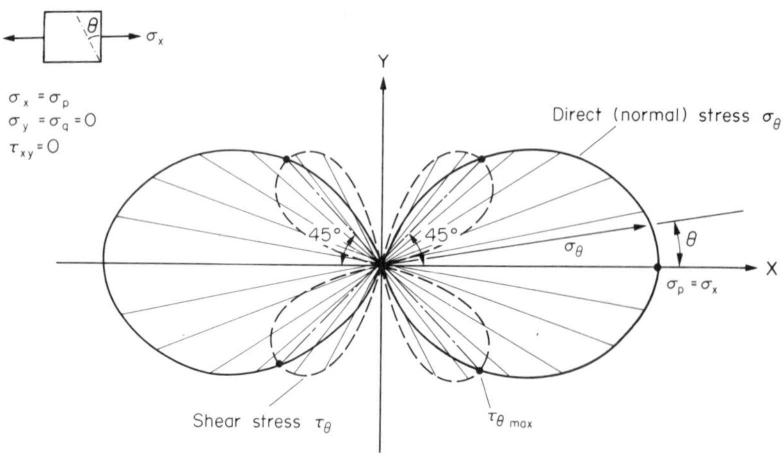

Fig. 13.13. Polar plot of stress distribution at a point under uniaxial applied stress.

in directions at right angles and maximum shearing stresses on planes at 45° with zero shear on the x and y (principal) axes.

(b) *Biaxial stresses*

In almost all modes of loading on structural members or engineering components the stresses produced are a maximum at the free (outside) surface. This is particularly evident for the cases of pure bending or torsion as shown by the stress diagrams of Figs. 4.4 and 8.4, respectively, but is also true for other more complex combined loading situations with the major exception of direct bearing loads where maximum stress conditions can be sub-surface. Additionally, at free surfaces the stress normal to the surface is always zero so that the most severe stress condition often reduces, at worst, to a two-dimensional plane stress system within the surface of the component. It should be evident, therefore that the biaxial stress system is of considerable importance to practical design considerations.

The Cartesian plot of a typical bi-axial stress state is shown in Fig. 13.14 whilst Fig. 13.15 shows the polar plot of stresses resulting from the bi-axial stress system present on the surface of a thin cylindrical pressure vessel for which $\sigma_p = \sigma_H$ and $\sigma_q = \sigma_L = \frac{1}{2}\sigma_H$ with $\tau_{xy} = 0$.

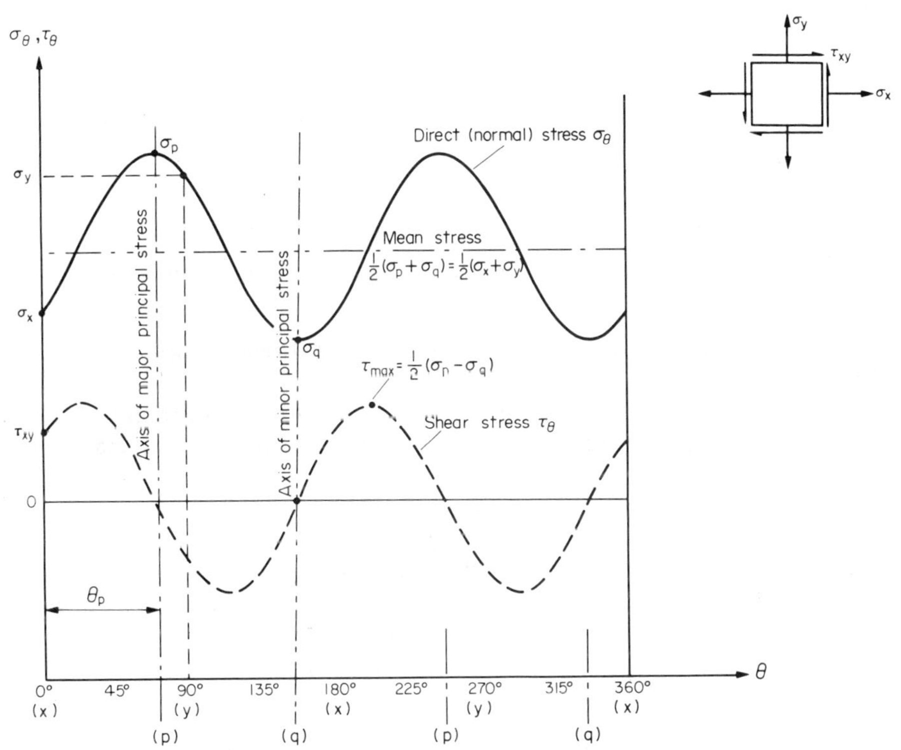

Fig. 13.14. Cartesian plot of stress distribution at a point under a typical biaxial applied stress system.

It should be noted that the whole of the information conveyed on these alternative representations is also available from the relevant Mohr circle which, additionally, is more amenable to quantitative analysis. They do not, therefore, replace Mohr's circle but are included merely to provide alternative pictorial representations which may aid a clearer understanding of the general problem of stress distribution at a point. The equivalent diagrams for strain are given in §14.16.

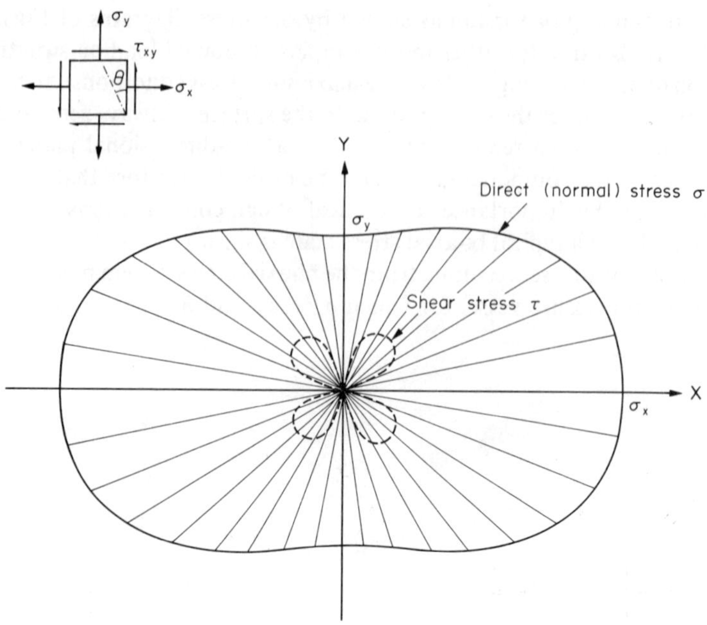

Fig. 13.15. Polar plot of stress distribution under typical biaxial applied stress system.

13.8. Three-dimensional stresses – graphical representation

Figure 13.16 shows the general *three-dimensional* state of stress at any point in a body, i.e. the body will be subjected to three mutually perpendicular direct stresses and three shear stresses. (See Chapter 23.)

Figure 13.17 shows a *principal element* at the same point, i.e. one in general rotated relative to the first until the stresses on the faces are principal stresses with no associated shear.

Figure 13.18 then represents true views on the various faces of the principal element, and for each two-dimensional stress condition so obtained a Mohr circle may be drawn. These

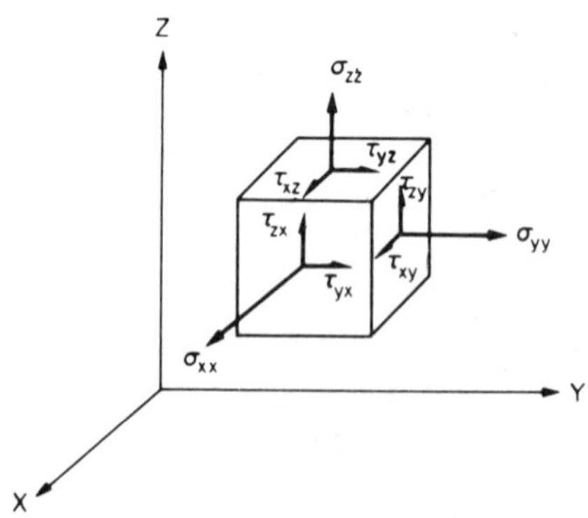

Fig. 13.16. Three-dimensional stress system.

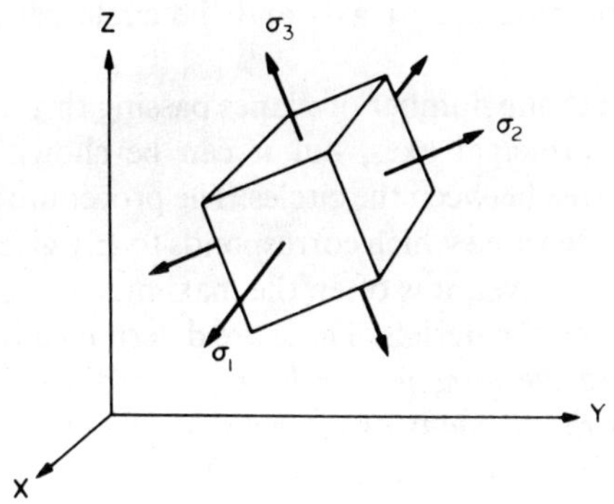

Fig. 13.17. Principal element.

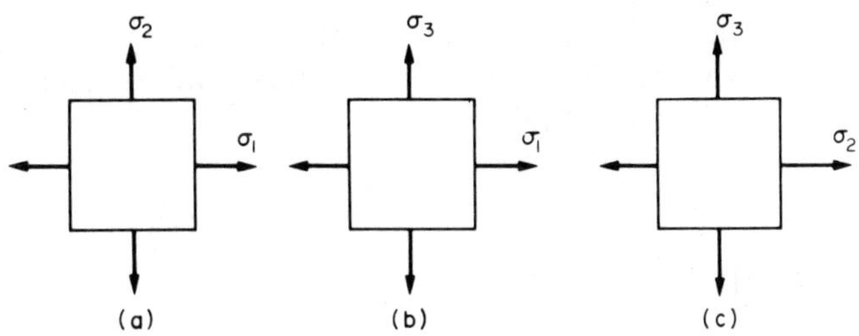

Fig. 13.18. True views on the various faces of the principal element

can then be combined to produce the complete three-dimensional Mohr circle representation shown in Fig. 13.19.

The large circle between points σ_1 and σ_3 represents stresses on all planes through the point in question containing the σ_2 axis. Likewise the small circle between σ_2 and σ_3 represents

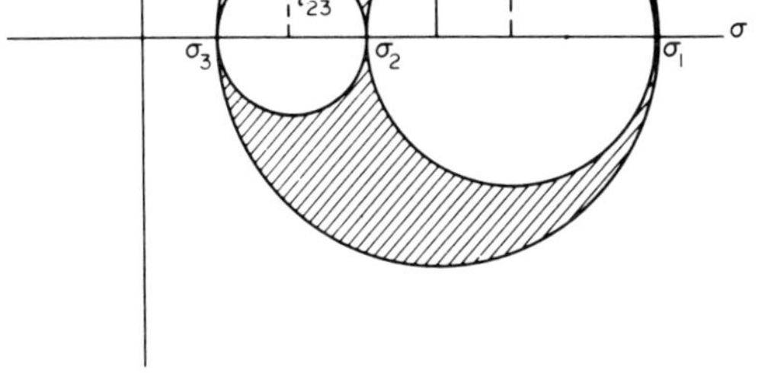

Fig. 13.19. Mohr circle representation of three-dimensional stress state showing the principal circle, the radius of which is equal to the greatest shear stress present in the system.

stresses on all planes containing the σ_1 axis and the circle between σ_1 and σ_2 all planes containing the σ_3 axis.

There are, of course, an infinite number of planes passing through the point which do not contain any of the three principal axes, but it can be shown that all such planes are represented by the shaded area between the circles. The procedure involved in the location of a particular point in the shaded area which corresponds to any given plane will be introduced in Chapter 23. In practice, however, it is often the maximum direct and shear stresses which will govern the elastic failure of materials. These are determined from the larger of the three circles which is thus termed the *principal circle* (τ_{max} = radius).

It is perhaps evident now that in many two-dimensional cases the maximum (greatest) shear stress value will be missed by not considering $\sigma_3 = 0$ and constructing the principal circle.

Consider the stress state shown in Fig. 13.20(a). If the principal stresses σ_1, σ_2 and σ_3 all have non-zero values the system will be termed "three-dimensional"; if one of the principal stresses is zero the system is said to be "two-dimensional" and with two principal stresses zero a "uniaxial" stress condition is obtained. In all cases, however, it is necessary to consider all three principal stress values in the determination of the maximum shear stress since out-of-plane shear stresses will be dependent on all three values and one will be a maximum – see Fig. 13.20(b), (c) and (d).

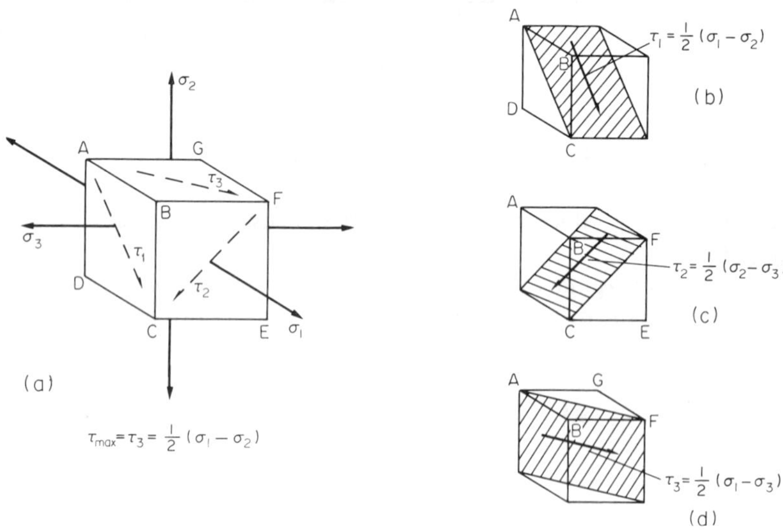

Fig. 13.20. Maximum shear stresses in a three-dimensional stress system.

Examples of the crucial effect of consideration of the third (zero) principal stress value in apparently "two-dimensional" stress states are given below:

(a) Thin cylinder.

An element in the surface of a thin cylinder subjected to internal pressure p will have principal stresses:

$$\sigma_1 = \sigma_H = pd/2t$$

$$\sigma_2 = \sigma_L = pd/4t$$

with the third, radial, stress σ_r assumed to be zero – see Fig. 13.21(a).

A two-dimensional Mohr circle representation of the stresses in the element will give Fig. 13.21(b) with a maximum shear stress:

$$\tau_{max} = \tfrac{1}{2}(\sigma_1 - \sigma_2)$$

$$= \tfrac{1}{2}\left(\frac{pd}{2t} - \frac{pd}{4t}\right) = \frac{pd}{8t}$$

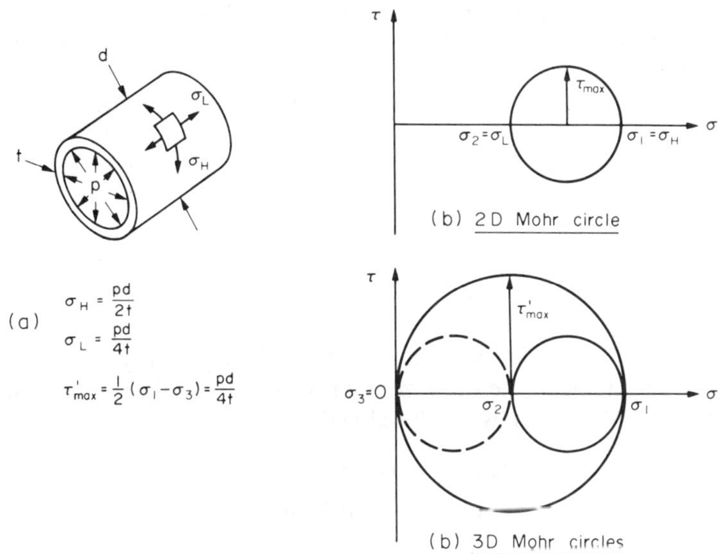

Fig. 13.21. Maximum shear stresses in a pressurised thin cylinder.

A three-dimensional Mohr circle construction, however, is shown in Fig. 13.21(c), the zero value of σ_3 producing a much larger principal circle and a maximum shear stress:

$$\tau_{max} = \tfrac{1}{2}(\sigma_1 - \sigma_3) = \tfrac{1}{2}\left(\frac{pd}{2t} - 0\right) = \frac{pd}{4t}$$

i.e. **twice** the value obtained from the two-dimensional circle.

(b) *Sphere*

Consider now an element in the surface of a sphere subjected to internal pressure p as shown in Fig. 13.22(a). Principal stresses on the element will then be $\sigma_1 = \sigma_2 = \dfrac{pd}{4t}$ with $\sigma_r = \sigma_3 = 0$ normal to the surface.

The two-dimensional Mohr circle is shown in Fig. 13.22(b), in this case reducing to a point since σ_1 and σ_2 are equal. The maximum shear stress, which always equals the radius of Mohr's, circle is thus zero and would seem to imply that, although the material of the vessel may well be ductile and susceptible to shear failure, no shear failure could ensue. However,

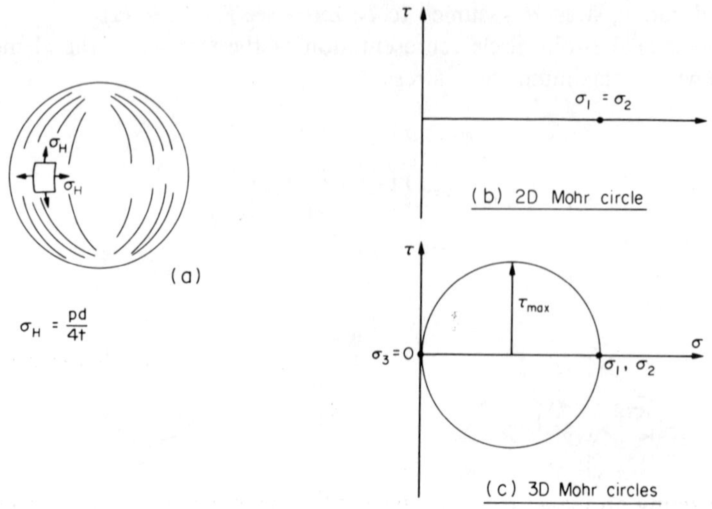

Fig. 13.22. Maximum shear stresses in a pressurised thin sphere.

this is far from the truth as will be evident when the full three-dimensional representation is drawn as in Fig. 13.22(c) with the third, zero, principal stress taken into account.

A maximum shear stress is now produced within the $\sigma_1\sigma_3$ plane of value:

$$\tau_{\mathbf{max}} = \tfrac{1}{2}(\sigma_1 - \sigma_3) = \boldsymbol{pd/8t}$$

The greatest value of τ can be obtained *analytically* by using the statement

$$\tau_{\max} = \tfrac{1}{2}(\text{greatest principal stress} - \text{least principal stress})$$

and considering separately the principal stress conditions as illustrated in Fig. 13.18.

Examples

Example 13.1 (A)

A circular bar 40 mm diameter carries an axial tensile load of 100 kN. What is the value of the shear stress on the planes on which the normal stress has a value of 50 MN/m^2 tensile?

Solution

Tensile stress $$\sigma_y = \frac{F}{A} = \frac{100 \times 10^3}{\pi \times (0.02)^2} = 79.6\ \text{MN/m}^2$$

Now the normal stress on an oblique plane is given by eqn. (13.1):

$$\sigma_\theta = \sigma_y \sin^2\theta$$

$$50 \times 10^6 = 79.6 \times 10^6 \sin^2\theta$$

$$\theta = 52^\circ\, 28'$$

The shear stress on the oblique plane is then given by eqn. (13.2):

$$\tau_\theta = \tfrac{1}{2}\sigma_y \sin 2\theta$$

$$= \tfrac{1}{2} \times 79.6 \times 10^6 \times \sin 104^\circ\ 56'$$

$$= 38.6 \times 10^6$$

The required shear stress is 38.6 MN/m².

Example 13.2 (A/B)

Under certain loading conditions the stresses in the walls of a cylinder are as follows:

(a) 80 MN/m² tensile;
(b) 30 MN/m² tensile at right angles to (a);
(c) shear stresses of 60 MN/m² on the planes on which the stresses (a) and (b) act; the shear couple acting on planes carrying the 30 MN/m² stress is clockwise in effect.

Calculate the principal stresses and the planes on which they act. What would be the effect on these results if owing to a change of loading (a) becomes compressive while stresses (b) and (c) remain unchanged?

Solution

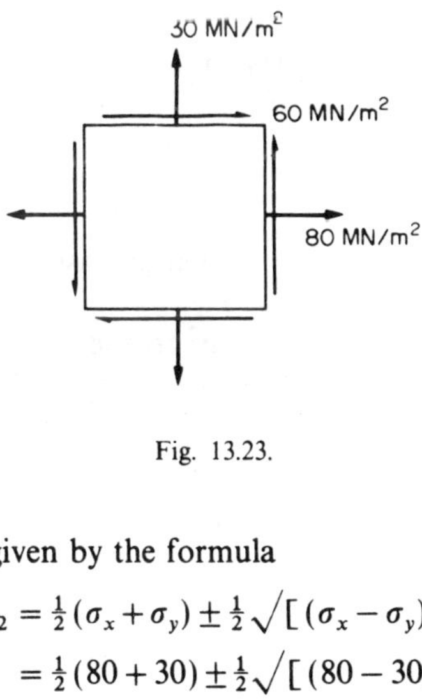

Fig. 13.23.

The principal stresses are given by the formula

$$\sigma_1 \text{ and } \sigma_2 = \tfrac{1}{2}(\sigma_x + \sigma_y) \pm \tfrac{1}{2}\sqrt{[(\sigma_x - \sigma_y)^2 + 4\tau_{xy}^2]}$$

$$= \tfrac{1}{2}(80 + 30) \pm \tfrac{1}{2}\sqrt{[(80 - 30)^2 + (4 \times 60^2)]}$$

$$= 55 \pm 5\sqrt{(25 + 144)}$$

$$= 55 \pm 65$$

$\therefore$ $\sigma_1 = \mathbf{120\ MN/m^2}$

and $\sigma_2 = \mathbf{-10\ MN/m^2}$ (i.e. compressive)

The planes on which these stresses act can be determined from eqn. (13.14),

i.e. $$\tan\theta_1 = \frac{\sigma_p - \sigma_x}{\tau_{xy}}$$

$\therefore$ $$\tan\theta_1 = \frac{120-80}{60} = 0.6667$$

$\therefore$ $$\theta_1 = \mathbf{33^\circ\ 41'}$$

Also $$\tan\theta_2 = = \frac{-10-80}{60} = 1.50$$

$\therefore$ $$\theta_2 = \mathbf{-56^\circ\ 19'} \quad \textbf{or} \quad \mathbf{123^\circ\ 41'}$$

N.B. – The resulting angles are at 90° to each other as expected.

If the loading is now changed so that the 80 MN/m² stress becomes compressive:

$$\sigma_1 = \tfrac{1}{2}(-80+30) + \tfrac{1}{2}\sqrt{[(-80-30)^2 + (4\times 60^2)]}$$
$$= -25 + 5\sqrt{(121+144)}$$
$$= -25 + 81.5 = \mathbf{56.5\ MN/m^2}$$

and $$\sigma_2 = -25 - 81.5 = \mathbf{-106.5\ MN/m^2}$$

Then $$\tan\theta_1 = \frac{56.5-(-80)}{60} = 2.28$$

$\therefore$ $$\theta_1 = \mathbf{66^\circ\ 19'}$$

and $$\theta_2 = 66^\circ\ 19' + 90 = \mathbf{156^\circ\ 19'}$$

Mohr's circle solutions

In the first part of the question the stress system and associated Mohr's circle are as drawn in Fig. 13.24.

By measurement: $\sigma_1 =$ **120 MN/m² tensile**

$\sigma_2 =$ **10 MN/m² compressive**

and $\theta_1 =$ **34° counterclockwise from *BC***

$\theta_2 =$ **124° counterclockwise from *BC***

When the 80 MN/m² stress is reversed, the stress system is that in Fig. 13.25, giving Mohr's circle as drawn.

The required values are then:

$\sigma_1 =$ **56.5 MN/m² tensile**

$\sigma_2 =$ **106.5 MN/m² compressive**

$\theta_1 =$ **66° 15′ counterclockwise to *BC***

and $\theta_2 =$ **156° 15′ counterclockwise to *BC***

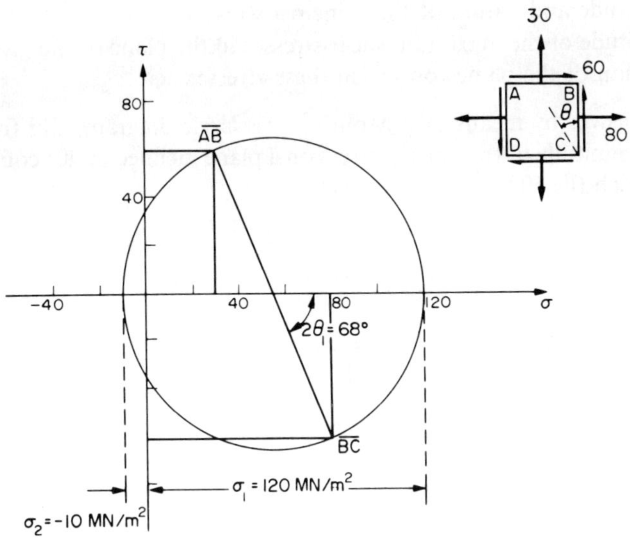

Fig. 13.24.

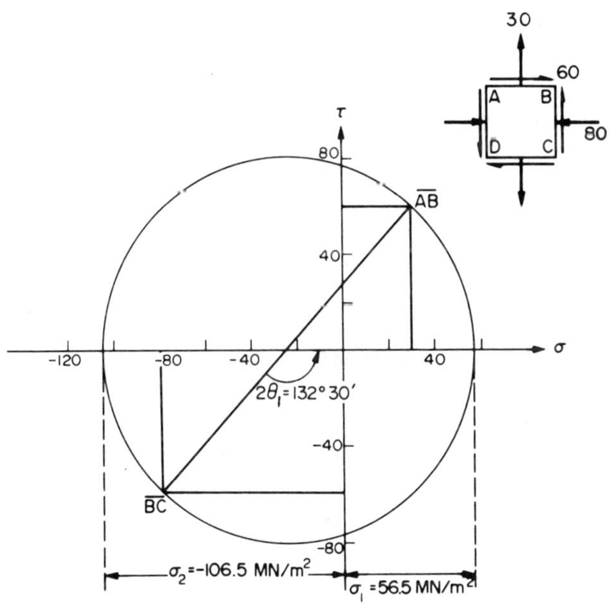

Fig. 13.25.

Example 13.3 (B)

A material is subjected to two mutually perpendicular direct stresses of 80 MN/m^2 tensile and 50 MN/m^2 compressive, together with a shear stress of 30 MN/m^2. The shear couple acting on planes carrying the 80 MN/m^2 stress is clockwise in effect. Calculate

(a) the magnitude and nature of the principal stresses;
(b) the magnitude of the maximum shear stresses in the plane of the given stress system;
(c) the direction of the planes on which these stresses act.

Confirm your answer by means of a Mohr's stress circle diagram, and from the diagram determine the magnitude of the normal stress on a plane inclined at 20° counterclockwise to the plane on which the 50 MN/m² stress acts.

Solution

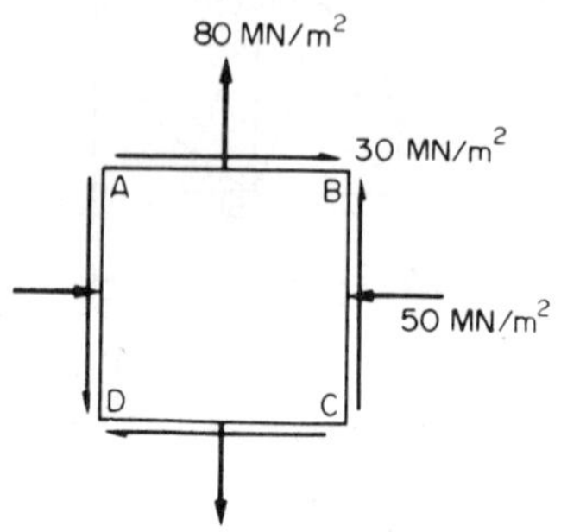

Fig. 13.26.

(a) To find the principal stresses:

$$\sigma_1 \text{ and } \sigma_2 = \tfrac{1}{2}(\sigma_x + \sigma_y) \pm \tfrac{1}{2}\sqrt{[(\sigma_x - \sigma_y)^2 + 4\tau_{xy}^2]}$$
$$= \tfrac{1}{2}(-50 + 80) \pm \tfrac{1}{2}\sqrt{[(-50 - 80)^2 + (4 \times 900)]}$$
$$= 5[3 \pm \sqrt{(169 + 36)}] = 5[3 \pm 14.31]$$

$$\therefore \quad \sigma_1 = 86.55 \text{ MN/m}^2$$
$$\sigma_2 = -56.55 \text{ MN/m}^2$$

The principal stresses are

86.55 MN/m² tensile and **56.55 MN/m² compressive**

(b) To find the maximum shear stress:

$$\tau_{max} = \frac{\sigma_1 - \sigma_2}{2} = \frac{86.55 - (-56.55)}{2} = \frac{143.1}{2} = 71.6 \text{ MN/m}^2$$

Maximum shear stress = **71.6 MN/m²**

(c) To find the directions of the principal planes:

$$\tan\theta_1 = \frac{\sigma_p - \sigma_x}{\tau_{xy}} = \frac{86.55 - (-50)}{30}$$
$$= \frac{136.55}{30} = 4.552$$

$$\therefore \qquad \theta_1 = \mathbf{77^\circ\,36'}$$

$$\therefore \qquad \theta_2 = 77^\circ\,36' + 90^\circ = \mathbf{167^\circ\,36'}$$

The principal planes are inclined at 77° 36′ to the plane on which the 50 MN/m² stress acts. The maximum shear planes are at 45° to the principal planes.

Mohr's circle solution

The stress system shown in Fig. 13.26 gives the Mohr's circle in Fig. 13.27.

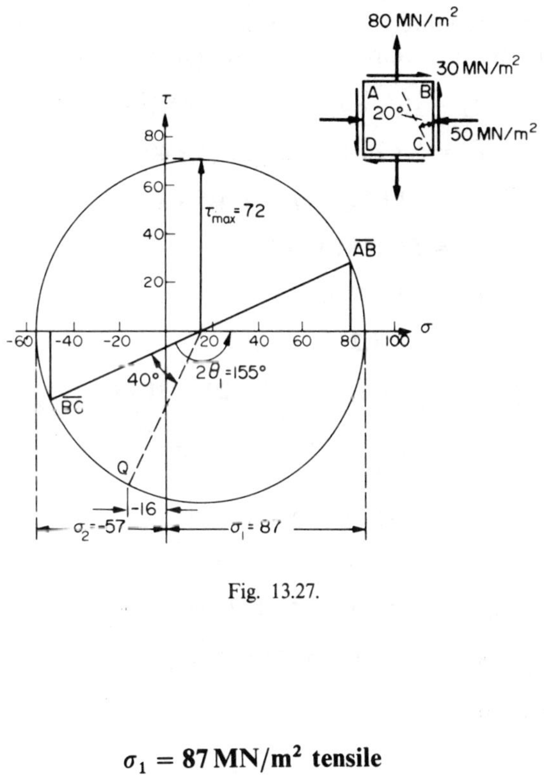

Fig. 13.27.

By measurement

$$\sigma_1 = \mathbf{87\,MN/m^2\ tensile}$$

$$\sigma_2 = \mathbf{57\,MN/m^2\ compressive}$$

$$\tau_{max} = \mathbf{72\,MN/m^2}$$

and $$\theta_1 = \frac{155^\circ}{2} = \mathbf{77^\circ\,30'}$$

The direct or normal stress on a plane inclined at 20° counterclockwise to *BC* is obtained by measuring from $\overline{BC}$ on the Mohr's circle through 2 × 20° = 40° in the same direction.

This gives $$\sigma = \mathbf{16\,MN/m^2\ compressive}$$

Example 13.4 (B)

At a given section a shaft is subjected to a bending stress of 20 MN/m^2 and a shear stress of 40 MN/m^2. Determine:

(a) the principal stresses;
(b) the directions of the principal planes;
(c) the maximum shear stress and the planes on which this acts;
(d) the tensile stress which, acting alone, would produce the same maximum shear stress;
(e) the shear stress which, acting alone, would produce the same maximum tensile principal stress.

Solution

(a) The bending stress is a direct stress and can be treated as acting on the x axis, so that $\sigma_x = 20$ MN/m^2; since no other direct stresses are given, $\sigma_y = 0$.

Principal stress
$$\sigma_1 = \tfrac{1}{2}(\sigma_x + \sigma_y) + \tfrac{1}{2}\sqrt{[(\sigma_x - \sigma_y)^2 + 4\tau_{xy}^2]}$$
$$= \tfrac{1}{2} \times 20 + \tfrac{1}{2}\sqrt{[20^2 + (4 \times 40^2)]}$$
$$= 10 + 5\sqrt{(68)} = 10 + 5 \times 8.246$$
$$= \mathbf{51.23\ MN/m^2}$$

and
$$\sigma_2 = 10 - 41.23$$
$$= \mathbf{-31.23\ MN/m^2}$$

(b) Then
$$\tan\theta_1 = \frac{\sigma_p - \sigma_x}{\tau_{xy}} = \frac{51.23 - 20}{40} = \frac{31.23}{40} = 0.7808$$

$\therefore$
$$\theta_1 = \mathbf{37^\circ\, 59'}$$

and
$$\tan\theta_2 = \frac{-31.23 - 20}{40} = \frac{-51.23}{40} = -1.2808$$

$\therefore$
$$\theta_2 = \mathbf{-52^\circ\, 1'} \quad \text{or} \quad \mathbf{127^\circ\, 59'}$$

both angles being measured counterclockwise from the plane on which the 20 MN/m^2 stress acts.

(c) Maximum shear stress

$$\tau_{max} = \frac{\sigma_1 - \sigma_2}{2} = \frac{51.23 - (-31.23)}{2}$$
$$= \frac{82.46}{2} = \mathbf{41.23\ MN/m^2}$$

This acts on planes at 45° to the principal planes,

i.e. **at 82° 59′ or −7° 1′**

(d) Maximum shear stress
$$\tau_{max} = \tfrac{1}{2}\sqrt{[(\sigma_x - \sigma_y)^2 + 4\tau_{xy}^2]}$$

Thus if a tensile stress is to act alone to give the same maximum shear stress ($\sigma_x = 0$ and $\tau_{xy} = 0$):

$$\text{maximum shear stress} = \tfrac{1}{2}\sqrt{(\sigma_x^2)} = \tfrac{1}{2}\sigma_x$$

$$41.23 = \tfrac{1}{2}\sigma_x$$

i.e. $$\sigma_x = \mathbf{82.46\ MN/m^2}$$

The required tensile stress is 82.46 MN/m².

(e) Principal stress

$$\sigma_1 = \tfrac{1}{2}(\sigma_x+\sigma_y)+\tfrac{1}{2}\sqrt{[(\sigma_x-\sigma_y)^2+4\tau_{xy}^2]}$$

Thus if a shear stress is to act alone to give the same principal stress ($\sigma_x = \sigma_y = 0$):

$$\sigma_1 = \tfrac{1}{2}\sqrt{(4\tau_{xy}^2)} = \tau_{xy}$$

$$51.23 = \tau_{xy}$$

The required shear stress is **51.23 MN/m²**.

Mohr's circle solutions

(a), (b), (c) The stress system and corresponding Mohr's circle are as shown in Fig. 13.28. By measurement:

(a) $\sigma_1 \simeq$ **51 MN/m² tensile**

$\sigma_2 \simeq$ **31 MN/m² compressive**

(b) $\theta_1 = \dfrac{76^\circ}{2} = \mathbf{38^\circ}$

$\theta_2 = 38^\circ + 90^\circ = \mathbf{128^\circ}$

(c) $\tau_{max} \simeq$ **41 MN/m²**

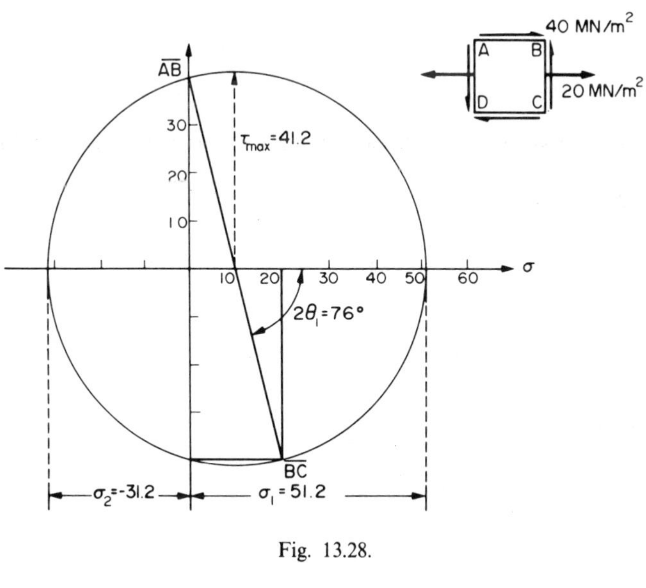

Fig. 13.28.

Angle of maximum shear plane

$$= \frac{166}{2} = \mathbf{83^\circ}$$

(d) If a tensile stress σ_x is to act alone to give the same maximum shear stress, then $\sigma_y = 0$, $\tau_{xy} = 0$ and $\tau_{max} = 41\ \text{MN/m}^2$. The Mohr's circle therefore has a radius of $41\ \text{MN/m}^2$ and passes through the origin (Fig. 13.29).

Hence the required tensile stress is **82 MN/m²**.

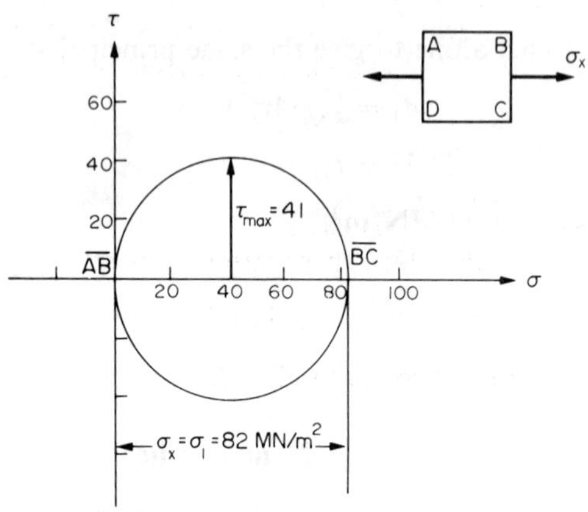

Fig. 13.29.

(e) If a shear stress is to act alone to produce the same principal stress, $\sigma_x = 0$, $\sigma_y = 0$ and $\sigma_1 = 51\ \text{MN/m}^2$. The Mohr's circle thus has its centre at the origin and passes through $\sigma = 51\ \text{MN/m}^2$ (Fig. 13.30).

Hence the required shear stress is **51 MN/m²** .

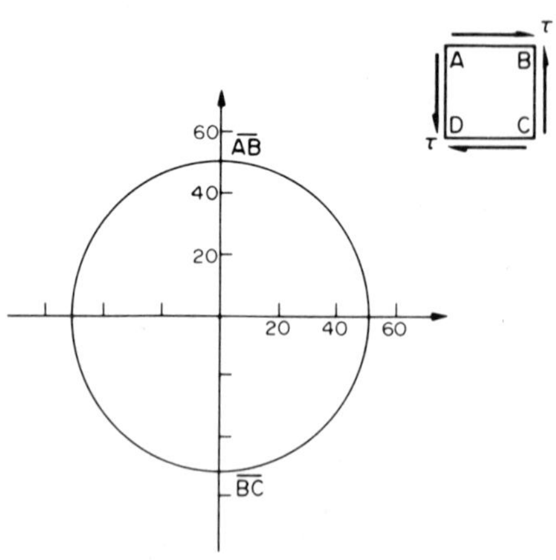

Fig. 13.30.

Example 13.5 (B)

At a point in a piece of elastic material direct stresses of 90 MN/m^2 tensile and 50 MN/m^2 compressive are applied on mutually perpendicular planes. The planes are also subjected to a shear stress. If the greater principal stress is limited to 100 MN/m^2 tensile, determine:

(a) the value of the shear stress;
(b) the other principal stress;
(c) the normal stress on the plane of maximum shear;
(d) the maximum shear stress.

Make a neat sketch showing clearly the positions of the principal planes and planes of maximum shear stress with respect to the planes of the applied stresses.

Solution

(a) Principal stress $\sigma_1 = \frac{1}{2}(\sigma_x + \sigma_y) + \frac{1}{2}\sqrt{[(\sigma_x - \sigma_y)^2 + 4\tau_{xy}^2]}$

This is limited to 100 MN/m^2; therefore shear stress τ_{xy} is given by

$$100 = \tfrac{1}{2}(90 - 50) + \tfrac{1}{2}\sqrt{[(90 + 50)^2 + 4\tau_{xy}^2]}$$

$$\therefore \quad 200 = 40 + 10\sqrt{[14^2 + 0.04\tau_{xy}^2]}$$

$$\therefore \quad \tau_{xy} = \sqrt{\left(\frac{16^2 - 14^2}{0.04}\right)} = \sqrt{\left(\frac{256 - 196}{0.04}\right)} = \frac{\sqrt{60}}{0.2}$$

$$= \mathbf{38.8\ MN/m^2}$$

The required shear stress is 38.8 MN/m^2.

(b) The other principal stress σ_2 is given by

$$\sigma_2 = \tfrac{1}{2}(\sigma_x + \sigma_y) - \tfrac{1}{2}\sqrt{[(\sigma_x - \sigma_y)^2 + 4\tau_{xy}^2]}$$

$$= \tfrac{1}{2}[(90 - 50) - 10\sqrt{(14^2 + 60)}] = \frac{40 - 10\sqrt{(256)}}{2}$$

$$= \frac{40 - 160}{2} = \mathbf{-60\ MN/m^2}$$

The other principal stress is 60 MN/m^2 compressive.

(c) The normal stress on the plane of maximum shear

$$= \frac{\sigma_1 + \sigma_2}{2} = \frac{100 - 60}{2}$$

$$= \mathbf{20\ MN/m^2}$$

The required normal stress is 20 MN/m^2 tensile.

(d) The maximum shear stress is given by

$$\tau_{max} = \frac{\sigma_1 - \sigma_2}{2} = \frac{100 + 60}{2}$$

$$= \mathbf{80\ MN/m^2}$$

The maximum shear stress is 80 MN/m^2.

In order to be able to draw the required sketch (Fig. 13.31) to indicate the relative positions of the planes on which the above stresses act, the angles of the principal planes are required. These are given by

$$\tan\theta = \frac{\sigma_p - \sigma_x}{\tau_{xy}} = \frac{100-(-50)}{38.8}$$

$$= \frac{150}{38.8} = 3.87$$

$$\therefore \qquad \theta_1 = \mathbf{75^\circ\ 30'}$$

to the plane on which the 50 MN/m² stress acts.

The required sketch is then shown in Fig. 13.31.

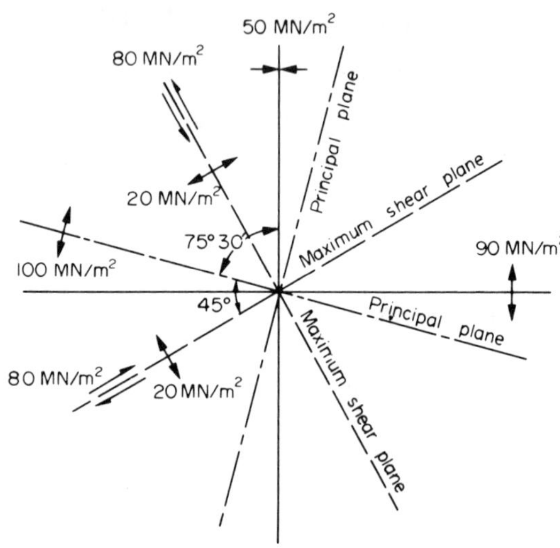

Fig. 13.31. Summary of principal planes and maximum shear planes.

Mohr's circle solution

The stress system is as shown in Fig. 13.32. The centre of the Mohr's circle is positioned midway between the two direct stresses given, and the radius is such that $\sigma_1 = 100\ \text{MN/m}^2$.

By measurement:

$$\tau = \mathbf{39\ MN/m^2}$$

$$\sigma_2 = \mathbf{60\ MN/m^2\ compressive}$$

$$\tau_{max} = \mathbf{80\ MN/m^2}$$

$$\theta_1 = \frac{151}{2} = \mathbf{75^\circ\ 30'\ to\ } \boldsymbol{BC}\mathbf{,\ the\ plane\ on\ which\ the\ 50\ MN/m^2\ stress\ acts}$$

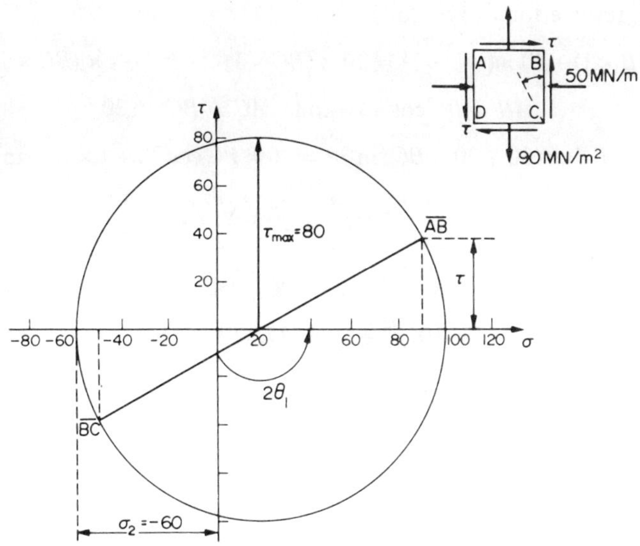

Fig. 13.32.

Example 13.6 (B)

In a certain material under load a plane *AB* carries a tensile direct stress of 30 MN/m^2 and a shear stress of 20 MN/m^2, while another plane *BC* carries a tensile direct stress of 20 MN/m^2 and a shear stress. If the planes are inclined to one another at 30° and plane *AC* at right angles to plane *AB* carries a direct stress unknown in magnitude and nature, find:

(a) the value of the shear stress on *BC*;
(b) the magnitude and nature of the direct stress on *AC*;
(c) the principal stresses.

Solution

Referring to Fig. 13.33 let the shear stress on *BC* be τ and the direct stress on *AC* be σ_x, assumed tensile. Consider the equilibrium of the elemental wedge *ABC*. Assume this wedge to be of unit depth. A complementary shear stress equal to that on *AB* will be set up on *AC*.

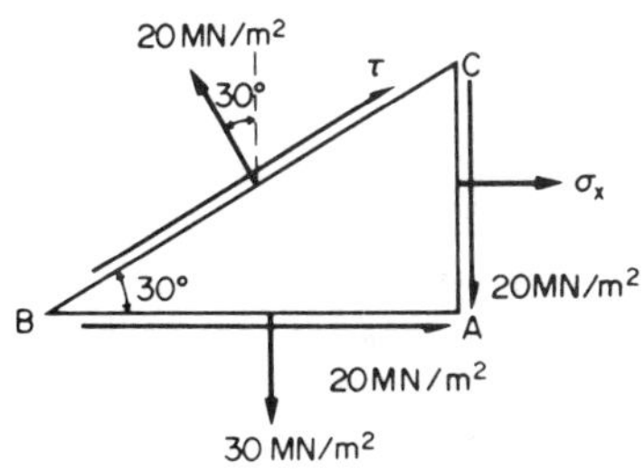

Fig. 13.33.

(a) To find τ, resolve forces vertically:

$$30 \times (AB \times 1) + 20 \times (AC \times 1) = 20 \times (BC \times 1)\cos 30^\circ + \tau \times (BC \times 1)\sin 30^\circ$$

Now $$AB = BC\cos 30 \quad \text{and} \quad AC = BC\sin 30$$

$$\therefore \quad 30 \times BC\cos 30 + 20 \times BC\sin 30 = 20 \times BC\cos 30 + \tau \times BC\sin 30$$

$$30\frac{\sqrt{3}}{2} + 20 \times \frac{1}{2} = 20 \times \frac{\sqrt{3}}{2} + \tau \times \frac{1}{2}$$

$$30\sqrt{3} + 20 = 20\sqrt{3} + \tau$$

$$\therefore \quad \tau = 10\sqrt{3} + 20 = \mathbf{37.32\ MN/m^2}$$

The required shear stress is $37.32\ \text{MN/m}^2$.

(b) To find σ_x, resolve forces horizontally:

$$20 \times (AB \times 1) + \sigma_x \times (AC \times 1) + \tau \times (BC \times 1)\cos 30^\circ = 20 \times (BC \times 1)\sin 30^\circ$$

$$20 \times BC\cos 30^\circ + \sigma_x \times BC\sin 30^\circ + \tau \times BC\cos 30^\circ = 20 \times BC\sin 30^\circ$$

$$20 \times \frac{\sqrt{3}}{2} + \sigma_x \times \frac{1}{2} + \tau \times \frac{\sqrt{3}}{2} = 20 \times \frac{1}{2}$$

$$20\sqrt{3} + \sigma_x + \sqrt{3} \times 37.32 = 10$$

$$\therefore \quad \sigma_x = 10 - \sqrt{3} \times 57.32 = 10 - 99.2$$

$$= \mathbf{-89.2\ MN/m^2, \text{ i.e. compressive}}$$

(c) The principal stresses are now given by

$$\sigma_{1,2} = \tfrac{1}{2}(\sigma_x + \sigma_y) \pm \tfrac{1}{2}\sqrt{[(\sigma_x - \sigma_y)^2 + 4\tau_{xy}^2]}$$

$$= \tfrac{1}{2}\{(-89.2 + 30) \pm \sqrt{[(-89.2 - 30)^2 + 4 \times 20^2]}\}$$

$$= 5\{-5.92 \pm \sqrt{[(-11.92)^2 + 16]}\}$$

$$= 5[-5.92 \pm \sqrt{158}] = 5[-5.92 \pm 12.57]$$

$$\therefore \quad \sigma_1 = \mathbf{33.25\ MN/m^2}$$

$$\sigma_2 = \mathbf{-92.45\ MN/m^2}$$

The principal stresses are $33.25\ \text{MN/m}^2$ tensile and $92.45\ \text{MN/m}^2$ compressive.

Example 13.7 (B)

A hollow steel shaft of 100 mm external diameter and 50 mm internal diameter transmits 0.75 MW at 500 rev/min and is also subjected to an axial end thrust of 50 kN. Determine the maximum bending moment which can be safely applied in conjunction with the applied torque and thrust if the maximum compressive principal stress is not to exceed $100\ \text{MN/m}^2$ compressive. What will then be the value of:

(a) the other principal stress;
(b) the maximum shear stress?

Solution

The torque on the shaft may be found from

$$\text{power} = T \times \omega$$

$$\therefore \quad T = \frac{0.75 \times 10^6 \times 60}{2\pi \times 500} = 14.3 \times 10^3 = 14.3\,\text{kN m}$$

The shear stress in the shaft at the surface is then given by the torsion theory

$$\frac{T}{J} = \frac{\tau}{R}$$

$$\tau = \frac{TR}{J} = \frac{14.3 \times 10^3 \times 50 \times 10^{-3} \times 2}{\pi(50^4 - 25^4)10^{-12}}$$

$$= 0.78 \times 10^8$$

$$= 78\,\text{MN/m}^2$$

The direct stress resulting from the end thrust is given by

$$\sigma_d = \frac{\text{load}}{\text{area}} = \frac{-50 \times 10^3}{\pi(50^2 - 25^2)} 10^{-6}$$

$$= -8.5 \times 10^6$$

$$= -8.5\,\text{MN/m}^2$$

The bending moment to be applied will produce a direct stress in the same direction as σ_d. Thus the total stress in the x direction is

$$\sigma_x = \sigma_b + \sigma_d$$

the greatest value of σ_x being obtained where the bending stress is of the same sign as the end thrust or, in other words, compressive. The stress system is therefore as shown in Fig. 13.34.

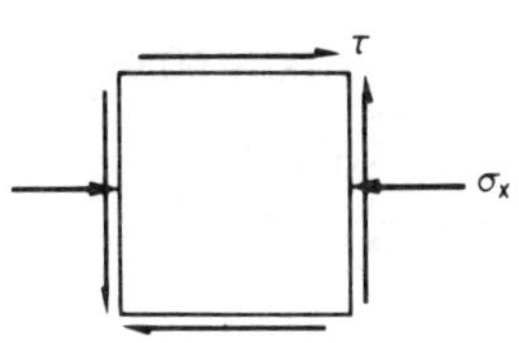

Fig. 13.34.

N.B. $\sigma_y = 0$; there is no stress in the y direction.

$$\sigma_1 = \tfrac{1}{2}(\sigma_x + \sigma_y) \pm \tfrac{1}{2}\sqrt{[(\sigma_x - \sigma_y)^2 + 4\tau_{xy}^2]}$$

Therefore substituting all stresses in units of MN/m^2,

$$-100 = \tfrac{1}{2}\sigma_x \pm \tfrac{1}{2}\sqrt{(\sigma_x^2 + 4\tau^2)}$$

$$\therefore \quad -200 - \sigma_x = \pm\sqrt{(\sigma_x^2 + 4\tau^2)}$$

$$\therefore \quad 4 \times 10^4 + 400\sigma_x + \sigma_x^2 = \sigma_x^2 + 4\tau^2$$

$$\therefore \quad 400\sigma_x = 4\tau^2 - 4 \times 10^4$$

$$= 24320 - 40000$$

$$\therefore \quad \sigma_x = -39.2\ \text{MN/m}^2$$

Therefore stress owing to bending

$$\sigma_b = \sigma_x - \sigma_d = -39.2 - (-8.5)$$

$$= -30.7\ \text{MN/m}^2 \quad \text{(i.e. compressive)}$$

But from bending theory

$$\sigma_b = \frac{My}{I}$$

$$\therefore \quad M = \frac{30.7 \times 10^6 \times \pi(50^4 - 25^4)10^{-12}}{50 \times 10^{-3} \times 4}$$

$$= 2830\ \text{N m}$$

$$= 2.83\ \text{kN m}$$

i.e. the bending moment which can be safely applied is **2.83 kN m.**

(a) The other principal stress

$$\sigma_2 = \tfrac{1}{2}\sigma_x + \tfrac{1}{2}\sqrt{(\sigma_x^2 + 4\tau^2)}$$

$$= -19.6 + \tfrac{1}{2}\sqrt{(39.2^2 + 24320)}$$

$$= -19.6 + 80.5$$

$$= \mathbf{60.9\ MN/m^2} \quad \text{(tensile)}$$

(b) The maximum shear stress is given by

$$\tau_{max} = \tfrac{1}{2}(\sigma_1 - \sigma_2)$$

$$= \tfrac{1}{2}(-100 - 60.9)$$

$$= \mathbf{-80.45\ MN/m^2}$$

i.e. the maximum shear stress is 80.45 MN/m^2.

Example 13.8

A beam of symmetrical I-section is simply supported at each end and loaded at the centre of its 3 m span with a concentrated load of 100 kN. The dimensions of the cross-section are: flanges 150 mm wide by 30 mm thick; web 30 mm thick; overall depth 200 mm.

For the transverse section at the point of application of the load, and considering a point at the top of the web where it meets the flange, calculate the magnitude and nature of the principal stresses. Neglect the self-mass of the beam.

Solution

At any section of the beam there will be two sets of stresses acting simultaneously:

(1) bending stresses $$\sigma_b = \frac{My}{I}$$

(2) shear stresses $$\tau = \frac{QA\bar{y}}{Ib}$$

together with their associated complementary shear stresses of the same value (Fig. 13.35a).

The stress system on any element of the beam can therefore be represented as in Fig. 13.36. The stress distribution diagrams are shown in Fig. 13.35b.

Bending stress

$$\sigma_b = \frac{My}{I}$$

$$M = \text{maximum bending moment}$$

$$= \frac{WL}{4} = \frac{100 \times 10^3 \times 3}{4} = 75\,\text{kN m}$$

and $$I = \frac{0.15 \times 0.2^3 - 0.12 \times 0.14^3}{12}\ \text{m}^4$$

$$= 72.56 \times 10^{-6}\,\text{m}^4$$

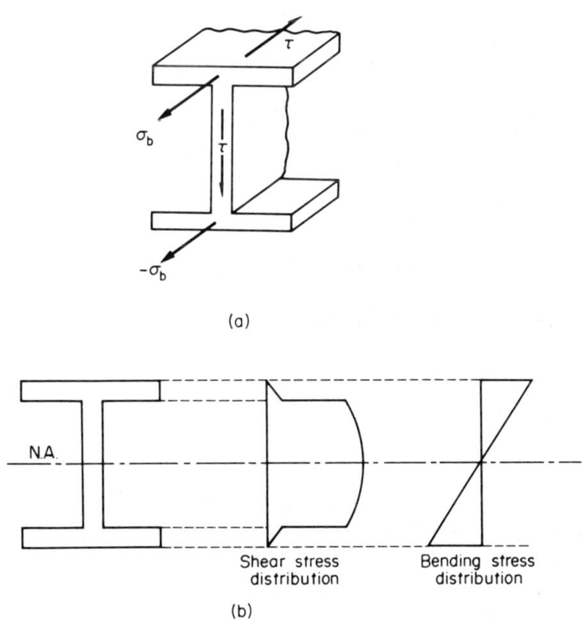

Fig. 13.35.

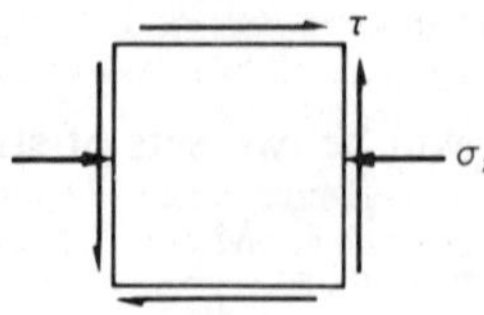

Fig. 13.36.

Therefore at the junction of web and flange

$$\sigma_b = \frac{75 \times 10^3 \times 0.07}{72.56 \times 10^{-6}}$$

$$= 72.35 \times 10^6 = 72.35\ \text{MN/m}^2 \text{ and is compressive}$$

Shear stress

$$\tau = \frac{QA\bar{y}}{Ib}$$

$$= \frac{50 \times 10^3 \times (150 \times 30) \times 85 \times 10^{-9}}{72.56 \times 10^{-4} \times 30 \times 10^{-3}}$$

$$= 8.79\ \text{MN/m}^2$$

The principal stresses are then given by

$$\sigma_1 \text{ or } \sigma_2 = \tfrac{1}{2}(\sigma_x + \sigma_y) \pm \tfrac{1}{2}\sqrt{[(\sigma_x - \sigma_y)^2 + 4\tau_{xy}^2]}$$

with $\sigma_x = -\sigma_b$ and $\sigma_y = 0$

$$\therefore \quad \sigma_1 \text{ or } \sigma_2 = \tfrac{1}{2}(-72.35) \pm \tfrac{1}{2}\sqrt{[(-72.35)^2 + 4 \times 8.79^2]}\ \text{MN/m}^2$$

$$= -36.2 \pm \sqrt{(5544)}$$

$$= -36.2 \pm 74.5$$

$$\therefore \quad \boldsymbol{\sigma_2 = -110.7\ \text{MN/m}^2}$$

$$\boldsymbol{\sigma_1 = +38.3\ \text{MN/m}^2}$$

i.e. the principal stresses are 110.7 MN/m^2 compressive and 38.3 MN/m^2 tensile in the top of the web. At the bottom of the web the stress values obtained would be of the same value but of opposite sign.

Problems

13.1 (A). An axial tensile load of 10 kN is applied to a 12 mm diameter bar. Determine the maximum shearing stress in the bar and the planes on which it acts. Find also the value of the normal stresses on these planes.
[44.1 MN/m^2 at 45° and 135°; $\pm$44.2 MN/m^2.]

13.2 (A). A compressive member of a structure is of 25 mm square cross-section and carries a load of 50 kN. Determine, from first principles, the normal, tangential and resultant stresses on a plane inclined at 60° to the axis of the bar.
[60, 34.6, 69.3 MN/m^2.]

13.3 (A). A rectangular block of material is subjected to a shear stress of 30 MN/m^2 together with its associated complementary shear stress. Determine the magnitude of the stresses on a plane inclined at 30° to the directions of the applied stresses, which may be taken as horizontal. [26, 15 MN/m^2.]

13.4 (A). A material is subjected to two mutually perpendicular stresses, one 60 MN/m^2 compressive and the other 45 MN/m^2 tensile. Determine the direct, shear and resultant stresses on a plane inclined at 60° to the plane on which the 45 MN/m^2 stress acts. [18.75, 45.5, 49.2 MN/m^2.]

13.5 (A/B). The material of Problem 13.4 is now subjected to an additional shearing stress of 10 MN/m^2. Determine the principal stresses acting on the material and the maximum shear stress. [46, −61, 53.5 MN/m^2.]

13.6 (A/B). At a certain section in a material under stress, direct stresses of 45 MN/m^2 tensile and 75 MN/m^2 tensile act on perpendicular planes together with a shear stress τ acting on these planes. If the maximum stress in the material is limited to 150 MN/m^2 tensile determine the value of τ. [88.7 MN/m^2.]

13.7 (A/B). At a point in a material under stress there is a compressive stress of 200 MN/m^2 and a shear stress of 300 MN/m^2 acting on the same plane. Determine the principal stresses and the directions of the planes on which they act. [216 MN/m^2 at 54.2° to 200 MN/m^2 plane; −416 MN/m^2 at 144.2°.]

13.8 (A/B). At a certain point in a material the following stresses act: a tensile stress of 150 MN/m^2, a compressive stress of 105 MN/m^2 at right angles to the tensile stress and a shear stress clockwise in effect of 30 MN/m^2. Calculate the principal stresses and the directions of the principal planes. [153.5, −108.5 MN/m^2; at 6.7° and 96.7° counterclockwise to 150 MN/m^2 plane.]

13.9 (B). The stresses across two mutually perpendicular planes at a point in an elastic body are 120 MN/m^2 tensile with 45 MN/m^2 clockwise shear, and 30 MN/m^2 tensile with 45 MN/m^2 counterclockwise shear. Find (i) the principal stresses, (ii) the maximum shear stress, and (iii) the normal and tangential stresses on a plane measured at 20° counterclockwise to the plane on which the 30 MN/m^2 stress acts. Draw sketches showing the positions of the stresses found above and the planes on which they act relative to the original stresses. [138.6, 11.4, 63.6, 69.5, −63.4 MN/m^2.]

13.10 (B). At a point in a strained material the stresses acting on planes at right angles to each other are 200 MN/m^2 tensile and 80 MN/m^2 compressive, together with associated shear stresses which may be assumed clockwise in effect on the 80 MN/m^2 planes. If the principal stress is limited to 320 MN/m^2 tensile, calculate:
(a) the magnitude of the shear stresses;
(b) the directions of the principal planes;
(c) the other principal stress;
(d) the maximum shear stress.
[219 MN/m^2, 28.7 and 118.7° counterclockwise to 200 MN/m^2 plane; −200 MN/m^2; 260 MN/m^2.]

13.11 (B). A solid shaft of 125 mm diameter transmits 0.5 MW at 300 rev/min. It is also subjected to a bending moment of 9 kN m and to a tensile end load. If the maximum principal stress is limited to 75 MN/m^2, determine the permissible end thrust. Determine the position of the plane on which the principal stress acts, and draw a diagram showing the position of the plane relative to the torque and the plane of the bending moment. [61.4 kN; 61° to shaft axis.]

13.12 (B). At a certain point in a piece of material there are two planes at right angles to one another on which there are shearing stresses of 150 MN/m^2 together with normal stresses of 300 MN/m^2 tensile on one plane and 150 MN/m^2 tensile on the other plane. If the shear stress on the 150 MN/m^2 planes is taken as clockwise in effect determine for the given point:
(a) the magnitudes of the principal stresses;
(b) the inclinations of the principal planes;
(c) the maximum shear stress and the inclinations of the planes on which it acts;
(d) the maximum strain if $E = 208\ GN/m^2$ and Poisson's ratio = 0.29.
[392.7, 57.3 MN/m^2; 31.7°, 121.7°; 167.7 MN/m^2, 76.7°, 166.7°; 1810 $\mu\varepsilon$.]

13.13 (B). A 250 mm diameter solid shaft drives a screw propeller with an output of 7 MW. When the forward speed of the vessel is 35 km/h the speed of revolution of the propeller is 240 rev/min. Find the maximum stress resulting from the torque and the axial compressive stress resulting from the thrust in the shaft; hence find for a point on the surface of the shaft (a) the principal stresses, and (b) the directions of the principal planes relative to the shaft axis. Make a diagram to show clearly the direction of the principal planes and stresses relative to the shaft axis. [U.L.] [90.8, 14.7, 98.4, −83.7 MN/m^2; 47° and 137°.]

13.14 (B). A hollow shaft is 460 mm inside diameter and 25 mm thick. It is subjected to an internal pressure of 2 MN/m^2, a bending moment of 25 kN m and a torque of 40 kN m. Assuming the shaft may be treated as a thin cylinder, make a neat sketch of an element of the shaft, showing the stresses resulting from all three actions. Determine the values of the principal stresses and the maximum shear stress. [21.5, 11.8, 16.6 MN/m^2.]

13.15 (B). In a piece of material a tensile stress σ_1 and a shearing stress τ act on a given plane. Show that the principal stresses are always of opposite sign. If an additional tensile stress σ_2 acts on a plane perpendicular to that of σ_1, find the condition that both principal stresses may be of the same sign. [U.L.] [$\tau = \sqrt{(\sigma_1 \sigma_2)}$.]

13.16 (B). A shaft 100 mm diameter is subjected to a twisting moment of 7 kN m, together with a bending moment of 2 kN m. Find, at the surface of the shaft, (a) the principal stresses, (b) the maximum shear stress. [47.3, -26.9 MN/m^2; 37.1 MN/m^2.]

13.17 (B). A material is subjected to a horizontal tensile stress of 90 MN/m^2 and a vertical tensile stress of 120 MN/m^2, together with shear stresses of 75 MN/m^2, those on the 120 MN/m^2 planes being counterclockwise in effect. Determine:

(a) the principal stresses;
(b) the maximum shear stress;
(c) the shear stress which, acting alone, would produce the same principal stress;
(d) the tensile stress which, acting alone, would produce the same maximum shear stress.

[181.5, 28.5 MN/m^2; 76.5 MN/m^2; 181.5 MN/m^2; 153 MN/m^2.]

13.18 (B). Two planes AB and BC in an elastic material under load are inclined at 45° to each other. The loading on the material is such that the stresses on these planes are as follows:

On AB, 150 MN/m^2 direct stress and 120 MN/m^2 shear.

On BC, 80 MN/m^2 shear and a direct stress σ.

Determine the value of the unknown stress σ on BC and hence determine the principal stresses which exist in the material. [190, 214, -74 MN/m^2.]

13.19 (B). A beam of I-section, 500 mm deep and 200 mm wide, has flanges 25 mm thick and web 12 mm thick. It carries a concentrated load of 300 kN at the centre of a simply supported span of 3 m. Calculate the principal stresses set up in the beam at the point where the web meets the flange. [83.4, -6.15 MN/m^2.]

13.20 (B). At a certain point on the outside of a shaft which is subjected to a torque and a bending moment the shear stresses are 100 MN/m^2 and the longitudinal direct stress is 60 MN/m^2 tensile. Find, by calculation from first principles or by graphical construction which must be justified:

(a) the maximum and minimum principal stresses;
(b) the maximum shear stress;
(c) the inclination of the principal stresses to the original stresses.

Summarize the answers clearly on a diagram, showing their relative positions to the original stresses.

[E.M.E.U.] [134.4, -74.4 MN/m^2; 104.4 MN/m^2; 35.5°.]

13.21 (B). A short vertical column is firmly fixed at the base and projects a distance of 300 mm from the base. The column is of I-section, 200 mm deep by 100 mm wide, flanges 10 mm thick, web 6 mm thick.

An inclined load of 80 kN acts on the top of the column in the centre of the section and in the plane containing the central line of the web; the line of action is inclined at 30 degrees to the vertical. Determine the position and magnitude of the greatest principal stress at the base of the column.

[U.L.] [48 MN/m^2 at junction of web and flange.]

CHAPTER 14

COMPLEX STRAIN AND THE ELASTIC CONSTANTS

Summary

The *relationships between the elastic constants* are

$$E = 2G(1 + \nu) \quad \text{and} \quad E = 3K(1 - 2\nu)$$

Poisson's ratio ν being defined as the ratio of lateral strain to longitudinal strain and bulk modulus K as the ratio of volumetric stress to volumetric strain.

The *strain in the x direction* in a material subjected to three mutually perpendicular stresses in the x, y and z directions is given by

$$\varepsilon_x = \frac{\sigma_x}{E} - \nu\frac{\sigma_y}{E} - \nu\frac{\sigma_z}{E} = \frac{1}{E}(\sigma_x - \nu\sigma_y - \nu\sigma_z)$$

Similar equations apply for ε_y and ε_z.

Thus the *principal strain* in a given direction can be found in terms of the principal stresses, since

$$\varepsilon_1 = \frac{\sigma_1}{E} - \nu\frac{\sigma_2}{E} - \nu\frac{\sigma_3}{E} = \frac{1}{E}(\sigma_1 - \nu\sigma_2 - \nu\sigma_3)$$

For a *two-dimensional* stress system (i.e. $\sigma_3 = 0$), *principal stresses* can be found from known principal strains, since

$$\sigma_1 = \frac{(\varepsilon_1 + \nu\varepsilon_2)}{(1 - \nu^2)}E \quad \text{and} \quad \sigma_2 = \frac{(\varepsilon_2 + \nu\varepsilon_1)}{(1 - \nu^2)}E$$

When the linear strains in two perpendicular directions are known, together with the associated shear strain, or when three linear strains are known, the principal strains are easily determined by the use of *Mohr's strain circle*.

14.1. Linear strain for tri-axial stress state

Consider an element subjected to three mutually perpendicular tensile stresses σ_x, σ_y and σ_z as shown in Fig. 14.1.

If σ_y and σ_z were not present the strain in the x direction would, from the basic definition of Young's modulus E, be

$$\varepsilon_x = \frac{\sigma_x}{E}$$

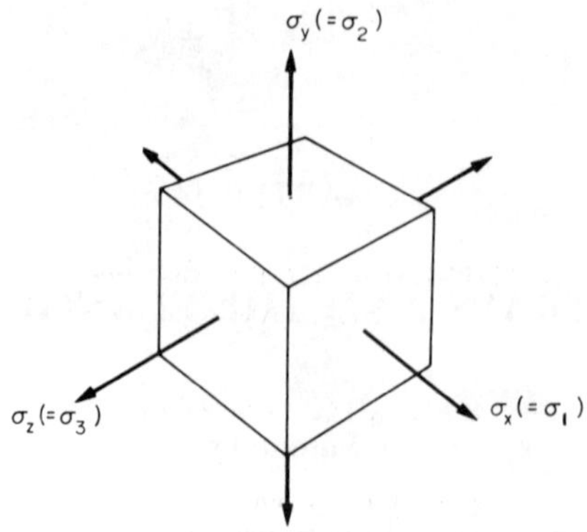

Fig. 14.1.

The effects of σ_y and σ_z in the x direction are given by the definition of Poisson's ratio ν to be

$$-\nu\frac{\sigma_y}{E} \quad \text{and} \quad -\nu\frac{\sigma_z}{E} \text{ respectively}$$

the negative sign indicating that if σ_y and σ_z are positive, i.e. tensile, then they tend to reduce the strain in the x direction.

Thus the total linear strain in the x direction is given by

$$\varepsilon_x = \frac{\sigma_x}{E} - \nu\frac{\sigma_y}{E} - \nu\frac{\sigma_z}{E}$$

i.e.

$$\varepsilon_x = \frac{1}{E}(\sigma_x - \nu\sigma_y - \nu\sigma_z) \tag{14.1}$$

Similarly the strains in the y and z directions would be

$$\varepsilon_y = \frac{1}{E}(\sigma_y - \nu\sigma_x - \nu\sigma_z)$$

$$\varepsilon_z = \frac{1}{E}(\sigma_z - \nu\sigma_x - \nu\sigma_y)$$

The three equations being known as the "generalised Hookes Law" from which the simple uniaxial form of §1.5 is obtained (when two of the three stresses are reduced to zero).

14.2. Principal strains in terms of stresses

In the absence of shear stresses on the faces of the element shown in Fig. 14.1 the stresses σ_x, σ_y and σ_z are in fact principal stresses. Thus the principal strain in a given direction is obtained from the principal stresses as

$$\varepsilon_1 = \frac{1}{E}(\sigma_1 - \nu\sigma_2 - \nu\sigma_3)$$

or
$$\varepsilon_2 = \frac{1}{E}(\sigma_2 - \nu\sigma_1 - \nu\sigma_3) \tag{14.2}$$

or
$$\varepsilon_3 = \frac{1}{E}(\sigma_3 - \nu\sigma_1 - \nu\sigma_2)$$

14.3. Principal stresses in terms of strains – two-dimensional stress system

For a two-dimensional stress system, i.e. $\sigma_3 = 0$, the above equations reduce to

$$\varepsilon_1 = \frac{1}{E}(\sigma_1 - \nu\sigma_2)$$

and
$$\varepsilon_2 = \frac{1}{E}(\sigma_2 - \nu\sigma_1)$$

with
$$\varepsilon_3 = \frac{1}{E}(-\nu\sigma_1 - \nu\sigma_2)$$

$\therefore$
$$E\varepsilon_1 = \sigma_1 - \nu\sigma_2$$
$$E\varepsilon_2 = \sigma_2 - \nu\sigma_1$$

Solving these equations simultaneously yields the following values for the principal stresses:

$$\sigma_1 = \frac{E}{(1-\nu^2)}(\varepsilon_1 + \nu\varepsilon_2)$$

and
$$\sigma_2 = \frac{E}{(1-\nu^2)}(\varepsilon_2 + \nu\varepsilon_1) \tag{14.3}$$

14.4. Bulk modulus K

It has been shown previously that Young's modulus E and the shear modulus G are defined as the ratio of stress to strain under direct load and shear respectively. Bulk modulus is similarly defined as a ratio of stress to strain under uniform pressure conditions. Thus if a material is subjected to a uniform pressure (or volumetric stress) σ in all directions then

$$\text{bulk modulus} = \frac{\text{volumetric stress}}{\text{volumetric strain}}$$

i.e.
$$K = \frac{\sigma}{\varepsilon_v} \tag{14.4}$$

the volumetric strain being defined below.

14.5. Volumetric strain

Consider a rectangular block of sides x, y and z subjected to a system of equal direct stresses σ on each face. Let the sides be changed in length by δx, δy and δz respectively under stress (Fig. 14.2).

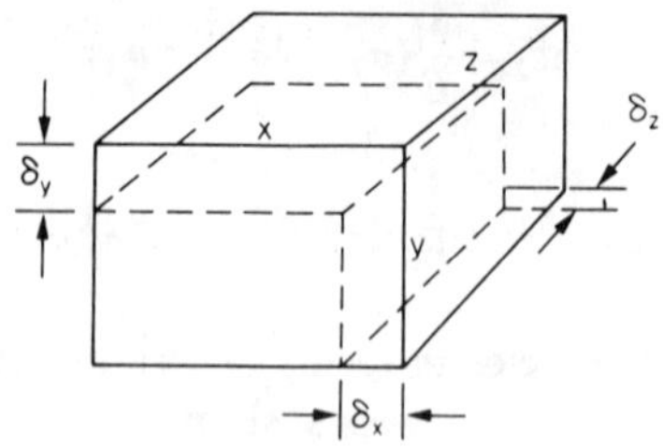

Fig. 14.2. Rectangular element subjected to uniform compressive stress on all faces producing decrease in size shown.

The volumetric strain is defined as follows:

$$\text{volumetric strain} = \frac{\text{change in volume}}{\text{original volume}} = \frac{\delta V}{V} = \frac{\delta V}{xyz}$$

The change in volume can best be found by calculating the volume of the strips to be cut off the original size of block to reduce it to the dotted block shown in Fig. 14.2.

Then

$$\delta V = \underset{\text{strip at back}}{xy\delta z} + \underset{\text{strip at side}}{y(z-\delta z)\delta x} + \underset{\text{strip on top}}{(x-\delta x)(z-\delta z)\delta y}$$

and neglecting the products of small quantities

$$\delta V = xy\delta z + yz\delta x + xz\delta y$$

$$\therefore \quad \text{volumetric strain} = \frac{(xy\delta z + yz\delta x + xz\delta y)}{xyz} = \varepsilon_v$$

$$\therefore \quad \varepsilon_v = \frac{\delta z}{z} + \frac{\delta x}{x} + \frac{\delta y}{y} = \varepsilon_z + \varepsilon_x + \varepsilon_y \qquad (14.5)$$

i.e. **volumetric strain = sum of the three mutually perpendicular linear strains**

14.6. Volumetric strain for unequal stresses

It has been shown above that the volumetric strain is the sum of the three perpendicular linear strains

$$\varepsilon_v = \varepsilon_x + \varepsilon_y + \varepsilon_z$$

Substituting for the strains in terms of stresses as given by eqn. (14.1),

$$\varepsilon_v = \frac{1}{E}(\sigma_x - \nu\sigma_y - \nu\sigma_z) + \frac{1}{E}(\sigma_y - \nu\sigma_x - \nu\sigma_z) + \frac{1}{E}(\sigma_z - \nu\sigma_x - \nu\sigma_y)$$

$$\boldsymbol{\varepsilon_v = \frac{1}{E}(\sigma_x + \sigma_y + \sigma_z)(1 - 2\nu)} \qquad (14.6)$$

It will be shown later that the following relationship applies between the elastic constants E, ν and K,

$$E = 3K(1 - 2\nu)$$

Thus the volumetric strain may be written in terms of the bulk modulus as follows:

$$\varepsilon_v = \frac{(\sigma_x + \sigma_y + \sigma_z)}{3K} \tag{14.7}$$

This equation applies to solid bodies only and cannot be used for the determination of internal volume (or capacity) changes of hollow vessels. It may be used, however, for changes in cylinder wall volume.

14.7. Change in volume of circular bar

A simple application of eqn. (14.6) is to the determination of volume changes of circular bars under direct load.

Consider, therefore, a circular bar subjected to a direct stress σ applied axially as shown in Fig. 14.3.

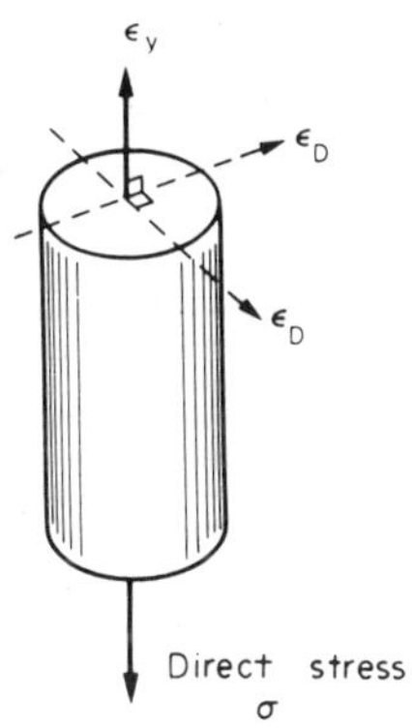

Fig. 14.3. Circular bar subjected to direct axial stress σ.

Here

$$\sigma_y = \sigma, \quad \sigma_x = 0 \quad \text{and} \quad \sigma_z = 0$$

Therefore from eqn. (14.6)

$$\varepsilon_v = \frac{\sigma}{E}(1 - 2\nu) = \frac{\delta V}{V}$$

$\therefore$

$$\text{change of volume} = \delta V = \frac{\sigma V}{E}(1 - 2\nu) \tag{14.8}$$

This formula could have been obtained from eqn. (14.5) with

$$\varepsilon_y = \frac{\sigma}{E} \quad \text{and} \quad \varepsilon_x = \varepsilon_z = \varepsilon_D = -\nu\frac{\sigma}{E}$$

then

$$\varepsilon_v = \varepsilon_x + \varepsilon_y + \varepsilon_z = \frac{\delta V}{V}$$

$\therefore$

$$\delta V = V\left(\frac{\sigma}{E} - 2\nu\frac{\sigma}{E}\right) = \frac{\sigma V}{E}(1 - 2\nu)$$

14.8. Effect of lateral restraint

(a) Restraint in one direction only

Consider a body subjected to a two-dimensional stress system with a rigid lateral restraint provided in the y direction as shown in Fig. 14.4. Whilst the material is free to contract laterally in the x direction the "Poisson's ratio" extension along the y axis is totally prevented.

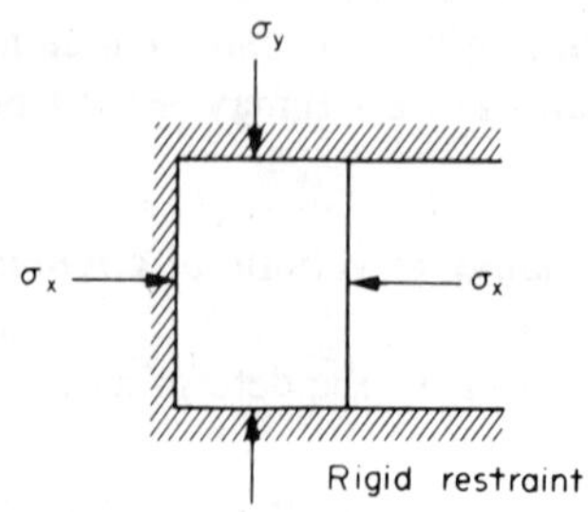

Fig. 14.4. Material subjected to lateral restraint in the y direction.

Therefore strain in the y direction with σ_x and σ_y both compressive, i.e. negative,

$$= \varepsilon_y = -\frac{1}{E}(\sigma_y - \nu\sigma_x) = 0$$

$$\therefore \qquad \sigma_y = \nu\sigma_x$$

Thus strain in the x direction

$$= \varepsilon_x = -\frac{1}{E}(\sigma_x - \nu\sigma_y)$$

$$= -\frac{1}{E}(\sigma_x - \nu^2\sigma_x)$$

$$= -\frac{\boldsymbol{\sigma_x}}{\boldsymbol{E}}(\mathbf{1} - \boldsymbol{\nu}^{\mathbf{2}}) \qquad (14.9)$$

Thus the introduction of a lateral restraint affects the stiffness and hence the load-carrying capacity of the material by producing an **effective change of Young's modulus** from

$$E \quad \text{to} \quad E/(1-\nu^2)$$

(b) Restraint in two directions

Consider now a material subjected to a three-dimensional stress system σ_x, σ_y and σ_z with restraint provided in both the y and z directions. In this case,

$$\varepsilon_y = -\frac{1}{E}(\sigma_y - \nu\sigma_x - \nu\sigma_z) = 0 \qquad (1)$$

and

$$\varepsilon_z = -\frac{1}{E}(\sigma_z - \nu\sigma_x - \nu\sigma_y) = 0 \qquad (2)$$

From (1),

$$\sigma_y = \nu\sigma_x + \nu\sigma_z$$

$$\therefore \qquad \sigma_z = (\sigma_y - \nu\sigma_x)\frac{1}{\nu} \qquad (3)$$

Substituting in (2),

$$\frac{1}{\nu}(\sigma_y - \nu\sigma_x) - \nu\sigma_x - \nu\sigma_y = 0$$

$$\therefore \qquad \sigma_y - \nu\sigma_x - \nu^2\sigma_x - \nu^2\sigma_y = 0$$

$$\sigma_y(1-\nu^2) = \sigma_x(\nu + \nu^2)$$

$$\sigma_y = \sigma_x\frac{\nu(1+\nu)}{(1-\nu^2)}$$

$$= \frac{\sigma_x\nu}{(1-\nu)}$$

and from (3),

$$\sigma_z = \frac{1}{\nu}\left[\frac{\nu\sigma_x}{(1-\nu)} - \nu\sigma_x\right]$$

$$= \sigma_x\left[\frac{1-(1-\nu)}{(1-\nu)}\right] = \frac{\nu\sigma_x}{(1-\nu)}$$

$$\therefore \qquad \text{strain in } x \text{ direction} = -\frac{\sigma_x}{E} + \nu\frac{\sigma_y}{E} + \nu\frac{\sigma_z}{E}$$

$$= -\frac{\sigma_x}{E}\left[1 - \frac{\nu^2}{(1-\nu)} - \frac{\nu^2}{(1-\nu)}\right]$$

$$= -\frac{\sigma_x}{E}\left[1 - \frac{2\nu^2}{(1-\nu)}\right] \qquad (14.10)$$

Again Young's modulus E is effectively changed, this time to

$$E\bigg/\left[1 - \frac{2\nu^2}{(1-\nu)}\right]$$

14.9. Relationship between the elastic constants E, G, K and ν

(a) E, G and ν

Consider a cube of material subjected to the action of the shear and complementary shear forces shown in Fig. 14.5 producing the strained shape indicated.

Assuming that the strains are small the angle ACB may be taken as 45°.

Therefore strain on diagonal OA

$$= \frac{BC}{OA} \simeq \frac{AC\cos 45^\circ}{a\sqrt{2}} = \frac{AC}{a\sqrt{2}} \times \frac{1}{\sqrt{2}} = \frac{AC}{2a}$$

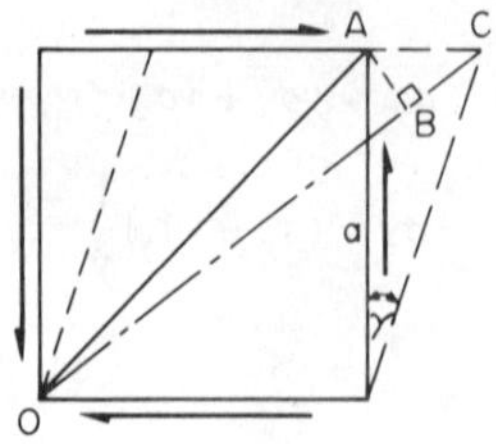

Fig. 14.5. Element subjected to shear and associated complementary shear.

But $AC = a\gamma$, where γ = angle of distortion or shear strain.

$$\therefore \qquad \text{strain on diagonal} = \frac{a\gamma}{2a} = \frac{\gamma}{2}$$

Now

$$\frac{\text{shear stress } \tau}{\text{shear strain } \gamma} = G$$

$$\therefore \qquad \gamma = \frac{\tau}{G}$$

$$\therefore \qquad \text{strain on diagonal} = \frac{\tau}{2G} \qquad (1)$$

From § 13.2 the shear stress system can be replaced by a system of direct stresses at 45°, as shown in Fig. 14.6. One set will be compressive, the other tensile, and both will be equal in value to the applied shear stresses.

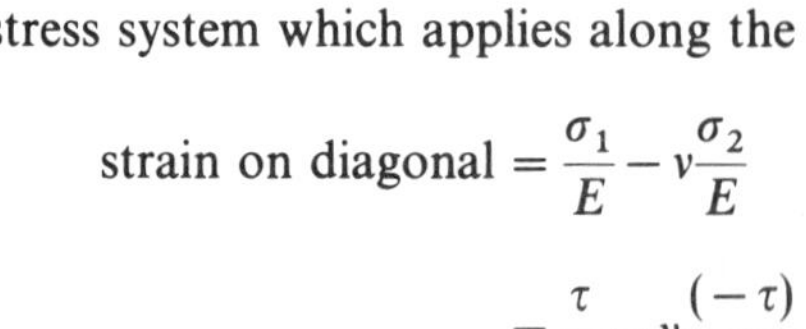

Fig. 14.6. Direct stresses due to shear.

Thus, from the direct stress system which applies along the diagonals:

$$\text{strain on diagonal} = \frac{\sigma_1}{E} - \nu\frac{\sigma_2}{E}$$

$$= \frac{\tau}{E} - \nu\frac{(-\tau)}{E}$$

$$= \frac{\tau}{E}(1+\nu) \qquad (2)$$

Combining (1) and (2),

$$\frac{\tau}{2G} = \frac{\tau}{E}(1+\nu)$$

$$\mathbf{E = 2G(1+\nu)} \qquad (14.11)$$

(*b*) *E, K and* ν

Consider a cube subjected to three equal stresses σ as in Fig. 14.7 (i.e. volumetric stress = σ).

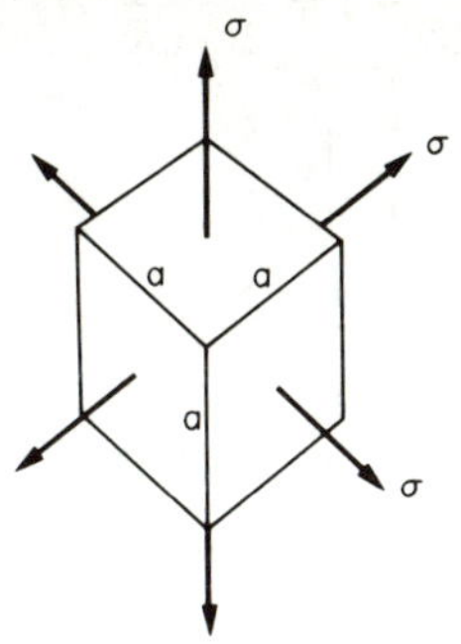

Fig. 14.7. Cubical element subjected to uniform stress σ on all faces ("volumetric" or "hydrostatic" stress).

$$\text{Total strain along one edge} = \frac{\sigma}{E} - \nu\frac{\sigma}{E} - \nu\frac{\sigma}{E}$$

$$= \frac{\sigma}{E}(1-2\nu)$$

But

$$\text{volumetric strain} = 3 \times \text{linear strain} \qquad \text{(see eqn. 14.5)}$$

$$= \frac{3\sigma}{E}(1-2\nu) \tag{3}$$

By definition:

$$\text{bulk modulus } K = \frac{\text{volumetric stress}}{\text{volumetric strain}}$$

$$\text{volumetric strain} = \frac{\sigma}{K} \tag{4}$$

Equating (3) and (4),

$$\frac{\sigma}{K} = \frac{3\sigma}{E}(1-2\nu)$$

$$\therefore \qquad E = \mathbf{3K(1-2\nu)} \tag{14.12}$$

(*c*) *G, K and* ν

Equations (14.11) and (14.12) can now be combined to give the final relationship as follows: From eqn. (14.11),

$$\nu = \frac{E}{2G} - 1$$

and from eqn. (14.12),

$$\nu = \frac{1}{2} - \frac{E}{6K}$$

Therefore, equating,

$$\frac{E}{2G} - 1 = \frac{1}{2} - \frac{E}{6K}$$

$$E\left[\frac{1}{2G} + \frac{1}{6K}\right] = \frac{3}{2}$$

$$\therefore \quad E\left[\frac{6K+2G}{12KG}\right] = \frac{3}{2}$$

i.e.

$$\boldsymbol{E = \frac{9KG}{(3K+G)}} \tag{14.13}$$

14.10. Strains on an oblique plane

(a) Linear strain

Consider a rectangular block of material $OLMN$ as shown in the xy plane (Fig. 14.8). The strains along Ox and Oy are ε_x and ε_y, and γ_{xy} is the shearing strain.

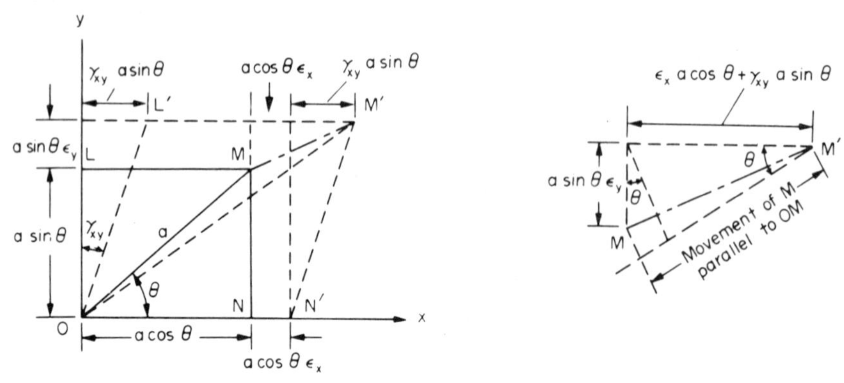

Fig. 14.8. Strains on an inclined plane.

Let the diagonal OM be of length a; then $ON = a\cos\theta$ and $OL = a\sin\theta$, and the increases in length of these sides under strain are $\varepsilon_x a\cos\theta$ and $\varepsilon_y a\sin\theta$ (i.e. strain × original length).

If M moves to M', the movement of M parallel to the x axis is

$$\varepsilon_x a\cos\theta + \gamma_{xy} a\sin\theta$$

and the movement parallel to the y axis is

$$\varepsilon_y a\sin\theta$$

Thus the movement of M parallel to OM, which since the strains are small is practically coincident with MM', is

$$(\varepsilon_x a\cos\theta+\gamma_{xy}a\sin\theta)\cos\theta+(\varepsilon_y a\sin\theta)\sin\theta$$

Then

$$\text{strain along } OM=\frac{\text{extension}}{\text{original length}}$$

$$=(\varepsilon_x\cos\theta+\gamma_{xy}\sin\theta)\cos\theta+(\varepsilon_y\sin\theta)\sin\theta$$

$$\therefore\qquad \varepsilon_\theta=\varepsilon_x\cos^2\theta+\varepsilon_y\sin^2\theta+\gamma_{xy}\sin\theta\cos\theta$$

$$\therefore\qquad \boldsymbol{\varepsilon_\theta=\tfrac{1}{2}(\varepsilon_x+\varepsilon_y)+\tfrac{1}{2}(\varepsilon_x-\varepsilon_y)\cos 2\theta+\tfrac{1}{2}\gamma_{xy}\sin 2\theta}\qquad(14.14)$$

This is identical in form with the equation defining the direct stress on any inclined plane θ with ε_x and ε_y replacing σ_x and σ_y and $\frac{1}{2}\gamma_{xy}$ replacing τ_{xy}, i.e. **the shear stress is replaced by HALF the shear strain.**

(b) Shear strain

To determine the shear strain in the direction OM consider the displacement of point P at the foot of the perpendicular from N to OM (Fig. 14.9).

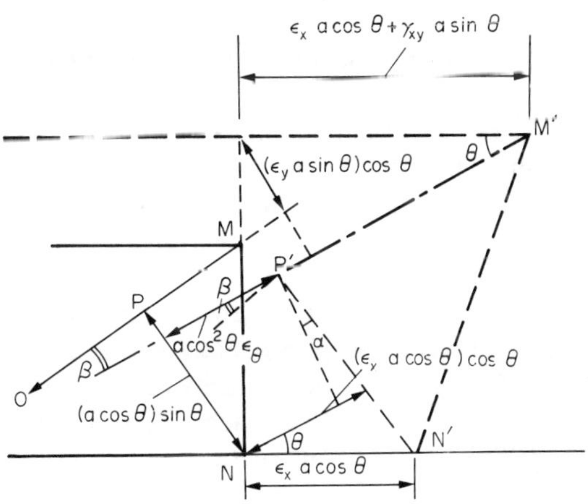

Fig. 14.9. Enlarged view of part of Fig. 14.8.

In the strained condition this point moves to P'.

Since $\qquad$ strain along $OM=\varepsilon_\theta$

$$\text{extension of } OM = OM\cdot\varepsilon_\theta$$

$$\therefore\qquad \text{extension of } OP = OP\cdot\varepsilon_\theta$$

But $\qquad OP=(a\cos\theta)\cos\theta$

$$\therefore\qquad \text{extension of } OP = a\cos^2\theta\,\varepsilon_\theta$$

During straining the line PN rotates counterclockwise through a small angle α.

$$\alpha = \frac{(\varepsilon_x a\cos\theta)\cos\theta - a\cos^2\theta\varepsilon_\theta}{a\cos\theta\sin\theta}$$

$$= (\varepsilon_x - \varepsilon_\theta)\cot\theta$$

The line OM also rotates, but clockwise, through a small angle

$$\beta = \frac{(\varepsilon_x a\cos\theta + \gamma_{xy} a\sin\theta)\sin\theta - (\varepsilon_y a\sin\theta)\cos\theta}{a}$$

Thus the required shear strain γ_θ in the direction OM, i.e. the amount by which the angle OPN changes, is given by

$$\gamma_\theta = \alpha + \beta = (\varepsilon_x - \varepsilon_\theta)\cot\theta + (\varepsilon_x\cos\theta + \gamma_{xy}\sin\theta)\sin\theta - \varepsilon_y\sin\theta\cos\theta$$

Substituting for ε_θ from eqn. (14.14) gives

$$\gamma_\theta = 2(\varepsilon_x - \varepsilon_y)\cos\theta\sin\theta - \gamma_{xy}(\cos^2\theta - \sin^2\theta)$$

$$\therefore \quad \tfrac{1}{2}\gamma_\theta = \tfrac{1}{2}(\varepsilon_x - \varepsilon_y)\sin 2\theta - \tfrac{1}{2}\gamma_{xy}\cos 2\theta$$

which again is similar in form to the expression for the shear stress τ on any inclined plane θ.

For consistency of sign convention, however (see §14.11 below), because OM' moves clockwise with respect to OM it is considered to be a negative shear strain, i.e.

$$\boldsymbol{\tfrac{1}{2}\gamma_\theta = -\left[\tfrac{1}{2}(\varepsilon_x - \varepsilon_y)\sin 2\theta - \tfrac{1}{2}\gamma_{xy}\cos 2\theta\right]} \qquad (14.15)$$

14.11. Principal strain – Mohr's strain circle

Since the equations for stress and strain on oblique planes are identical in form, as noted above, it is evident that Mohr's stress circle construction can be used equally well to represent strain conditions using the horizontal axis for linear strains and the vertical axis for *half* the shear strain. It should be noted, however, that angles given by Mohr's stress circle refer to the directions of the planes on which the stresses act and not to the direction of the stresses themselves. The directions of the stresses and hence the associated strains are therefore normal (i.e. at 90°) to the directions of the planes. Since angles are doubled in Mohr's circle construction it follows therefore that for true similarity of working a relative rotation of the axes of 2 × 90 = 180° must be introduced. This is achieved by plotting positive shear strains vertically *downwards* on the strain circle construction as shown in Fig. 14.10.

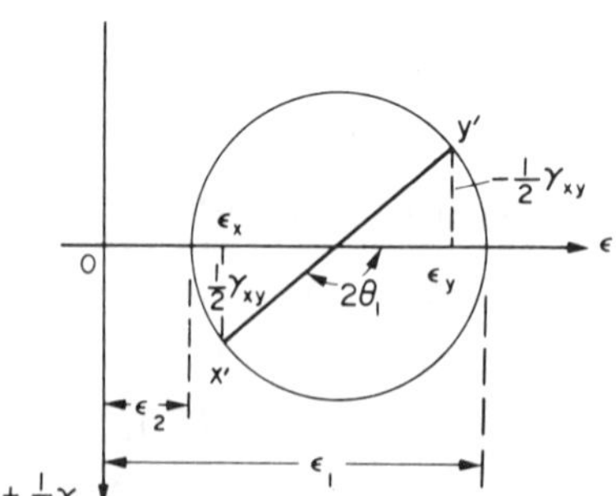

Fig. 14.10. Mohr's strain circle.

The **sign convention** adopted for strains is as follows:

Linear strains: extension positive
compression negative.

Shear strains:

The convention for shear strains is a little more difficult. The first subscript in the symbol γ_{xy} usually denotes the shear strain associated with that direction, i.e. with Ox. Similarly, γ_{yx} is the shear strain associated with Oy. If, under strain, the line associated with the first subscript moves counterclockwise with respect to the other line, the shearing strain is said to be positive, and if it moves clockwise it is said to be negative. It will then be seen that positive shear strains are associated with planes carrying positive shear stresses and negative shear strains with planes carrying negative shear stresses.

Thus, $$\gamma_{xy} = -\gamma_{yx}$$

Mohr's circle for strains ε_x, ε_y and shear strain γ_{xy} (positive referred to x direction) is therefore constructed as for the stress circle with $\frac{1}{2}\gamma_{xy}$ replacing τ_{xy} and the axis of shear reversed, as shown in Fig. 14.10.

The maximum principal strain is then ε_1 at an angle θ_1 to ε_x in the same angular direction as that in Mohr's circle (Fig. 14.11).

Again, **angles are doubled on Mohr's circle.**

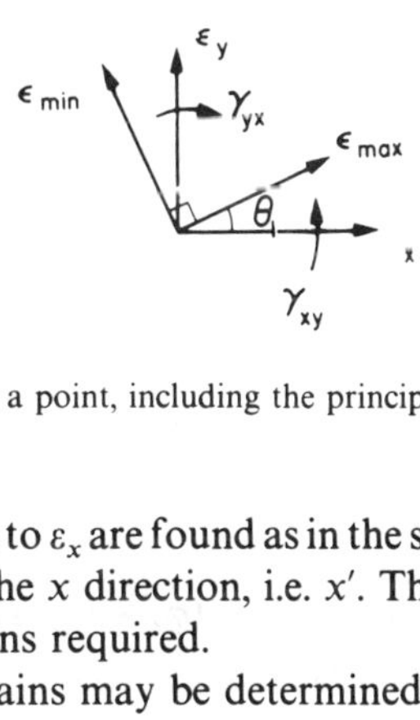

Fig. 14.11. Strain system at a point, including the principal strains and their inclination.

Strain conditions at any angle α to ε_x are found as in the stress circle by marking off an angle 2α from the point representing the x direction, i.e. x'. The coordinates of the point on the circle thus obtained are the strains required.

Alternatively, the principal strains may be determined *analytically* from eqn. (14.14),

i.e. $$\varepsilon_\theta = \tfrac{1}{2}(\varepsilon_x + \varepsilon_y) + \tfrac{1}{2}(\varepsilon_x - \varepsilon_y)\cos 2\theta + \tfrac{1}{2}\gamma_{xy}\sin 2\theta$$

As for the derivation of the principal stress equations on page 331, the principal strains, i.e. the maximum and minimum values of strain, occur at values of θ obtained by equating $d\varepsilon_\theta/d\theta$ to zero.

The procedure is identical to that of page 331 for the stress case and will not be repeated here. The values obtained are

$$\varepsilon_1 \text{ or } \varepsilon_2 = \tfrac{1}{2}(\varepsilon_x + \varepsilon_y) \pm \tfrac{1}{2}\sqrt{[(\varepsilon_x - \varepsilon_y)^2 + \gamma_{xy}^2]} \qquad (14.16)$$

i.e. once again identical in form to the principal stress equation with ε replacing σ and $\frac{1}{2}\gamma$ replacing τ.

Similarly, $$\tfrac{1}{2}\gamma_{max} = \pm\tfrac{1}{2}\sqrt{[(\varepsilon_x - \varepsilon_y)^2 + \gamma_{xy}^2]} \qquad (14.17)$$

14.12. Mohr's strain circle – alternative derivation from the general stress equations

The direct stress on any plane within a material inclined at an angle θ to the xy axes is given by eqn. (13.8) as:

$$\sigma_\theta = \tfrac{1}{2}(\sigma_x + \sigma_y) + \tfrac{1}{2}(\sigma_x - \sigma_y)\cos 2\theta + \tau_{xy}\sin 2\theta$$

$$\therefore \quad \sigma_{\theta+90} = \tfrac{1}{2}(\sigma_x + \sigma_y) + \tfrac{1}{2}(\sigma_x - \sigma_y)\cos(2\theta + 180^\circ) + \tau_{xy}\sin(2\theta + 180^\circ)$$

$$= \tfrac{1}{2}(\sigma_x + \sigma_y) - \tfrac{1}{2}(\sigma_x - \sigma_y)\cos 2\theta - \tau_{xy}\sin 2\theta$$

Also, from eqn. (13.9),

$$\tau_\theta = \tfrac{1}{2}(\sigma_x - \sigma_y)\sin 2\theta - \tau_{xy}\cos 2\theta \qquad (1)$$

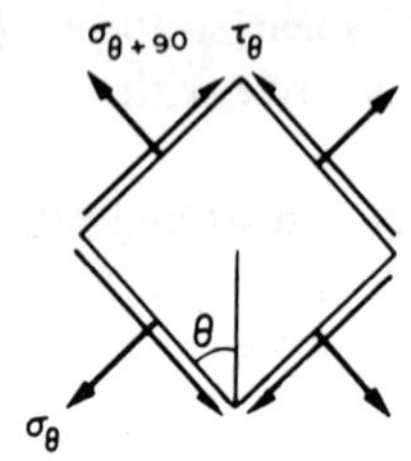

Fig. 14.12.

Now for the two-dimensional stress system shown in Fig. 14.12,

$$\varepsilon_\theta = \frac{1}{E}(\sigma_\theta - \nu\sigma_{\theta+90})$$

$$= \frac{1}{E}\{[\tfrac{1}{2}(\sigma_x + \sigma_y) + \tfrac{1}{2}(\sigma_x - \sigma_y)\cos 2\theta + \tau_{xy}\sin 2\theta]$$

$$- \nu[\tfrac{1}{2}(\sigma_x + \sigma_y) - \tfrac{1}{2}(\sigma_x - \sigma_y)\cos 2\theta - \tau_{xy}\sin 2\theta]\}$$

$$= \frac{1}{E}[\tfrac{1}{2}(1-\nu)(\sigma_x + \sigma_y) + \tfrac{1}{2}(1+\nu)(\sigma_x - \sigma_y)\cos 2\theta + (1+\nu)\tau_{xy}\sin 2\theta]$$

But
$$\varepsilon_y = \frac{1}{E}(\sigma_y - \nu\sigma_x)$$

and
$$\varepsilon_x = \frac{1}{E}(\sigma_x - \nu\sigma_y)$$

from which
$$\sigma_y = \frac{E}{(1-\nu^2)}[\varepsilon_y + \nu\varepsilon_x]$$

and
$$\sigma_x = \frac{E}{(1-\nu^2)}[\varepsilon_x + \nu\varepsilon_y]$$

$$\therefore \quad \tfrac{1}{2}(\sigma_x + \sigma_y) = \frac{E(1+\nu)}{2(1-\nu^2)}(\varepsilon_y + \varepsilon_x)$$

and $$\tfrac{1}{2}(\sigma_x - \sigma_y) = \frac{E(1-\nu)}{2(1-\nu^2)}(\varepsilon_x - \varepsilon_y)$$

$\therefore$ $$\varepsilon_\theta = \frac{1}{E}\left[\frac{E(1-\nu)(1+\nu)(\varepsilon_y+\varepsilon_x)}{2(1-\nu^2)} + \frac{E(1+\nu)(1-\nu)(\varepsilon_x-\varepsilon_y)}{2(1-\nu^2)}\cos 2\theta + (1+\nu)\tau_{xy}\sin 2\theta\right]$$

$$= \tfrac{1}{2}(\varepsilon_y+\varepsilon_x) + \tfrac{1}{2}(\varepsilon_x-\varepsilon_y)\cos 2\theta + \tau_{xy}\frac{(1+\nu)}{E}\sin 2\theta$$

Now $$\frac{\tau}{\gamma} = G \quad \therefore \quad \tau = G\gamma \quad \text{and} \quad E = 2G(1+\nu)$$

$\therefore$ $$\tau_{xy} = \frac{E}{2(1+\nu)}\gamma_{xy}$$

$\therefore$ $$\boldsymbol{\varepsilon_\theta = \tfrac{1}{2}(\varepsilon_x+\varepsilon_y) + \tfrac{1}{2}(\varepsilon_x-\varepsilon_y)\cos 2\theta + \tfrac{1}{2}\gamma_{xy}\sin 2\theta} \qquad (14.14)$$

Similarly, substituting for $\tfrac{1}{2}(\sigma_x - \sigma_y)$ and τ_{xy} in (1),

$$\tau_\theta = \frac{E(\varepsilon_x-\varepsilon_y)(1-\nu)}{2(1-\nu^2)}\sin 2\theta - \frac{E}{2(1+\nu)}\gamma_{xy}\cos 2\theta$$

But $$\tau_\theta = \frac{E}{2(1+\nu)}\gamma_\theta$$

$\therefore$ $$\frac{E}{2(1+\nu)}\gamma_\theta = \frac{E(\varepsilon_x-\varepsilon_y)}{2(1+\nu)}\sin 2\theta - \frac{E}{2(1+\nu)}\gamma_{xy}\cos 2\theta$$

$$\gamma_\theta = (\varepsilon_x-\varepsilon_y)\sin 2\theta - \gamma_{xy}\cos 2\theta$$

$\therefore$ $$\tfrac{1}{2}\gamma_\theta = \tfrac{1}{2}(\varepsilon_x-\varepsilon_y)\sin 2\theta - \tfrac{1}{2}\gamma_{xy}\cos 2\theta$$

Again, for consistency of sign convention, since OM will move clockwise under strain, the above shear strain must be considered negative,

i.e. $$\boldsymbol{\tfrac{1}{2}\gamma_\theta = -[\tfrac{1}{2}(\varepsilon_x-\varepsilon_y)\sin 2\theta - \tfrac{1}{2}\gamma_{xy}\cos 2\theta]} \qquad (14.15)$$

Equations (14.14) and (14.15) are similar in form to eqns. (13.8) and (13.9) which are the basis of Mohr's circle solution for stresses provided that $\tfrac{1}{2}\gamma_{xy}$ is used in place of τ_{xy} and linear stresses σ are replaced by linear strains ε. **These equations will therefore provide a graphical solution known as Mohr's strain circle if axes of ε and $\tfrac{1}{2}\gamma$ are used.**

14.13. Relationship between Mohr's stress and strain circles

Consider now a material subjected to the two-dimensional principal stress system shown in Fig. 14.13a. The stress and strain circles are then as shown in Fig. 14.13(b) and (c).

For Mohr's stress circle (Fig. 14.13b),

$$OA \times \text{stress scale} = \frac{(\sigma_1+\sigma_2)}{2}$$

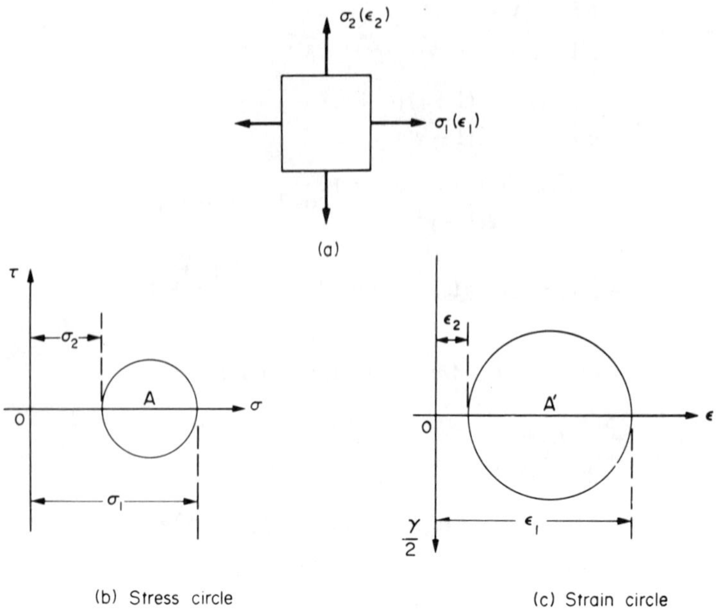

Fig. 14.13.

$$\therefore \qquad OA = \frac{(\sigma_1 + \sigma_2)}{2 \times \text{stress scale}} \tag{1}$$

and radius of stress circle × stress scale $= \frac{1}{2}(\sigma_1 - \sigma_2)$ (2)

For Mohr's strain circle (Fig. 14.13c),

$$OA' \times \text{strain scale} = \frac{(\varepsilon_1 + \varepsilon_2)}{2}$$

But

$$\varepsilon_1 = \frac{1}{E}(\sigma_1 - \nu\sigma_2)$$

and

$$\varepsilon_2 = \frac{1}{E}(\sigma_2 - \nu\sigma_1)$$

$$\therefore \qquad \varepsilon_1 + \varepsilon_2 = \frac{1}{E}[(\sigma_1 + \sigma_2) - \nu(\sigma_1 + \sigma_2)]$$

$$= \frac{1}{E}(\sigma_1 + \sigma_2)(1 - \nu)$$

$$\therefore \qquad OA' = \frac{(\sigma_1 + \sigma_2)(1 - \nu)}{2E \times \text{strain scale}} \tag{3}$$

Thus, in order that the circles shall be concentric (Fig. 14.14),

$$OA = OA'$$

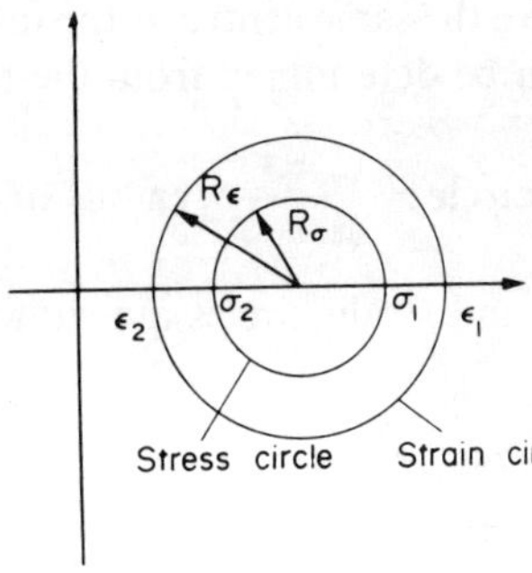

Fig. 14.14. Combined stress and strain circles.

Therefore from (1) and (3)

$$\frac{(\sigma_1+\sigma_2)}{2\times\text{stress scale}} = \frac{(\sigma_1+\sigma_2)(1-\nu)}{2E\times\text{strain scale}}$$

$$\therefore \qquad \textbf{stress scale} = \frac{\boldsymbol{E}}{\boldsymbol{(1-\nu)}}\times\textbf{strain scale} \tag{14.18}$$

Now radius of strain circle × strain scale

$$= \tfrac{1}{2}(\varepsilon_1-\varepsilon_2)$$

$$= \frac{1}{2E}[(\sigma_1-\nu\sigma_2)-(\sigma_2-\nu\sigma_1)]$$

$$= \frac{1}{2E}(\sigma_1-\sigma_2)(1+\nu) \tag{4}$$

$$\therefore \qquad \frac{\text{radius of stress circle}\times\text{stress scale}}{\text{radius of strain circle}\times\text{strain scale}} = \frac{\tfrac{1}{2}(\sigma_1-\sigma_2)}{\dfrac{1}{2E}(\sigma_1-\sigma_2)(1+\nu)}$$

$$= \frac{E}{(1+\nu)}$$

$$\frac{\text{radius of stress circle}}{\text{radius of strain circle}} = \frac{E}{(1+\nu)}\times\frac{\text{strain scale}}{\text{stress scale}}$$

i.e.

$$\frac{R_\sigma}{R_\varepsilon} = \frac{E}{(1+\nu)}\times\frac{(1-\nu)}{E}$$

$$= \frac{\boldsymbol{(1-\nu)}}{\boldsymbol{(1+\nu)}} \tag{14.19}$$

In other words, provided suitable scales are chosen so that

$$\textbf{stress scale} = \frac{\boldsymbol{E}}{\boldsymbol{(1-\nu)}}\times\textbf{strain scale}$$

the stress and strain circles will have the same centre. If the radius of one circle is known the radius of the other circle can then be determined from the relationship

$$\textbf{radius of stress circle} = \frac{(1-\nu)}{(1+\nu)} \times \textbf{radius of strain circle}$$

Other relationships for the stress and strain circles are shown in Fig. 14.15.

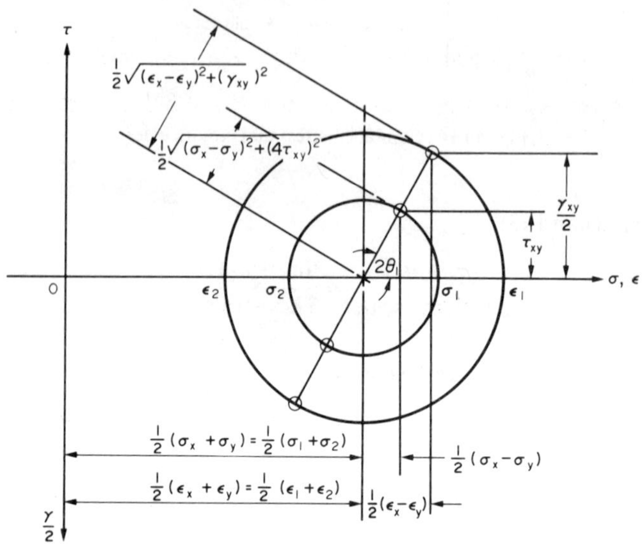

Fig. 14.15. Other relationships for Mohr's stress and strain circles.

14.14. Construction of strain circle from three known strains (McClintock method)–rosette analysis

In order to measure principal strains on the surface of engineering components the normal experimental technique involves the bonding of a strain gauge rosette (see Chapter 21) at the point under consideration. This gives the values of strain in three known directions and enables Mohr's strain circle to be constructed as follows.

Consider the three-strain system shown in Fig. 14.16, the known directions of strain being at angles α_a, α_b and α_c to a principal strain direction (this being one of the primary requirements of such readings). The construction sequence is then:

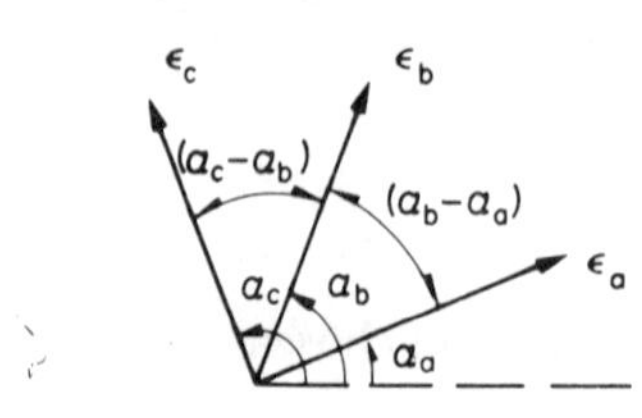

Fig. 14.16. System of three known strains.

(1) On a horizontal line *da* mark off the known strains ε_a, ε_b and ε_c to the same scale to give points *a*, *b* and *c* (see Fig. 14.18).

(2) From *a*, *b* and *c* draw perpendiculars to the line *da*.

(3) From a convenient point *X* on the perpendicular through *b* mark off lines corresponding to the known strain directions of ε_a and ε_c to intersect (projecting back if necessary) perpendiculars through *c* and *a* at *C* and *A*.

Note that these directions must be identical relative to *Xb* as they are relative to ε_b in Fig. 14.16,

i.e. *XC* is $\alpha_c - \alpha_b$ counterclockwise from *Xb* and *XA* is $\alpha_b - \alpha_a$ clockwise from *Xb*

(4) Construct perpendicular bisectors of the lines *XA* and *XC* to meet at the point *Y*, which is then the centre of Mohr's strain circle (Fig. 14.18).

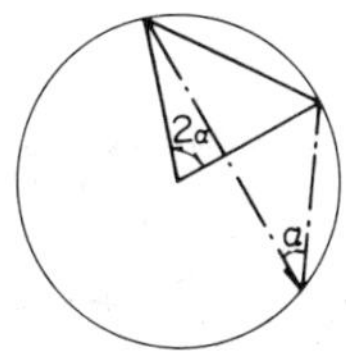

Fig. 14.17. Useful relationship for development of Mohr's strain circle (see Fig. 14.18).

(5) With centre *Y* and radius *YA* or *YC* draw the strain circle to cut *Xb* in the point *B*.

(6) The vertical shear strain axis can now be drawn through the zero of the strain scale *da*; the horizontal linear strain axis passes through *Y*.

(7) Join points *A*, *B* and *C* to *Y*. These radii must then be in the same angular order as the original strain directions. As in Mohr's stress circle, however, angles between them will be double in value, as shown in Fig. 14.18.

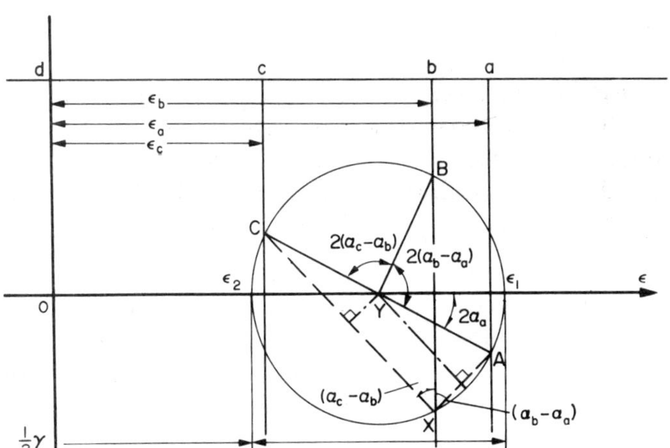

Fig. 14.18. Construction of strain circle from three known strains – McClintock construction. (Strain gauge rosette analysis.)

The principal strains are then ε_1 and ε_2 as indicated. Principal stresses can now be determined either from the relationships

$$\sigma_1 = \frac{E}{(1-\nu^2)}[\varepsilon_1 + \nu\varepsilon_2] \quad \text{and} \quad \sigma_2 = \frac{E}{(1-\nu^2)}[\varepsilon_2 + \nu\varepsilon_1]$$

or by superimposing the stress circle using the relationships established in §14.13.

The above construction applies whatever the values of strain and whatever the angles between the individual gauges of the rosette. The process is simplified, however, if the rosette axes are arranged:

(a) in sequence, in order of ascending or descending strain magnitude,

(b) so that the included angle between axes of maximum and minimum strain is less than 180°.

For example, consider three possible results of readings from the rosette of Fig. 14.16 as shown in Fig. 14.19(i), (ii) and (iii).

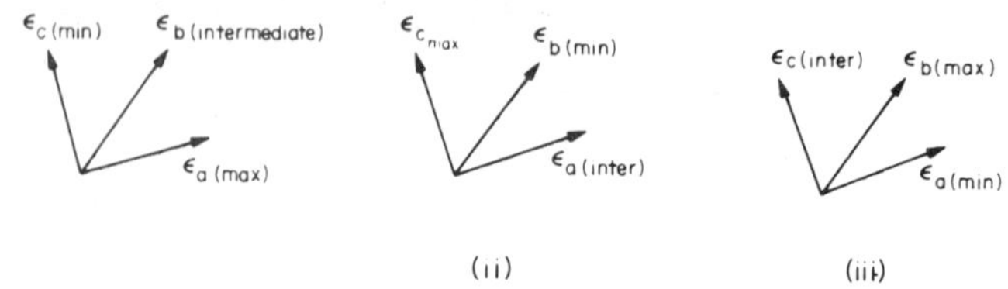

Fig. 14.19. Three possible orders of results from any given strain gauge rosette.

These may be rearranged as suggested above by projecting axes where necessary as shown in Fig. 14.20(i), (ii) and (iii).

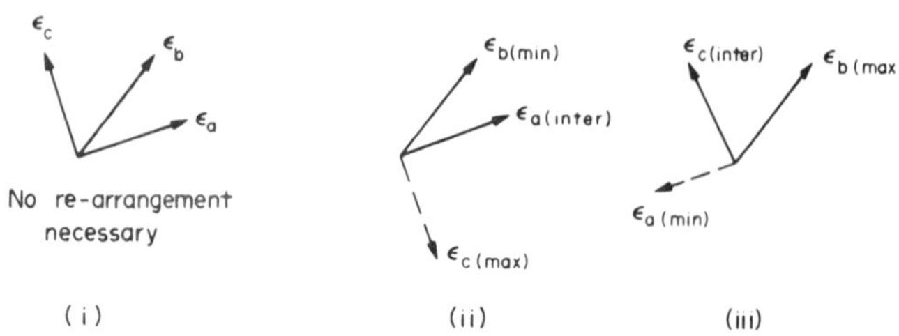

Fig. 14.20. Suitable rearrangement of Fig. 14.19 to facilitate the McClintock construction.

In all the above cases, the most convenient construction still commences with the starting point X on the vertical through the intermediate strain value, and will appear similar in form to the construction of Fig. 14.18.

Mohr's strain circle solution of rosette readings is strongly recommended because of its simplicity, speed and the ease with which principal stresses may be obtained by superimposing Mohr's stress circle. In addition, when one becomes familiar with the construction procedure, there is little opportunity for arithmetical error. As stated in the previous chapter, the advent of cheap but powerful calculators and microcomputers may reduce the effectiveness of Mohr's circle as a quantitative tool. It remains, however, a very powerful medium for the teaching and understanding of complex stress and strain systems and a valuable "aide-memoire" for some of the complex formulae which may be required for solution by other means. For example Fig. 14.21 shows the use of a free-hand sketch of the Mohr circle given by rectangular strain gauge rosette readings to obtain, from simple geometry, the corresponding principal strain equations.

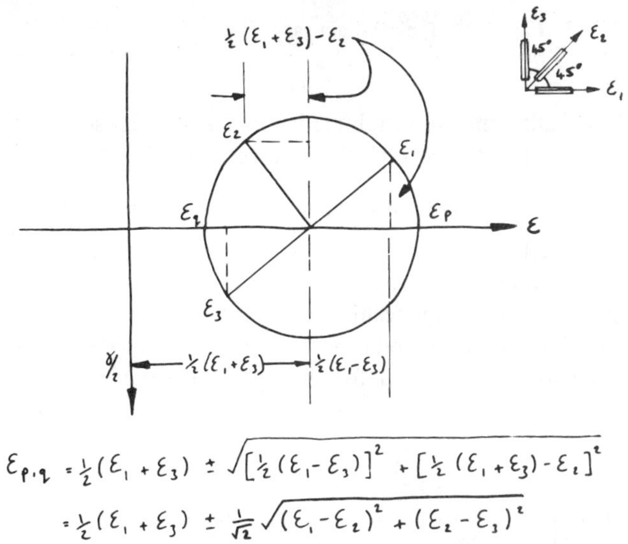

Fig. 14.21. Free-hand sketch of Mohr's strain circle.

14.15. Analytical determination of principal strains from rosette readings

The values of the principal strains associated with the three strain readings taken from a strain gauge rosette may be found by calculation using eqn. (14.14),

i.e.
$$\varepsilon_\theta = \tfrac{1}{2}(\varepsilon_x+\varepsilon_y)+\tfrac{1}{2}(\varepsilon_x-\varepsilon_y)\cos 2\theta+\tfrac{1}{2}\gamma_{xy}\sin 2\theta$$

This equation can be applied three times for the three values of θ of the rosette gauges. Thus with three known values of ε_θ for three known values of θ, three simultaneous equations will give the unknown strains ε_x, ε_y and γ_{xy}.

The principal strains can then be determined from eqn. (14.16).

$$\varepsilon_1 \text{ or } \varepsilon_2 = \tfrac{1}{2}(\varepsilon_x+\varepsilon_y)\pm\tfrac{1}{2}\sqrt{[(\varepsilon_x-\varepsilon_y)^2+{\gamma_{xy}}^2]}$$

The direction of the principal strain axes are then given by the equivalent strain expression to that derived for stresses [eqn. (13.10)],

i.e.
$$\tan 2\theta = \frac{\gamma_{xy}}{(\varepsilon_x-\varepsilon_y)} \qquad (14.20)$$

angles being given relative to the X axis.

The majority of rosette gauges in common use today are either rectangular rosettes with $\theta = 0°$, 45° and 90° or delta rosettes with $\theta = 0°$, 60° and 120° (Fig. 14.22).

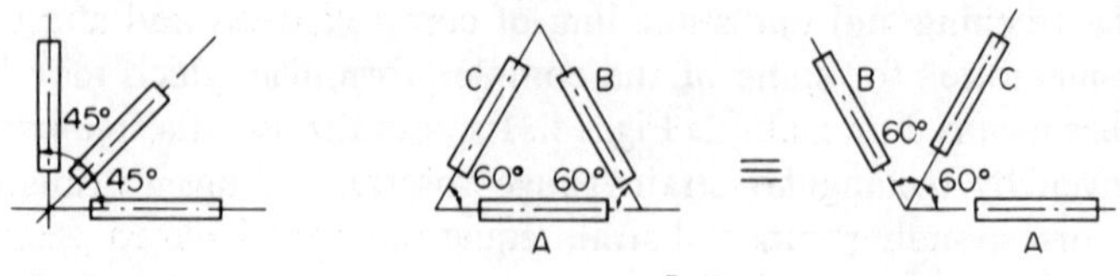

Fig. 14.22. Typical strain gauge rosette configurations.

In each case the calculations are simplified if the X axis is chosen to coincide with $\theta = 0$. Then, for both types of rosette, eqn. (14.14) reduces (for $\theta = 0$) to

$$\varepsilon_0 = \tfrac{1}{2}(\varepsilon_x + \varepsilon_y) + \tfrac{1}{2}(\varepsilon_x - \varepsilon_y) = \varepsilon_x$$

and ε_x is obtained directly from the ε_0 strain gauge reading. Similarly, for the rectangular rosette ε_y is obtained directly from the ε_{90° reading.

If a large number of rosette gauge results have to be analysed, the calculation process may be computerised. In this context the relationship between the rosette readings and resulting principal stresses shown in Table 14.1 for three standard types of strain gauge rosette is recommended.

TABLE 14.1. Principal strains and stresses from strain gauge rosettes*
(Gauge readings = σ_1, σ_2 and σ_3; Principal stresses = σ_p and σ_q.)

Rectangular (45°) Rosette——Arbitrarily oriented with respect to principal axes.

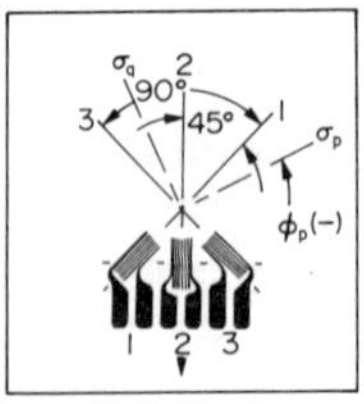

$$\epsilon_{p,q} = \frac{\epsilon_1+\epsilon_3}{2} \pm \frac{1}{\sqrt{2}}\sqrt{(\epsilon_1-\epsilon_2)^2+(\epsilon_2-\epsilon_3)^2}$$

$$\sigma_{p,q} = \frac{E}{2}\left(\frac{\epsilon_1+\epsilon_3}{1-\nu} \pm \frac{\sqrt{2}}{1+\nu}\sqrt{(\epsilon_1-\epsilon_2)^2+(\epsilon_2-\epsilon_3)^2}\right)$$

$$\phi_{p,q} = \frac{1}{2}\tan^{-1}\left(\frac{(\epsilon_2-\epsilon_3)-(\epsilon_1-\epsilon_2)}{\epsilon_1-\epsilon_3}\right) \left(\text{if } \epsilon_1 > \frac{\epsilon_1+\epsilon_3}{2}, \phi_{p,q} = \phi_p\right.$$

$$\text{if } \epsilon_1 < \frac{\epsilon_1+\epsilon_3}{2}, \phi_{p,q} = \phi_q$$

$$\left.\text{if } \epsilon_1 = \frac{\epsilon_1+\epsilon_3}{2}, \phi_p = \pm 45^\circ\right)$$

Delta (equiangular) Rosette——Arbitrarily oriented with respect to principal axes.

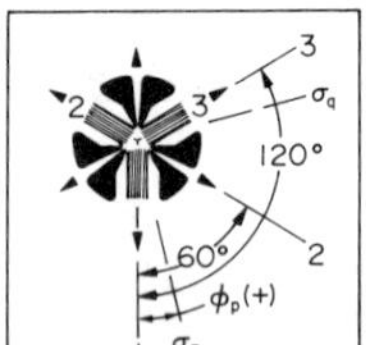

$$\epsilon_{p,q} = \frac{\epsilon_1+\epsilon_2+\epsilon_3}{3} \pm \frac{\sqrt{2}}{3}\sqrt{(\epsilon_1-\epsilon_2)^2+(\epsilon_2-\epsilon_3)^2+(\epsilon_1-\epsilon_3)^2}$$

$$\sigma_{p,q} = \frac{E}{3}\left(\frac{\epsilon_1+\epsilon_2+\epsilon_3}{1-\nu} \pm \frac{\sqrt{2}}{1+\nu}\sqrt{(\epsilon_1-\epsilon_2)^2+(\epsilon_2-\epsilon_3)^2+(\epsilon_1-\epsilon_3)^2}\right)$$

$$\phi_{p,q} = \frac{1}{2}\tan^{-1}\left(\frac{\sqrt{3}(\epsilon_2-\epsilon_3)}{(\epsilon_1-\epsilon_2)+(\epsilon_1-\epsilon_3)}\right) \left(\text{if } \epsilon_1 > \frac{\epsilon_1+\epsilon_2+\epsilon_3}{3}, \phi_{p,q} = \phi_p\right.$$

$$\text{if } \epsilon_1 < \frac{\epsilon_1+\epsilon_2+\epsilon_3}{3}, \phi_{p,q} = \phi_q$$

$$\left.\text{if } \epsilon_1 = \frac{\epsilon_1+\epsilon_2+\epsilon_3}{3}, \phi_p = \pm 45^\circ\right)$$

Tee Rosette——Gage elements must be aligned with principal axes.

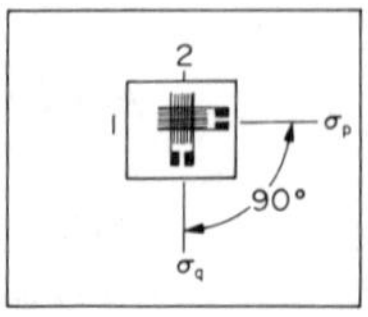

$$\epsilon_p = \epsilon_1$$

$$\epsilon_q = \epsilon_2$$

$$\sigma_p = \frac{E}{1-\nu^2}(\epsilon_1+\nu\epsilon_2)$$

$$\sigma_q = \frac{E}{1-\nu^2}(\epsilon_2+\nu\epsilon_1)$$

Rosette Gage-Numbering Considerations

The equations at the left for calculating principal strains and stresses from rosette strain measurements assume that the gage elements are numbered in a particular manner. Improper numbering of the gage elements will lead to ambiguity in the interpretation of $\phi_{p,q}$; and, in the case of the rectangular rosette, can also cause errors in the calculated principal strains and stresses.

Treating the latter situation first, it is always necessary in a rectangular rosette that gage numbers 1 and 3 be assigned to the two mutually perpendicular gages. Any other numbering arrangement will produce incorrect principal strains and stresses.

Ambiguities in the interpretation of $\phi_{p,q}$ for both rectangular and delta rosettes can be eliminated by numbering the gage elements as follows:

In a rectangular rosette, Gage 2 must be 45° away from Gage 1 : and Gage 3 must be 90° away, in the same direction . Similarly, in a delta rosette. Gages 2 and 3 must be 60° and 120° away respectively, in the same direction from Gage 1. By definition, $\phi_{p,q}$ is the angle from the axis of Gage 1 to the nearest principal axis. When $\phi_{p,q}$ is positive, the direction is the same as that of the gage numbering; and, when negative, the opposite.

* Reproduced with permission from Vishay Measurements Ltd wall chart.

14.16. Alternative representations of strain distributions at a point

Alternative forms of representation for the distribution of stress at a point were presented in §13.7; the directly equivalent representations for strain are given below.

The values of the direct strain ε_θ and shear strain γ_θ for any inclined plane θ are given by equations (14.14) and (14.15) as

$$\varepsilon_\theta = \tfrac{1}{2}(\varepsilon_x + \varepsilon_y) + \tfrac{1}{2}(\varepsilon_x - \varepsilon_y)\cos 2\theta + \tfrac{1}{2}\gamma_{xy}\sin 2\theta$$

$$\tfrac{1}{2}\gamma_\theta = -[\tfrac{1}{2}(\varepsilon_x - \varepsilon_y)\sin 2\theta - \tfrac{1}{2}\gamma_{xy}\cos 2\theta]$$

Plotting these values for the uniaxial stress state on Cartesian axes yields the curves of Fig. 14.23 which can then be compared directly to the equivalent stress distributions of Fig. 13.12. Again the shear curves are "shifted" by 45° from the normal strain curves.

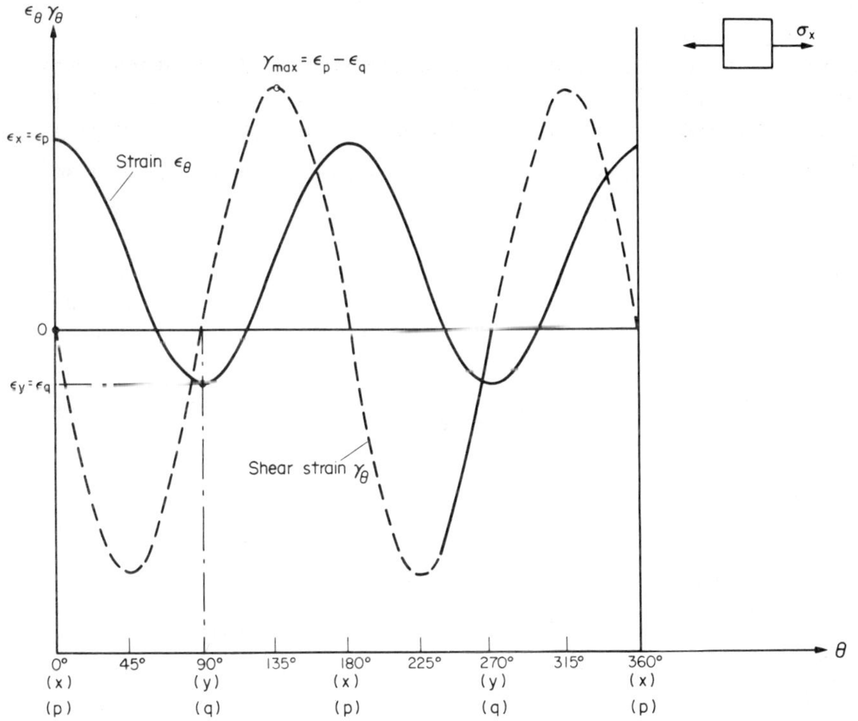

Fig. 14.23. Cartesian plot of strain distribution at a point under uniaxial applied stress.

Comparison with Fig. 13.12 shows that the normal stress and shear stress curves are each in phase with their respective normal strain and shear strain curves. Other relationships between the shear strain and normal strain curves are identical to those listed on page 335 for the normal stress and shear stress distributions.

The alternative polar strain representation for the uniaxial stress system is shown in Fig. 14.24 whilst the Cartesian and polar diagrams for the same biaxial stress systems used for Figs. 13.14 and 13.15 are shown in Figs. 14.25 and 14.26.

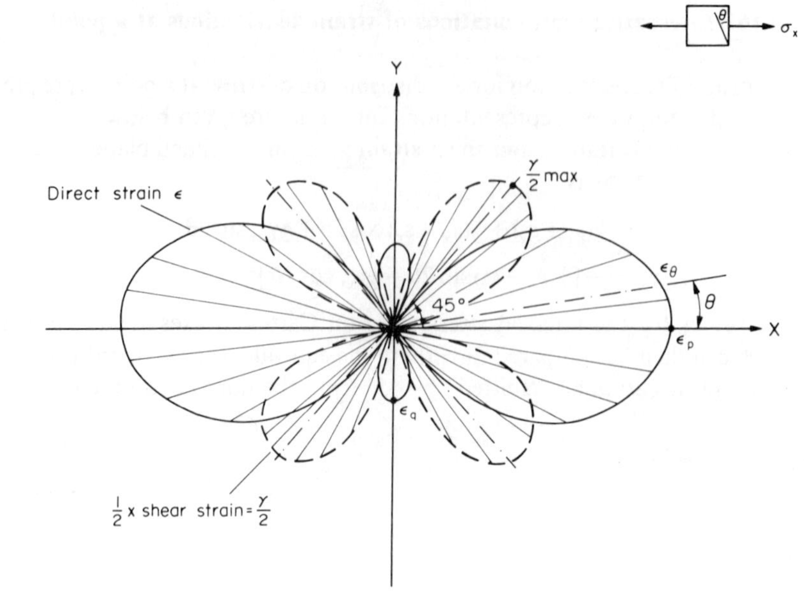

Fig. 14.24. Polar plot of strain distribution at a point under uniaxial applied stress.

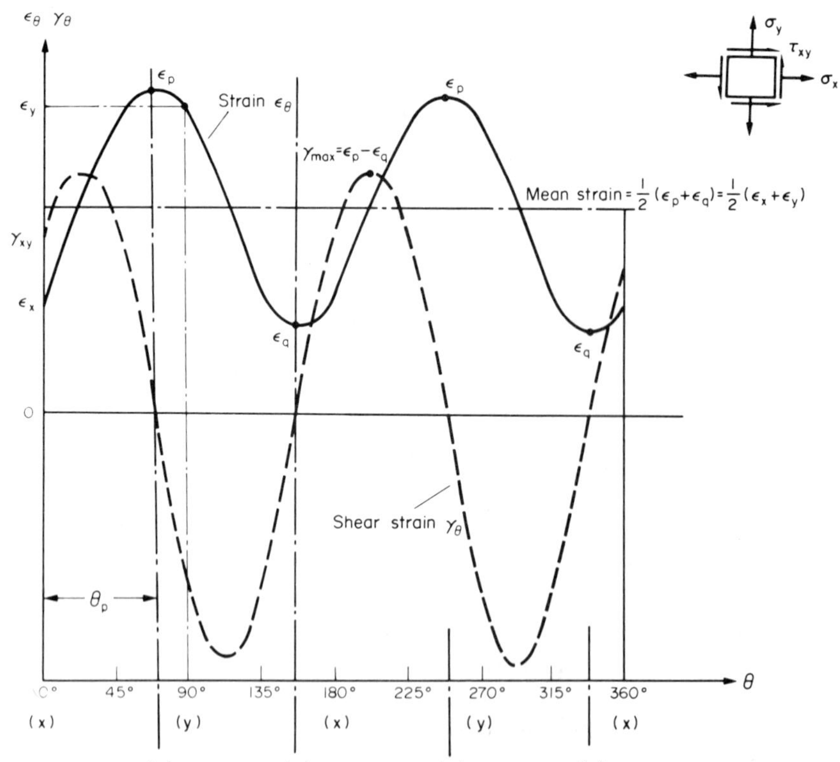

Fig. 14.25. Cartesian plot of strain distribution at a point under a typical biaxial applied stress system.

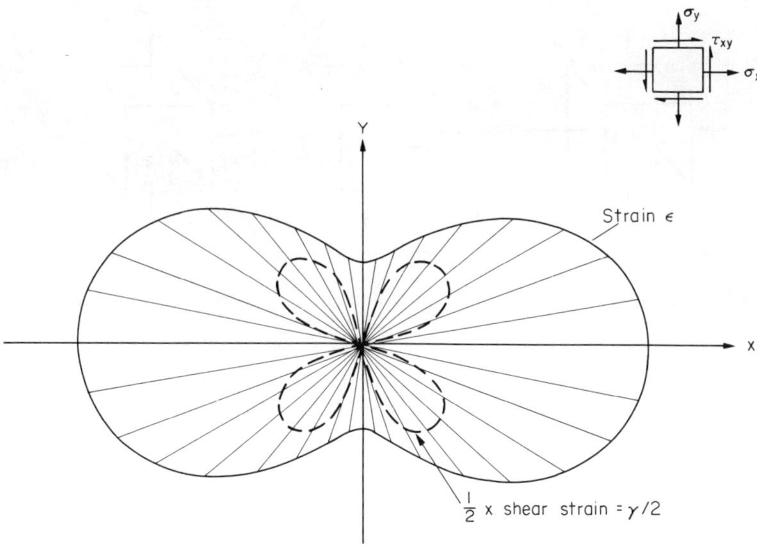

Fig. 14.26. Polar plot of strain distribution at a point under a typical biaxial applied stress system.

14.17. Strain energy of three-dimensional stress system

(a) Total strain energy

Any three-dimensional stress system may be reduced to three principal stresses σ_1, σ_2 and σ_3 acting on a unit cube, the faces of which are principal planes and, therefore, by definition, subjected to zero shear stress. If the corresponding principal strains are ε_1, ε_2 and ε_3, then the total strain energy U_t per unit volume is equal to the total work done by the system and given by the equation

$$U_t = \Sigma \tfrac{1}{2}\sigma\varepsilon$$

since the stresses are applied gradually from zero (see page 258).

$$\therefore \qquad U_t = \tfrac{1}{2}\sigma_1\varepsilon_1 + \tfrac{1}{2}\sigma_2\varepsilon_2 + \tfrac{1}{2}\sigma_3\varepsilon_3$$

Substituting for the principal strains using eqn. (14.2),

$$U_t = \frac{1}{2E}[\sigma_1(\sigma_1 - \nu\sigma_2 - \nu\sigma_3) + \sigma_2(\sigma_2 - \nu\sigma_3 - \nu\sigma_1) + \sigma_3(\sigma_3 - \nu\sigma_2 - \nu\sigma_1)]$$

$$\therefore \qquad \boldsymbol{U_t = \frac{1}{2E}[\sigma_1^2 + \sigma_2^2 + \sigma_3^2 - 2\nu(\sigma_1\sigma_2 + \sigma_2\sigma_3 + \sigma_3\sigma_1)]} \textbf{ per unit volume} \qquad (14.21)$$

(b) Shear (or "distortion") strain energy

As above, consider the three-dimensional stress system reduced to principal stresses σ_1, σ_2 and σ_3 acting on a unit cube as in Fig. 14.27. For convenience the principal stresses may be

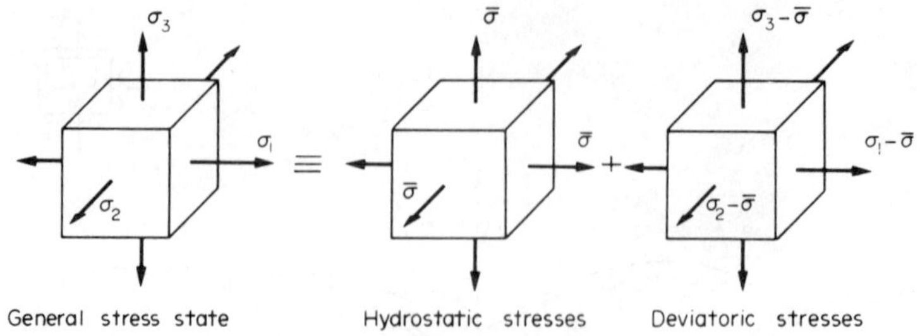

Fig. 14.27. Resolution of general three-dimensional principal stress state into "hydrostatic" and "deviatoric" components.

written in terms of a mean stress $\bar{\sigma} = \frac{1}{3}(\sigma_1 + \sigma_2 + \sigma_3)$ and additional shear stress terms,

i.e.
$$\sigma_1 = \tfrac{1}{3}(\sigma_1 + \sigma_2 + \sigma_3) + \tfrac{1}{3}(\sigma_1 - \sigma_2) + \tfrac{1}{3}(\sigma_1 - \sigma_3)$$
$$\sigma_2 = \tfrac{1}{3}(\sigma_1 + \sigma_2 + \sigma_3) + \tfrac{1}{3}(\sigma_2 - \sigma_1) + \tfrac{1}{3}(\sigma_2 - \sigma_3)$$
$$\sigma_3 = \tfrac{1}{3}(\sigma_1 + \sigma_2 + \sigma_3) + \tfrac{1}{3}(\sigma_3 - \sigma_1) + \tfrac{1}{3}(\sigma_3 - \sigma_2)$$

The mean stress term may be considered as a *hydrostatic* tensile stress, equal in all directions, the strains associated with this giving rise to no distortion, i.e. the unit cube under the action of the hydrostatic stress alone would be strained into a cube. The hydrostatic stresses are sometimes referred to as the *spherical* or *dilatational* stresses.

The strain energy associated with the hydrostatic stress is termed the *volumetric strain energy* and is found by substituting

$$\sigma_1 = \sigma_2 = \sigma_3 = \tfrac{1}{3}(\sigma_1 + \sigma_2 + \sigma_3)$$

into eqn. (14.21),

i.e.
$$\text{volumetric strain energy} = \frac{3}{2E}\left[\left(\frac{\sigma_1 + \sigma_2 + \sigma_3}{3}\right)^2\right](1 - 2\nu)$$

$\therefore$
$$\boldsymbol{U_v = \frac{(1 - 2\nu)}{6E}[(\sigma_1 + \sigma_2 + \sigma_3)^2]} \textbf{ per unit volume} \qquad (14.22)$$

The remaining terms in the modified principal stress equations are shear stress terms (i.e. functions of principal stress differences in the various planes) and these are the only stresses which give rise to distortion of the stressed element. They are therefore termed *distortional* or *deviatoric* stresses.

Now

total strain energy per unit volume = shear strain energy per unit volume + volumetric strain energy per unit volume

i.e.
$$U_t = U_s + U_v$$

Therefore shear strain energy per unit volume is given by:

$$U_s = U_t - U_v$$

i.e.
$$U_s = \frac{1}{2E}[\sigma_1^2 + \sigma_2^2 + \sigma_3^2 - 2\nu(\sigma_1\sigma_2 + \sigma_2\sigma_3 + \sigma_3\sigma_1)] - \frac{(1 - 2\nu)}{6E}[(\sigma_1 + \sigma_2 + \sigma_3)^2]$$

This simplifies to

$$U_s = \frac{(1+\nu)}{6E}[(\sigma_1-\sigma_2)^2+(\sigma_2-\sigma_3)^2+(\sigma_3-\sigma_1)^2]$$

and, since $E = 2G(1+\nu)$,

$$U_s = \frac{1}{12G}[(\sigma_1-\sigma_2)^2+(\sigma_2-\sigma_3)^2+(\sigma_3-\sigma_1)^2] \qquad (14.23a)$$

or, alternatively,

$$U_s = \frac{1}{6G}[\sigma_1^2+\sigma_2^2+\sigma_3^2-(\sigma_1\sigma_2+\sigma_2\sigma_3+\sigma_3\sigma_1)] \qquad (14.23b)$$

It is interesting to note here that even a uniaxial stress condition may be divided into hydrostatic (dilatational) and deviatoric (distortional) terms as shown in Fig. 14.28.

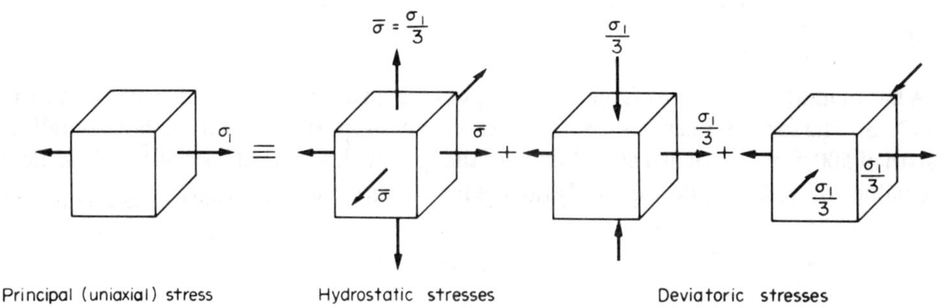

Fig. 14.28. Resolution of uniaxial stress into hydrostatic and deviatoric components.

Examples

Example 14.1

When a bar of 25 mm diameter is subjected to an axial pull of 61 kN the extension on a 50 mm gauge length is 0.1 mm and there is a decrease in diameter of 0.013 mm. Calculate the values of E, ν, G, and K.

Solution

$$\text{Longitudinal stress} = \frac{\text{load}}{\text{area}} = \frac{61\times10^3}{\frac{1}{4}\pi(0.025)^2} = 124.2\ \text{MN/m}^2$$

$$\text{Longitudinal strain} = \frac{\text{extension}}{\text{original length}} = \frac{0.1\times10^3}{10^3\times50} = 2\times10^{-3}$$

$$\text{Young's modulus } E = \frac{\text{stress}}{\text{strain}} = \frac{124.2\times10^6}{2\times10^{-3}} = \mathbf{62.1\ GN/m^2}$$

$$\text{Lateral strain} = \frac{\text{change in diameter}}{\text{original diameter}} = \frac{0.013\times10^3}{10^3\times25} = 0.52\times10^{-3}$$

$$\text{Poisson's ratio } (\nu) = \frac{\text{lateral strain}}{\text{longitudinal strain}} = \frac{0.52\times10^{-3}}{2\times10^{-3}} = \mathbf{0.26}$$

Now $$E = 2G(1+\nu) \quad \therefore \quad G = \frac{E}{2(1+\nu)}$$

$$G = \frac{62.1 \times 10^9}{2(1+0.26)} = \mathbf{24.6\ GN/m^2}$$

Also $$E = 3K(1-2\nu) \quad \therefore \quad K = \frac{E}{3(1-2\nu)}$$

$$K = \frac{62.1 \times 10^9}{3 \times 0.48} = \mathbf{43.1\ GN/m^2}$$

Example 14.2

A bar of mild steel 25 mm diameter twists 2 degrees in a length of 250 mm under a torque of 430 N m. The same bar deflects 0.8 mm when simply supported at each end horizontally over a span of 500 mm and loaded at the centre of the span with a vertical load of 1.2 kN. Calculate the values of E, G, K and Poisson's ratio ν for the material.

Solution

$$J = \frac{\pi}{32} D^4 = \frac{\pi}{32}(0.025)^4 = 0.0383 \times 10^{-6}\ \text{m}^4$$

Angle of twist $$\theta = 2 \times \frac{\pi}{180} = 0.0349 \text{ radian}$$

From the simple torsion theory $$\frac{T}{J} = \frac{G\theta}{L} \quad \therefore \quad G = \frac{TL}{J\theta}$$

$$G = \frac{430 \times 250 \times 10^6}{0.0349 \times 10^3 \times 0.0383} = 80.3 \times 10^9\ \text{N/m}^2$$

$$= \mathbf{80.3\ GN/m^2}$$

For a simply supported beam the deflection at mid-span with central load W is

$$\delta = \frac{WL^3}{48EI}$$

Then $$E = \frac{WL^3}{48\delta I} \quad \text{and} \quad I = \frac{\pi}{64}D^4 = \frac{\pi}{64}(0.025)^4 = 0.0192 \times 10^{-6}\ \text{m}^4$$

$\therefore$ $$E = \frac{1.2 \times 10^3 \times (0.5)^3 \times 10^6 \times 10^3}{48 \times 0.0192 \times 0.8} = 203 \times 10^9\ \text{N/m}^2$$

$$= \mathbf{203\ GN/m^2}$$

Now $$E = 2G(1+\nu) \quad \therefore \quad \nu = \frac{E}{2G} - 1$$

$\therefore$ $$\nu = \frac{203}{2 \times 80.3} - 1 = \mathbf{0.268}$$

Also $$E = 3K(1-2\nu) \quad \therefore \quad K = \frac{E}{3(1-2\nu)}$$

$$K = \frac{203 \times 10^9}{3(1-0.536)} = 146 \times 10^9 \text{ N/m}^2$$

$$= \mathbf{146 \text{ GN/m}^2}$$

Example 14.3

A rectangular bar of metal 50 mm × 25 mm cross-section and 125 mm long carries a tensile load of 100 kN along its length, a compressive load of 1 MN on its 50 × 125 mm faces and a tensile load of 400 kN on its 25 × 125 mm faces. If $E = 208$ GN/m² and $\nu = 0.3$, find

(a) the change in volume of the bar;
(b) the increase required in the 1 MN load to produce no change in volume.

Solution

(a) $$\sigma_x = \frac{\text{load}}{\text{area}} = \frac{100 \times 10^3 \times 10^6}{50 \times 25} = 80 \text{ MN/m}^2$$

$$\sigma_y = \frac{400 \times 10^3 \times 10^6}{125 \times 25} = 128 \text{ MN/m}^2$$

$$\sigma_z = \frac{-1 \times 10^6 \times 10^6}{125 \times 50} = -160 \text{ MN/m}^2 \quad \text{(Fig. 14.29)}$$

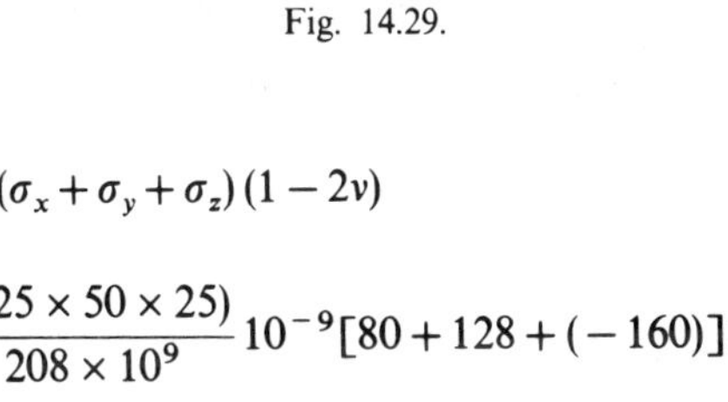

Fig. 14.29.

From §14.6

$$\text{change in volume} = \frac{V}{E}(\sigma_x + \sigma_y + \sigma_z)(1-2\nu)$$

$$= \frac{(125 \times 50 \times 25)}{208 \times 10^9} 10^{-9}[80 + 128 + (-160)]10^6 \times 0.4$$

$$= \frac{125 \times 50 \times 25 \times 48 \times 0.4}{208 \times 10^{12}} \text{m}^3 = \mathbf{14.4 \text{ mm}^3}$$

i.e. the bar increases in volume by 14.4 mm³.

(b) If the 1 MN load is to be changed, then σ_z will be changed; therefore the equation for the change in volume becomes

$$\text{change in volume} = 0 = \frac{(125 \times 50 \times 25)}{208 \times 10^9} 10^{-9}(80 + 128 + \sigma_z)10^6 \times 0.4$$

Then

$$0 = 80 + 128 + \sigma_z$$

$$\sigma_z = -208 \text{ MN/m}^2$$

Now

$$\text{load} = \text{stress} \times \text{area}$$

$\therefore$

$$\text{new load required} = -208 \times 10^6 \times 125 \times 50 \times 10^{-6}$$

$$= -1.3 \text{ MN}$$

Therefore the compressive load of 1 MN must be **increased by 0.3 MN** for no change in volume to occur.

Example 14.4

A steel bar *ABC* is of circular cross-section and transmits an axial tensile force such that the total change in length is 0.6 mm. The total length of the bar is 1.25 m, *AB* being 750 mm and 20 mm diameter and *BC* being 500 mm long and 13 mm diameter (Fig. 14.30). Determine for the parts *AB* and *BC* the changes in (a) length, and (b) diameter. Assume Poisson's ratio ν for the steel to be 0.3 and Young's modulus *E* to be 200 GN/m².

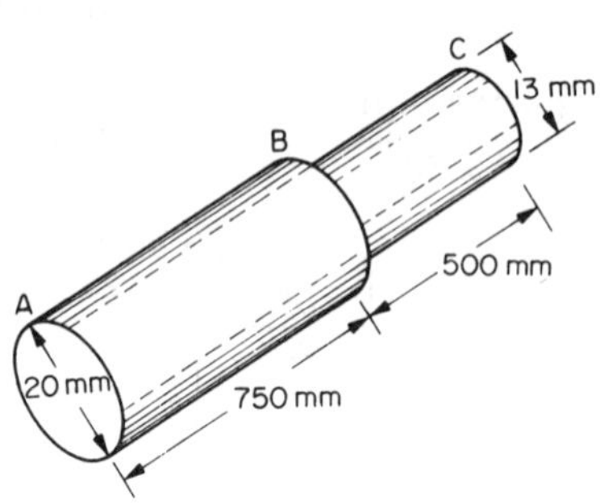

Fig. 14.30.

Solution

(a) Let the tensile force be *P* newtons.

Then

$$\text{stress in } AB = \frac{\text{load}}{\text{area}} = \frac{P}{\frac{1}{4}\pi(0.02)^2} = \frac{P}{100\pi} \text{ MN/m}^2$$

$$\text{stress in } BC = \frac{P}{\frac{1}{4}\pi(0.013)^2} = \frac{P}{42\pi} \text{ MN/m}^2$$

Then $\quad \text{strain in } AB = \dfrac{\text{stress}}{E} = \dfrac{P \times 10^6}{100\pi \times 200 \times 10^9} = \dfrac{P}{20\pi} \times 10^{-6}$

and $\quad \text{strain in } BC = \dfrac{P \times 10^6}{42\pi \times 200 \times 10^9} = \dfrac{P}{8.4\pi} \times 10^{-6}$

$$\text{change in length of } AB = \frac{P \times 10^{-6}}{20\pi} \times 750 \times 10^{-3} = 11.95P \times 10^{-9}$$

$$\text{change in length of } BC = \frac{P \times 10^{-6}}{8.4\pi} \times 500 \times 10^{-3} = 18.95P \times 10^{-9}$$

$$\text{total change in length} = (11.95P + 18.95P)10^{-9} = 0.6 \times 10^{-3}$$

$\therefore \quad P(11.95 + 18.95)10^{-9} = 0.6 \times 10^{-3}$

$\therefore \quad P = \dfrac{0.6 \times 10^9}{10^3 \times 30.9} = \mathbf{19.4\ kN}$

Then $\quad \text{change in length of } AB = 19.4 \times 10^3 \times 11.95 \times 10^{-9}$

$$= 0.232 \times 10^{-3}\ \text{m} = \mathbf{0.232\ mm}$$

and $\quad \text{change in length of } BC = 19.4 \times 10^3 \times 18.95 \times 10^{-9}$

$$= 0.368 \times 10^{-3} = \mathbf{0.368\ mm}$$

(b) The lateral (in this case "diametral") strain can be found from the definition of Poisson's ratio ν.

$$\nu = \frac{\text{lateral strain}}{\text{longitudinal strain}}$$

$\therefore \quad \text{lateral strain} = \text{strain on the diameter } (= \text{diametral strain})$

$$= \nu \times \text{longitudinal strain}$$

$$\text{Lateral strain on } AB = \frac{\nu P \times 10^{-6}}{20\pi} = \frac{0.3 \times 19.4 \times 10^3}{20\pi \times 10^6}$$

$$= 92.7 \times 10^{-6} (= 92.7\ \mu\varepsilon) \quad \text{compressive}$$

$$\text{Lateral strain on } BC = \frac{\nu P \times 10^{-6}}{8.4 \times \pi} = \frac{0.3 \times 19.4 \times 10^3}{8.4\pi \times 10^6}$$

$$= 220.5 \times 10^{-6}\ (= 220.5\ \mu\varepsilon) \quad \text{compressive}$$

Then, $\text{change in diameter of } AB = 92.7 \times 10^{-6} \times 20 \times 10^{-3}$

$$= 1.854 \times 10^{-6} = \mathbf{0.00185\ mm}$$

and $\quad \text{change in diameter of } BC = 220.5 \times 10^{-6} \times 13 \times 10^{-3}$

$$= 2.865 \times 10^{-6} = \mathbf{0.00286\ mm}$$

Both these changes are *decreases*.

Example 14.5

At a certain point a material is subjected to the following strains:

$$\varepsilon_x = 400 \times 10^{-6}; \quad \varepsilon_y = 200 \times 10^{-6}; \quad \gamma_{xy} = 350 \times 10^{-6} \text{ radian}$$

Determine the magnitudes of the principal strains, the directions of the principal strain axes and the strain on an axis inclined at 30° clockwise to the x axis.

Solution

Mohr's strain circle is as shown in Fig. 14.31.

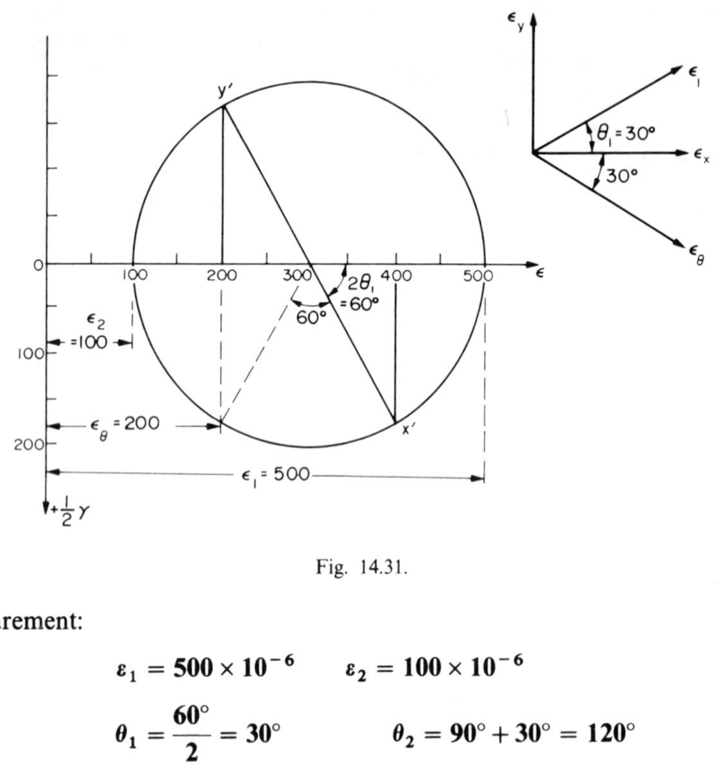

Fig. 14.31.

By measurement:

$$\varepsilon_1 = 500 \times 10^{-6} \qquad \varepsilon_2 = 100 \times 10^{-6}$$

$$\theta_1 = \frac{60^\circ}{2} = 30^\circ \qquad \theta_2 = 90^\circ + 30^\circ = 120^\circ$$

$$\varepsilon_{30} = 200 \times 10^{-6}$$

the angles being measured counterclockwise from the direction of ε_x.

Example 14.6

A material is subjected to two mutually perpendicular strains, $\varepsilon_x = 350 \times 10^{-6}$ and $\varepsilon_y = 50 \times 10^{-6}$, together with an unknown shear strain γ_{xy}. If the principal strain in the material is 420×10^{-6}, determine:

(a) the magnitude of the shear strain;
(b) the other principal strain;
(c) the direction of the principal strain axes;
(d) the magnitudes of the principal stresses if $E = 200\ \text{GN/m}^2$ and $\nu = 0.3$.

Solution

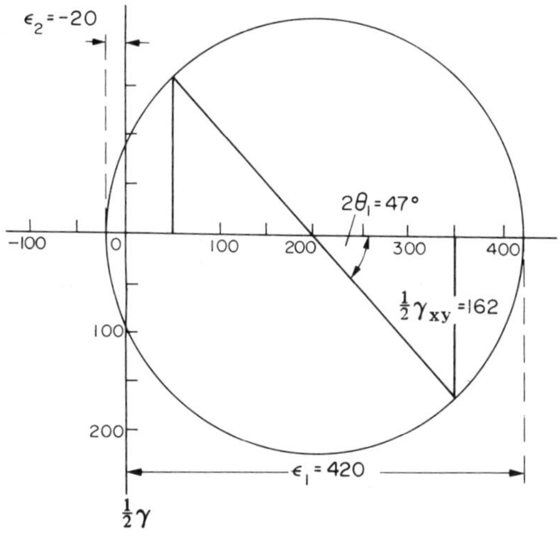

Fig. 14.32.

Mohr's strain circle is as shown in Fig. 14.32. The centre has been positioned half-way between ε_x and ε_y, and the radius is such that the circle passes through the ε axis at 420×10^{-6}. Then, by measurement:

(a) Shear strain $\gamma_{xy} = 2 \times 162 \times 10^{-6} = \mathbf{324 \times 10^{-6}}$ **radian.**
(b) Other principal strain $= \mathbf{-20 \times 10^{-6}}$ **(compressive).**
(c) Direction of principal strain $\varepsilon_1 = \dfrac{47°}{2} = \mathbf{23°30'}$.

Direction of principal strain $\varepsilon_2 = 90° + 23°30' = \mathbf{113°30'}$.

(d) The principal stresses may then be determined from the equations

$$\sigma_1 = \frac{(\varepsilon_1 + \nu\varepsilon_2)}{1-\nu^2}E \quad \text{and} \quad \sigma_2 = \frac{(\varepsilon_2 + \nu\varepsilon_1)}{1-\nu^2}E$$

$$\therefore \qquad \sigma_1 = \frac{[420 + 0.3(-20)]\,10^{-6} \times 200 \times 10^9}{1-(0.3)^2}$$

$$= \frac{414 \times 200 \times 10^3}{0.91} = \mathbf{91\ MN/m^2\ tensile}$$

and
$$\sigma_2 = \frac{(-20+0.3\times 420)10^{-6}\times 200\times 10^9}{1-(0.3)^2}$$

$$= \frac{106\times 200\times 10^3}{0.91} = \mathbf{23.3\ MN/m^2\ tensile}$$

Thus the principal stresses are 91 MN/m^2 and 23.3 MN/m^2, both tensile.

Example 14.7

The following strain readings were recorded at the angles stated relative to a given horizontal axis:

$$\varepsilon_a = -2.9\times 10^{-5} \text{ at } 20^\circ$$

$$\varepsilon_b = 3.1\times 10^{-5} \text{ at } 80^\circ$$

$$\varepsilon_c = -0.5\times 10^{-5} \text{ at } 140^\circ$$

as shown in Fig. 14.33. Determine the magnitude and direction of the principal stresses. $E = 200$ GN/m^2; $\nu = 0.3$.

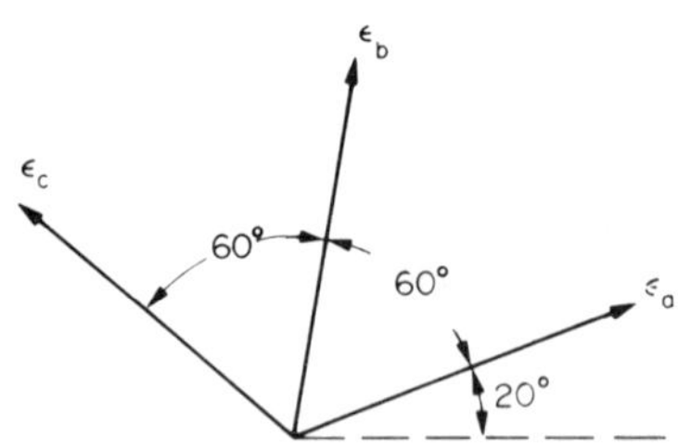

Fig. 14.33.

Solution

Consider now the construction shown in Fig. 14.34 giving the strain circle for the strain values in the question and illustrated in Fig. 14.33.

For a strain scale of 1 cm = 1×10^{-5} strain, in order to superimpose a stress circle concentric with the strain circle, the necessary scale is

$$1\text{ cm} = \frac{E}{(1-\nu)}\times 1\times 10^{-5} = \frac{200\times 10^9}{0.7}\times 10^{-5}$$

$$= 2.86\ MN/m^2$$

Also radius of strain circle = 3.5 cm

$\therefore$ radius of stress circle $= 3.5\times\dfrac{(1-\nu)}{(1+\nu)} = \dfrac{3.5\times 0.7}{1.3}$

$$= 1.886\text{ cm}$$

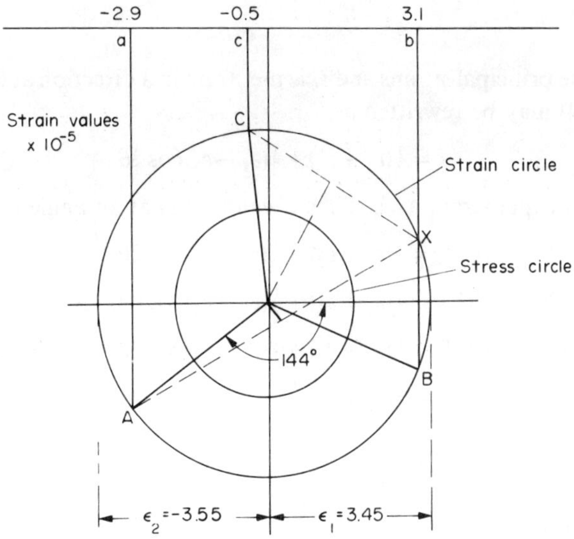

Fig. 14.34.

Superimposing the stress circle of radius 1.886 cm concentric with the strain circle, the principal stresses to a scale 1 cm = 2.86 MN/m^2 are found to be

$$\sigma_1 = 1.8 \times 2.86 \times 10^6 = \mathbf{5.15\ MN/m^2}$$

$$\sigma_2 = -2.0 \times 2.86 \times 10^6 = \mathbf{-5.72\ MN/m^2}$$

The principal strains will be at an angle $\frac{144}{2} = 72^\circ$ and 162° counterclockwise from the direction of ε_a, i.e. **92°** and **182°** counterclockwise from the given horizontal axis. The principal stresses will therefore also be in these directions.

Example 14.8

A rectangular rosette of strain gauges on the surface of a material under stress recorded the following readings of strain:

gauge A	$+450 \times 10^{-6}$
gauge B, at 45° to A	$+200 \times 10^{-6}$
gauge C, at 90° to A	-200×10^{-6}

the angles being counterclockwise from A.

Determine:

(a) the magnitudes of the principal strains,

(b) the directions of the principal strain axes, both by calculation and by Mohr's strain circle.

Solution

If ε_1 and ε_2 are the principal strains and ε_θ is the strain in a direction at θ to the direction of ε_1, then eqn. (14.14) may be rewritten as

$$\varepsilon_\theta = \tfrac{1}{2}(\varepsilon_1+\varepsilon_2)+\tfrac{1}{2}(\varepsilon_1-\varepsilon_2)\cos 2\theta$$

since $\gamma_{xy} = 0$ on principal strain axes. Thus if gauge A is at an angle θ to ε_1:

$$450\times10^{-6} = \tfrac{1}{2}(\varepsilon_1+\varepsilon_2)+\tfrac{1}{2}(\varepsilon_1-\varepsilon_2)\cos 2\theta \qquad (1)$$

$$200\times10^{-6} = \tfrac{1}{2}(\varepsilon_1+\varepsilon_2)+\tfrac{1}{2}(\varepsilon_1-\varepsilon_2)\cos(90^\circ+2\theta) \qquad (2)$$

$$-200\times10^{-6} = \tfrac{1}{2}(\varepsilon_1+\varepsilon_2)+\tfrac{1}{2}(\varepsilon_1-\varepsilon_2)\cos(180^\circ+2\theta)$$

$$= \tfrac{1}{2}(\varepsilon_1+\varepsilon_2)-\tfrac{1}{2}(\varepsilon_1-\varepsilon_2)\cos 2\theta \qquad (3)$$

Adding (1) and (3),

$$250\times10^{-6} = \varepsilon_1+\varepsilon_2 \qquad (4)$$

Substituting (4) in (2),

$$200\times10^{-6} = 125\times10^{-6}+\tfrac{1}{2}(\varepsilon_1-\varepsilon_2)\cos(90^\circ+2\theta)$$

$$\therefore \quad 75\times10^{-6} = -\tfrac{1}{2}(\varepsilon_1-\varepsilon_2)\sin 2\theta \qquad (5)$$

Substituting (4) in (1),

$$450\times10^{-6} = 125\times10^{-6}+\tfrac{1}{2}(\varepsilon_1-\varepsilon_2)\cos 2\theta$$

$$\therefore \quad 325\times10^{-6} = \tfrac{1}{2}(\varepsilon_1-\varepsilon_2)\cos 2\theta \qquad (6)$$

Dividing (5) by (6),

$$\frac{75\times10^{-6}}{325\times10^{-6}} = \frac{-\tfrac{1}{2}(\varepsilon_1-\varepsilon_2)\sin 2\theta}{\tfrac{1}{2}(\varepsilon_1-\varepsilon_2)\cos 2\theta}$$

$$\therefore \quad \tan 2\theta = -0.231$$

$$\therefore \quad 2\theta = -13^\circ \quad \text{or} \quad 180^\circ-13^\circ = 167^\circ$$

$$\therefore \quad \theta = \mathbf{83^\circ 30'}$$

Thus gauge A is $83^\circ 30'$ counterclockwise from the direction of ε_1.

Therefore from (6),

$$325\times10^{-6} = \tfrac{1}{2}(\varepsilon_1-\varepsilon_2)\cos 167^\circ$$

$$\therefore \quad \varepsilon_1-\varepsilon_2 = \frac{2\times325\times10^{-6}}{-0.9744} = -667\times10^{-6}$$

But from eqn. (4),

$$\varepsilon_1+\varepsilon_2 = 250\times10^{-6}$$

Therefore adding,

$$2\varepsilon_1 = -417\times10^{-6} \quad \therefore \quad \varepsilon_1 = \mathbf{-208.5\times10^{-6}}$$

and subtracting,

$$2\varepsilon_2 = 917\times10^{-6} \quad \therefore \quad \varepsilon_2 = \mathbf{458.5\times10^{-6}}$$

Thus the principal strains are -208×10^{-6} and 458.5×10^{-6}, the former being on an axis **83°30′** clockwise from gauge A.

Alternatively, these results may be obtained using Mohr's strain circle as shown in Fig. 14.35. The circle has been drawn using the construction procedure of §14.14 and gives principal strains of

$\varepsilon_1 = \mathbf{458.5 \times 10^6}$, tensile, at **6°30′** counterclockwise from gauge A
$\varepsilon_2 = \mathbf{208.5 \times 10^6}$, compressive, at **83°30′** clockwise from gauge A

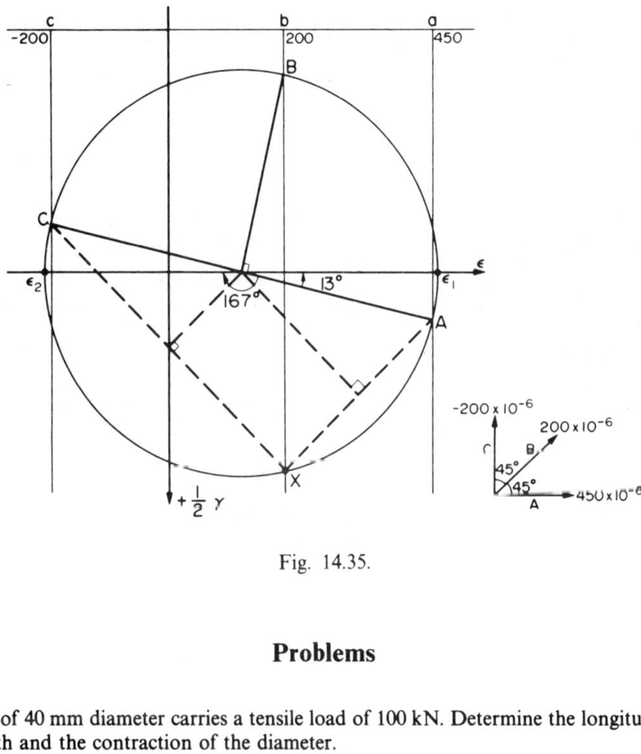

Fig. 14.35.

Problems

14.1 (A). A bar of 40 mm diameter carries a tensile load of 100 kN. Determine the longitudinal extension of a 50 mm gauge length and the contraction of the diameter.
Young's modulus $E = 210$ GN/m^2 and Poisson's ratio $\nu = 0.3$. [0.019, 0.0045 mm.]

14.2 (A). Establish the relationship between Young's modulus E, the modulus of rigidity G and the bulk modulus K in the form

$$E = \frac{9KG}{3K+G}$$

14.3 (A). The extension of a 100 mm gauge length of 14.33 mm diameter bar was found to be 0.15 mm when a tensile load of 50 kN was applied. A torsion specimen of the same specification was made with a 19 mm diameter and a 200 mm gauge length. On test it twisted 0.502 degree under the action of a torque of 45 N m. Calculate E, G, K and ν. [206.7, 80.9, 155 GN/m^2; 0.278.]

14.4 (A). A rectangular steel bar of 25 mm × 12 mm cross-section deflects 6 mm when simply supported on its 25 mm face over a span of 1.2 m and loaded at the centre with a concentrated load of 126 N. If Poisson's ratio for the material is 0.28 determine the values of (a) the rigidity modulus, and (b) the bulk modulus. [82, 159 GN/m^2.]

14.5 (A). Calculate the changes in dimensions of a 37 mm × 25 mm rectangular bar when loaded with a tensile load of 600 kN.
Take $E = 210$ GN/m^2 and $\nu = 0.3$. [0.034, 0.023 mm.]

14.6 (A). A rectangular block of material 125 mm × 100 mm × 75 mm carries loads normal to its faces as follows: 1 MN tensile on the 125 × 100 mm faces; 0.48 MN tensile on the 100 × 75 mm faces; zero load on the 125 × 75 mm faces.

If Poisson's ratio = 0.3 and $E = 200 \text{ GN/m}^2$, determine the changes in dimensions of the block under load. What is then the change in volume? [0.025, 0.0228, −0.022 mm; 270 mm³.]

14.7 (A). A rectangular bar consists of two sections, *AB* 25 mm square and 250 mm long and *BC* 12 mm square and 250 mm long. For a tensile load of 20 kN determine:

(a) the change in length of the complete bar;
(b) the changes in dimensions of each portion.

Take $E = 80 \text{ GN/m}^2$ and Poisson's ratio $\nu = 0.3$. [0.534 mm; 0.434, −0.003, 0.1, −0.0063 mm.]

14.8 (A). A cylindrical brass bar is 50 mm diameter and 250 mm long. Find the change in volume of the bar when an axial compressive load of 150 kN is applied.

Take $E = 100 \text{ GN/m}^2$ and $\nu = 0.27$. [172.5 mm³.]

14.9 (A). A certain alloy bar of 32 mm diameter has a gauge length of 100 mm. A tensile load of 25 kN produces an extension of 0.014 mm on the gauge length and a torque of 2.5 kN m produces an angle of twist of 1.63 degrees. Calculate *E*, *G*, *K* and ν. [222, 85.4, 185 GN/m², 0.3.]

14.10 (A/B). Derive the relationships which exist between the elastic constants (a) *E*, *G* and ν, and (b) *E*, *K* and ν. Find the change in volume of a steel cube of 150 mm side immersed to a depth of 3 km in sea water.

Take *E* for steel = 210 GN/m², $\nu = 0.3$ and the density of sea water = 1025 kg/m³. [580 mm³.]

14.11 (B). Two steel bars have the same length and the same cross-sectional area, one being circular in section and the other square. Prove that when axial loads are applied the changes in volume of the bars are equal.

14.12 (B). Determine the percentage change in volume of a bar 50 mm square and 1 m long when subjected to an axial compressive load of 10 kN. Find also the restraining pressure on the sides of the bar required to prevent all lateral expansion.

For the bar material, $E = 210 \text{ GN/m}^2$ and $\nu = 0.27$. [0.876×10^{-3}%, 1.48 MN/m².]

14.13 (B). Derive the formula for longitudinal strain due to axial stress σ_x when all lateral strain is prevented. A piece of material 100 mm long by 25 mm square is in compression under a load of 60 kN. Determine the change in length of the material if all lateral strain is prevented by the application of a uniform external lateral pressure of a suitable intensity.

For the material, $E = 70 \text{ GN/m}^2$ and Poisson's ratio $\nu = 0.25$. [0.114 mm.]

14.14 (B). Describe briefly an experiment to find Poisson's ratio for a material.

A steel bar of rectangular cross-section 40 mm wide and 25 mm thick is subjected to an axial tensile load of 100 kN. Determine the changes in dimensions of the sides and hence the percentage decrease in cross-sectional area if $E = 200 \text{ GN/m}^2$ and Poisson's ratio = 0.3. [-6×10^{-3}, -3.75×10^{-3}, −0.03%.]

14.15 (B). A material is subjected to the following strain system:

$$\varepsilon_x = 200 \times 10^{-6}; \; \varepsilon_y = -56 \times 10^{-6}; \quad \gamma_{xy} = 230 \times 10^{-6} \text{ radian}$$

Determine:

(a) the principal strains;
(b) the directions of the principal strain axes;
(c) the linear strain on an axis inclined at 50° counterclockwise to the direction of ε_x.

[244, -100×10^{-6}; 21°; 163×10^{-6}.]

14.16 (B). A material is subjected to two mutually perpendicular linear strains together with a shear strain. Given that this system produces principal strains of 0.0001 compressive and 0.0003 tensile and that one of the linear strains is 0.00025 tensile, determine the magnitudes of the other linear strain and the shear strain.

[-50×10^{-6}, 265×10^{-6}.]

14.17 (C). A 50 mm diameter cylinder is subjected to an axial compressive load of 80 kN. The cylinder is partially enclosed by a well-fitted casing covering almost the whole length, which reduces the lateral expansion by half. Determine the ratio between the axial strain when the casing is fitted and that when it is free to expand in diameter. Take $\nu = 0.3$. [0.871.]

14.18 (C). A thin cylindrical shell has hemispherical ends and is subjected to an internal pressure. If the radial change of the cylindrical part is to be equal to that of the hemispherical ends, determine the ratio between the thickness necessary in the two parts. Take $\nu = 0.3$. [2.43 : 1.]

14.19 (B). Determine the values of the principal stresses present in the material of Problem 14.16. Describe an experimental technique by which the directions and magnitudes of these stresses could be determined in practice. For the material, take $E = 208 \text{ GN/m}^2$ and $\nu = 0.3$. [61.6, 2.28 MN/m².]

14.20 (B). A rectangular prism of steel is subjected to purely normal stresses on all six faces (i.e. the stresses are principal stresses). One stress is 60 MN/m^2 tensile, and the other two are denoted by σ_x and σ_y and may be either tensile or compressive, their magnitudes being such that there is no strain in the direction of σ_y and that the maximum shearing stress in the material does not exceed 75 MN/m^2 on any plane. Determine the range of values within which σ_x may lie and the corresponding values of σ_y. Make sketches to show the two limiting states of stress, and calculate the strain energy per cubic metre of material in the two limiting conditions. Assume that the stresses are not sufficient to cause elastic failure. For the prism material $E = 208$ GN/m^2; $\nu = 0.286$.
[U.L.] [-90 to 210; -8.6, 77.2 MN/m^2.]

For the following problems on the application of strain gauges additional information may be obtained in §21.2 (*Vol.* 2).

14.21 (A/B). The following strains are recorded by two strain gauges, their axes being at right angles: $\varepsilon_x = 0.00039$; $\varepsilon_y = -0.00012$ (i.e. one tensile and one compressive). Find the values of the stresses σ_x and σ_y acting along these axes if the relevant elastic constants are $E = 208$ GN/m^2 and $\nu = 0.3$. [80.9, -0.69 MN/m^2.]

14.22 (B). Explain how strain gauges can be used to measure shear strain and hence shear stresses in a material.
Find the value of the shear stress present in a shaft subjected to pure torsion if two strain gauges mounted at 45° to the axis of the shaft record the following values of strain: 0.00029; -0.00029. If the shaft is of steel, 75 mm diameter, $G = 80$ GN/m^2 and $\nu = 0.3$, determine the value of the applied torque. [46.4 MN/m^2, 3.84 kN m.]

14.23 (B). The following strains were recorded on a rectangular strain rosette: $\varepsilon_a = 450 \times 10^{-6}$; $\varepsilon_b = 230 \times 10^{-6}$; $\varepsilon_c = 0$.
Determine:
(a) the principal strains and the directions of the principal strain axes;
(b) the principal stresses if $E = 200$ GN/m^2 and $\nu = 0.3$.
[451×10^{-6} at 1° clockwise from A, -1×10^{-6} at 91° clockwise from A; 98, 29.5 MN/m^2.]

14.24 (B). The values of strain given in Problem 14.23 were recorded on a 60° rosette gauge. What are now the values of the principal strains and the principal stresses?
[484×10^{-6}, -27×10^{-6}; 104 MN/m^2, 25.7 MN/m^2.]

14.25 (B). Describe briefly how you would proceed, with the aid of strain gauges, to find the principal stresses present on a material under the action of a complex stress system.
Find, by calculation, the principal stresses present in a material subjected to a complex stress system given that strain readings in directions at 0°, 45° and 90° to a given axis are $+240 \times 10^{-6}$, $+170 \times 10^{-6}$ and $+40 \times 10^{-6}$ respectively.
For the material take $E = 210$ GN/m^2 and $\nu = 0.3$. [59, 25 MN/m^2.]

14.26 (B). Check the calculation of Problem 14.25 by means of Mohr's strain circle.

14.27 (B). A closed-ended steel pressure vessel of diameter 2.5 m and plate thickness 18 mm has electric resistance strain gauges bonded on the outer surface in the circumferential and axial directions. These gauges have a resistance of 200 ohms and a gauge factor of 2.09. When the pressure is raised to 9 MN/m^2 the change of resistance is 1.065 ohms for the circumferential gauge and 0.265 ohm for the axial gauge. Working from first principles calculate the value of Young's modulus and Poisson's ratio. [I.Mech.E.] [0.287, 210 GN/m^2.]

14.28 (B). Briefly describe the mode of operation of electric resistance strain gauges, and a simple circuit for the measurement of a static change in strain.
The torque on a steel shaft of 50 mm diameter which is subjected to pure torsion is measured by a strain gauge bonded on its outer surface at an angle of 45° to the longitudinal axis of the shaft. If the change of the gauge resistance is 0.35 ohm in 200 ohms and the strain gauge factor is 2, determine the torque carried by the shaft. For the shaft material $E = 210$ GN/m^2 and $\nu = 0.3$. [I.Mech.E.] [3.47 kN.]

14.29 (A/B). A steel test bar of diameter 11.3 mm and gauge length 56 mm was found to extend 0.08 mm under a load of 30 kN and to have a contraction on the diameter of 0.00452 mm. A shaft of 80 mm diameter, made of the same quality steel, rotates at 420 rev/min. An electrical resistance strain gauge bonded to the outer surface of the shaft at an angle of 45° to the longitudinal axis gave a recorded resistance change of 0.189 Ω. If the gauge resistance is 100 Ω and the gauge factor is 2.1 determine the maximum power transmitted. [650 kW.]

14.30 (B). A certain equiangular strain gauge rosette is made up of three separate gauges. After it has been installed it is found that one of the gauges has, in error, been taken from an odd batch; its gauge factor is 2.0, that of the other two being 2.2. As the three gauges appear identical it is impossible to say which is the rogue and it is decided to proceed with the test. The following strain readings are obtained using a gauge factor setting on the strain gauge equipment of 2.2:

Gauge direction	0°	60°	120°
Strain $\times 10^{-6}$	$+1$	-250	$+200$

Taking into account the various gauge factor values evaluate the greatest possible shear stress value these readings can represent.

For the specimen material $E = 207\ \text{GN/m}^2$ and $\nu = 0.3$. [City U.] [44 MN/m^2.]

14.31 (B). A solid cylindrical shaft is 250 mm long and 50 mm diameter and is made of aluminium alloy. The periphery of the shaft is constrained in such a way as to prevent lateral strain. Calculate the axial force that will compress the shaft by 0.5 mm.

Determine the change in length of the shaft when the lateral constraint is removed but the axial force remains unaltered.

Calculate the required reduction in axial force for the non-constrained shaft if the axial strain is not to exceed 0.2 %

Assume the following values of material constants, $E = 70\ \text{GN/m}^2$; $\nu = 0.3$.

[C.E.I.] [370 kN; 0.673 mm; 95.1 kN.]

14.32 (C). An electric resistance strain gauge rosette is bonded to the surface of a square plate, as shown in Fig. 14.36. The orientation of the rosette is defined by the angle gauge A makes with the X direction. The angle between gauges A and B is 120° and between A and C is 120°. The rosette is supposed to be orientated at 45° to the X direction. To check this orientation the plate is loaded with a uniform tension in the X direction only (i.e. $\sigma_y = 0$), unloaded and then loaded with a uniform tension stress of the same magnitude, in the Y direction only (i.e. $\sigma_x = 0$), readings being taken from the strain gauges in both loading cases.

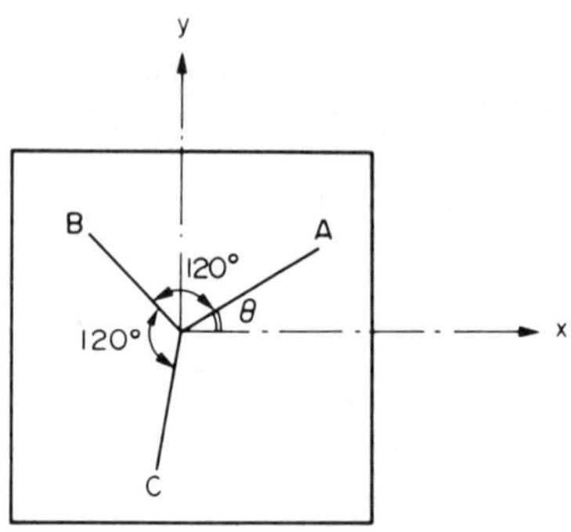

Fig. 14.36.

Denoting the greater principal strain in both loading cases by ε_1, show that if the rosette is correctly orientated, then

(a) the strain shown by gauge A should be

$$\varepsilon_A = \frac{(1-\nu)}{2}\varepsilon_1$$

for both load cases, and

(b) that shown by gauge B should be

$$\varepsilon_B = \frac{(1-\nu)}{2}\varepsilon_1 + \frac{(1+\nu)}{2}\frac{\sqrt{3}}{2}\varepsilon_1 \quad \text{for the } \sigma_x \text{ case}$$

or

$$\varepsilon_B = \frac{(1-\nu)}{2}\varepsilon_1 - \frac{(1+\nu)}{2}\frac{\sqrt{3}}{2}\varepsilon_1 \quad \text{for the } \sigma_y \text{ case}$$

Hence obtain the corresponding expressions for ε_c. [As for B, but reversed.]

CHAPTER 15

THEORIES OF ELASTIC FAILURE

Summary

TABLE 15.1

Theory	Value in tension test at failure	Value in complex stress system	Criterion for failure
Maximum principal stress (*Rankine*)	σ_y	σ_1	$\sigma_1 = \sigma_y$
Maximum shear stress (*Guest–Tresca*)	$\frac{1}{2}\sigma_y$	$\frac{1}{2}(\sigma_1 - \sigma_3)$	$\sigma_1 - \sigma_3 = \sigma_y$
Maximum principal strain (*Saint-Venant*)	$\frac{\sigma_y}{E}$	$\frac{\sigma_1}{E} - \nu\frac{\sigma_2}{E} - \nu\frac{\sigma_3}{E}$	$\sigma_1 - \nu\sigma_2 - \nu\sigma_3 = \sigma_y$
Total strain energy per unit volume (*Haigh*)	$\frac{\sigma_y^2}{2E}$	$\frac{1}{2E}[\sigma_1^2 + \sigma_2^2 + \sigma_3^2 - 2\nu(\sigma_1\sigma_2 + \sigma_2\sigma_3 + \sigma_3\sigma_1)]$	$\sigma_1^2 + \sigma_2^2 + \sigma_3^2 - 2\nu(\sigma_1\sigma_2 + \sigma_2\sigma_3 + \sigma_3\sigma_1) = \sigma_y^2$
Shear strain energy per unit volume Distortion energy theory (*Maxwell–Huber–von Mises*)	$\frac{\sigma_y^2}{6G}$	$\frac{1}{12G}[(\sigma_1 - \sigma_2)^2 + (\sigma_2 - \sigma_3)^2 + (\sigma_3 - \sigma_1)^2]$	$\frac{1}{2}[(\sigma_1 - \sigma_2)^2 + (\sigma_2 - \sigma_3)^2 + (\sigma_3 - \sigma_1)^2] = \sigma_y^2$
Modified shear stress Internal friction theory (*Mohr*)			$\frac{\sigma_1}{\sigma_{y_t}} + \frac{\sigma_2}{\sigma_{y_c}} = 1$

Introduction

When dealing with the design of structures or components the physical properties of the constituent materials are usually found from the results of laboratory experiments which have only subjected the materials to the simplest stress conditions. The most usual test is the simple tensile test in which the value of the stress at yield or at fracture (whichever occurs first) is easily determined. The strengths of materials under complex stress systems are not generally known except in a few particular cases. In practice it is these complicated systems of stress which are more often encountered, and therefore it is necessary to have some basis for determining allowable working stresses so that failure will not occur. Thus the function of the theories of elastic failure is to predict from the behaviour of materials in a simple tensile test when elastic failure will occur under *any* condition of applied stress.

A number of theoretical criteria have been proposed each seeking to obtain adequate correlation between estimated component life and that actually achieved under service load conditions for both brittle and ductile material applications. The five main theories are:

(a) Maximum principal stress theory (Rankine).
(b) Maximum shear stress theory (Guest–Tresca).
(c) Maximum principal strain (Saint-Venant).
(d) Total strain energy per unit volume (Haigh).
(e) Shear strain energy per unit volume (Maxwell–Huber–von Mises).

In each case the value of the selected critical property implied in the title of the theory is determined for both the simple tension test and a three-dimensional complex stress system. These values are then equated to produce the so-called *criterion for failure* listed in the last column of Table 15.1.

In Table 15.1 σ_y is the stress at the yield point in the simple tension test, and σ_1, σ_2 and σ_3 are the three principal stresses in the three-dimensional complex stress system in order of magnitude. Thus in the case of the maximum shear stress theory $\sigma_1 - \sigma_3$ is the greatest numerical difference between two principal stresses taking into account signs and the fact that one principal stress may be zero.

Each of the first five theories listed in Table 15.1 will be introduced in detail in the following text, as will a sixth theory, (f) **Mohr's modified shear stress theory**. Whereas the previous theories (a) to (e) assume equal material strength in tension and compression, the Mohr's modified theory attempts to take into account the additional strength of brittle materials in compression.

15.1. Maximum principal stress theory

This theory assumes that when the maximum principal stress in the complex stress system reaches the elastic limit stress in simple tension, failure occurs. The criterion of failure is thus

$$\sigma_1 = \sigma_y$$

It should be noted, however, that failure could also occur in compression if the least principal stress σ_3 were compressive and its value reached the value of the yield stress in compression for the material concerned before the value of σ_{y_t} was reached in tension. An additional criterion is therefore

$$\sigma_3 = \sigma_y \quad \textbf{(compressive)}$$

Whilst the theory can be shown to hold fairly well for brittle materials, there is considerable experimental evidence that the theory should not be applied for ductile materials. For example, even in the case of the pure tension test itself, failure for ductile materials takes place not because of the direct stresses applied but in shear on planes at 45° to the specimen axis. Also, truly homogeneous materials can withstand very high hydrostatic pressures without failing, thus indicating that maximum direct stresses alone do not constitute a valid failure criteria for all loading conditions.

15.2. Maximum shear stress theory

This theory states that failure can be assumed to occur when the maximum shear stress in the complex stress system becomes equal to that at the yield point in the simple tensile test.

Since the maximum shear stress is half the greatest difference between two principal stresses the criterion of failure becomes

$$\tfrac{1}{2}(\sigma_1-\sigma_3)=\tfrac{1}{2}(\sigma_y-0)$$

i.e.

$$\boldsymbol{\sigma_1-\sigma_3=\sigma_y} \qquad (15.1)$$

the value of σ_3 being algebraically the smallest value, i.e. taking account of sign *and the fact that one stress may be zero.* This produces fairly accurate correlation with experimental results particularly for ductile materials, and is often used for ductile materials in machine design. The criterion is often referred to as the "Tresca" theory and is one of the widely used laws of plasticity.

15.3. Maximum principal strain theory

This theory assumes that failure occurs when the maximum strain in the complex stress system equals that at the yield point in the tensile test,

i.e.

$$\frac{\sigma_1}{E}-\nu\frac{\sigma_2}{E}-\nu\frac{\sigma_3}{E}=\frac{\sigma_y}{E}$$

$$\boldsymbol{\sigma_1-\nu\sigma_2-\nu\sigma_3=\sigma_y} \qquad (15.2)$$

This theory is contradicted by the results obtained from tests on flat plates subjected to two mutually perpendicular tensions. The Poisson's ratio effect of each tension reduces the strain in the perpendicular direction so that according to this theory failure should occur at a higher load. This is not always the case. The theory holds reasonably well for cast iron but is not generally used in design procedures these days.

15.4. Maximum total strain energy per unit volume theory

The theory assumes that failure occurs when the total strain energy in the complex stress system is equal to that at the yield point in the tensile test.

From the work of §14.17 the criterion of failure is thus

$$\frac{1}{2E}[\sigma_1^2+\sigma_2^2+\sigma_3^2-2\nu(\sigma_1\sigma_2+\sigma_2\sigma_3+\sigma_3\sigma_1)]=\frac{\sigma_y^2}{2E}$$

i.e.

$$\boldsymbol{\sigma_1^2+\sigma_2^2+\sigma_3^2-2\nu(\sigma_1\sigma_2+\sigma_2\sigma_3+\sigma_3\sigma_1)=\sigma_y^2} \qquad (15.3)$$

The theory gives fairly good results for ductile materials but is seldom used in preference to the theory below.

15.5. Maximum shear strain energy per unit volume (or distortion energy) theory

Section 14.17 again indicates how the strain energy of a stressed component can be divided into volumetric strain energy and shear strain energy components, the former being associated with volume change and no distortion, the latter producing distortion of the stressed elements. This theory states that failure occurs when the maximum shear strain energy component in the complex stress system is equal to that at the yield point in the tensile test,

i.e. $$\frac{1}{12G}[(\sigma_1-\sigma_2)^2+(\sigma_2-\sigma_3)^2+(\sigma_3-\sigma_1)^2]=\frac{\sigma_y^2}{6G} \quad \text{(eqn. (14.23a))}$$

or $$\frac{1}{6G}[\sigma_1^2+\sigma_2^2+\sigma_3^2-(\sigma_1\sigma_2+\sigma_2\sigma_3+\sigma_3\sigma_1)=\frac{\sigma_y^2}{6G}$$

$$\therefore \quad (\sigma_1-\sigma_2)^2+(\sigma_2-\sigma_3)^2+(\sigma_3-\sigma_1)^2=2\sigma_y^2 \tag{15.4}$$

This theory has received considerable verification in practice and is widely regarded as the most reliable basis for design, particularly when dealing with ductile materials. It is often referred to as the "von Mises" or "Maxwell" criteria and is probably the best theory of the five. It is also sometimes referred to as the **distortion energy** or **maximum octahedral shear stress theory** (see §23.19).

In the above theories it has been assumed that the properties of the material in tension and compression are similar. It is well known, however, that certain materials, notably concrete, cast iron, soils, etc., exhibit vastly different properties depending on the nature of the applied stress. For brittle materials this has been explained by Griffith,† who has introduced the principle of surface energy at microscopic cracks and shown that an existing crack will propagate rapidly if the available elastic strain energy release is greater than the surface energy of the crack.‡ In this way Griffith indicates the greater seriousness of tensile stresses compared with compressive ones with respect to failure, particularly in fatigue environments. A further theory has been introduced by Mohr to predict failure of materials whose strengths are considerably different in tension and shear; this is introduced below.

15.6. Mohr's modified shear stress theory for brittle materials (sometimes referred to as the internal friction theory)

Brittle materials in general show little ability to deform plastically and hence will usually fracture at, or very near to, the elastic limit. Any of the so-called "yield criteria" introduced above, therefore, will normally imply fracture of a brittle material. It has been stated previously, however, that brittle materials are usually considerably stronger in compression than in tension and to allow for this Mohr has proposed a construction based on his stress circle in the application of the maximum shear stress theory. In Fig. 15.1 the circle on diameter OA is that for pure tension, the circle on diameter OB that for pure compression and the circle centre O and diameter CD is that for pure shear. Each of these types of test can be performed to failure relatively easily in the laboratory. An envelope to these curves, shown dotted, then represents the failure envelope according to the Mohr theory. A failure condition is then indicated when the stress circle for a particular complex stress condition is found to cut the envelope.

† A. A. Griffith, The phenomena of rupture and flow of solids, *Phil. Trans. Royal Soc.*, London, 1920.
‡ J. F. Knott, *Fundamentals of Fracture Mechanics* (Butterworths, London), 1973.

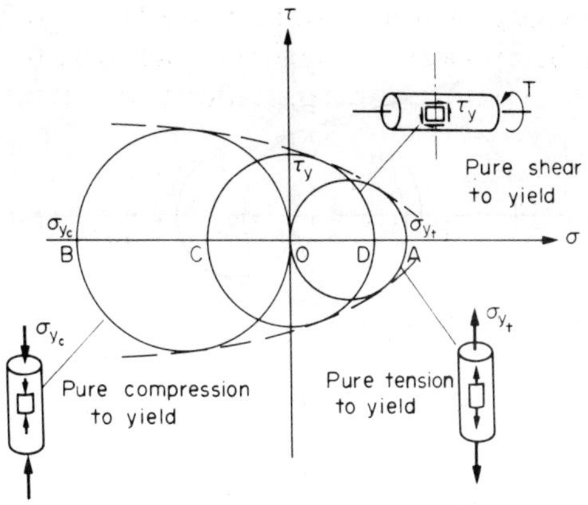

Fig. 15.1. Mohr theory on σ–τ axes.

As a close approximation to this procedure Mohr suggests that only the pure tension and pure compression failure circles need be drawn with OA and OB equal to the yield or fracture strengths of the brittle material. Common tangents to these circles may then be used as the failure envelope as shown in Fig. 15.2. Circles drawn tangent to this envelope then represent the condition of failure at the point of tangency.

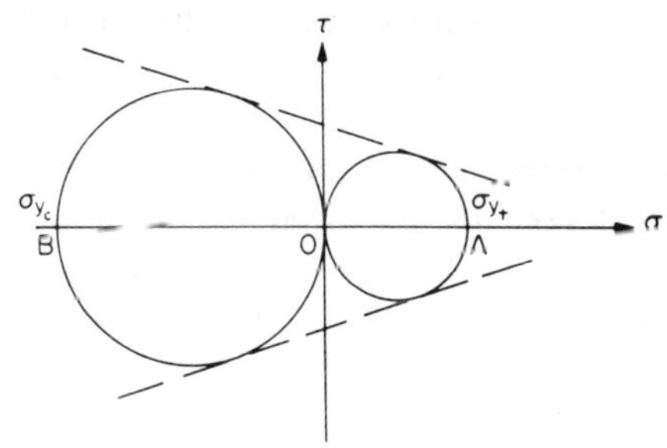

Fig. 15.2. Simplified Mohr theory on σ–τ axes.

In order to develop a theoretical expression for the failure criterion, consider a general stress circle with principal stresses of σ_1 and σ_2. It is then possible to develop an expression relating σ_1, σ_2, the principal stresses, and $\sigma_{y_t}, \sigma_{y_c}$, the yield strengths of the brittle material in tension and compression respectively.

From the geometry of Fig. 15.3,

$$\frac{KL}{KM} = \frac{JL}{MH}$$

Now, in terms of the stresses,

$$KL = \tfrac{1}{2}(\sigma_1 + \sigma_2) - \sigma_1 + \tfrac{1}{2}\sigma_{y_t} = \tfrac{1}{2}(\sigma_{y_t} - \sigma_1 + \sigma_2)$$

$$KM = \tfrac{1}{2}\sigma_{y_t} + \tfrac{1}{2}\sigma_{y_c} = \tfrac{1}{2}(\sigma_{y_t} + \sigma_{y_c})$$

$$JL = \tfrac{1}{2}(\sigma_1 + \sigma_2) - \tfrac{1}{2}\sigma_{y_t} = \tfrac{1}{2}(\sigma_1 + \sigma_2 - \sigma_{y_t})$$

$$MH = \tfrac{1}{2}\sigma_{y_c} - \tfrac{1}{2}\sigma_{y_t} = \tfrac{1}{2}(\sigma_{y_c} - \sigma_{y_t})$$

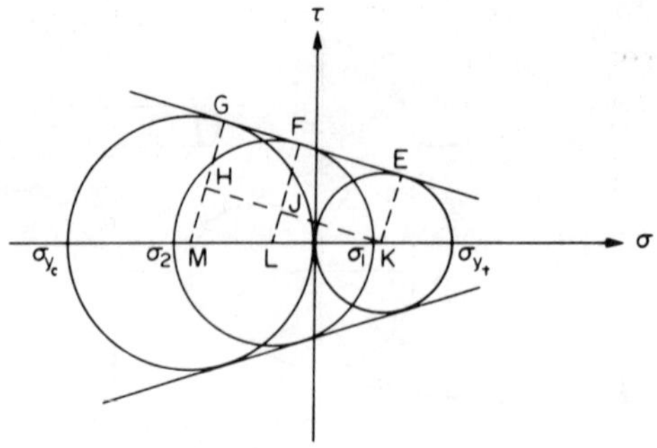

Fig. 15.3.

Substituting,

$$\frac{\sigma_{y_t} - \sigma_1 + \sigma_2}{\sigma_{y_t} + \sigma_{y_c}} = \frac{\sigma_1 + \sigma_2 - \sigma_{y_t}}{\sigma_{y_c} - \sigma_{y_t}}$$

Cross-multiplying and simplifying this reduces to

$$\frac{\sigma_1}{\sigma_{y_t}} + \frac{\sigma_2}{\sigma_{y_c}} = 1 \qquad (15.5)$$

which is then the Mohr's modified shear stress criterion for brittle materials.

15.7. Graphical representation of failure theories for two-dimensional stress systems (one principal stress zero)

Having obtained the equations for the elastic failure criteria above in the general three-dimensional stress state it is relatively simple to obtain the corresponding equations when one of the principal stresses is zero.

Each theory may be represented graphically as described below, the diagrams often being termed *yield loci.*

(a) *Maximum principal stress theory*

For simplicity of treatment, ignore for the moment the normal convention for the principal stresses, i.e. $\sigma_1 > \sigma_2 > \sigma_3$ and consider the two-dimensional stress state shown in Fig. 15.4

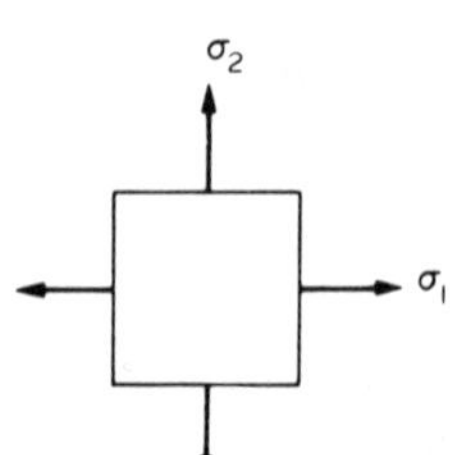

Fig. 15.4. Two-dimensional stress state ($\sigma_3 = 0$).

where σ_3 is zero and σ_2 may be tensile or compressive as appropriate, i.e. σ_2 may have a value less than σ_3 for the purpose of this development.

The maximum principal stress theory then states that failure will occur when σ_1 or $\sigma_2 = \sigma_{y_t}$ or σ_{y_c}. Assuming $\sigma_{y_t} = \sigma_{y_c} = \sigma_y$, these conditions are represented graphically on σ_1, σ_2 coordinates as shown in Fig. 15.5. If the point with coordinates (σ_1, σ_2) representing any complex two-dimensional stress system falls outside the square, then failure will occur according to the theory.

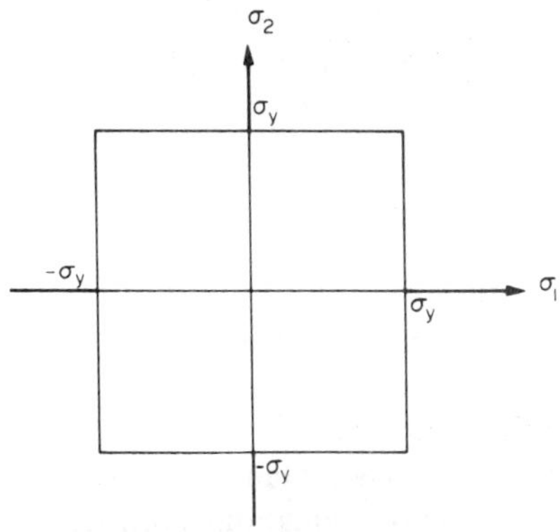

Fig. 15.5. Maximum principal stress failure envelope (locus).

(b) *Maximum shear stress theory*

For *like stresses*, i.e. σ_1 and σ_2, both tensile or both compressive (first and third quadrants), the maximum shear stress criterion is

$$\tfrac{1}{2}(\sigma_1 - 0) = \tfrac{1}{2}\sigma_y \quad \text{or} \quad \tfrac{1}{2}(\sigma_2 - 0) = \tfrac{1}{2}\sigma_y$$

i.e.

$$\sigma_1 = \sigma_y \quad \text{or} \quad \sigma_2 = \sigma_y$$

thus producing the same result as the previous theory in the first and third quadrants.

For *unlike stresses* the criterion becomes

$$\tfrac{1}{2}(\sigma_1 - \sigma_2) = \tfrac{1}{2}\sigma_y$$

since consideration of the third stress as zero will not produce as large a shear as that when σ_2 is negative. Thus for the second and fourth quadrants,

$$\frac{\sigma_1}{\sigma_y} - \frac{\sigma_2}{\sigma_y} = 1 \quad \left(\text{or} \quad \frac{\sigma_2}{\sigma_y} - \frac{\sigma_1}{\sigma_y} = 1\right)$$

These are straight lines and produce the failure envelope of Fig. 15.6. Again, any point outside the failure envelope represents a condition of potential failure.

(c) *Maximum principal strain theory*

For yielding in tension the theory states that

$$\sigma_1 - \nu\sigma_2 = \sigma_y$$

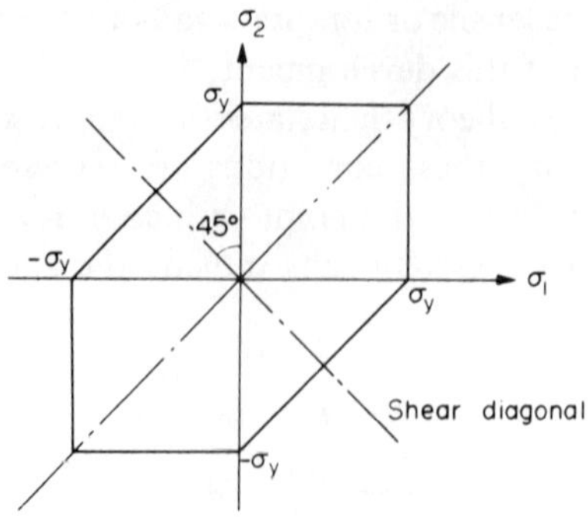

Fig. 15.6. Maximum shear stress failure envelope.

and for compressive yield, with σ_2 compressive,

$$\sigma_2 - \nu\sigma_1 = \sigma_y$$

Since this theory does not find general acceptance in any engineering field it is sufficient to note here, without proof, that the above equations produce the rhomboid failure envelope shown in Fig. 15.7.

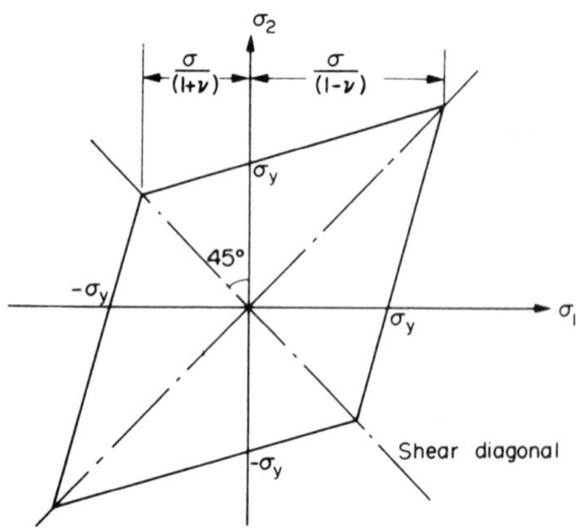

Fig. 15.7. Maximum principal strain failure envelope.

(*d*) *Maximum strain energy per unit volume theory*

With $\sigma_3 = 0$ this failure criterion reduces to

$$\sigma_1^2 + \sigma_2^2 - 2\nu\sigma_1\sigma_2 = \sigma_y^2$$

i.e. $$\left(\frac{\sigma_1}{\sigma_y}\right)^2 + \left(\frac{\sigma_2}{\sigma_y}\right)^2 - 2\nu\left(\frac{\sigma_1}{\sigma_y}\right)\left(\frac{\sigma_2}{\sigma_y}\right) = 1$$

This is the equation of an ellipse with major and minor semi-axes

$$\frac{\sigma_y}{\sqrt{(1-\nu)}} \quad \text{and} \quad \frac{\sigma_y}{\sqrt{(1+\nu)}}$$

respectively, each at 45° to the coordinate axes as shown in Fig. 15.8.

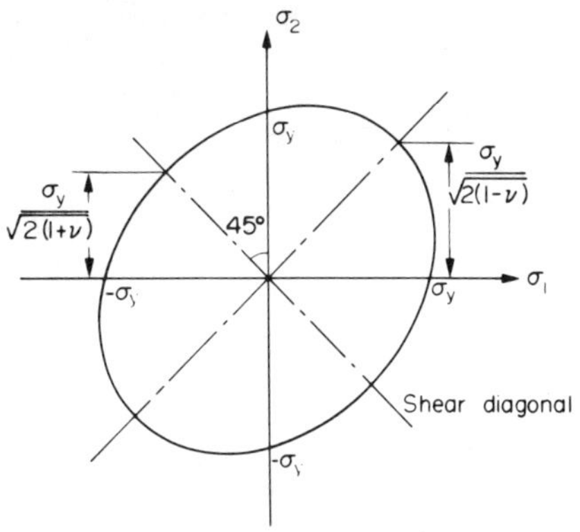

Fig. 15.8. Failure envelope for maximum strain energy per unit volume theory.

(e) Maximum shear strain energy per unit volume theory

With $\sigma_3 = 0$ the criteria of failure for this theory reduces to

$$\tfrac{1}{2}[(\sigma_1-\sigma_2)^2+\sigma_2^2+\sigma_1^2] = \sigma_y^2$$

$$\sigma_1^2+\sigma_2^2-\sigma_1\sigma_2 = \sigma_y^2$$

$$\left(\frac{\sigma_1}{\sigma_y}\right)^2+\left(\frac{\sigma_2}{\sigma_y}\right)^2-\left(\frac{\sigma_1}{\sigma_y}\right)\left(\frac{\sigma_2}{\sigma_y}\right) = 1$$

again an ellipse with semi-axes $\sqrt{(2)}\sigma_y$ and $\sqrt{(\frac{2}{3})}\sigma_y$ at 45° to the coordinate axes as shown in Fig. 15.9. The ellipse will circumscribe the maximum shear stress hexagon.

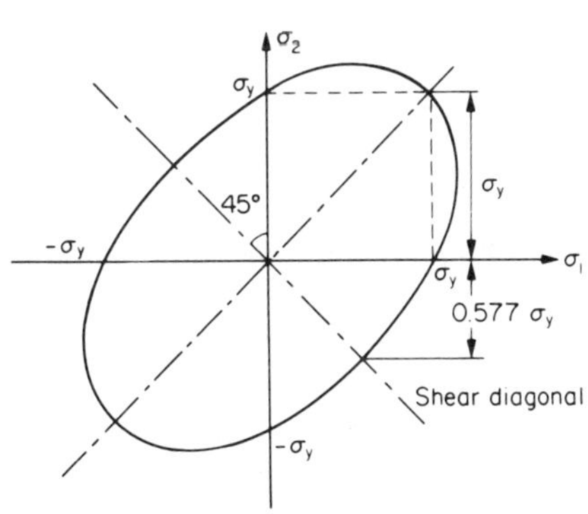

Fig. 15.9. Failure envelope for maximum shear strain energy per unit volume theory.

(*f*) *Mohr's modified shear stress theory* ($\sigma_{y_c} > \sigma_{y_t}$)

For the original formulation of the theory based on the results of pure tension, pure compression and pure shear tests the Mohr failure envelope is as indicated in Fig. 15.10.

In its simplified form, however, based on just the pure tension and pure compression results, the failure envelope becomes that of Fig. 15.11.

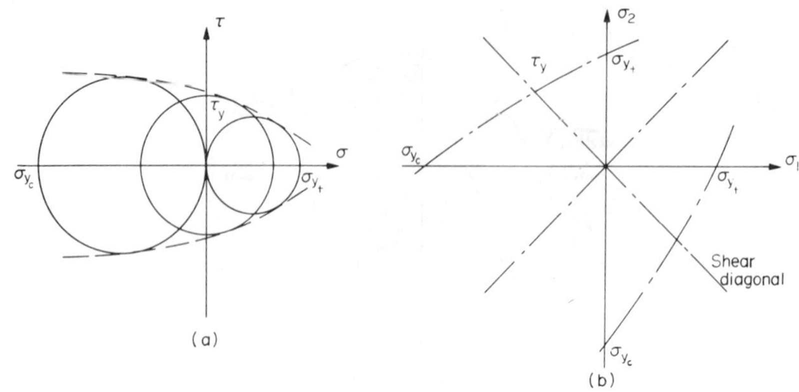

Fig. 15.10. (a) Mohr theory on σ–τ axes. (b) Mohr theory failure envelope on σ_1–σ_2 axes.

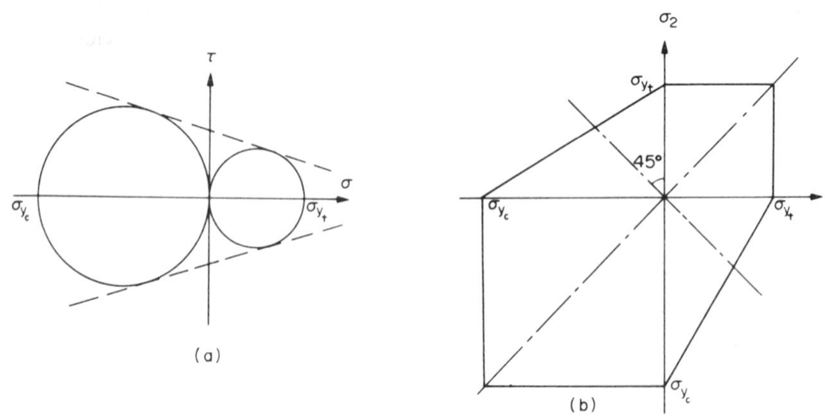

Fig. 15.11. (a) Simplified Mohr theory on σ–τ axes. (b) Failure envelope for simplified Mohr theory.

15.8. Graphical solution of two-dimensional theory of failure problems

The graphical representations of the failure theories, or yield loci, may be combined onto a single set of σ_1 and σ_2 coordinate axes as shown in Fig. 15.12. Inside any particular locus or failure envelope elastic conditions prevail whilst points outside the loci suggest that yielding or fracture will occur. It will be noted that in most cases the maximum shear stress criterion is the most conservative of the theories. The combined diagram is particularly useful since it allows experimental points to be plotted to give an immediate assessment of failure

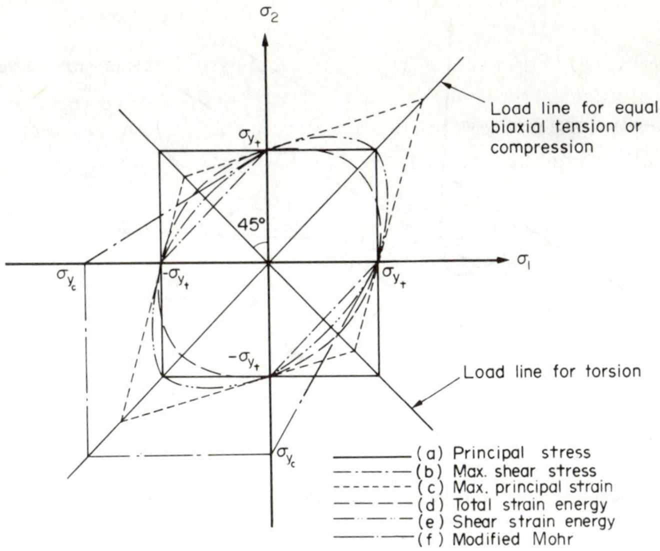

Fig. 15.12. Combined yield loci for the various failure theories.

probability according to the various theories. In the case of equal biaxial tension or compression for example $\sigma_1/\sigma_2 = 1$ and a so-called *load line* may be drawn through the origin with a slope of unity to represent this loading case. This line cuts the yield loci in the order of theories *d*; (*a*, *b*, *e*, *f*); and *c*. In the case of pure torsion, however, $\sigma_1 = \tau$ and $\sigma_2 = -\tau$, i.e. $\sigma_1/\sigma_2 = -1$. This load line will therefore have a slope of -1 and the order of yield according to the various theories is now changed considerably to (*b*; *e*, *f*, *d*, *c*, *a*). The load line procedure may be used to produce rapid solutions of failure problems as shown in Example 15.2.

15.9. Graphical representation of the failure theories for three-dimensional stress systems

15.9.1. Ductile materials

(*a*) *Maximum shear strain energy or distortion energy* (*von Mises*) *theory*

It has been stated earlier that the failure of most ductile materials is most accurately governed by the distortion energy criterion which states that, at failure,

$$(\sigma_1-\sigma_2)^2+(\sigma_2-\sigma_3)^2+(\sigma_3-\sigma_1)^2 = 2\sigma_y^2 = \text{constant}$$

In the special case where $\sigma_3 = 0$, this has been shown to give a yield locus which is an ellipse symmetrical about the shear diagonal. For a three-dimensional stress system the above equation defines the surface of a regular prism having a circular cross-section, i.e. a cylinder with its central axis along the line $\sigma_1 = \sigma_2 = \sigma_3$. The axis thus passes through the origin of the principal stress coordinate system shown in Fig. 15.13 and is inclined at equal angles to each

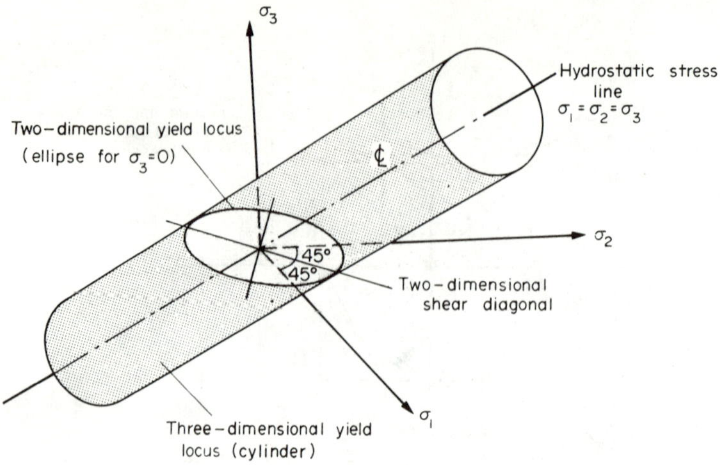

Fig. 15.13. Three-dimensional yield locus for Maxwell–von Mises distortion energy (shear strain energy per unit volume) theory.

axis. It will be observed that when $\sigma_3 = 0$ the failure condition reverts to the ellipse mentioned above, i.e. that produced by intersection of the (σ_1, σ_2) plane with the inclined cylinder.

The yield locus for the von Mises theory in a three-dimensional stress system is thus the *surface* of the inclined cylinder. Points within the cylinder represent safe conditions, points outside indicate failure conditions. It should be noted that the cylinder axis extends indefinitely along the $\sigma_1 = \sigma_2 = \sigma_3$ line, this being termed the *hydrostatic stress line*. It can be shown that hydrostatic stress alone cannot cause yielding and it is presumed that all other stress conditions which fall within the cylindrical boundary may be considered equally safe.

(*b*) *Maximum shear stress (Tresca) theory*

With a few exceptions, e.g. aluminium alloys and certain steels, the yielding of most ductile materials is adequately governed by the Tresca maximum shear stress condition, and because of its relative simplicity it is often used in preference to the von Mises theory. For the Tresca theory the three-dimensional yield locus can be shown to be a regular prism with hexagonal cross-section (Fig. 15.14). The central axis of this figure is again on the line $\sigma_1 = \sigma_2 = \sigma_3$ (the hydrostatic stress line) and again extends to infinity.

Points representing stress conditions plotted on the principal stress coordinate axes indicate safe conditions if they lie within the surface of the hexagonal cylinder. The two-dimensional yield locus of Fig. 15.6 is obtained as before by the intersection of the σ_1, σ_2 plane $(\sigma_3 = 0)$ with this surface.

15.9.2. Brittle materials

Failure of brittle materials has been shown previously to be governed by the maximum principal tensile stress present in the three-dimensional stress system. This is thought to be

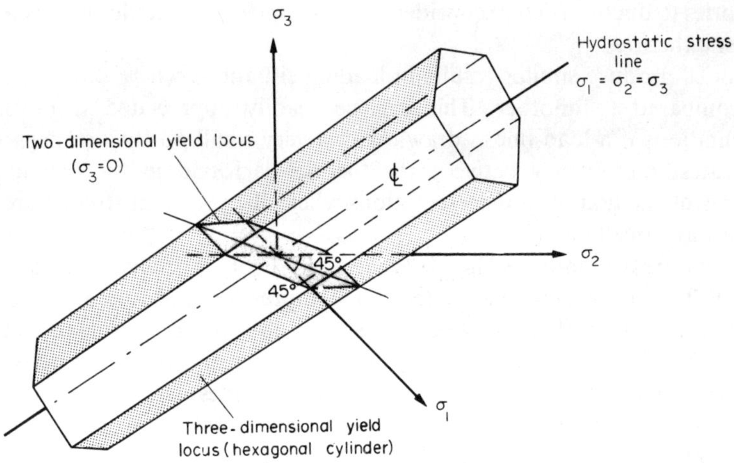

Fig. 15.14. Three-dimensional yield locus for Tresca (maximum shear stress) theory.

due to the microscopic cracks, flaws or discontinuities which are present in most brittle materials and which act as local stress raisers. These stress raisers, or *stress concentrations*, have a much greater adverse effect in tension and hence produce the characteristic weaker behaviour of brittle materials in tension than in compression.

Thus if the greatest tensile principal stress exceeds the yield stress then failure occurs, and such a simple condition does not require a graphical representation.

15.10. Limitations of the failure theories

It is important to remember that the theories introduced above are those of *elastic* failure, i.e. they relate to the "failure" which is assumed to occur under elastic loading conditions at an equivalent stage to that of yielding in a simple tensile test. If it is anticipated that loading conditions are such that the component may fail in service in a way which cannot easily be related to standard simple loading tests (e.g. under fatigue, creep, buckling, impact loading, etc.) then the above "classical" elastic failure theories should not be applied. A good example of this is the brittle fracture failure of steel under low temperature or very high strain rate (impact) conditions compared with simple ductile failure under normal ambient conditions. If any doubt exists about the relevance of the failure theories then, ideally, specially designed tests should be carried out on the component with loading conditions as near as possible to those expected in service. If, however, elastic failure can be assumed to be relevant it is necessary to consider which of the theories is the most appropriate for the material in question and for the service loading condition expected.

In most cases the Von Mises "distortion energy" theory is considered to be the most reliable and relevant theory with the following exceptions:

(a) For brittle materials the maximum principal stress or Mohr "internal friction" theories are most suitable. (It must be noted, however, that the former is definitely unsafe for ductile materials.) Some authorities also recommend the Mohr theory for extension of the theories to ductile *fracture* consideration as opposed to ductile yielding as assumed in the elastic theories.

(b) All theories produce similar results in loading situations where one principal stress is large compared to another. This can be readily appreciated from the graphical representations if a load-line is drawn with a very small positive or negative slope.

(c) The greatest discrepancy between the theories is found in the second and fourth quadrants of the graphical representations where the principal stresses are of opposite sign but numerically equal.

(d) For bi-axial stress conditions, the Mohr modified theory is often preferred, provided that reliable test data are available for tension, compression and torsion.

(e) In most general bi-axial and tri-axial stress conditions the Tresca maximum shear stress theory is the most conservative (i.e. the safest) theory and this, together with its easily applied and simple formula, probably explains its widespread use in industry.

(f) The St. Venant maximum principal strain and Haigh total strain energy per unit volume theories are now rarely, if ever, used in general engineering practice.

15.11. Effect of stress concentrations

Whilst stress concentrations have their most significant effect under fatigue loading conditions and impact situations, nevertheless, there are also some important considerations for static loading applications, namely:

(a) In the presence of ductile yielding, stress concentrations are relatively unimportant since the yielding which will occur at the concentration, e.g. the tip of a notch, will merely redistribute the stresses and not necessarily lead to failure. If, however, there is only marginal ductility, or in the presence of low temperatures, then stress concentrations become more significant as the likelihood of brittle failure increases. It is wise, therefore, to keep stress concentration factors as low as possible.

(b) For brittle materials like cast iron, internal stress concentrations arise within the material due to the presence of, e.g., flaws, impurities or graphite flakes. These produce stress increases at least as large as those given by surface stress concentrations which, therefore, may have little or no effect on failure. A cast iron bar with a small transverse hole, for example, may not fracture at the hole when a tensile load is applied!

15.12. Safety factors

When using elastic design procedures incorporating any of the failure theories introduced in this chapter it is normal to incorporate safety factors to take account of various imponderables which arise when one attempts to forecast accurately service loads or operating conditions or to make allowance for variations in material properties or behaviour from those assumed by the acceptance of "standard" values. "Ideal" application of the theories, i.e. a rigorous mathematical analysis, is thus rarely possible and the following factors indicate in a little more detail the likely sources of inaccuracy:

1. Whilst design may have been based up nominally static loading, changing service conditions or misuse by operators can often lead to dynamic, fluctuating or impact loading situations which will produce significant increases in maximum stress levels.

2. A precise knowledge of the mechanical properties of the material used in the design is seldom available. Standard elastic values found in reference texts assume ideal homogeneous and isotropic materials with equal "strengths" in all directions. This is rarely true in practice and the effect of internal flaws, inclusions or other weaknesses in the material may be quite significant.
3. The method of manufacture or construction of the component can have a significant effect on service life, particularly if residual stresses are introduced by, e.g., welding or straining beyond the elastic limit during the assembly stages.
4. Complex designs often give rise to difficult analysis problems which even after time-consuming and expensive theoretical procedures, at best yield only a reasonable estimate of maximum service stresses.

Despite these problems and the assumptions which are often required to overcome them, it has been shown that elastic design procedures can be made to agree with experimental results within a reasonable margin of error provided that appropriate safety factors are applied.

It has been shown in § 1.16 that alternative definitions are used for the safety factor depending upon whether it is based on the tensile strength of the material used or its yield strength, i.e., either

$$\text{safety factor, } n = \frac{\text{tensile strength}}{\text{allowable working stress}}$$

or

$$\text{safety factor, } n = \frac{\text{yield stress (or proof stress)}}{\text{allowable working stress}}$$

Clearly, *it is important when quoting safety factors to state which definition has been used.*

Safety values vary depending on the type of industry and the area of application of the component being designed. National codes of practice (e.g. British Standards) or other external authority regulations often quote mandatory values to be applied and some companies produce their own guideline values.

Table 15.2 shows the way in which the various factors outlined above contribute to the overall factor of safety for some typical service conditions. These values are based on the yield stress of the materials concerned.

TABLE 15.2. Typical safety factors.

Application	(a) Nature of stress	(b) Nature of load	(c) Type of service	Overall safety factor (a) × (b) × (c)
Steelwork in buildings	1	1	2	2
Pressure vessels	1	1	3	3
Transmission shafts	3	1	2	6
Connecting rods	3	2	1.5	9

It should be noted, however, that the values given in the "type of service" column can be considered to be conservative and severe misuse or overload could increase these (and, hence, the overall factors) by as much as five times.

Recent legislative changes such as "Product Liability" and "Health and Safety at Work" will undoubtedly cause renewed concern that appropriate safety factors are applied, and may lead to the adoption of higher values. Since this could well result in uneconomic utilisation of materials, such a trend would be regrettable and a move to enhanced product testing and service load monitoring is to be preferred.

15.13. Modes of failure

Before concluding this chapter, the first which looks at design procedures to overcome possible failure (in this case elastic overload), it is appropriate to introduce the reader to the many other ways in which components may fail in order that an appreciation is gained of the complexities often facing designers of engineering components. Sub-classification and a certain amount of cross-referencing does make the list appear to be formidably long but even allowing for these it is evident that the designer, together with his supporting materials and stress advisory teams, has an unenviable task if satisfactory performance and reliability of components is to be obtained in the most complex loading situations. The list below is thus a summary of the so-called "*modes* (*or methods*) *of failure*"

1. Mechanical overload/under-design
2. Elastic yielding – force and/or temperature induced.
3. Fatigue
    - high cycle
    - low cycle
    - thermal
    - corrosion
    - fretting
    - impact
    - surface
4. Brittle fracture
5. Creep
6. Combined creep and fatigue
7. Ductile rupture
8. Corrosion
    - direct chemical
    - galvanic
    - pitting
    - cavitation
    - stress
    - intergranular
    - crevice
    - erosion
    - hydrogen damage
    - selective leaching
    - biological
    - corrosion fatigue
9. Impact
    - fracture
    - fatigue

deformation
wear
fretting

10. Instability
    buckling
    creep buckling
    torsional instability
11. Wear
    adhesive
    abrasive
    corrosive
    impact
    deformation
    surface fatigue
    fretting
12. Vibration
13. Environmental
    thermal shock
    radiation damage
    lubrication failure
14. Contact
    spalling
    pitting
    galling and seizure
15. Stress rupture
16. Thermal relaxation

Examples

Example 15.1

A material subjected to a simple tension test shows an elastic limit of 240 MN/m^2. Calculate the factor of safety provided if the principal stresses set up in a complex two-dimensional stress system are limited to 140 MN/m^2 tensile and 45 MN/m^2 compressive. The appropriate theories of failure on which your answer should be based are:

(a) the maximum shear stress theory;
(b) the maximum shear strain energy theory.

Solution

(*a*) *Maximum shear stress theory*

This theory states that failure will occur when the maximum shear stress in the material equals the maximum shear stress value at the yield point in a simple tension test, i.e. when

$$\tfrac{1}{2}(\sigma_1 - \sigma_3) = \tfrac{1}{2}\sigma_y$$

or

$$\sigma_1 - \sigma_3 = \sigma_y$$

In this case the system is two-dimensional, i.e. the principal stress in one plane is zero. However, since one of the given principal stresses is a compressive one, it follows that the zero value is that of σ_2 since the negative value of σ_3 associated with the compressive stress will produce a numerically greater value of stress difference $\sigma_1 - \sigma_3$ and hence must be used in the above criterion.

Thus $\sigma_1 = 140\,\text{MN/m}^2$, $\sigma_2 = 0$ and $\sigma_3 = -45\,\text{MN/m}^2$.

Now with a factor of safety applied the design yield point becomes σ_y/n and this must replace σ_y in the yield criterion which then becomes

$$\frac{\sigma_y}{n} = \sigma_1 - \sigma_3$$

$$\therefore \quad \frac{240}{n} = 140 - (-45) \quad \text{units of MN/m}^2 \text{ throughout}$$

$$n = \frac{240}{185} = \mathbf{1.3}$$

The required factor of safety is 1.3.

(b) *Maximum shear strain energy theory*

Once again equating the values of the quantity concerned in the tensile test and in the complex stress system,

$$\frac{\sigma_y^2}{6G} = \frac{1}{12G}[(\sigma_1 - \sigma_2)^2 + (\sigma_2 - \sigma_3)^2 + (\sigma_3 - \sigma_1)^2]$$

$$\sigma_y^2 = \tfrac{1}{2}[(\sigma_1 - \sigma_2)^2 + (\sigma_2 - \sigma_3)^2 + (\sigma_3 - \sigma_1)^2]$$

With the three principal stress values used above and with σ_y/n replacing σ_y

$$\left(\frac{240}{n}\right)^2 = \tfrac{1}{2}\{(140-0)^2 + [0-(-45)]^2 + (-45-140)^2\}$$

$$\frac{5.76 \times 10^4}{n^2} = \tfrac{1}{2}[1.96 + 0.203 + 3.42]10^4$$

$$n^2 = \frac{2 \times 5.76 \times 10^4}{5.583 \times 10^4} = 2.063$$

$$\therefore \quad n = \mathbf{1.44}$$

The required factor of safety is now 1.44.

Example 15.2

A steel tube has a mean diameter of 100 mm and a thickness of 3 mm. Calculate the torque which can be transmitted by the tube with a factor of safety of 2.25 if the criterion of failure is (a) maximum shear stress; (b) maximum strain energy; (c) maximum shear strain energy. The elastic limit of the steel in tension is $225\,\text{MN/m}^2$ and Poisson's ratio ν is 0.3.

Solution

From the torsion theory

$$\frac{T}{J} = \frac{\tau}{R} \quad \therefore \quad \tau = \frac{TR}{J}$$

Now mean diameter of tube = 100 mm and thickness = 3 mm.

$$\therefore \qquad J = \pi dt \times r^2 = \frac{\pi d^3 t}{4} \quad \text{(approximately)}$$

$$= \frac{\pi \times 0.1^3 \times 0.003}{4} = 2.36 \times 10^{-6}\,\text{m}^4$$

$$\therefore \qquad \text{shear stress } \tau = \frac{T \times 51.5 \times 10^{-3}}{2.36 \times 10^{-6}} = (2.18 \times 10^4)T\ \text{N/m}^2$$

$$= 21.8T\ \text{kN/m}^2$$

(*a*) *Maximum shear stress*

Torsion introduces pure shear onto elements within the tube material and it has been shown in § 13.2 that pure shear produces an equivalent principal direct stress system, one tensile and one compressive and both equal in value to the applied shear stress,

i.e. $$\sigma_1 = \tau, \quad \sigma_3 = -\tau \quad (\text{and } \sigma_2 = 0)$$

Thus for the maximum shear stress criterion, taking account of the safety factor,

$$\frac{\sigma_y}{n} = \sigma_1 - \sigma_3 = \tau - (-\tau)$$

$$\therefore \qquad \frac{225 \times 10^6}{2.25} = 2\tau = 2 \times 21.8T \times 10^3$$

$$\therefore \qquad T = \frac{100 \times 10^6}{2 \times 21.8 \times 10^3} = \mathbf{2.3 \times 10^3\ N\,m}$$

The torque which can be safely applied = 2.3 kN m.

(*b*) *Maximum strain energy*

From eqn. (15.3) the relevant criterion of failure is

$$\sigma_y^2 = \sigma_1^2 + \sigma_2^2 + \sigma_3^2 - 2\nu(\sigma_1\sigma_2 + \sigma_2\sigma_3 + \sigma_3\sigma_1)$$

Taking account of the safety factor

$$\left(\frac{225 \times 10^6}{2.25}\right)^2 = \tau^2 + 0 + (-\tau)^2 - 2 \times 0.3[\tau \times (-\tau)]$$

$$= 2.6\tau^2$$

$$= 2.6(21.8 \times 10^3 T)^2$$

$$\therefore \quad T = \frac{100 \times 10^6}{\sqrt{(2.6)} \times 21.8 \times 10^3} = \mathbf{2.84 \times 10^3\ N\,m}$$

The safe torque is now 2.84 kN m.

(c) *Maximum shear strain energy*

From eqn. (15.4) the criterion of failure is

$$\sigma_y^2 = \tfrac{1}{2}[(\sigma_1 - \sigma_2)^2 + (\sigma_2 - \sigma_3)^2 + (\sigma_3 - \sigma_1)^2]$$

$$\therefore \quad \left(\frac{225 \times 10^6}{2.25}\right)^2 = \tfrac{1}{2}\{(\tau - 0)^2 + [0 - (-\tau)]^2 + (-\tau - \tau)^2\}$$

$$= 3\tau^2$$

$$\therefore \quad \tau = \frac{100 \times 10^6}{\sqrt{3}} = 21.8 \times 10^3 T$$

$$\therefore \quad T = \frac{100 \times 10^6}{21.8 \times 10^3 \times \sqrt{3}} = \mathbf{2.65 \times 10^3\ N\,m}$$

The safe torque is now 2.65 kN m.

Example 15.3

A structure is composed of circular members of diameter *d*. At a certain position along one member the loading is found to consist of a shear force of 10 kN together with an axial tensile load of 20 kN. If the elastic limit in tension of the material of the members is 270 MN/m^2 and there is to be a factor of safety of 4, estimate the magnitude of *d* required according to (a) the maximum principal stress theory, and (b) the maximum shear strain energy per unit volume theory. Poisson's ratio $\nu = 0.283$.

Solution

The stress system at the point concerned is as shown in Fig. 15.15, the principal stress normal to the surface of the member being zero.

Now the direct stress along the axis of the bar is tensile, i.e. positive, and given by

$$\sigma_x = \frac{\text{load}}{\text{area}} = \frac{20}{\pi d^2/4} = \frac{80}{\pi d^2} \quad \text{kN/m}^2$$

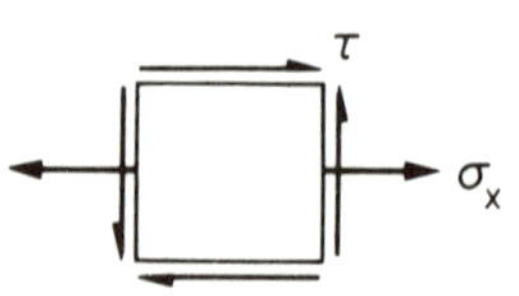

Fig. 15.15.

and the shear stress is

$$\tau = \frac{\text{shear load}}{\text{area}} = \frac{10}{\pi d^2/4} = \frac{40}{\pi d^2} \quad \text{kN/m}^2$$

The principal stresses are given by Mohr's circle construction (πd^2 being a common denominator) or from

$$\sigma_1 \text{ and } \sigma_3 = \tfrac{1}{2}(\sigma_x + \sigma_y) \pm \tfrac{1}{2}\sqrt{[(\sigma_x - \sigma_y)^2 + 4\tau_{xy}^2]}$$

with σ_y zero,

i.e.
$$\sigma_1 \text{ or } \sigma_3 = \frac{1}{2}\left\{\frac{80}{\pi d^2} \pm \sqrt{\left[\left(\frac{80}{\pi d^2}\right)^2 + 4\left(\frac{40}{\pi d^2}\right)^2\right]}\right\}$$

$$= \frac{40}{\pi d^2}(1 \pm \sqrt{2})$$

$\therefore$
$$\sigma_1 = \frac{40 \times 2.414}{\pi d^2} = \frac{30.7}{d^2} \quad \text{kN/m}^2$$

$$\sigma_3 = -\frac{40 \times 0.414}{\pi d^2} = -\frac{5.27}{d^2} \quad \text{kN/m}^2$$

and
$$\sigma_2 = 0$$

Since the elastic limit in tension is 270 MN/m^2 and the factor of safety is 4, the working stress or effective yield stress is

$$\sigma_y = \frac{270}{4} = 67.5\,\text{MN/m}^2$$

(a) *Maximum principal stress theory*

Failure is assumed to occur when

$$\sigma_1 = \sigma_y$$

$\therefore$
$$\frac{30.7 \times 10^3}{d^2} = 67.5 \times 10^6$$

$\therefore$
$$d^2 = \frac{30.7}{67.5} \times 10^{-3} = 4.55 \times 10^{-4}\,\text{m}^2$$

$\therefore$
$$d = 2.13 \times 10^{-2}\,\text{m} = \mathbf{21.3\,mm}$$

(b) *Maximum shear strain energy*

From eqn. (15.4) the criterion of failure is

$$2\sigma_y^2 = (\sigma_1 - \sigma_2)^2 + (\sigma_2 - \sigma_3)^2 + (\sigma_3 - \sigma_1)^2$$

Therefore taking account of the safety factor

$$2(67.5 \times 10^6)^2 = \left[\left(\frac{30.7}{d^2}\right)^2 + \left(-\frac{5.27}{d^2}\right)^2 + \left(\frac{-5.27 - 30.7}{d^2}\right)^2\right] \times 10^6$$

$$= \frac{2264 \times 10^6}{d^4}$$

$$\therefore \quad d^4 = \frac{1132 \times 10^6}{(67.5 \times 10^6)^2}$$

$$\therefore \quad d^2 = \frac{33.6 \times 10^3}{67.5 \times 10^6} = 4.985 \times 10^{-4}\,\text{m}^2$$

$$\therefore \quad d = \mathbf{22.3\,mm}$$

Example 15.4

Assuming the formulae for the principal stresses and the maximum shear stress induced in a material owing to combined stresses and the fundamental formulae for pure bending, derive a formula in terms of the bending moment M and the twisting moment T for the equivalent twisting moment on a shaft subjected to combined bending and torsion for

(a) the maximum principal stress criterion;
(b) the maximum shear stress criterion.

Solution

The *equivalent torque*, or turning moment, is defined as that torque which, acting alone, will produce the same conditions of stress as the combined bending and turning moments.

At failure the stress produced by the equivalent torque T_E is given by the torsion theory

$$\frac{T}{J} = \frac{\tau}{R}$$

$$\therefore \quad \tau_{max} = \frac{T_E R}{J} = \frac{T_E \times D}{2J}$$

The direct stress owing to bending is

$$\sigma_x = \frac{My_{max}}{I} = \frac{MD}{2I} = \frac{MD}{J}$$

and the shear stress due to torsion is

$$\tau = \frac{TD}{2J}$$

The principal stresses are then given by

$$\sigma_{1,3} = \tfrac{1}{2}(\sigma_x + \sigma_y) \pm \tfrac{1}{2}\sqrt{[(\sigma_x - \sigma_y)^2 + 4\tau_{xy}^2]} \quad \text{with} \quad \sigma_y = 0 \quad \text{and} \quad \sigma_2 = 0$$

$$= \frac{1}{2}\left(\frac{MD}{J}\right) \pm \frac{1}{2}\sqrt{\left[\left(\frac{MD}{J}\right)^2 + 4\left(\frac{TD}{2J}\right)^2\right]}$$

$$= \frac{D}{2J}(M \pm \sqrt{[M^2 + T^2]})$$

$$\therefore \qquad \sigma_1 = \frac{D}{2J}(M + \sqrt{[M^2 + T^2]})$$

$$\sigma_3 = \frac{D}{2J}(M - \sqrt{[M^2 + T^2]})$$

(a) *For maximum principal stress criterion*

$$\frac{T_E D}{2J} = \sigma_1 = \frac{D}{2J}(M + \sqrt{[M^2 + T^2]})$$

$$\therefore \qquad T_E = M + \sqrt{(M^2 + T^2)}$$

(b) *For maximum shear stress criterion*

$$\frac{T_E D}{2J} = \tfrac{1}{2}(\sigma_1 - \sigma_3)$$

$$= \frac{1}{2}\left\{\frac{D}{2J}(M + \sqrt{[M^2 + T^2]}) - \frac{D}{2J}(M - \sqrt{[M^2 + T^2]})\right\}$$

$$T_E = \sqrt{(M^2 + T^2)}$$

Example 15.5

The test strengths of a material under pure compression and pure tension are $\sigma_{y_c} = 350\ \text{MN/m}^2$ and $\sigma_{y_t} = 300\ \text{MN/m}^2$. In a certain design of component the material may be subjected to each of the five biaxial stress states shown in Fig. 15.16. Assuming that failure is deemed to occur when yielding takes place, arrange the five stress states in order of diminishing factor of safety according to the maximum principal or normal stress, maximum shear stress, maximum shear strain energy (or distortion energy) and modified Mohr's (or internal friction) theories.

Solution

A graphical solution of this problem can be employed by constructing the combined yield loci for the criteria mentioned in the question. Since σ_1 the maximum principal stress is $+100\ \text{MN/m}^2$ in each of the stress states only half the combined loci diagram is required, i.e. the positive σ_1 half.

Here it must be remembered that for stress condition (e) pure shear is exactly equivalent to two mutually perpendicular direct stresses – one tensile, the other compressive, acting on 45° planes and of equal value to the applied shear, i.e. for condition (e) $\sigma_1 = 100\ \text{MN/m}^2$ and $\sigma_2 = -100\ \text{MN/m}^2$ (see §13.2).

It is now possible to construct the "load lines" for each stress state with slopes of σ_2/σ_1. An immediate solution is then obtained by considering the intersection of each load line with the failure envelopes.

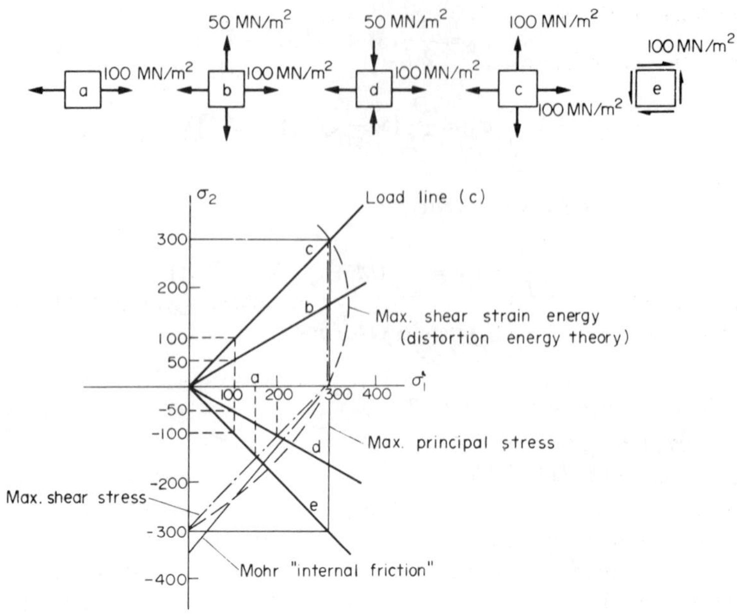

Fig. 15.16.

Maximum principal stress theory

All five load lines cut the failure envelope for this theory at $\sigma_1 = 300\ \text{MN/m}^2$. According to this theory, therefore, all the stress states will produce failure when the maximum direct stress reaches $300\ \text{MN/m}^2$. Since the maximum principal stress present in each stress state is $100\ \text{MN/m}^2$ it therefore follows that the safety factor for each state according to the maximum principal stress theory is $\dfrac{300}{100} = 3$.

Maximum shear stress theory

The load lines *a*, *b* and *c* cut the failure envelope for this theory at $\sigma_1 = 300\ \text{MN/m}^2$ whilst *d* and *e* cut it at $\sigma_1 = 200\ \text{MN/m}^2$ and $\sigma_1 = 150\ \text{MN/m}^2$ respectively as shown in Fig. 15.16. The safety factors are, therefore,

$$a, b, c = \frac{300}{100} = \mathbf{3}, \quad d = \frac{200}{100} = \mathbf{2}, \quad e = \frac{150}{100} = \mathbf{1.5}$$

Maximum shear strain energy theory

In decreasing order, the factors of safety for this theory, found as before from the points where each load line crosses the failure envelope, are

$$b = \frac{347}{100} = \mathbf{3.47}, \quad a, c = \frac{300}{100} = \mathbf{3}, \quad d = \frac{227}{100} = \mathbf{2.27}, \quad e = \frac{173}{100} = \mathbf{1.73}$$

Mohr's modified or internal friction theory (with $\sigma_{y_c} = 350\ MN/m^2$)

In this case the safety factors are:

$$a, b, c = \frac{300}{100} = \mathbf{3}, \quad d = \frac{210}{100} = \mathbf{2.1}, \quad e = \frac{162}{100} = \mathbf{1.62}$$

Example 15.6

The cast iron used in the manufacture of an engineering component has tensile and compressive strengths of 400 MN/m^2 and 1.20 GN/m^2 respectively.

(a) If the maximum value of the tensile principal stress is to be limited to one-quarter of the tensile strength, determine the maximum value and nature of the other principal stress using Mohr's modified yield theory for brittle materials.
(b) What would be the values of the principal stresses associated with a maximum shear stress of 450 MN/m^2 according to Mohr's modified theory?
(c) At some point in a component principal stresses of 100 MN/m^2 tensile and 100 MN/m^2 compressive are found to be present. Estimate the safety factor with respect to initial yield using the maximum principal stress, maximum shear stress, distortion energy and Mohr's modified theories of elastic failure.

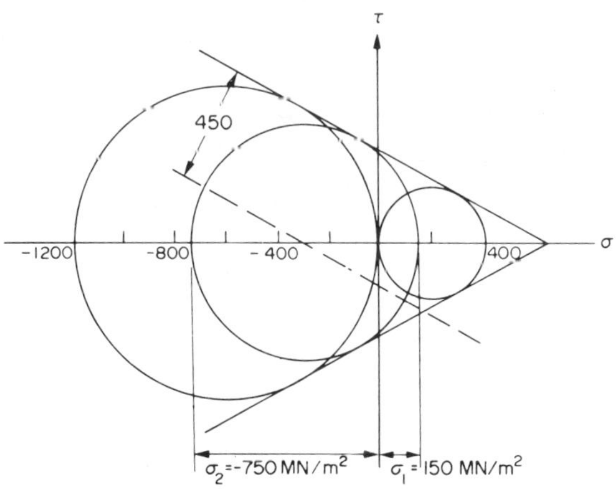

Fig. 15.17.

Solution

(a) Maximum principal stress $= \dfrac{400}{4} = 100\ \text{MN/m}^2$

According to Mohr's theory

$$\frac{\sigma_1}{\sigma_{y_t}} + \frac{\sigma_2}{\sigma_{y_c}} = 1$$

$$\therefore \qquad \frac{100 \times 10^6}{400 \times 10^6} + \frac{\sigma_2}{-1.2 \times 10^9} = 1$$

$$\therefore \qquad \sigma_2 = -1.2 \times 10^9 (1 - \tfrac{1}{4}) = \mathbf{-900\ MN/m^2}$$

(b) In any Mohr circle construction the radius of the circle equals the maximum shear stress value. In order to answer this part of the question, therefore, it is necessary to draw the Mohr failure envelope on σ–τ axes as shown in Fig. 15.17 and to construct the circle which is tangential to the envelope and has a radius of 450 MN/m^2. This is achieved by drawing a line parallel to the failure envelope and a distance of 450 MN/m^2 (to scale) from it. Where this line cuts the σ axis is then the centre of the required circle. The desired principal stresses are then, as usual, the extremities of the horizontal diameter of the circle.

Thus from Fig. 15.17

$$\boldsymbol{\sigma_1 = 150\ \mathrm{MN/m^2}} \quad \text{and } \boldsymbol{\sigma_2 = -750\ \mathrm{MN/m^2}}$$

(c) The solution here is similar to that used for Example 15.5. The yield loci are first plotted for the given failure theories and the required safety factors determined from the points of intersection of the loci and the load line with a slope of $100/-100 = -1$.

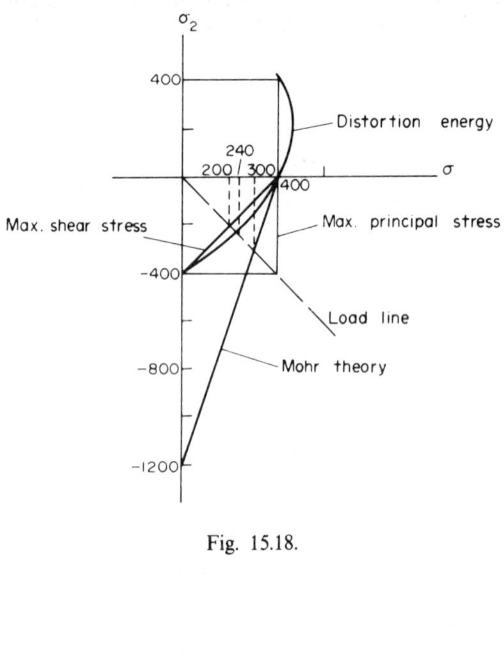

Fig. 15.18.

Thus from Fig. 15.18 the safety factors are:

$$\text{Maximum principal stress} = \frac{400}{100} = \mathbf{4}$$

$$\text{Maximum shear stress} = \frac{200}{100} = \mathbf{2}$$

$$\text{Distortion energy} \quad = \frac{240}{100} = 2.4$$

$$\text{Mohr theory} \quad = \frac{300}{100} = 3$$

Problems

15.1 (B). If the principal stresses at a point in an elastic material are 120 MN/m^2 tensile, 180 MN/m^2 tensile and 75 MN/m^2 compressive, find the stress at the limit of proportionality expected in a simple tensile test assuming:
(a) the maximum shear stress theory;
(b) the maximum shear strain energy theory;
(c) the maximum principal strain theory.
Assume $\nu = 0.294$. [255, 230.9, 166.8 MN/m^2.]

15.2 (B). A horizontal shaft of 75 mm diameter projects from a bearing, and in addition to the torque transmitted the shaft carries a vertical load of 8 kN at 300 mm from the bearing. If the safe stress for the material, as determined in a simple tensile test, is 135 MN/m^2, find the safe torque to which the shaft may be subjected using as the criterion (a) the maximum shearing stress; (b) the maximum strain energy. Poisson's ratio $\nu = 0.29$.
[U.L.] [5.05, 6.3 kN m.]

15.3 (B). Show that the strain energy per unit volume of a material under a single direct stress is given by $\frac{1}{2}$ (stress × strain). Hence show that for a material under the action of the principal stresses σ_1, σ_2 and σ_3 the strain energy per unit volume becomes

$$\frac{1}{2E}[\sigma_1^2+\sigma_2^2+\sigma_3^2-2\nu(\sigma_1\sigma_2+\sigma_1\sigma_3+\sigma_2\sigma_3)]$$

A thin cylinder 1 m diameter and 3 m long is filled with a liquid to a pressure of 2 MN/m^2. Assuming a yield stress for the material of 240 MN/m^2 in simple tension and a safety factor of 4, determine the necessary wall thickness of the cylinder, taking the maximum shear strain energy as the criterion of failure.
For the cylinder material, $E = 207$ GN/m^2 and $\nu = 0.286$. [14.4 mm.]

15.4 (B). An aluminium-alloy tube of 25 mm outside diameter and 22 mm inside diameter is to be used as a shaft. It is 500 mm long, in self-aligning bearings, and supports a load of 0.5 kN at mid-span. In order to find the maximum allowable shear stress a length of tube was tested in tension and reached the limit of proportionality at 21 kN. Assuming the criterion for elastic failure to be the maximum shear stress, find the greatest torque to which the shaft could be subjected. [98.2 N m.]

15.5 (B). A bending moment of 4 kN m is found to cause elastic failure of a solid circular shaft. An exactly similar shaft is now subjected to a torque T. Determine the value of T which will cause failure of the shaft according to the following theories:
(a) maximum principal stress;
(b) maximum principal strain;
(c) maximum shear strain energy. ($\nu = 0.3$.)
Which of these values would you expect to be the most reliable and why? [8, 6.15, 4.62 kN m.]

15.6 (B). A thin cylindrical pressure vessel with closed ends is required to withstand an internal pressure of 4 MN/m^2. The inside diameter of the vessel is to be 500 mm and a factor of safety of 4 is required. A sample of the proposed material tested in simple tension gave a yield stress of 360 MN/m^2.
Find the thickness of the vessel, assuming the criterion of elastic failure to be (a) the maximum shear stress, (b) the shear strain energy. [E.M.E.U.] [11.1, 9.62 mm.]

15.7 (B). Derive an expression for the strain energy stored in a material when subjected to three principal stresses.
A material is subjected to a system of three mutually perpendicular stresses as follows: f tensile, $2f$ tensile and f compressive. If this material failed in simple tension at a stress of 180 MN/m^2, determine the value of f if the criterion of failure is:

(a) maximum principal stress;
(b) maximum shear stress;
(c) maximum strain energy.
Take Poisson's ratio $\nu = 0.3$. [90, 60, 70 MN/m^2.]

15.8 (B). The external and internal diameters of a hollow steel shaft are 150 mm and 100 mm. A power transmission test with a torsion dynamometer showed an angle of twist of 0.13 degree on a 250 mm length when the speed was 500 rev/min. Find the power being transmitted and the torsional strain energy per metre length.

If, in addition to this torque, a bending moment of 15 kN m together with an axial compressive force of 80 kN also acted upon the shaft, find the value of the equivalent stress in simple tension corresponding to the maximum shear strain energy theory of elastic failure. Take $G = 80\ \text{GN/m}^2$.

[I.Mech.E.] [1.52 MW; 13.13 J/m; 113 MN/m^2.]

15.9 (B/C). A close-coiled helical spring has a wire diameter of 2.5 mm and a mean coil diameter of 40 mm. The spring is subjected to a combined axial load of 60 N and a torque acting about the axis of the spring. Determine the maximum permissible torque if (a) the material is brittle and ultimate failure is to be avoided, the criterion of failure is the maximum tensile stress, and the ultimate tensile stress is 1.2 GN/m^2, (b) the material is ductile and failure by yielding is to be avoided, the criterion of failure is the maximum shear stress, and the yield in tension is 0.9 GN/m^2.

[I.Mech.E.] [1.645, 0.68 N m.]

15.10 (C). A closed-ended thick-walled steel cylinder with a diameter ratio of 2 is subjected to an internal pressure. If yield occurs at a pressure of 270 MN/m^2 find the yield strength of the steel used and the diametral strain at the bore at yield. Yield can be assumed to occur at a critical value of the maximum shear stress. It can be assumed that the stresses in a thick-walled cylinder are:

$$\text{hoop stress } \sigma_H = A + \frac{B}{r^2}$$

$$\text{radial stress } \sigma_r = A - \frac{B}{r^2}$$

$$\text{axial stress } \sigma_L = \tfrac{1}{2}(\sigma_H + \sigma_r)$$

where A and B are constants and r is any radius.

For the cylinder material $E = 210\ \text{GN/m}^2$ and $\nu = 0.3$. [I.Mech.E.] [721 MN/m^2; 2.4×10^{-3}.]

15.11 (C). For a certain material subjected to plane stress it is assumed that the criterion of elastic failure is the shear strain energy per unit volume. By considering co-ordinates relative to two axes at 45° to the principal axes, show that the limiting values of the two principal stresses can be represented by an ellipse having semi-diameters $\sigma_e\sqrt{2}$ and $\sigma_e\sqrt{\tfrac{2}{3}}$, where σ_e is the equivalent simple tension. Hence show that for a given value of the major principal stress the elastic factor of safety is greatest when the minor principal stress is half the major, both stresses being of the same sign. [U.L.]

15.12 (C). A horizontal circular shaft of diameter d and second moment of area I is subjected to a bending moment $M \cos\theta$ in a vertical plane and to an axial twisting moment $M \sin\theta$. Show that the principal stresses at the ends of a vertical diameter are $\tfrac{1}{2}Mk(\cos\theta \pm 1)$, where

$$k = \frac{d}{2I}$$

If strain energy is the criterion of failure, show that

$$\tau_{max} = \frac{\tau_0\sqrt{2}}{[\cos^2\theta(1-\nu)+(1+\nu)]^{\frac{1}{2}}}$$

where τ_{max} = maximum shearing stress,
τ_0 = maximum shearing stress in the special case when $\theta = 0$,
ν = Poisson's ratio. [U.L.]

15.13 (C). What are meant by the terms "yield criterion" and "yield locus" as related to ductile metals and why, in general, are principal stresses involved?

Define the maximum shear stress and shear strain energy theories of yielding. Describe the three-dimensional loci and sketch the plane stress loci for the above theories. [C.E.I.]

15.14 (B). The maximum shear stress theory of elastic failure is sometimes criticised because it makes no allowance for the magnitude of the intermediate principal stress. On these grounds a theory is preferred which predicts that yield will not occur provided that

$$(\sigma_1-\sigma_2)^2+(\sigma_2-\sigma_3)^2+(\sigma_3-\sigma_1)^2 < 2\sigma_y^2$$

What is the criterion of failure implied here?

Assuming that σ_1 and σ_3 are fixed and unequal, find the value of σ_2 which will be most effective in preventing failure according to this theory. If this theory is correct, by what percentage does the maximum shear stress theory underestimate the strength of a material in this case? [City U.] [$\tfrac{1}{2}(\sigma_1+\sigma_3)$; 13.4%.]

15.15 (B). The cast iron used in the manufacture of an engineering component has tensile and compressive strengths of 400 MN/m^2 and 1.20 GN/m^2 respectively.

(a) If the maximum value of the tensile principal stress is to be limited to one-quarter of the tensile strength, determine the maximum value and nature of the other principal stress using Mohr's modified yield theory for brittle materials.

(b) What would be the values of the principal stresses associated with a maximum shear stress of 450 MN/m^2 according to Mohr's modified theory?

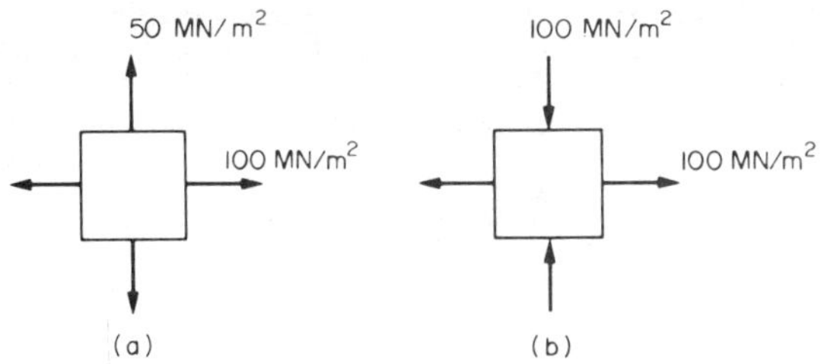

Fig. 15.19.

(c) Estimate the safety factor with respect to initial failure for the stress conditions of Fig. 15.19 using the maximum principal stress, maximum shear stress, distortion energy and Mohr's modified theories of elastic failure. [B.P.] [-900 MN/m^2; 150, -750 MN/m^2; 4, 4, 4.7, 4 and 4, 2, 2.4, 3.]

15.16 (B). Show that for a material subjected to two principal stresses, σ_1 and σ_2, the strain energy per unit volume is

$$\frac{1}{2E}(\sigma_1^2 + \sigma_2^2 - 2\sigma_1\sigma_2)$$

A thin-walled steel tube of internal diameter 150 mm, closed at its ends, is subjected to an internal fluid pressure of 3 MN/m^2. Find the thickness of the tube if the criterion of failure is the maximum strain energy. Assume a factor of safety of 4 and take the elastic limit in pure tension as 300 MN/m^2. Poisson's ratio $\nu = 0.28$.

[I.Mech.E.] [2.95 mm]

15.17 (B). A circular shaft, 100 mm diameter is subjected to combined bending moment and torque, the bending moment being 3 times the torque. If the direct tension yield point of the material is 300 MN/m^2 and the factor of safety on yield is to be 4, calculate the allowable twisting moment by the three following theories of failure:

(a) Maximum principal stress theory
(b) Maximum shear stress theory
(c) Maximum shear strain energy theory. [U.L.] [2.86, 2.79, 2.83 kNm]

15.18 (B). A horizontal shaft of 75 mm diameter projects from a bearing and, in addition to the torque transmitted, the shaft carries a vertical load of 8 kN at 300 mm from the bearing. If the safe stress for the material, as determined in a simple tension test, is 135 MN/m^2 find the safe torque to which the shaft may be subjected using as a criterion

(a) the maximum shearing, stress,
(b) the maximum strain energy per unit volume.

Poisson's ratio $\nu = 0.29$. [U.L.] [5.59, 8.3 kNm.]

APPENDIX 1

TYPICAL MECHANICAL AND PHYSICAL PROPERTIES FOR ENGINEERING METALS

Material	Young's modulus of elasticity E (GN/m^2)	Shear modulus G (GN/m^2)	"Elastic" limit σ_y (MN/m^2)	Shear yield strength τ_y (MN/m^2)	Tensile strength (MN/m^2)	Ultimate strength in shear (MN/m^2)	Percentage elongation (%)	Density (kg/m^3)	Linear coefficient of thermal expansion ($\times 10^{-6}$/°C)
Aluminium alloy	69	26	230	—	390	240	23	2770	23
Brass	102	38	—	—	350	—	40	8350	18.9
Bronze	115	45	210	—	310	—	20	7650	18
Cast iron: Grey	90	41	—	—	210	—	8	7640	10.5
Malleable	170	83	248	166	370	330	12	7640	12
Low carbon (mild) steel	207	80	280	175	480	350	25	7800	11.7
Nickel-chrome steel	208	82	1200	650	1700	950	12	7800	11.7
Titanium	107	40	480	—	551	—	—	4507	9.5
Magnesium	45	17	262	—	379	165	—	1791	28.8

APPENDIX 2

TYPICAL MECHANICAL PROPERTIES OF NON-METALS

Material	Young's modulus of elasticity E (GN/m^2)	Tensile strength (MN/m^2)	Compressive strength (MN/m^2)	Elongation (maximum) %
Acetals	—	69	124	75
Cellulose acetate	1.4	41	207	20
Cellulose nitrate	1.4	48	138	40
Epoxy (glass filler)	—	145	234	—
Hard rubber	3.0	48	—	—
Melamine	8.0	55	227	0.7
Nylon filaments	4.1	340	—	—
Polycarbonate – unreinforced Makralon	2.3	70	83	100
Reinforced Makralon	6.0	90	—	8
Polyester (unfilled)	2.0	41	—	2
Polyethylene H.D.	—	28	22	100
Polyethylene L.D.	—	10	—	800
Polypropylene	—	34	510	250
Polystyrene	3.4	20	76	1.2
Polystyrene – impact resistant	1.4	38	41	80
P.T.F.E.	—	34	248	70
P.V.C. (rigid)	3.4	50–60	69	40
P.V.C. (plasticised)	—	20	0.7	200
Rubber (natural-vulcanised)	—	7–34	—	—
Silicones (elastomeric)	—	1.5–6	—	—
Timber	9.0	70	—	—
Urea (cellulose filler)	10.0	62	241	0.7

* Data taken in part from *Design Engineering Handbook on Plastics* (Product Journals Ltd).

APPENDIX 3

OTHER PROPERTIES OF NON-METALS*

Material	Chemical resistance					Max. useful temp. (°C)
	Organic Solvents	Acids		Alkalis		
		Weak	Strong	Weak	Strong	
Acetal	×	×	00	×	×	90
Acrylic	Varies	×	×–0	×	×	90
Nylon 66	×	×	00	×	×	150
Polycarbonate	Varies	×	0	×–0	00	120
Polyethylene LD	×	×	×–00	×	×	90
Polyethylene HD	×	×	×–0	×	×	120
Polypropylene	×	×	×–0	×	×	150
Polystyrene	Varies	×	×–0	×	×	95
PTFE	×	×	×	×	×	240
PVC	Varies	×	×–0	×	×	80
Epoxy	×	×	×	×	0	430
Melamine	×	×	00	×	0	100–200
Phenolic	×	×–0	0–00	0–00	00	200
Polyester/glass	×–0	0	00	0	00	250
Silicone	×–0	×–0	00	0–00	00	180
Urea	×–0	×–0	00	0–00	00	90
× – Resistant, 0 – slightly attacked, 00 – markedly attacked						

* Data taken from Design Engineering Handbook on Plastics. (Product Journals Ltd).

INDEX